£10

PERGAMON INTERNATIONAL
of Science, Technology, Engineering and
The 1000-volume original paperback library in
industrial training and the enjoyment
Publisher: Robert Maxwell, M.C.

D1128864

CHEMICAL ENGINEERING

VOLUME THREE

Second Edition (SI units)

THE PERGAMON TEXTBOOK
INSPECTION COPY SERVICE

An inspection copy of any book published in the Pergamon International Library will
gladly be sent to academic staff without obligation for their consideration for course
adoption or recommendation. Copies may be retained for a period of 60 days from
receipt and returned if not suitable. When a particular title is adopted or recommended
for adoption for class use and the recommendation results in a sale of 12 or more copies,
the inspection copy may be retained with our compliments. The Publishers will be
pleased to receive suggestions for revised editions and new titles to be published in this
important International Library.

Other Titles in the Series

Chemical Engineering, Volume 1, Third edition (in SI units)
Fluid Flow, Heat Transfer and Mass Transfer
  (With Editorial Assistance from J. R. Backhurst and J. H. Harker)

Chemical Engineering, Volume 2, Third edition (in SI units)
Unit Operations
  (With Editorial Assistance from J. R. Backhurst and J. H. Harker)

Chemical Engineering, Volume 4 (in SI units)
Solutions to the Problems in Volume 1
  (J. R. Backhurst and J. H. Harker)

Chemical Engineering, Volume 5 (in SI units)
Solutions to the Problems in Volume 2
  (J. R. Backhurst and J. H. Harker)

* Chemical Engineering, Volume 6 (in SI units)
  Design
    (R. K. Sinnott)

The above-listed titles may be of interest to readers and can
be supplied as inspection copies for 60-day approval by
writing to the nearest Pergamon office

* In preparation. Full details of this title available on request

CHEMICAL ENGINEERING

VOLUME THREE

Chemical Reactor Design, Biochemical Reaction Engineering including Computational Techniques and Control

by

J. M. COULSON
University of Newcastle upon Tyne

and

J. F. RICHARDSON
University College of Swansea

EDITORS OF VOLUME THREE

J. F. RICHARDSON

and

D. G. PEACOCK
The School of Pharmacy, London

SECOND EDITION

(SI units)

PERGAMON PRESS

OXFORD · NEW YORK · TORONTO · SYDNEY · PARIS · FRANKFURT

U.K.	Pergamon Press Ltd., Headington Hill Hall, Oxford OX3 0BW, England
U.S.A.	Pergamon Press Inc., Maxwell House, Fairview Park, Elmsford, New York 10523, U.S.A.
CANADA	Pergamon of Canada, Suite 104, 150 Consumers Road, Willowdale, Ontario M2J 1P9, Canada
AUSTRALIA	Pergamon Press (Aust.) Pty. Ltd., P.O. Box 544, Potts Point, N.S.W. 2011, Australia
FRANCE	Pergamon Press SARL, 24 rue des Ecoles, 75240 Paris, Cedex 05, France
FEDERAL REPUBLIC OF GERMANY	Pergamon Press GmbH, 6242 Kronberg/Taunus, Pferdstrasse 1, Federal Republic of Germany

First edition 1971

Reprinted 1975

Second edition 1979

British Library Cataloguing in Publication Data

Chemical engineering.
Vol. 3: Chemical reactor design, biochemical reaction engineering including computational techniques and control.—2nd ed. (SI units).
1. Chemical engineering
I. Coulson, John Metcalfe   II. Richardson, John Francis   III. Peacock, Donald G.
660.2   TP155   79-40177

ISBN 0-08-023818-1 (Hardcover)
ISBN 0-08-023819-X (Flexicover)

Printed in Great Britain by Page Bros (Norwich) Ltd

Contents

Preface to the Second Edition

APART from general updating and correction, the main alterations in the second edition of Volume 3 are additions to Chapter 1 on Reactor Design and the inclusion of a Table of Error Functions in the Appendix.

In Chapter 1 two new sections have been added. In the first of these is a discussion of non-ideal flow conditions in reactors and their effect on residence time distribution and reactor performance. In the second section an important class of chemical reactions—that in which a solid and a gas react non-catalytically—is treated. Together, these two additions to the chapter considerably increase the value of the book in this area.

All quantities are expressed in SI units, as in the second impression, and references to earlier volumes of the series take account of the modifications which have recently been made in the presentation of material in the third editions of these volumes.

Foreword to the Second Impression

THE opportunity has been taken to correct a number of errors and to introduce SI units wherever practicable. The SI system is still undergoing modification in detail and it will be some while before methods of presentation become completely standardised. It should be stressed that the main advantages of the system lie in calculations, but problems will frequently be stated in terms of everyday units. In some cases, data are available only in one of the older sets of units and it is necessary for chemical engineers to be able to tackle problems in whatever units they are formulated.

Preface to the First Edition

CHEMICAL ENGINEERING, as we know it today, developed as a major engineering discipline in the United Kingdom in the interwar years and has grown rapidly since that time. The unique contribution of the subject to the industrial scale development of processes in the chemical and allied industries was initially attributable to the improved understanding it gave to the transport processes—fluid flow, heat transfer and mass transfer—and to the development of design principles for the unit operations, nearly all of which are concerned with the physical separation of complex mixtures, both homogeneous and heterogeneous, into their components. In this context the chemical engineer was concerned much more closely with the separation and purification of the products from a chemical reactor than with the design of the reactor itself.

The situation is now completely changed. With a fair degree of success achieved in the physical separation processes, interest has moved very much towards the design of the reactor, and here too the processes of fluid flow, heat transfer and mass transfer can be just as important. Furthermore, many difficult separation problems can be obviated by correct choice of conditions in the reactor. Chemical manufacture has become more demanding with a high proportion of the economic rewards to be obtained in the production of sophisticated chemicals, pharmaceuticals, antibiotics and polymers, to name a few, which only a few years earlier were unknown even in the laboratory. Profit margins have narrowed too, giving a far greater economic incentive to obtain the highest possible yield from raw materials. Reactor design has therefore become a vital ingredient of the work of the chemical engineer.

Volumes 1 and 2, though no less relevant now, reflected the main areas of interest of the chemical engineer in the early 1950s. In Volume 3 the coverage of chemical engineering is brought up to date with an emphasis on the design of systems in which chemical and even biochemical reactions occur. It includes chapters on adsorption, on the general principles of the design of reactors, on the design and operation of reactors employing heterogeneous catalysts, and on the special features of systems exploiting biochemical and microbiological processes. Many of the materials which are processed in chemical and biochemical reactors are complex in physical structure and the flow properties of non-Newtonian materials are therefore considered worthy of special treatment. With the widespread use of computers, many of the design problems which are too complex to solve analytically or graphically are now capable of numerical solution, and their application to chemical engineering problems forms the subject of a chapter. Parallel with the growth in complexity of chemical plants has developed the need for much closer control of their operation, and a chapter on process control is therefore included.

Each chapter of Volume 3 is the work of a specialist in the particular field, and the authors are present or past members of the staff of the Chemical Engineering Department of the University College of Swansea. W. J. Thomas is now at the Bath University of Technology and J. M. Smith is at the Technische Hogeschool, Delft.

J.M.C.
J.F.R.
D.G.P.

Acknowledgments

THE authors wish to acknowledge permission to copy diagrams and data as listed below.

Fig. 2.22   Cambridge University Press

Fig. 5.1   Publishers of the *Journal of Cell Biology*

Fig. 5.6   The A.P.V. Company Limited

Fig. 6.16   Ferranti Limited

Fig. 6.18   Brookfield Engineering Laboratories

Fig. 6.60   Lightnin Mixers Limited

Fig. 6.61   Sesquehanna Corporation

Fig. 7.5   B.P. Chemicals (U.K.) Limited

Fig. 7.10   Sutcliffe, Speakman and Company Limited

Fig. 7.19   Sutcliffe, Speakman and Company Limited

Fig. 7.20   Courtaulds Limited

Table 7.2   Courtaulds Limited

Table 7.3   Sutcliffe, Speakman and Company Limited

Table A.1   (Laplace Transforms)   McGraw-Hill Limited

List of Contributors

Reactor design—general principles
   J. C. LEE (*University College of Swansea*)

The design of catalytic reactors
   W. J. THOMAS (*Bath University of Technology*)

Process control
   A. P. WARDLE (*University College of Swansea*)

Computers and methods for computation
   D. J. GUNN (*University College of Swansea*)

Biochemical reaction engineering
   B. ATKINSON (*University of Manchester Institute of Science and Technology*)

Non-Newtonian technology
   J. M. SMITH (*Technische Hogeschool, Delft*)

Sorption processes
   J. H. BOWEN (*University College of Swansea*)

CHAPTER 1

Reactor design—general principles

BASIC OBJECTIVES IN DESIGN OF A REACTOR

In chemical engineering physical operations such as fluid flow, heat transfer, mass transfer and separation processes play a very large part; these have been discussed in Volumes 1 and 2. In any manufacturing process where there is a chemical change taking place, however, the chemical reactor is at the heart of the plant.

In size and appearance it may often seem to be one of the least impressive items of equipment, but its demands and performance are usually the most important factors in the design of the whole plant.

When a new chemical process is being developed, at least some indication of the performance of the reactor is needed before any economic assessment of the project as a whole can be made. As the project develops and its economic viability becomes established, so further work is carried out on the various chemical engineering operations involved. Thus, when the stage of actually designing the reactor in detail has been reached, the project as a whole will already have acquired a fairly definite form. Among the major decisions which will have been taken is the rate of production of the desired product. This will have been determined from a market forecast of the demand for the product in relation to its estimated selling price. The reactants to be used to make the product and their chemical purity will have been established. The basic chemistry of the process will almost certainly have been investigated, and information about the composition of the products from the reaction, including any byproducts, should be available.

On the other hand, a reactor may have to be designed as part of a modification to an existing process. Because the new reactor has then to tie in with existing units, its duties can be even more clearly specified than when the whole process is new. Naturally, in practice, detailed knowledge about the performance of the existing reactor would be incorporated in the design of the new one.

As a general statement of the basic objectives in designing a reactor, we can say therefore that the aim is to produce a *specified product* at a *given rate* from *known reactants*. In proceeding further however there are a number of important decisions to be made and there may be scope for considerable ingenuity in order to achieve the best result. At the outset the two most important questions to be settled are:

(a) The type of reactor to be used and its method of operation. Will the

1

reaction be carried out as a batch process, a continuous flow process, or possibly as a hybrid of the two? Will the reactor operate isothermally, adiabatically or in some intermediate manner?

(b) The physical condition of the reactants at the inlet to the reactor. Thus, the basic processing conditions in terms of pressure, temperature and compositions of the reactants on entry to the reactor have to be decided, if not already specified as part of the original process design.

Subsequently, the aim is to reach logical conclusions concerning the following principal features of the reactor:

(a) The overall size of the reactor, its general configuration and the more important dimensions of any internal structures.

(b) The exact composition and physical condition of the products emerging from the reactor. The composition of the products must of course lie within any limits set in the original specification of the process.

(c) The temperatures prevailing within the reactor and any provision which must be made for heat transfer.

(d) The operating pressure within the reactor and any pressure drop associated with the flow of the reaction mixture.

Byproducts and their economic importance

Before taking up the design of reactors in detail, let us first consider the very important question of whether any byproducts are formed in the reaction. Obviously, consumption of reactants to give unwanted, and perhaps unsaleable, byproducts is wasteful and will directly affect the operating costs of the process. Apart from this, however, the nature of any byproducts formed and their amounts must be known so that plant for separating and purifying the products from the reaction may be correctly designed. The appearance of unforeseen byproducts on start-up of a full-scale plant can be utterly disastrous. Economically, although the cost of the reactor may sometimes not appear to be great compared with that of the associated separation equipment such as distillation columns, etc., it is the composition of the mixture of products issuing from the reactor which determines the capital and operating costs of the separation processes.

For example, in producing ethylene[14] together with several other valuable hydrocarbons like butadiene from the thermal cracking of naphtha, the design of the whole complex plant is determined by the composition of the mixture formed in a tubular reactor in which the conditions are very carefully controlled. As we shall see later, the design of a reactor itself can affect the amount of byproducts formed and therefore the size of the separation equipment required. The design of a reactor and its mode of operation can thus have profound repercussions on the remainder of the plant.

Preliminary appraisal of a reactor project

In the following pages we shall see that reactor design involves all the basic principles of chemical engineering with the addition of chemical kinetics. Mass transfer, heat transfer and fluid flow are all concerned and complications arise when, as so often is the case, interaction occurs between these transfer processes and the reaction itself. In designing a reactor it is essential to weigh up all the various factors involved and, by an exercise of judgement, to place them in their proper order of importance. Often the basic design of the reactor is determined by what is seen to be the most troublesome step. It may be the chemical kinetics; it may be mass transfer between phases; it may be heat transfer; or it may even be the need to ensure safe operation. For example, in oxidising naphthalene or o-xylene to phthalic anhydride with air, the reactor must be designed so that ignitions, which are not infrequent, may be rendered harmless. The theory of reactor design is being extended rapidly and more precise methods for detailed design and optimisation are being evolved. However, if the final design is to be successful, *the major decisions taken at the outset must be correct.* Initially, a careful appraisal of the basic role and functioning of the reactor is required and at this stage the application of a little chemical engineering common sense may be invaluable.

CLASSIFICATION OF REACTORS AND CHOICE OF REACTOR TYPE

Homogeneous and heterogeneous reactors

Chemical reactors may be divided into two main categories, homogeneous and heterogeneous. In homogeneous reactors only one phase, usually a gas or a liquid, is present. If more than one reactant is involved, provision must of course be made for mixing them together to form a homogenous whole. Often, mixing the reactants is the way of starting off the reaction, although sometimes the reactants are mixed and then brought to the required temperature.

In heterogeneous reactors two, or possibly three, phases are present, common examples being gas–liquid, gas–solid, liquid–solid and liquid–liquid systems. In cases where one of the phases is a solid, it is quite often present as a catalyst; gas–solid catalytic reactors particularly form an important class of heterogeneous chemical reaction systems. It is worth noting that, in a heterogeneous reactor, the chemical reaction itself may be truly heterogeneous, but this is not necessarily so. In a gas-solid catalytic reactor, the reaction takes place on the surface of the solid and is thus heterogeneous. However, bubbling a gas through a liquid may serve just to dissolve the gas in the liquid where it then reacts homogeneously; the reaction is thus homogeneous but the reactor is heterogeneous in that it is required to effect contact between two phases—gas and liquid. Generally, heterogeneous reactors exhibit a greater variety of configuration and contacting pattern than homogeneous reactors. Initially, therefore, we shall be concerned mainly with the simpler homogeneous reactors, although parts of the treatment that follows can be extended to heterogeneous reactors with little modification.

Batch reactors and continuous reactors

Another kind of classification which cuts across the homogeneous–heterogeneous division is the mode of operation—batchwise or continuous.

Batchwise operation, shown in Fig. 1.1*a*, is familiar to anybody who has carried out small-scale preparative reactions in the laboratory. There are many situations, however, especially in large-scale operation, where considerable advantages accrue by carrying out a chemical reaction continuously in a flow reactor.

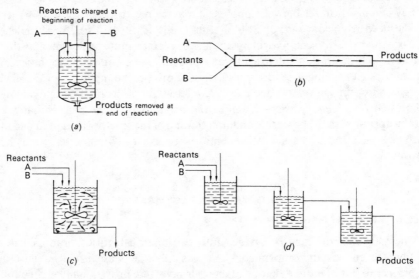

FIG. 1.1. Basic types of chemical reactors

(*a*) Batch reactor
(*b*) Tubular flow reactor
(*c*) Continuous stirred tank reactor (C.S.T.R.) or "Back-mix reactor"
(*d*) C.S.T.R.s in series as frequently used

Figure 1.1 illustrates the two basic types of flow reactor which may be employed. In the *tubular flow reactor* the aim is to pass the reactants along a tube so that there is as little intermingling as possible between the reactants entering the tube and the products leaving at the far end. In the *continuous stirred tank reactor* (C.S.T.R.) an agitator is deliberately introduced to disperse the reactants thoroughly into the reaction mixture immediately they enter the tank. The product stream is drawn off continuously and, in the ideal state of perfect mixing, will have the same composition as the contents of the tank. In some ways, using a C.S.T.R., or *backmix reactor* as it is sometimes called, seems a curious method of conducting a reaction because as soon as the reactants enter the tank they are mixed and a portion leaves in the product stream flowing out. To reduce this effect, it is often advantageous to employ a number of stirred tanks connected in series as shown in Fig. 1.1*d*.

The stirred tank reactor is by its nature well suited to liquid-phase reactions. The tubular reactor, although sometimes used for liquid-phase reactions, is the natural choice for gas-phase reactions, even on a small scale. Usually the temperature or catalyst is chosen so that the rate of reaction is high, in which case a comparatively small tubular reactor is sufficient to handle a high volumetric flowrate of gas. A few gas-phase reactions, examples being partial combustion and certain chlorinations, are carried out in reactors which resemble the stirred tank reactor; rapid mixing is usually brought about by arranging for the gases to enter with a vigorous swirling motion instead of by mechanical means.

Variations in contacting pattern—semi-batch operation

Another question which should be asked in assessing the most suitable type of reactor is whether there is any advantage to be gained by varying the contacting pattern. Figure 1.2a illustrates the *semi-batch* mode of operation. The reaction

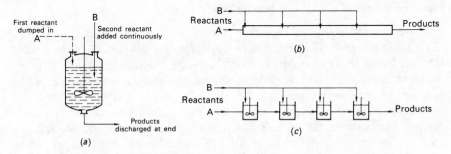

FIG. 1.2. Examples of possible variations in reactant contacting pattern

    (a) Semi-batch operation
    (b) Tubular reactor with divided feed
    (c) Stirred tank reactors with divided feed
    (in each case the concentration of **B**, C_B, is low throughout)

vessel here is essentially a batch reactor, and at the start of a batch it is charged with one of the reactants **A**. However, the second reactant **B** is not all added at once, but continuously over the period of the reaction. This is the natural and obvious way to carry out many reactions. For example, if a liquid has to be treated with a gas, perhaps in a chlorination or hydrogenation reaction, the gas is normally far too voluminous to be charged all at once to the reactor; instead it is fed continuously at the rate at which it is used up in the reaction. Another case is where the reaction is too violent if both reactants are mixed suddenly together. Organic nitration, for example, can be conveniently controlled by regulating the rate of addition of the nitrating acid. The maximum rate of addition of the second reactant in such a case will be determined by the rate of heat transfer.

A characteristic of semi-batch operation is that the concentration C_B of the reactant added slowly, **B** in Fig. 1.2, is low throughout the course of the reaction. This may be an advantage if more than one reaction is possible, and if the desired reaction is favoured by a low value of C_B. Thus, the semi-batch method may be chosen for a further reason, that of improving the yield of the desired product, as shown on page 68.

Summarising, a semi-batch reactor may be chosen:
(a) to react a gas with a liquid,
(b) to control a highly exothermic reaction, and
(c) to improve product yield in suitable circumstances.

In semi-batch operation, when the initial charge of **A** has been consumed, the flow of **B** is interrupted, the products discharged, and the cycle begun again with a fresh charge of **A**. If required, however, the advantages of semi-batch operation may be retained but the reactor system designed for continuous flow of both reactants. In the tubular flow version (Fig. 1.2b) and the stirred tank version (Fig. 1.2c), the feed of **B** is divided between several points. These are known as *cross-flow* reactors. In both cases C_B is low throughout.

Influence of heat of reaction on reactor type

Associated with every chemical change there is a heat of reaction, and only in a few cases is this so small that it can be neglected. The magnitude of the heat of reaction has often a major influence on the design of a reactor. With a strongly exothermic reaction, for example, a substantial rise in temperature of the reaction mixture will take place unless provision is made for heat to be transferred as the reaction proceeds. It is important to try to appreciate clearly the relation between the enthalpy of reaction, the heat transferred, and the temperature change of the reaction mixture; quantitatively this is expressed by an enthalpy balance (e.g. page 27). If the temperature of the reaction mixture is to remain constant (isothermal operation), the heat equivalent to the heat of reaction at the operating temperature must be transferred to or from the reactor. If no heat is transferred (adiabatic operation), the temperature of the reaction mixture will rise or fall as the reaction proceeds. In practice, it may be most convenient to adopt a policy intermediate between these two extremes: in the case of a strongly exothermic reaction, some heat-transfer from the reactor may be necessary in order to keep the reaction under control, but a moderate temperature rise may be quite acceptable, especially if strictly isothermal operation would involve an elaborate and costly control scheme.

In setting out to design a reactor, therefore, two very important questions to ask are:
(a) What is the heat of reaction?
(b) What is the acceptable range over which the temperature of the reaction mixture may be permitted to vary?

The answers to these questions may well dominate the whole design. Usually, the temperature range can only be roughly specified; often the lower temperature limit is determined by the slowing down of the reaction, and the upper temperature limit by the onset of undesirable side reactions.

Adiabatic reactors

If it is feasible, adiabatic operation is to be preferred for simplicity of design. Figure 1.3 shows the reactor section of a plant for the catalytic reforming of petroleum naphtha; this is an important process for improving the octane number of gasoline. The reforming reactions are mostly endothermic so that in adiabatic operation the temperature would fall during the course of the reaction.

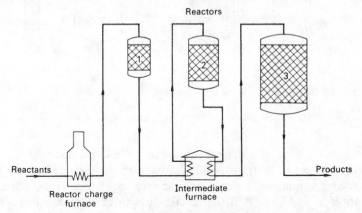

Fig. 1.3. Reactor system of a petroleum naphtha catalytic reforming plant. (The reactor is divided into three units each of which operates *adiabatically*, the heat required being supplied at intermediate stages via an external furnace)

If the reactor were made as one single unit, this temperature fall would be too large, i.e. either the temperature at the inlet would be too high and undesired reactions would occur, or the reaction would be incomplete because the temperature near the outlet would be too low. The problem is conveniently solved by dividing the reactor into three sections. Heat is supplied externally between the sections, and the intermediate temperatures are raised so that each section of the reactor will operate adiabatically. Dividing the reactor into sections also has the advantage that the intermediate temperature can be adjusted independently of the inlet temperature; thus an optimum temperature distribution can be determined. In this example we can see that the furnaces where heat is transferred and the catalytic reactors are quite separate units, each designed specifically for the one function. This separation of function generally provides ease of control, flexibility of operation and often leads to a good overall engineering design.

Reactors with heat transfer

If the reactor does not operate adiabatically, then its design must include provision for heat transfer. Figure 1.4 shows some of the ways in which the contents of a batch reactor may be heated or cooled. In *a* and *b* the jacket and the coils form part of the reactor itself, whereas in *c* an external heat exchanger is used with a recirculating pump. If one of the constituents of the reaction mixture, possibly a solvent, is volatile at the operating temperature, the external heat exchanger may be a reflux condenser, just as in the laboratory.

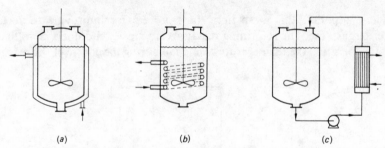

FIG. 1.4. Batch reactors showing different methods of heating or cooling

(a) Jacket
(b) Internal coils
(c) External heat exchangers

Figure 1.5 shows ways of designing tubular reactors to include heat transfer. If the amount of heat to be transferred is large, then the ratio of heat transfer surface to reactor volume will be large, and the reactor will look very much like a heat exchanger as in Fig. 1.5b. If the reaction has to be carried out at a high temperature and is strongly endothermic (for example, the production of ethylene by the thermal cracking of naphtha or ethane—see also page 40), the reactor will be directly fired by the combustion of oil or gas and will look like a pipe furnace (Fig. 1.5c).

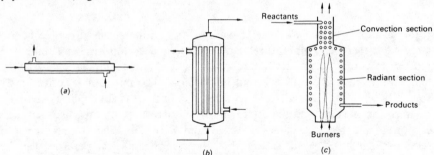

FIG. 1.5. Methods of heat transfer to tubular reactors
(a) Jacketed pipe
(b) Multitube reactor (tubes in parallel)
(c) Pipe furnace (pipes mainly in series although some pipe runs may be in parallel)

Autothermal reactor operation

If a reaction requires a relatively high temperature before it will proceed at a reasonable rate, the products of the reaction will leave the reactor at a high temperature and, in the interests of economy, heat will normally be recovered from them. Since heat must be supplied to the reactants to raise them to the reaction temperature, a common arrangement is to use the hot products to heat the incoming feed as shown in Fig. 1.6a. If the reaction is sufficiently exothermic, enough heat will be produced in the reaction to overcome any losses in the system and to provide the necessary temperature difference in the heat exchanger. The term *autothermal* is used to describe such a system which is completely self-supporting in its thermal energy requirements.

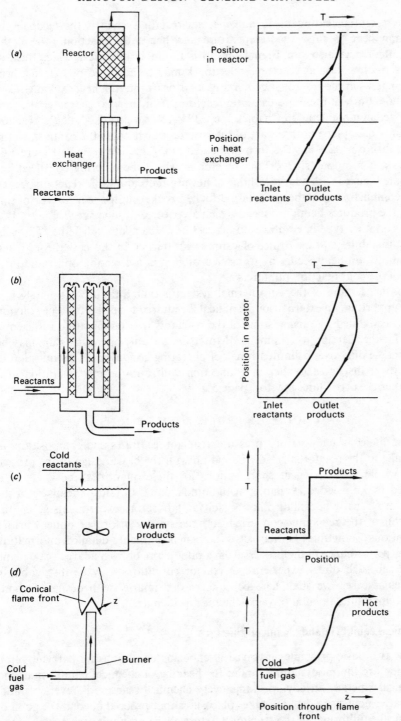

FIG. 1.6. Autothermal reactor operation

The essential feature of an autothermal reactor system is the feedback of reaction heat to raise the temperature and hence the reaction rate of the incoming reactant stream. Figure 1.6 shows a number of ways in which this can occur. With a tubular reactor the feedback may be achieved by external heat exchange, as in the reactor shown in Fig. 1.6a, or by internal heat exchange as in Fig. 1.6b. Both of these are catalytic reactors; their thermal characteristics are discussed in more detail in Chapter 2, pp. 171 *et seq.* Being catalytic the reaction can only take place in that part of the reactor which holds the catalyst, so the temperature profile has the form indicated alongside the reactor. Figure 1.6c shows a continuous stirred tank reactor in which the entering cold feed immediately mixes with a large volume of hot products and rapid reaction occurs. The combustion chamber of a liquid fuelled rocket motor is a reactor of this type, the products being hot gases which are ejected at high speed. Figure 1.6d shows another type of combustion process in which a laminar flame of conical shape is stabilised at the orifice of a simple gas burner. In this case the feedback of combustion heat occurs by transfer upstream in a direction opposite to the flow of the cold reaction mixture.

Another feature of the autothermal system is that, although ultimately it is self-supporting, an external source of heat is required to start it up. The reaction has to be ignited by raising some of the reactants to a temperature sufficiently high for the reaction to commence. Moreover, a stable operating state may be obtainable only over a limited range of operating conditions. This question of stability is discussed further in connection with autothermal operation of a continuous stirred tank reactor (page 55).

CHOICE OF PROCESS CONDITIONS

The choice of temperature, pressure, reactant feed rates and compositions at the inlet to the reactor is closely bound up with the basic design of the process as a whole. In arriving at specifications for these quantities, the engineer is guided by knowledge available on the fundamental physical chemistry of the reaction. Usually he will also have results of laboratory experiments giving the fraction of the reactants converted and the products formed under various conditions. Sometimes he may have the benefit of highly detailed information on the performance of the process from a pilot plant, or even a large-scale plant. Although such direct experience of reactor conditions may be invaluable in particular cases, we shall here be concerned primarily with design methods based upon fundamental physico-chemical principles.

Chemical equilibria and chemical kinetics

The two basic principles involved in choosing conditions for carrying out a reaction are thermodynamics, under the heading of chemical equilibrium, and chemical kinetics. Strictly speaking, every chemical reaction is reversible and, no matter how fast a reaction takes place, it cannot proceed beyond the point of chemical equilibrium in the reaction mixture at the particular temperature and pressure concerned. Thus, under any prescribed conditions, the principle of chemical equilibrium, through the equilibrium constant, determines *how far* the reaction can possibly proceed given sufficient time for equilibrium to be reached. On the other hand, the principle of chemical kinetics determines at what *rate* the reaction will proceed towards this maximum extent. If the equilibrium constant is very large, then for all practical purposes the reaction may be said to be *irreversible*. However, even when a reaction is claimed to be *irreversible* an engineer would be very unwise not to calculate the equilibrium constant and check the position of equilibrium, especially if high conversions are required.

In deciding process conditions, the two principles of thermodynamic equilibrium and kinetics need to be considered together; indeed, any complete rate equation for a reversible reaction will include the equilibrium constant or its equivalent (see page 21) but complete rate equations are not always available to the engineer. The first question to ask is: in what temperature range will the chemical reaction take place at a reasonable rate (in the presence, of course, of any catalyst which may have been developed for the reaction)? The next step is to calculate values of the equilibrium constant in this temperature range using the principles of chemical thermodynamics. (Such methods are beyond the scope of this chapter and any reader unfamiliar with this subject should consult a standard textbook[10].) The equilibrium constant K_p of a reaction depends only on the temperature as indicated by the relation:

$$\frac{d \ln K_p}{dT} = \frac{\Delta H}{RT^2}$$

where $-\Delta H$ is the heat of reaction. The equilibrium constant is then used to determine the limit to which the reaction can proceed under the conditions of temperature, pressure and reactant compositions which appear to be most suitable.

Calculation of equilibrium conversion

Whereas the equilibrium constant itself depends on the temperature only, the conversion at equilibrium depends on the composition of the original reaction mixture and, in general, on the pressure. If the equilibrium constant is very high, the reaction may be treated as being irreversible. If the equilibrium constant is low, however, it may be possible to obtain acceptable conversions only by using high or low pressures. Two important examples are the reactions:

$$C_2H_4 + H_2O \rightleftharpoons C_2H_5OH$$
$$N_2 + 3H_2 \rightleftharpoons 2NH_3$$

both of which involve a decrease in the number of mols as the reaction proceeds, and therefore high pressures are used to obtain satisfactory equilibrium conversions.

Thus, in those cases in which reversibility of the reaction imposes a serious limitation, the equilibrium conversion must be calculated in order that the most advantageous conditions to be employed in the reactors may be chosen; this may be seen in detail in the following example of the styrene process.

A study of the design of this process is also very instructive in showing how the basic features of the reaction, namely equilibrium, kinetics, and suppression of byproducts, have all been satisfied in quite a clever way by using steam as a diluent.

Example: *A process for the manufacture of styrene by the dehydrogenation of ethylbenzene*

Let us suppose that we are setting out from first principles to investigate the dehydrogenation of ethylbenzene which is a well established process for manufacturing styrene:

$$C_6H_5 \cdot CH_2 \cdot CH_3 = C_6H_5 \cdot CH:CH_2 + H_2$$

There is available a catalyst which will give a suitable rate of reaction at 560°C. At this temperature the equilibrium constant for the reaction above is:

$$\frac{P_{St} \times P_H}{P_{Et}} = K_p = 100 \text{ mbar} \qquad \dots (1.1)$$

where P_{Et}, P_{St} and P_H are the partial pressures of ethylbenzene, styrene and hydrogen respectively.

Part (i)

Feed pure ethylbenzene: If a feed of pure ethylbenzene is used at 1 bar pressure, determine the fractional conversion at equilibrium.

Solution

This calculation requires not only the use of the equilibrium constant, but also a material balance over the reactor. To avoid confusion, it is as well to set out this material balance quite clearly even in this comparatively simple case.

First it is necessary to choose a basis; let this be 1 mol of ethylbenzene fed into the reactor: a fraction α_e of this will be converted at equilibrium. Then, from the above stoichiometric equation, α_e mol styrene and α_e mol hydrogen are formed, and $(1 - \alpha_e)$ mol ethylbenzene remains unconverted. Let the total pressure at the outlet of the reactor be P which we shall later set equal to 1 bar.

$$C_6H_5 \cdot C_2H_5 \longrightarrow \boxed{\text{REACTOR}} \longrightarrow \begin{array}{l} C_6H_5 \cdot C_2H_5 \\ C_6H_5 \cdot C_2H_3 \\ H_2 \end{array}$$

Temperature 560°C = 833 K
Pressure P (1 bar = $1\cdot0 \times 10^5$ N/m²)

	IN	*a*	*b*	*c*
	mol	mol	mol fraction	partial pressure
$C_6H_5 \cdot C_2H_5$	1	$1 - \alpha_e$	$\dfrac{1 - \alpha_e}{1 + \alpha_e}$	$\dfrac{1 - \alpha_e}{1 + \alpha_e} P$
$C_6H_5 \cdot C_2H_3$	—	α_e	$\dfrac{\alpha_e}{1 + \alpha_e}$	$\dfrac{\alpha_e}{1 + \alpha_e} P$
H_2	—	α_e	$\dfrac{\alpha_e}{1 + \alpha_e}$	$\dfrac{\alpha_e}{1 + \alpha_e} P$
TOTAL		$1 + \alpha_e$		

Since for 1 mol of ethylbenzene entering, the total number of mols increases to $1 + \alpha_e$, the mol fractions of the various species in the reaction mixture at the reactor outlet are shown in column b above. At a total pressure P, the partial pressures are given in column c (assuming ideal gas behaviour). If the reaction mixture is at chemical equilibrium, these partial pressures must satisfy equation 1.1 above:

$$K_p = \frac{P_{St} \times P_H}{P_{Et}} = \frac{\dfrac{\alpha_e}{(1 + \alpha_e)}P\ \dfrac{\alpha_e}{(1 + \alpha_e)}P}{\dfrac{(1 - \alpha_e)}{(1 + \alpha_e)}P} = \frac{\alpha_e^2}{1 - \alpha_e^2}P$$

i.e.

$$\frac{\alpha_e^2}{1 - \alpha_e^2}P = 1 \cdot 0 \times 10^4 \ \text{N/m}^3 \qquad \ldots (1.2)$$

Thus, when $P = 1$ bar, $\alpha_e = 0 \cdot 30$; i.e. the maximum possible conversion using pure ethylbenzene at 1 bar is only 30 per cent; this is not very satisfactory (although it is possible in some processes to operate at low conversions by separating and recycling reactants). Ways of improving this figure are now sought.

Note that equation 1.2 above shows that as P decreases α_e increases; this is the quantitative expression of Le Chatelier's principle that, because the total number of mols increases in the reaction, the decomposition of ethylbenzene is favoured by a reduction in pressure. There are, however, disadvantages in operating such a process at sub-atmospheric pressures. One disadvantage is that any ingress of air through leaks might result in ignition. A better solution in this instance is to reduce the partial pressure by diluting the ethylbenzene with an inert gas, while maintaining the total pressure slightly in excess of atmospheric. The inert gas most suitable for this process is steam: one reason for this is that it can be condensed easily in contrast to a gas such as nitrogen which would introduce greater problems in separation.

Part (ii)

Feed ethylbenzene with steam: If the feed to the process consists of ethylbenzene diluted with steam in the ratio 15 mol steam: 1 mol ethylbenzene, determine the new fractional conversion at equilibrium α'_e.

Solution

Again we set out the material balance in full, the basis being 1 mol ethylbenzene into the reactor.

$C_6H_5 \cdot C_2H_5$ ——→ REACTOR ——→ $C_6H_5 \cdot C_2H_5$ Temperature 560°C = 833 K

H_2O $C_6H_5 \cdot C_2H_3$

                            H_2 Pressure P (1 bar = $1 \cdot 0 \times 10^5$ N/m²)

                            H_2O

	IN mol	a mol	OUT b mol fraction	c partial pressure
$C_6H_5 \cdot C_2H_5$	1	$1 - \alpha'_e$	$\dfrac{1 - \alpha'_e}{16 + \alpha'_e}$	$\dfrac{1 - \alpha'_e}{16 + \alpha'_e}P$
$C_6H_5 \cdot C_2H_3$	—	α'_e	$\dfrac{\alpha'_e}{16 + \alpha'_e}$	$\dfrac{\alpha'_e}{16 + \alpha'_e}P$
H_2	—	α'_e	$\dfrac{\alpha'_e}{16 + \alpha'_e}$	$\dfrac{\alpha'_e}{16 + \alpha'_e}P$
H_2O	15	15	$\dfrac{15}{16 + \alpha'_e}$	

TOTAL $16 + \alpha'_e$

$$K_p = \frac{P_{St} \times P_H}{P_{Et}} = \frac{\dfrac{\alpha'_e}{(16 + \alpha'_e)}P \dfrac{\alpha'_e}{(16 + \alpha'_e)}P}{\dfrac{(1 - \alpha'_e)}{(16 + \alpha'_e)}P}$$

$$= \frac{\alpha'^2_e}{(16 + \alpha'_e)(1 - \alpha'_e)}P$$

i.e.

$$\frac{\alpha'^2_e}{(16 + \alpha'_e)(1 - \alpha'_e)}P = 1 \cdot 0 \times 10^{-4} \qquad \ldots (1.3)$$

Thus when $P = 1$ bar, $\alpha'_e = 0 \cdot 70$; i.e. the maximum possible conversion has now been raised to 70 per cent. Inspection of equation 1.3 shows that the equilibrium conversion increases as the ratio of steam to ethylbenzene increases. However, as more steam is used, its cost increases and offsets the value of the increase in ethylbenzene conversion. The optimum steam:ethylbenzene ratio is thus determined by an economic balance.

Part (iii)

Final choice of reaction conditions in the styrene process:

Solution

The use of steam has a number of other advantages in the styrene process. The most important of these is that it acts as a source of internal heat supply so that the reactor can be operated adiabatically. The dehydrogenation reaction is strongly endothermic, the heat of reaction at 560°C being $(-\Delta H) = -125,000$ kJ/kmol. It is instructive to look closely at the conditions which were originally worked out for this process (Fig. 1.7). Most of the steam, 90 per cent of the total

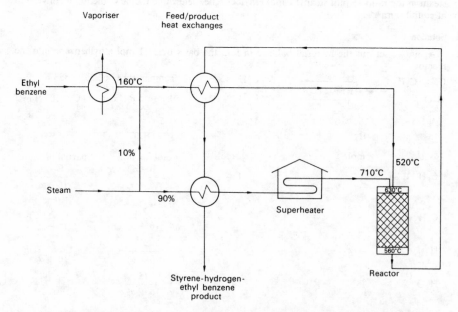

FIG. 1.7. A process for styrene from ethylbenzene using 15 mol steam: 1 mol ethylbenzene. Operating pressure 1 bar. Conversion per pass 0·40. Overall relative yield 0·90

used, is heated separately from the ethylbenzene stream, and to a higher temperature (710°C) than is required at the inlet to the reactor. The ethylbenzene is heated in the heat exchangers to only 520°C and is then rapidly mixed with the hotter steam to give a temperature of 630°C at the inlet to the catalyst bed. If the ethylbenzene were heated to 630°C more slowly by normal heat exchange decomposition and coking of the heat transfer surfaces would tend to occur. Moreover, the tubes of this heat exchanger would have to be made of a more expensive alloy to resist the more severe working conditions. To help avoid coking, 10 per cent of the steam used is passed through the heat exchanger with the ethylbenzene. The presence of a large proportion of steam in the reactor also prevents coke deposition on the catalyst. By examining the equilibrium constant of reactions involving carbon such as:

$$C_6H_5 \cdot CH_2 \cdot CH_3 \rightleftharpoons 8C + 5H_2$$
$$C + H_2O \rightleftharpoons CO + H_2$$

it may be shown that coke formation is not possible at high steam:ethylbenzene ratios.

The styrene process operates with a fractional conversion of ethylbenzene per pass of 0·40 compared with the equilibrium conversion of 0·70. This actual conversion of 0·40 is determined by the *rate* of the reaction over the catalyst at the temperature prevailing in the reactor. (Adiabatic operation means that the temperature falls with increasing conversion and the reaction tends to be quenched at the outlet). The unreacted ethylbenzene is separated and recycled to the reactor. The overall yield in the process, i.e. mols of ethylbenzene transformed into styrene per mol of ethylbenzene supplied, is 0·90, the remaining 0·10 being consumed in unwanted side reactions. Notice that the conversion per pass could be increased by increasing the temperature at the inlet to the catalyst bed beyond 630°C, but the undesirable side reactions would increase, and the overall yield of the process would fall. The figure of 630°C for the inlet temperature is thus determined by an economic balance between the cost of separating unreacted ethylbenzene (which is high if the inlet temperature and conversion per pass are low), and the cost of ethylbenzene consumed in wasteful side reactions (which is high if the inlet temperature is high).

Ultimate choice of reactor conditions

The use of steam in the styrene process above is an example of how an engineer can exercise a degree of ingenuity in reactor design. The advantages conferred by the steam may be summarised as follows:

(a) it lowers the partial pressure of the ethylbenzene without the need to operate at sub-atmospheric pressures;

(b) it provides an internal heat source for the endothermic heat of reaction, making adiabatic operation possible; and

(c) it prevents coke formation on the catalyst and coking problems in the ethylbenzene heaters.

As the styrene process shows, it is not generally feasible to operate a reactor with a conversion per pass equal to the equilibrium conversion. The rate of a chemical reaction decreases as equilibrium is approached, so that the equilibrium conversion can only be attained if either the reactor is very large or the reaction unusually fast. The size of reactor required to give any particular conversion, which of course cannot exceed the maximum conversion predicted from the equilibrium constant, is calculated from the kinetics of the reaction. For this purpose we need quantitative data on the rate of reaction, and the rate equations which describe the kinetics are considered in the following section.

If there are two or more reactants involved in the reaction, both can be converted completely in a single pass only if they are fed to the reactor in the stoichiometric proportion. In many cases, the stoichiometric ratio of reactants may be the best, but in some instances, where one reactant (especially water or air) is very much cheaper than the other, it may be economically advantageous to use it in excess. For a given size of reactor, the object is to increase the conversion of the more costly reactant, possibly at the expense of a substantial decrease in the fraction of the cheaper reactant converted. Examination of the kinetics of the reaction is required to determine whether this can be achieved, and to calculate quantitatively the effects of varying the reactant ratio. Another and perhaps more common reason for departing from the stoichiometric proportions of reactants is to minimise the amount of byproducts formed. This question is discussed further on page 68.

Ultimately, the final choice of the temperature, pressure, reactant ratio and conversion at which the reactor will operate depends on an assessment of the overall economics of the process. This will take into account the cost of the reactants, the cost of separating the products and the costs associated with any recycle streams. It should include all the various operating costs and capital costs of reactor and plant. In the course of making this economic assessment, a whole series of calculations of operating conditions, final conversion and reactor size may be performed with the aid of a computer, provided that the data are available. Each of these sets of conditions may be technically feasible, but the one chosen will be that which gives the maximum profitability for the project as a whole.

CHEMICAL KINETICS AND RATE EQUATIONS

When a homogeneous mixture of reactants is passed into a reactor, either batch or tubular, the concentrations of the reactants fall as the reaction proceeds. Experimentally it has been found that, in general, the rate of the reaction decreases as the concentrations of the reactants decrease. In order to calculate the size of the reactor required to manufacture a particular product at a desired overall rate of production, the design engineer therefore needs to know how the rate of reaction at any time or at any point in the reactor depends on the concentrations of the reactants. Since the reaction rate varies also with temperature, generally increasing rapidly with increasing temperature, a *rate equation*, expressing the rate of reaction as a function of concentrations and temperature, is required in order to design a reactor.

Definition of reaction rate, order of reaction and rate constant

Let us consider a homogeneous irreversible reaction

$$v_A A + v_B B + v_C C \rightarrow \quad \text{Products} \qquad \dots (1.4)$$

where A, B, C are the reactants and v_A, v_B, v_C the corresponding coefficients in the stoichiometric equation. The rate of reaction can be measured as the mols of A transformed per unit volume and unit time. Thus, if n_A is the number of mols of **A** present in a volume V of reaction mixture, the *rate of reaction* with respect to **A** is defined as:

$$\mathscr{R}_A = -\frac{1}{V}\frac{dn_A}{dt} \qquad \ldots (1.5)$$

However, the rate of reaction can also be measured as the mols of **B** transformed per unit volume and unit time, in which case:

$$\mathscr{R}_B = -\frac{1}{V}\frac{dn_B}{dt}$$

and $\mathscr{R}_B = (v_B/v_A)\mathscr{R}_A$; similarly $\mathscr{R}_C = (v_C/v_A)\mathscr{R}_A$ and so on. Obviously, when quoting a reaction rate, care must be taken to specify which reactant is being considered, otherwise ambiguity may arise. Another common source of confusion is the units in which the rate of reaction is measured. Appropriate units for $\mathscr{R}_A$ can be seen quite clearly from equation 1.5; they are kmol of A/m^3s or lb mol of A/ft^3 s.

At constant temperature, the rate of reaction $\mathscr{R}_A$ is a function of the concentrations of the reactants. Experimentally, it has been found that often (but not always) the function has the mathematical form:

$$\mathscr{R}_A \left(= -\frac{1}{V}\frac{dn_A}{dt} \right) = k\, C_A^p C_B^q C_C^r \qquad \ldots (1.6)$$

$C_A (= n_A/V)$ being the molar concentration of **A**, etc. The exponents p, q, r in this expression are quite often (but not necessarily) whole numbers. When the functional relationship has the form of equation 1.6, the reaction is said to be of order p with respect to reactant **A**, q with respect to **B** and r with respect to **C**. The order of the reaction overall is $(p + q + r)$.

The coefficient k in equation 1.6 is by definition the *rate constant* of the reaction. Its dimensions depend on the exponents p, q, r (i.e. on the order of the reaction); the units in which it is to be expressed may be inferred from the defining equation 1.6. For example, if a reaction:

$$\mathbf{A} \rightarrow \text{Products}$$

behaves as a simple first order reaction, it has a rate equation:

$$\mathscr{R}_A = k_1 C_A$$

If the rate of reaction $\mathscr{R}_A$ is measured in units of kmol m^3 s and the concentration C_A in kmol/m^3, then k_1 has the units s^{-1}. On the other hand, if the reaction above behaved as a second order reaction with a rate equation:

$$\mathscr{R}_A = k_2 C_A^2$$

the units of this rate constant, with $\mathscr{R}_A$ in kmol/m^3 s and C_A in kmol/m^3, are $m^3(kmol)^{-1} s^{-1}$. A possible source of confusion is that in some instances in the chemical literature, the rate equation, for say a second order gas phase reaction may be written $\mathscr{R}_A = k_p P_A^2$, where P_A is the partial pressure of **A** and may be measured in N/m^2, bar or even in mm Hg. This form of expression results in rather confusing hybrid units for k_p and is not to be recommended.

If a large excess of one or more of the reactants is used, such that the concentration of that reactant changes hardly at all during the course of the reaction, the effective order of the reaction is reduced. Thus if in carrying out a reaction which is normally second order with a rate equation $\mathscr{R}_A = k_2 C_A C_B$, an excess of **B** is used, then C_B remains constant and equal to the initial value C_{B0}. The rate equation may then be written $\mathscr{R}_A = k_1 C_A$ where $k_1 = k_2 C_{B0}$ and the reaction is now said to be *pseudo first order*.

Influence of temperature. Activation energy

Experimentally, the influence of temperature on the rate constant of a reaction is well represented by the original equation of Arrhenius:

$$k = \mathscr{A} \exp\left(-E/\mathbf{R}T\right) \qquad \qquad \dots (1.7)$$

where T is the absolute temperature and $\mathbf{R}$ the gas constant. In this equation E is termed the *activation energy*, and $\mathscr{A}$ the *frequency factor*. There are theoretical reasons to suppose that temperature dependence should be more exactly described by an equation of the form $k = \mathscr{A}'T^m \exp\left(-E/\mathbf{R}T\right)$, with m usually in the range 0 to 2. However, the influence of the exponential term in equation 1.7 is in practice so strong as to mask any variation in $\mathscr{A}$ with temperature, and this simple form of the relationship (equation 1.7) is therefore quite adequate. E is called the *activation energy* because in the molecular theory of chemical kinetics it is associated with an energy barrier which the reactants must surmount to form an activated complex in the transition state. Similarly, $\mathscr{A}$ is associated with the frequency with which the activated complex breaks down into products; or, in terms of the simple collision theory, it is associated with the frequency of collisions.

Values of the activation energy E are usually quoted in kJ/kmol, $\mathbf{R}$ being expressed as kJ/kmol K. For most reactions the activation energy lies in the range 50,000–250,000 kJ/kmol, which implies a very rapid increase in rate constant with temperature. Thus, for a reaction which is occurring at a temperature in the region of 100°C and has an activation energy of 100,000 kJ/kmol, the reaction rate will be doubled for a temperature rise of only 10°C.

Thus, the complete rate equation for an irreversible reaction normally has the form:

$$\mathscr{R}_A = \mathscr{A} \exp\left(-E/\mathbf{R}T\right) C_A^p C_B^q C_C^r \qquad \qquad \dots (1.8)$$

Unfortunately, the exponential temperature term $\exp(-E/\mathbf{R}T)$ is rather troublesome to handle mathematically, both by analytical methods and numerical techniques. In reactor design this means that calculations for reactors which are not operated isothermally tend to become complicated. In a few cases, useful results can be obtained by abandoning the exponential term altogether and substituting a linear variation of reaction rate with temperature, but this approach is quite inadequate unless the temperature range is very small.

Rate equations and reaction mechanism

One of the reasons why chemical kinetics is an important branch of physical chemistry is that the rate of a chemical reaction may be a significant guide to its mechanism. The engineer concerned with reactor design and development is not interested in reaction mechanism *per se*, but should be aware that an insight into the mechanism of the reaction can provide a valuable clue to the kind of rate equation to be used in a design problem. In the present chapter, it will be possible to make only a few observations on the subject, and for further information the excellent text of FROST and PEARSON[4] should be consulted.

The first point which must be made is that the *overall* stoichiometry of a reaction is *no guide whatsoever* to its rate equation or to the mechanism of reaction. A stoichiometric equation is no more than a material balance; thus the reaction:

$$KClO_3 + 6FeSO_4 + 3H_2SO_4 \rightarrow KCl + 3Fe_2(SO_4)_3 + 3H_2O$$

is in fact second order in dilute solution with the rate of reaction proportional to the concentrations of ClO_3^- and Fe^{2+} ions. In the general case, given by equation 1.4, the stoichiometric coefficients v_A, v_B, v_C, are not necessarily related to the orders p, q, r for the reaction.

However, if it is known from kinetic or other evidence that a reaction $M + N \rightarrow$ Product is a simple *elementary reaction*, i.e., if it is known that its mechanism is simply the interaction between a molecule of M and a molecule of N, then the molecular theory of reaction rates predicts that the rate of this elementary step is proportional to the concentration of species M and the concentration of species N, i.e. it is second order overall. The reaction is also said to be *bimolecular* since two molecules are involved in the actual chemical transformation.

Thus, the reaction between H_2 and I_2 is known to be an elementary bimolecular reaction:

$$H_2 + I_2 \rightarrow 2HI$$

It is in fact virtually the only known example of a simple bimolecular reaction between stable species in the gas phase.

The rate of the forward reaction is thus given by:

$$\mathscr{R}_{I_2} = k_f C_{H_2} C_{I_2}$$

The reaction between H_2 and Br_2, however, is in reality quite different although the stoichiometric equation:

$$H_2 + Br_2 \rightarrow 2HBr$$

looks similar. It has a chain mechanism consisting of the elementary steps:

$$Br_2 \rightleftharpoons 2Br\cdot \quad \textit{chain initiation and termination}$$
$$\left. \begin{array}{l} Br\cdot + H_2 \rightleftharpoons HBr + H\cdot \\ H\cdot + Br_2 \rightarrow HBr + Br\cdot \end{array} \right\} \textit{chain propagation}$$

The rate of the last reaction, for example, is proportional to the concentration of $H\cdot$ and the concentration of Br_2, i.e. it is second order. When the rates of these elementary steps are combined into an overall rate equation, this becomes:

$$\mathscr{R}_{Br_2} = \frac{k' C_{H_2} C_{Br_2}^{\frac{1}{2}}}{1 + k'' \dfrac{C_{HBr}}{C_{Br_2}}}$$

where k' and k'' are constants, which are combinations of the rate constants of the elementary steps. This rate equation has a different form from the usual type given by equation 1.6, and cannot therefore be said to have any order because the definition of order applies only to the usual form.

We shall find that the rate equations of gas–solid heterogeneous catalytic reactions (Chapter 2) also do not, in general, have the same form as equation 1.6.

However, many reactions, although their mechanism may be quite complex, do conform to simple first or second order rate equations. This is because the rate of the overall reaction is limited by just one of the elementary reactions which is then said to be rate-determining. The kinetics of the overall reaction thus reflect the kinetics of this particular step. An example is the pyrolysis of ethane[9] which is important industrially as a source of ethylene[14] (see also page 40). The main overall reaction is:

$$C_2H_6 \rightarrow C_2H_4 + H_2$$

Although there are complications concerning this reaction, under most circumstances it is first order, the kinetics being largely determined by the first step in a chain mechanism:

$$C_2H_6 \rightarrow 2CH_3\cdot$$

which is followed by the much faster reactions

$$\left.\begin{array}{l} CH_3\cdot + C_2H_6 \rightarrow C_2H_5\cdot + CH_4 \\ C_2H_5\cdot \qquad\quad \rightarrow C_2H_4 + H\cdot \\ H\cdot + C_2H_6 \quad \rightarrow C_2H_5\cdot + H_2 \end{array}\right\} \textit{ chain propagation}$$

Eventually the reaction chains are broken by termination reactions. Other free radical reactions also take place to a lesser extent leading to the formation of CH_4 and some higher hydrocarbons among the products.

Reversible reactions

For reactions which do not proceed virtually to completion, it is necessary to include the kinetics of the reverse reaction, or the equilibrium constant, in the rate equation.

The equilibrium state in a chemical reaction can be considered from two distinct points of view. The first is from the standpoint of classical thermodynamics, and leads to relationships between the equilibrium constant and thermodynamic quantities such as free energy and heat of reaction, from which we can very usefully calculate equilibrium conversion. The second is a kinetic viewpoint, in which the state of chemical equilibrium is regarded as a dynamic balance between forward and reverse reactions; at equilibrium the rates of the forward reactions and of the reverse reaction are just equal to each other, making the net rate of transformation zero.

Consider a reversible reaction:

$$A + B \underset{k_r}{\overset{k_f}{\rightleftharpoons}} M + N$$

which is second order overall in each direction, and first order with respect to each species. The hydrolysis of an ester such as ethyl acetate is an example

$$CH_3COOC_2H_5 + NaOH \rightleftharpoons CH_3COO\,Na + C_2H_5OH$$

The rate of the forward reaction expressed with respect to A, $\mathscr{R}_{+A}$ is given by $\mathscr{R}_{+A} = k_f C_A C_B$, and the rate of the reverse reaction (again expressed with respect to A and written $\mathscr{R}_{-A}$) is given by $\mathscr{R}_{-A} = k_r C_M C_N$. The net rate of reaction in the direction left to right is thus:

$$\mathscr{R}_A = \mathscr{R}_{+A} - \mathscr{R}_{-A} = k_f C_A C_B - k_r C_M C_N \qquad \dots (1.9)$$

At equilibrium, when $C_A = C_{Ae}$ etc., $\mathscr{R}_A$ is zero and we have:

$$k_f C_{Ae} C_{Be} = k_r C_{Me} C_{Ne}$$

or

$$\frac{C_{Me} C_{Ne}}{C_{Ae} C_{Be}} = \frac{k_f}{k_r}$$

But $C_{Me} C_{Ne} / C_{Ae} C_{Be}$ is the equilibrium constant K_c and hence $k_f/k_r = K_c$. Often it is convenient to substitute for k_r in equation 1.9 so that we have as a typical example of a *rate equation for a reversible reaction*:

$$\mathscr{R}_A = k_f \left(C_A C_B - \frac{C_M C_N}{K_c} \right)$$

We see from the above example that the forward and reverse rate constants are not completely independent, but are related by the equilibrium constant, which in turn is related to the thermodynamic free energy, etc. More detailed examination of the kinds of kinetic equations which might be used to describe the forward and reverse reactions shows that, to be consistent with the thermodynamic equilibrium constant, the form of the rate equation for the reverse reaction cannot be completely independent of the forward rate equation. A good example is the formation of phosgene:

$$CO + Cl_2 \rightleftharpoons CO\,Cl_2$$

The rate of the forward reaction is given by $\mathscr{R}_{+CO} = k_f C_{CO} C_{Cl_2}^{\frac{3}{2}}$. This rate equation indicates that the chlorine concentration must also appear in the reverse rate equation. Let this be $\mathscr{R}_{-CO} = k_r C_{COCl_2}^{p} C_{Cl_2}^{q}$; then at equilibrium, when $\mathscr{R}_{+CO} = \mathscr{R}_{-CO}$, we must have:

$$\frac{\mathscr{R}_{-CO}}{\mathscr{R}_{+CO}} = 1 = \frac{k_r}{k_f} \left(\frac{C_{COCl_2}^p C_{Cl_2}^q}{C_{CO} C_{Cl_2}^{\frac{1}{2}}} \right)_{eq}$$

But we know from the thermodynamic equilibrium constant that:

$$K_c = \left(\frac{C_{COCl_2}}{C_{CO} C_{Cl_2}} \right)_{eq}$$

Therefore it follows that $p = 1$ and $q = \frac{1}{2}$. The complete rate equation is therefore:

$$\mathscr{R}_{CO} = k_f C_{CO} C_{Cl_2}^{\frac{3}{2}} - k_r C_{COCl_2} C_{Cl_2}^{\frac{1}{2}}$$

or

$$\mathscr{R}_{CO} = k_f \left(C_{CO} C_{Cl_2}^{\frac{3}{2}} - \frac{C_{COCl_2} C_{Cl_2}^{\frac{1}{2}}}{K_c} \right)$$

Rate equations for constant volume batch reactors

In applying a rate equation to a situation where the volume of a given reaction mixture (i.e. the density) remains constant throughout the reaction, the treatment is very much simplified if the equation is expressed in terms of a variable χ, which is defined as the number of mols of a particular reactant transformed per unit volume of reaction mixture (e.g. $C_{A0} - C_A$) at any instant of time t. The quantity χ is very similar to a molar concentration and has the same units. By simple stoichiometry, the mols of the other reactants transformed and products generated can also be expressed in terms of χ, and the rate of the reaction can be expressed as the rate of increase in χ with time. Thus, by definition,

$$\mathscr{R}_A = -\frac{1}{V} \frac{dn_A}{dt}$$

and if V is constant this becomes:

$$\mathscr{R}_A = -\frac{d(n_A/V)}{dt} = -\frac{dC_A}{dt} = \frac{d\chi}{dt} \qquad \dots (1.10)$$

χ being the mols of **A** which have reacted. The general rate equation 1.6 may then be written:

$$\frac{d\chi}{dt} = k(C_{A0} - \chi)^p \left(C_{B0} - \frac{v_B}{v_A} \chi \right)^q \left(C_{C0} - \frac{v_C}{v_A} \chi \right)^r \qquad \dots (1.11)$$

where C_{A0} etc. are the initial concentrations. This equation may then in general, under constant temperature conditions, be integrated to give χ as a function of time, so that the reaction time for any particular conversion can be readily calculated.

The equations which result when these integrations are carried out for reactions of various orders are discussed in considerable detail in most texts dealing with the physico-chemical aspects of chemical kinetics[4, 9]. Table 1.1 shows a summary of some of the simpler cases; the integrated forms can be easily verified by the reader if desired.

One particular point of interest is the expression for the *half-life* of a reaction $t_{\frac{1}{2}}$; this is the time required for one half of the reactant in question to disappear. A first order reaction is unique in that the *half-life* is independent of the initial concentration of the reactant. This characteristic is sometimes used as a test of whether a reaction really is first order. Also since $t_{\frac{1}{2}} = \dfrac{2 \cdot 3}{k_1} \log_{10} 2$, a first order rate constant can be readily converted into a *half-life* which one can easily remember as characteristic of the reaction.

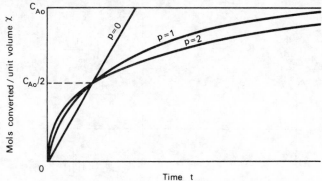

FIG. 1.8. Batch reactions at constant volume: Comparison of curves for zero, first and second order reactions

A further point of interest about the equations shown in Table 1.1 is to compare the shapes of graphs of χ (or fractional conversion $\chi/C_{A0} = \alpha_A$) vs. time for reactions of different orders. Figure 1.8 shows a comparison between first and second order reactions involving a single reactant only, together with the straight line for a zero order reaction. The rate constants have been taken so that the curves coincide at 50 per cent conversion. The rate of reaction at any time is given by the slope of the curve (as indicated by equation 1.10). It may be seen that the rate of the second order reaction is high at first but falls rapidly with increasing time and, compared with first order reactions, longer reaction times are required for high conversions. The zero order reaction is the only one where the reaction rate does not decrease with increasing conversion. Many biological systems have apparent reaction orders between 0 and 1 and will have a behaviour intermediate between the curves shown.

Experimental determination of kinetic constants

The interpretation of laboratory scale experiments to determine order and rate constant is another subject which is considered at length in physical chemistry texts[4, 9]. Essentially, it is a process of fitting a rate equation of the

TABLE 1.1. *Rate equations for constant volume batch reactors*

Reaction type	Rate equation	Integrated form
Irreversible reactions		
First order $A \rightarrow$ products	$\dfrac{d\chi}{dt} = k_1(C_{A0} - \chi)$	$t = \dfrac{1}{k_1} \ln \dfrac{C_{A0}}{(C_{A0} - \chi)}$ $t_{\frac{1}{2}} = \dfrac{\ln 2}{k_1}$
Second order $A + B \rightarrow$ products	$\dfrac{d\chi}{dt} = k_2(C_{A0} - \chi)(C_{B0} - \chi)$	$t = \dfrac{1}{k_2(C_{B0} - C_{A0})} \ln \dfrac{C_{A0}(C_{B0} - \chi)}{C_{B0}(C_{A0} - \chi)}$ $C_{A0} \neq C_{B0}$
$2A \rightarrow$ products	$\dfrac{d\chi}{dt} = k_2(C_{A0} - \chi)^2$	$t = \dfrac{1}{k_2} \left(\dfrac{1}{C_{A0} - \chi} - \dfrac{1}{C_{A0}} \right)$ $t_{\frac{1}{2}} = \dfrac{1}{k_2 C_{A0}}$
Order p, one reactant $A \rightarrow$ products	$\dfrac{d\chi}{dt} = k(C_{A0} - \chi)^p$	$t = \dfrac{1}{k(p-1)} \left[\dfrac{1}{(C_{A0} - \chi)^{p-1}} - \dfrac{1}{C_{A0}^{p-1}} \right]$ $t_{\frac{1}{2}} = \dfrac{2^{p-1} - 1}{k(p-1)C_{A0}^{p-1}} \quad p \neq 1$
Reversible reactions		
First order both directions $A \underset{k_r}{\overset{k_f}{\rightleftharpoons}} M$	$\dfrac{d\chi}{dt} = k_f(C_{A0} - \chi) - k_r(C_{M0} + \chi)$	$t = \dfrac{1}{(k_f + k_r)} \ln \dfrac{k_f C_{A0} - k_r C_{M0}}{k_f(C_{A0} - \chi) - k_r(C_{M0} + \chi)}$ or since $K_c = k_f/k_r$ $t = \dfrac{K_c}{k_f(1 + K_c)} \ln \dfrac{K_c C_{A0} - C_{M0}}{K_c(C_{A0} - \chi) - (C_{M0} + \chi)}$ If $C_{M0} = 0$ $t = \dfrac{1}{(k_f + k_r)} \ln \dfrac{\chi_e}{(\chi_e - \chi)}$
Second order both directions $A + B \underset{k_r}{\overset{k_f}{\rightleftharpoons}} M + N$	If $C_{A0} = C_{B0}$ and $C_{M0} = C_{N0} = 0$ $\dfrac{d\chi}{dt} = k_f(C_{A0} - \chi)^2 - k_r \chi^2$	$t = \dfrac{\sqrt{K_c}}{2k_f C_{A0}} \ln \dfrac{C_{A0} + \chi[1/(\sqrt{K_c}) - 1]}{C_{A0} - \chi[1/(\sqrt{K_c}) + 1]}$

24

general form given by equation 1.6 to a set of numerical data. The experiments which are carried out to obtain the kinetic constants may be of two kinds, depending on whether the rate equation is to be used in its original (*differential*) form, or in its *integrated* form (see Table 1.1). If the differential form is to be used, the experiments must be designed so that the rate of disappearance of reactant A, $\mathscr{R}_A$, can be measured without its concentration changing appreciably. With batch or tubular reactors this has the disadvantage in practice that very accurate measurements of C_A must be made so that, when differences in concentration ΔC_A are taken to evaluate $\mathscr{R}_A$ (e.g. for a batch reactor, equation 1.10 in finite difference form is $\mathscr{R}_A = -\Delta C_A/\Delta t$), the difference may be obtained with sufficient accuracy. Continuous stirred tank reactors do not suffer from this disadvantage; by operating in the steady state, steady concentrations of the reactants are maintained and the rate of reaction is determined readily.

If the rate equation is to be employed in its integrated form, the problem of determining kinetic constants from experimental data from batch or tubular reactors is in many ways equivalent to taking the design equations and working backwards. Thus, for a batch reactor with constant volume of reaction mixture at constant temperature, the equations listed in Table 1.1 apply. For example, if a reaction is suspected of being second order overall, the experimental results are plotted in the form:

$$\frac{1}{C_{B0} - C_{A0}} \ln \left\{ \frac{C_{A0}(C_{B0} - \chi)}{C_{B0}(C_{A0} - \chi)} \right\} \text{ versus } t$$

If the points lie close to a straight line, this is taken as confirmation that a second order equation satisfactorily describes the kinetics, and the value of the rate constant k_2 is found by fitting the best straight line to the points by linear regression. Experiments using tubular and continuous stirred tank reactors to determine kinetic constants are discussed in the sections describing these reactors (pages 45, 57).

Unfortunately, many of the chemical processes which are important industrially are quite complex. A complete description of the kinetics of a process, including byproduct formation as well as the main chemical reaction, may involve several individual reactions, some occurring simultaneously, some proceeding in a consecutive manner. Often the results of laboratory experiments in such cases are ambiguous, and even if complete elucidation of such a complex reaction pattern is possible, it may take several man-years of experimental effort. Whereas ideally the design engineer would like to have a complete set of rate equations for all the reactions involved in a process, in practice the data available to him often fall far short of this.

GENERAL MATERIAL AND THERMAL BALANCES

The starting point for the design of any type of reactor is the general material balance. This material balance can be carried out with respect to one of the reactants or to one of the products. However, if we are dealing with a single reaction such as:

$$v_A A + v_B B = v_M M + v_N N$$

then, in the absence of any separation of the various components by diffusion, it is not necessary to write separate material balance equations for each of the reactants and products. The stoichiometric equation shows that if v_A mols of **A** react, v_B mols of **B** must also have disappeared, v_M mols of **M** must have been formed together with v_N mols of **N**. In such a case the extent to which the reaction has proceeded at any stage can be expressed in terms of the fractional conversion α of any selected reactant, for example **A**. (See also page 22 where similarly the rate of the reaction could be expressed by considering one reactant only.) Alternatively, one of the products **M** or **N** could be chosen as the entity for the material balance equation; however, it is usual to use one reactant as a basis because there may be some uncertainty about just what products are present when the procedure is extended to more complex reactions in which several byproducts are formed, whereas usually the chemical nature of the reactants is known for certain.

Basically, the general material balance for a reactor follows the same pattern as all material and energy balances, namely:

$$\text{Input} - \text{Output} = \text{Accumulation}$$

but with the important difference that the reactant in question can disappear through chemical reaction. The material balance must therefore be written:

$$\text{Input} - \text{Output} - \text{Reaction} = \text{Accumulation}$$
$$\quad (1) \qquad\quad (2) \qquad\quad (3) \qquad\qquad (4)$$

In setting out this equation in an exact form for any particular reactor, the material balance has to be carried out
   (a) over a certain element of volume, and
   (b) over a certain element of time.
If the compositions vary with position in the reactor, which is the case with a tubular reactor, a differential element of volume δV_t must be used, and the equation integrated at a later stage. Otherwise, if the compositions are uniform, e.g. a well-mixed batch reactor or a continuous stirred tank reactor, then the size of the volume element is immaterial; it may conveniently be unit volume (1 m³) or it may be the whole reactor. Similarly, if the compositions are changing with time as in a batch reactor, the material balance must be made over a differential element of time. Otherwise for a tubular or a continuous stirred tank reactor operating in a steady state, where compositions do not vary with time, the time interval used is immaterial and may conveniently be unit time (1 s). Bearing in mind these considerations the general material balance may be written:

Rate of flow of reactant into volume element		Rate of flow of reactant out of volume element		Rate of reactant removal by reaction within volume element		Rate of accumulation of reactant within volume element	
	$-$		$-$		$=$		$\quad \ldots\ldots (1.12)$
(1)		(2)		(3)		(4)	

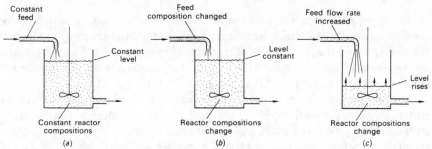

FIG. 1.9. Continuous stirred tank reactor showing steady state operation (*a*) and two modes of unsteady state operation (*b*) and (*c*)

For each of the three basic types of chemical reactor this equation may be reduced to a simplified form. For a *batch reactor* terms (1) and (2) are zero and the *Rate of Accumulation*, i.e. the rate of disappearance of the reactant, is equal to the rate of reaction. For a *tubular reactor* or a *continuous stirred tank reactor*, if operating in a *steady state*, the *Rate of Accumulation* term (4) is by definition zero, and the *Rate of reactant removal by reaction* is just balanced by the difference between inflow and outflow.

For unsteady state operation of a flow reactor, it is important to appreciate the distinction between the *Reaction* term (3) above and the *Accumulation* term (4), which are equal for a batch reactor. Transient operation of a flow reactor occurs during start-up and in response to disturbances in the operating conditions. The nature of transients induced by disturbances and the differences between terms (3) and (4) above can best be visualised for the case of a *continuous stirred tank reactor* (Fig. 1.9). In Fig. 1.9*a*, the reactor is operating in a steady state. In Fig. 1.9*b* it is subject to an increase in the input of reactant owing to a disturbance in the feed composition. This results in a rise in the concentration of the reactant within the reaction vessel corresponding to the *Accumulation* term (4) which is quite distinct from, and additional to, the *Reactant removal by the reaction* term (3). Figure 1.9*c* shows another kind of transient which will cause compositions in the reactor to change, namely a change in the volume of reaction mixture contained in the reactor. Other variables which must be controlled, apart from feed composition and flowrates in and out, are temperature and, for gas reactions particularly, pressure. Variations of any of these quantities with time will cause a change in the composition levels of the reactant in the reactor, and these will appear in the *Accumulation* term (4) in the material balance.

The heat balance for a reactor has a form very similar to the general material balance, i.e.

Rate of heat inflow into volume element	−	Rate of heat outflow from volume element	−	Rate of absorption of heat by chemical reaction in volume element	=	Rate of heat accumulation in volume element	.... (1.13)
(1)		(2)		(3)		(4)	

In the *Inflow* and *Outflow* terms (1) and (2), the heat flow may be of two kinds: the first is transfer of sensible heat or enthalpy by the fluid entering and leaving the element; and the second is heat transferred to or from the fluid across heat transfer surfaces, such as cooling coils situated in the reactor. The *Heat absorbed in the chemical reaction*, term (3), depends on the rate of reaction, which in turn depends on the concentration levels in the reactor as determined by the general material balance equation. Since the rate of reaction depends also on the temperature levels in the reactor as determined by the heat balance equation, the material balance and the heat balance interact with each other, and the two equations have to be solved simultaneously. The types of solutions obtained are discussed further under the headings of the various types of reactor—batch, tubular and continuous stirred tank.

BATCH REACTORS

There is a tendency in chemical engineering to try to make all processes continuous. Whereas continuous flow reactors are likely to be most economic for large scale production, the very real advantages of batch reactors, especially for smaller scale production, should not be overlooked. Small batch reactors generally require less auxiliary equipment, such as pumps, and their control systems are less elaborate and costly than those for continuous reactors, although manpower needs are greater. However, large batch reactors may sometimes be fitted with highly complex control systems. A big advantage of batch reactors in the dyestuff, fine chemical and pharmaceutical industries is their versatility. A corrosion resistant batch reactor such as an enamel or rubber-lined jacketed vessel (Fig. 1.4*a*) or a stainless steel vessel with heating and cooling coils (Fig. 1.4*b*) can be used for a wide variety of similar kinds of reaction. Sometimes only a few batches per year are required to meet the demand for an unusual product. In some processes, such as polymerisations and fermentations, batch reactors are traditionally preferred because the interval between batches provides an opportunity to clean the system thoroughly and ensure that no deleterious intermediates such as foreign bacteria build up and spoil the product. Moreover, it must not be forgotten that a squat tank is the most economical shape for holding a given volume of liquid, and for slow reactions a tubular flow reactor with a diameter sufficiently small to prevent backmixing, would be more costly than a simple batch reactor. Although at present we are concerned mainly with homogeneous reactions, we should note that the batch reactor has many advantages for heterogeneous reactions; the agitator can be designed to suspend solids in the liquid, and to disperse a second immiscible liquid or a gas.

In calculating the volume required for a batch reactor, we shall be specifying the volume of liquid which must be processed. In designing the vessel itself the heights should be increased by about 10 per cent to allow freeboard for waves and disturbances on the surface of the liquid; additional freeboard may have to be provided if foaming is anticipated.

Calculation of reaction time; basic design equation

Calculation of the time required to reach a particular conversion is the main objective in the design of batch reactors. Knowing the amount of reactant converted, i.e. the amount of the desired product formed per unit volume in this reaction time, the volume of reactor required for a given production rate can be found by simple scale-up as shown in the example on ethyl acetate below.

The reaction time t_r is determined by applying the general material balance equation 1.12. In the most general case, when the volume of the reaction mixture is not constant throughout the reaction, it is convenient to make the material balance over the whole volume of the reactor V_b. For the reactant **A**, if n_{A0} mols are charged initially, the number of mols remaining when the fraction of **A** converted is α_A is $n_{A0}(1 - \alpha_A)$ and, using a differential element of time, the rate at which this is changing, i.e. the *Accumulation* term (4) in equation 1.12 is:

$$\frac{\mathrm{d}}{\mathrm{d}t} [n_{A0}(1 - \alpha_A)] = -n_{A0} \frac{\mathrm{d}\alpha_A}{\mathrm{d}t}$$

The rate at which **A** is removed by reaction term (3) is $\mathscr{R}_A V_b$ and, since the *Flow* terms (1) and (2) are zero, we have:

$$-\mathscr{R}_A V_b = -n_{A0} \frac{\mathrm{d}\alpha_A}{\mathrm{d}t}$$
$$\text{Reaction} \qquad \text{Accumulation}$$

Thus, integrating over the period of the reaction to a final conversion α_{Af}, we obtain the basic design equation:

$$t_r = n_{A0} \int_0^{\alpha_{Af}} \frac{\mathrm{d}\alpha_A}{\mathscr{R}_A V_b} \qquad \dots (1.14)$$

For many liquid phase reactions it is reasonable to neglect any change in volume of the reaction mixture. Equation 1.14 then becomes:

$$t_r = \frac{n_{A0}}{V_b} \int_0^{\alpha_{Af}} \frac{\mathrm{d}\alpha_A}{\mathscr{R}_A} = C_{A0} \int_0^{\alpha_{Af}} \frac{\mathrm{d}\alpha_A}{\mathscr{R}_A} \qquad \dots (1.15)$$

This form of the equation is convenient if there is only one reactant. For more than one reactant and for reversible reactions, it is more convenient to write the equation in terms of χ, the mols of **A** converted per unit volume $\chi = C_{A0}\alpha_A$, and obtain:

$$t_r = \int_0^{\chi_f} \frac{\mathrm{d}\chi}{\mathscr{R}_A} \qquad \dots (1.16)$$

This is the integrated form of equation 1.10 obtained previously; it may be derived formally by applying the general material balance to unit volume under conditions of constant density, when the *Rate of reaction* term (3) is simply $\mathscr{R}_A$ and the *Accumulation* term (4) is:

$$\frac{dC_A}{dt}, \text{ i.e. } -\frac{d\chi}{dt}$$

Thus:

$$-\mathscr{R}_A = -\frac{d\chi}{dt}$$

which is equation 1.10.

Reaction time—isothermal operation

If the reactor is to be operated isothermally, the rate of reaction $\mathscr{R}_A$ can be expressed as a function of concentrations only, and the integration in equation 1.15 or 1.16 carried out. The integrated forms of equation 1.16 for a variety of the simple rate equations are shown in Table 1.1 and Fig. 1.8. We now consider an example with a rather more complicated rate equation involving a reversible reaction, and show also how the volume of the batch reactor required to meet a particular production requirement is calculated.

Example: *Production of ethyl acetate in a batch reactor*

Ethyl acetate is to be manufactured by the esterification of acetic acid with ethanol in an isothermal batch reactor. A production rate of 10 tonne/day of ethyl acetate is required.

$$CH_3 \cdot COOH + C_2H_5OH \rightleftharpoons CH_3 \cdot COOC_2H_5 + H_2O$$

$$\quad\quad A \quad\quad\quad\quad B \quad\quad\quad\quad\quad M \quad\quad\quad\quad N$$

The reactor will be charged with a mixture containing 500 kg/m³ ethanol and 250 kg/m³ acetic acid, the remainder being water and a small quantity of hydrochloric acid to act as a catalyst. The density of this mixture is 1045 kg/m³ which will be assumed constant throughout the reaction. The reaction is reversible with a rate equation which, over the concentration range of interest, can be written:

$$\mathscr{R}_A = k_f C_A C_B - k_r C_M C_N$$

At the operating temperature of 100°C the rate constants have the values:

$$k_f = 8 \cdot 0 \times 10^{-6} \text{ m}^3/\text{kmol s}$$
$$k_r = 2 \cdot 7 \times 10^{-6} \text{ m}^3/\text{kmol s}$$

The reaction mixture will be discharged when the conversion of the acetic acid is 30 per cent. A time of 30 min is required between batches for discharging, cleaning, and recharging. Determine the volume of the reactor required.

Solution

After a time t, if χ kmol/m³ of acetic acid (**A**) has reacted, its concentration will be $(C_{A0} - \chi)$ where C_{A0} is the initial concentration. From the stoichiometry of the reaction, if χ kmol/m³ of acetic acid has reacted, χ kmol/m^3 of ethanol also will have reacted and the same number of mols of ester and of water will have been formed. The rate equation may thus be written:

$$\mathscr{R}_A = k_f(C_{A0} - \chi)(C_{B0} - \chi) - k_r\chi(\dot{C}_{N0} + \chi).$$

From the original composition of the mixture, its density, and the molecular weights of acetic acid, ethanol and water which are 60, 46 and 18 respectively, $C_{A0} = 4\cdot2$ kmol/m^3; $C_{B0} = 10\cdot9$ kmol/m^3; $C_{N0} = 16\cdot4$ kmol/m^3. Thus, from equation 1.16, t_r in seconds is given by:

$$t_r = \int_0^{\chi_f} \frac{d\chi}{8\cdot0 \times 10^{-6}(4\cdot2 - \chi)(10\cdot9 - \chi) - 2\cdot7 \times 10^{-6}\chi(16\cdot4 + \chi)}$$

This integral may be evaluated either by splitting into partial fractions, or by graphical or numerical means. Using the method of partial fractions, we obtain after some fairly lengthy manipulation:

$$t_r = 16200\left[\log_{10}\frac{29 - \chi_f}{2\cdot4 - \chi_f} - 1\cdot082\right]$$

Since the final conversion of acetic acid is to be 30 per cent, $\chi_f = 0\cdot30$, $C_{A0} = 1\cdot26$ kmol/m^3; whence the reaction time t_r is, from the above equation, 4920 s.

Thus 1 m^3 of reactor volume produces $1\cdot26$ kmol of ethyl acetate (molecular weight 88) in a total batch time of 6720 s, i.e. in 4920 s reaction time and 1800 s shut-down time. This is an average production rate of:

$$1\cdot26 \times 88 \times \frac{24 \times 60^2}{6720}$$

i.e. 1420 kg/day per m^3 of reactor volume. Since the required production rate is 10,000 kg/day the required reactor volume is 10,000/1420 = $\underline{7\cdot1 \text{ m}^3}$.

This example on ethyl acetate is useful also in directing attention to an important point concerning reversible reactions in general. A reversible reaction will not normally go to completion, but will slow down as equilibrium is reached. This progress towards equilibrium can however sometimes be disturbed by continuously removing one or more of the products as formed. In the actual manufacture of ethyl acetate, the ester is removed as the reaction proceeds by distilling off a ternary azeotrope of molar composition ethyl acetate $60\cdot1$ per cent, ethanol $12\cdot4$ per cent and water $27\cdot5$ per cent. The net rate of reaction is thereby increased as the rate equation above shows; because C_M is always small the term for the rate of the reverse reaction $k_rC_MC_N$ is always small and the net rate of reaction is virtually equal to the rate of the forward reaction above, i.e. $k_fC_AC_B$.

Maximum production rate

For most reactions, the rate decreases as the reaction proceeds (important exceptions being a number of biological reactions which are autocatalytic). For a reaction with no volume change, the rate is represented by the slope of the curve of χ (mols converted per unit volume) versus time (Fig. 1.10), which decreases steadily with increasing time. The maximum reaction rate occurs at zero time, and, if our sole concern were to obtain maximum output from the reactor and the shutdown time were zero, it appears that the best course would be to discharge the reactor after only a short reaction time t_r, and refill with fresh reactants. It would then be necessary, of course, to separate a large amount

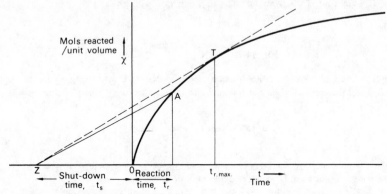

FIG. 1.10. Maximum production rate in a batch reactor with a shut-down time t_s

of reactant from a small amount of product. However, if the shut-down time is appreciable and has a value t_s then as we have seen in the example on ethyl acetate above, the average production rate per unit volume is:

$$\frac{\chi}{t_r + t_s}$$

The maximum production rate is therefore given by the maximum value of:

$$\frac{\chi}{t_r + t_s}$$

This maximum can be most conveniently found graphically (Fig. 1.10). The average production rate is given by the slope of the line ZA; this is obviously a maximum when the line is tangent to the curve of χ versus t, i.e. ZT as shown. The reaction time obtained $t_{r\,max}$ is not necessarily the optimum for the process as a whole, however. This will depend in addition upon the costs involved in feed preparation, separation of the products, and storage.

Reaction time—non-isothermal operation

If the temperature is not constant but varies during the course of the reaction, then the rate of reaction $\mathscr{R}_A$ in equation 1.15 or 1.16 will be a function of temperature as well as concentration (equation 1.8). The temperature at any stage is determined by a heat balance, the general form of which is given by equation 1.13. Since there is no material flow into or out of a batch reactor during reaction, the enthalpy changes associated with such flows in continuous reactors are absent. However, there may be a flow of heat to or from the reactor by heat transfer using the type of equipment shown in Fig. 1.4. In the case of a jacketed vessel or one with an internal coil, the heat transfer coefficient will be largely dependent on the agitator speed, which is usually held constant. Thus, assuming that the viscosity of the liquid does not change appreciably, which is reasonable in many cases (except for some polymerisations), the heat transfer

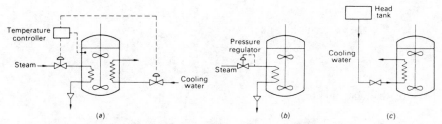

FIG. 1.11. Methods of operating batch reactors

(a) Isothermal operation of an exothermic reaction: heating to give required initial temperature, cooling to remove heat of reaction

(b) and (c) Non-isothermal operation: simple schemes having constant heat transfer coefficient U

coefficient may be taken as constant. If heating is effected by condensing saturated steam at constant pressure, as in Fig. 1.11b, the temperature on the coil side T_C is constant. If cooling is carried out with water (Fig. 1.11c), the rise in temperature of the water may be small if the flow rate is large, and T_C again taken as constant. Thus, we may write the rate of heat transfer to cooling coils of area A_t as:

$$Q = UA_t(T - T_C)$$

where T is the temperature of the reaction mixture. The heat balance taken over the whole reactor thus becomes:

$$-UA_t(T - T_C) + (-\Delta H_A)V_b\mathcal{R}_A = (\Sigma m_j c_j)\frac{dT}{dt} \qquad \dots (1.17)$$

Rate of heat flow out by heat transfer

Rate of heat release by chemical reaction

Rate of heat accumulation

where ΔH_A is the enthalpy change in the reaction per mol of A reacting, and $(\Sigma m_j c_j)$ is the sum of the heat capacities (i.e. mass × specific heat) of the reaction mixture and the reactor itself, including all the various internal components such as the agitator, whose temperatures also change.

Finding the time required for a particular conversion involves the solution of two simultaneous equations, i.e. 1.15 or 1.16 for the material balance and 1.17

for the heat balance. Generally a solution in analytical form is unobtainable and numerical methods or analogue simulation must be used. Taking, for example, a first order reaction with constant volume:

$$\mathscr{R}_A = \frac{d\chi}{dt} = k(C_{A0} - \chi)$$

and

$$k = \mathscr{A} \exp\left(\frac{-E}{RT}\right)$$

we have for the material balance:

$$\frac{d\chi}{dt} = \mathscr{A} \exp\left(\frac{-E}{RT}\right)(C_{A0} - \chi) \qquad \ldots (1.18)$$

and for the heat balance:

$$(-\Delta H_A) V_b \mathscr{A} \exp\left(\frac{-E}{RT}\right)(C_{A0} - \chi) - U A_t(T - T_C) = (\Sigma m_j c_j)\frac{dT}{dt} \qquad \ldots (1.19)$$

With t as the independent variable, the solution will be:

(a) χ as $f(t)$
(b) T as $F(t)$.

A typical requirement is that the temperature shall not rise above T_{mx} in order to avoid byproducts or hazardous operation. The forms of the solutions obtained are sketched in Fig. 1.12.

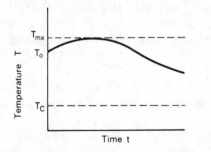
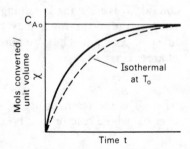

FIG. 1.12. Non-isothermal batch reactor: Typical curves for an exothermic reaction with just sufficient cooling (constant U and T_C) to prevent temperature rising above T_{mx}

Adiabatic operation

If the reaction is carried out adiabatically (i.e. without heat transfer, so that $Q = 0$), the heat balance shows that the temperature at any stage in the reaction can be expressed in terms of the conversion only. This is because, however fast or slow the reaction, the heat released by the reaction is retained as sensible heat in the reactor. Thus, for reaction at constant volume, putting $Q = 0$ and $\mathscr{R}_A = d\chi/dt$ in equation 1.17:

$$(-\Delta H_A) V_b \, d\chi = (\Sigma m_j c_j) \, dT \qquad \ldots (1.20)$$

Equation 1.20 may be solved to give the temperature as a function of χ. Usually the change in temperature $(T - T_0)$, where T_0 is the initial temperature, is proportional to χ, since $\Sigma m_j c_p$, the total heat capacity, does not vary appreciably with temperature or conversion. The appropriate values of the rate constant are then used to carry out the integration of equation 1.15 or 1.16 numerically, as shown in the following example.

Example: *Adiabatic batch reactor*

Acetic anhydride is hydrolysed by water in accordance with the equation:

$$(CH_3 \cdot CO)_2O + H_2O \rightleftharpoons 2CH_3 \cdot COOH$$

In a dilute aqueous solution where a large excess of water is present, the reaction is irreversible and pseudo-first order with respect to the acetic anhydride. The variation of the pseudo-first order rate constant with temperature is as follows:

Temperature	(°C)	15	20	25	30
	(K)	288	293	298	303
Rate constant	(s^{-1})	0·00134	0·00188	0·00263	0·00351

A batch reactor for carrying out the hydrolysis is charged with an anhydride solution containing 0·30 kmol/m^3 at 15°C. The specific heat and density of the reaction mixture are 3·8 kJ/kgK and 1070 kg/m^3, and may be taken as constant throughout the course of the reaction. The reaction is exothermic, the heat of reaction per kmol of anhydride being 210,000 kJ/kmol. If the reactor is operated adiabatically, estimate the time required for the hydrolysis of 80 per cent of the anhydride.

Solution

For the purposes of this example we shall neglect the heat capacity of the reaction vessel. Since the anticipated temperature rise is small, the heat of reaction will be taken as independent of temperature. Because the heat capacity of the reactor is neglected, we may most conveniently take the adiabatic heat balance (equation 1.20) over the unit volume, i.e. 1 m^3 of reaction mixture. Thus, integrating equation 1.20 with the temperature T_0 when $\chi = 0$:

$$(-\Delta H_A)\chi = mc(T - T_0)$$

i.e.

$$210,000 \, \chi = 1070 \times 3 \cdot 8 \, (T - T_0)$$

$\therefore$

$$(T - T_0) = 52\chi$$

Writing

$$\chi = C_{A0}\alpha = 0 \cdot 3 \, \alpha$$

$$(T - T_0) = 15 \cdot 6 \, \alpha$$

Thus, if the reaction went to completion ($\alpha = 1$), the adiabatic temperature rise would be 15·6°C. For a pseudo-first order reaction, the rate equation is:

$$\mathscr{R}_A = k_1(C_{A0} - \chi) = C_{A0}k_1(1 - \alpha)$$

and from equation 1.15:

$$t_r = \int_0^{0 \cdot 8} \frac{d\alpha}{k_1(1 - \alpha)}$$

To evaluate this integral graphically, we need to plot:

$$\frac{1}{k_1(1-\alpha)} \text{ versus } \alpha,$$

remembering that k_1 is a function of T (i.e. of α). Interpolating the values of k_1 given, we evaluate:

$$\frac{1}{k_1(1-\alpha)}$$

for various values of α, some of which are shown in Table 1.2.

TABLE 1.2. *Evaluation of integral to determine* t_r

Fractional conversion α	Temperature rise $(T-T_0)$ deg K	Temperature K	Rate constant $k_1(s^{-1})$	$\frac{1}{k_1(1-\alpha)}$
0	0	288	0·00134	740
0·2	3·1	291·1	0·00167	750
0·4	6·2	294·2	0·00205	810
0·6	9·4	297·4	0·00253	990
0·8	12·5	300·5	0·00305	1630

From the area under the graph up to $\alpha = 0.80$, we find that the required reaction time is approximately 720 s = 12 min.

TUBULAR FLOW REACTORS

The tubular flow reactor (Fig. 1.1b) is chosen when it is desired to operate the reactor continuously but without back-mixing of reactants and products. In the case of an *ideal* tubular reactor, the reaction mixture passes through in a state of *plug flow* which, as the name suggests, means that the fluid moves like a solid plug or piston. Furthermore, in the ideal reactor it is assumed that not only the local mass flow rate but also the fluid properties, temperature, pressure, and compositions are uniform across any section normal to the fluid motion. Of course the compositions, and possibly the temperature and pressure also, change between inlet and outlet of the reactor in the longitudinal direction. In the elementary treatment of tubular reactors, *longitudinal dispersion*, i.e. mixing by diffusion and other processes in the direction of flow, is also neglected.

Thus, in the idealised tubular reactor all elements of fluid take the same time to pass through the reactor and experience the same sequence of temperature, pressure and composition changes. In calculating the size of such a reactor, we are concerned with its volume only; its shape does not affect the reaction so long as plug flow occurs.

The flow pattern of the fluid is, however, only one of the criteria which determine the shape eventually chosen for a tubular reactor. The factors which must be taken into account are:

(a) whether plug flow can be attained,
(b) heat transfer requirements,
(c) pressure drop in the reactor,
(d) support of catalyst, if present, and
(e) ease and cheapness of construction.

Figure 1.13 shows various configurations which might be chosen. One of the cheapest ways of enclosing a given volume is to use a cylinder of height approximately equal to its diameter. In Fig. 1.13a the reactor is a simple cylinder of this kind. Without packing, however, swirling motions in the fluid would cause serious departures from plug flow. With packing in the vessel, such movements are damped out and the simple cylinder is then quite suitable for catalytic reactions where no heat transfer is required. If pressure drop is a problem, the depth of the cylinder may be reduced and its diameter increased as in Fig. 1.13b; to avoid serious departures from plug flow in such circumstances, the catalyst must be uniformly distributed and baffles are often used near the inlet and outlet.

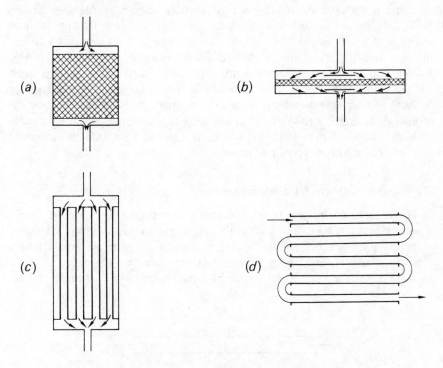

FIG. 1.13. Various configurations for tubular reactors

(a) Simple cylindrical shell: suitable only if packed with catalyst
(b) Shallow cylinder giving low pressure drop through catalyst bed
(c) Tubes in parallel: relatively low tube velocity
(d) Tubes in series: high tube velocity

When heat transfer to the reactor is required, a configuration with a high surface to volume ratio is employed. In the reactors shown in Fig. 1.13c and 1.13d the reaction volume is made up of a number of tubes. In c they are arranged in parallel, whereas in d they are in series. The parallel arrangement gives a lower velocity of the fluid in the tubes, which in turn results in a lower pressure drop, but also a lower heat transfer coefficient (which affects the temperature of the reactant mixture and must be taken into account in calculating the reactor volume). The parallel arrangement is very suitable if a second fluid outside the tubes is used for heat transfer; parallel tubes can be arranged between tube sheets in a compact bundle fitted into a shell, as in a shell and tube heat exchanger. On the other hand, with tubes in series, a high fluid velocity is obtained inside the tubes and a higher heat transfer coefficient results. The series arrangement is therefore often the more suitable if heat transfer is by radiation, when the high heat transfer coefficient helps to prevent overheating of the tubes and coke formation in the case of organic materials.

In practice, there is always some degree of departure from the ideal plug flow condition of uniform velocity, temperature, and composition profiles. If the reactor is not packed and the flow is turbulent, the velocity profile is reasonably flat in the region of the turbulent core (Volume 1, Chapter 3), but in laminar flow, the velocity profile is parabolic. More serious however than departures from a uniform velocity profile are departures from a uniform temperature profile. If there are variations in temperature across the reactor, there will be local variations in reaction rate and therefore in the composition of the reaction mixture. These transverse variations in temperature may be particularly serious in the case of strongly exothermic catalytic reactions which are cooled at the wall (Chapter 2, page 163 *et seq.*). An excellent discussion on how deviations from plug flow arise is given by DENBIGH[11].

Basic design equations for a tubular reactor

The basic equation for a tubular reactor is obtained by applying the general material balance, equation 1.12, with the plug flow assumptions. In steady state operation, which is usually the aim, the *Rate of accumulation* term (4) is zero. The material balance is taken with respect to a reactant **A** over a differential element of volume δV_t (Fig. 1.14). The fractional conversion of **A** in the mixture

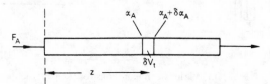

FIG. 1.14. Differential element of a tubular reactor

entering the element is α_A and leaving it is $(\alpha_A + \delta\alpha_A)$. If F_A is the feed rate of **A** into the reactor (mols per unit time) the material balance over δV_t gives:

$$F_A(1 - \alpha_A) - F_A(1 - \alpha_A - \delta\alpha_A) - \mathscr{R}_A\delta V_t = 0$$

$$\quad\text{Inflow}\qquad\qquad\text{Outflow}\qquad\qquad\text{Reaction (in steady state)}$$

$$\therefore \qquad F_A\delta\alpha_A = \mathscr{R}_A\delta V_t$$

Integrating:

$$\frac{V_t}{F_A} = \int_0^{\alpha_{Af}} \frac{d\alpha_A}{\mathscr{R}_A} \qquad \dots (1.21)$$

If the reaction mixture is a fluid whose density remains constant throughout the reaction, equation 1.21 may be written in terms of χ mols of reactant converted per unit volume of fluid. Since $\chi = \alpha_A C_{A0}$, equation 1.21 becomes:

$$\frac{V_t}{F_A} = \frac{1}{C_{A0}} \int_0^{\chi_f} \frac{d\chi}{\mathscr{R}_A}$$

However, $\dfrac{F_A}{C_{A0}} = v$, the volumetric flow rate of the reaction mixture. Hence:

$$\frac{V_t}{v} = \int_0^{\chi_f} \frac{d\chi}{\mathscr{R}_A} \qquad \dots (1.22)$$

This equation can be derived directly from the general material balance equation above (Fig. 1.14) by expressing the flow of reactant **A** into and out of the reactor element δV_t in terms of the volumetric rate of flow of mixture v, which of course is only valid if v is constant throughout the reactor, i.e.:

$$v(C_{A0} - \chi) - v(C_{A0} - \chi - \delta\chi) - \mathscr{R}_A \delta V_t = 0$$
$$\text{Inflow} \qquad\qquad \text{Outflow} \qquad\quad \text{Reaction} \quad \text{(in steady state)}$$

i.e.

$$v\delta\chi = \mathscr{R}_A \delta V_t$$

Hence this leads to equation 1.22.

For many tubular reactors, the pressure drop due to flow of the reaction mixture is relatively small, so that the reactor operates at almost constant pressure. An assumption of constant density and the use of equation 1.22 is usually acceptable for liquids and for gas reactions in which there is no change in the total numbers of mols on either side of the stoichiometric equation. If a gas phase reaction does involve a change in the number of mols or if there is a large temperature change, a volume element of reaction mixture will undergo expansion or contraction in passing through the reactor; an assumption of constant v is then unsatisfactory and equation 1.21 must be used. In these circumstances care must be exercised in how the concentrations which appear in the rate equation for $\mathscr{R}_A$ are expressed in terms of α_A, especially when inerts are present; this point is illustrated in the following example. Generally, if the reactor is operated isothermally, $\mathscr{R}_A$ depends on concentrations alone; if it is not operated isothermally, $\mathscr{R}_A$ is a function of temperature also, and the heat balance equation must be introduced. The example is concerned with an isothermal gas-phase reaction in which expansion of a volume element does occur.

Example: *Production of ethylene by pyrolysis of ethane in an isothermal tubular reactor*

Ethylene is manufactured on a very large scale[14] by the thermal cracking of ethane in the gas phase:

$$C_2H_6 \rightleftharpoons C_2H_4 + H_2$$

Significant amounts of CH_4 and C_2H_2 are also formed but will be ignored for the purposes of this example. The ethane is diluted with steam and passed through a tubular furnace. Steam is used for reasons very similar to those in the case of ethyl benzene pyrolysis (page 12): in particular it reduces the amounts of undesired byproducts. The economic optimum proportion of steam is, however, rather less than in the case of ethylbenzene. We will suppose that the reaction is to be carried out in an isothermal tubular reactor which will be maintained at 900°C. Ethane will be supplied to the reactor at a rate of 20 tonne/h; it will be diluted with steam in the ratio 0·5 mol steam: 1 mol ethane. The required fractional conversion of ethane is 0·6 (the conversion per pass is relatively low to reduce byproduct formation; unconverted ethane is separated and recycled). The operating pressure is 1·4 bar total, and will be assumed constant, i.e. the pressure drop through the reactor will be neglected.

Laboratory experiments, confirmed by data from large scale operations, have shown that ethane decomposition is a homogeneous first order reaction, the rate constant (s^{-1}) being given by the equation[20]:

$$k_1 = 1·535 \times 10^{14} \exp(-294,000/\mathbf{R}T)$$

Thus at 900°C (1173 K) the value of k_1 is 12·8 s^{-1}

We are required to determine the volume of the reactor.

Solution

The ethane decomposition reaction is in fact reversible, but in the first instance, to avoid undue complication, we shall neglect the reverse reaction; a more complete and satisfactory treatment is given below. For a simple first order reaction, the rate equation is:

$$\mathcal{R}_A \quad = \quad k_1 C_A$$

$$kmol/m^3\,s \qquad\qquad s^{-1}\,kmol/m^3$$

These units for $\mathcal{R}_A$ and C_A will eventually lead to the units m^3 for the volume of the reactor when we come to use equation 1.21. When we substitute for $\mathcal{R}_A$ in equation 1.21, however, to integrate, we must express C_A in terms of α_A, where the reactant $\mathbf{A}$ is C_2H_6. To do this we first note that $C_A = y_A C$ where y_A is the mol fraction and C is the molar density of the gas mixture $(kmol/m^3)$. Assuming ideal gas behaviour, C is the same for any gas mixture, being dependent only on pressure and temperature in accordance with the ideal gas laws. Thus 1 kmol of gas occupies 22·41 m^3 at stp. Therefore at 1·4 bar = $1·4 \times 10^5$ N/m^2 and 1173 K it will occupy:

$$22·41 \times \frac{1·013 \times 10^5}{1·4 \times 10^5} \times \frac{1173}{273} = 69·8\ m^3$$

and the molar density C is therefore:

$$1/69·8 = 0·0143\ kmol/m^3$$

We next determine y_A in terms of α_A by making a subsidiary stoichiometric balance over the reactor between the inlet and the point at which the fractional conversion of the ethane is α_A. Taking as a basis 1 mol of ethane entering the reactor we have:

	IN	AT CONVERSION α_A	
	(mol)	(a) mol	(b) mol fraction y
C_2H_6	1	$1 - \alpha_A$	$\left(\dfrac{1 - \alpha_A}{1\cdot 5 + \alpha_A}\right)$
C_2H_4	—	α_A	$\dfrac{\alpha_A}{1\cdot 5 + \alpha_A}$
H_2	—	α_A	$\dfrac{\alpha_A}{1\cdot 5 + \alpha_A}$
H_2O	0·5	0·5	$\dfrac{0\cdot 5}{1\cdot 5 + \alpha_A}$
TOTAL		$1\cdot 5 + \alpha_A$	1

Thus

$$y_A = \frac{1 - \alpha_A}{1\cdot 5 + \alpha_A}$$

Hence, to find the volume of the reactor, using the basic design equation 1.21 and the first order rate equation:

$$\frac{V_t}{F_A} = \int_0^{\alpha_{Af}} \frac{d\alpha_A}{\mathscr{R}_A} = \int_0^{\alpha_{Af}} \frac{d\alpha_A}{k_1 C_A} = \int_0^{\alpha_{Af}} \frac{d\alpha_A}{k_1 y_A C}$$

$$= \frac{1}{k_1 C} \int_0^{\alpha_{Af}} \frac{(1\cdot 5 + \alpha_A)}{1 - \alpha_A} \, d\alpha_A = \frac{1}{k_1 C} \int_0^{\alpha_{Af}} \left[\frac{2\cdot 5}{(1 - \alpha_A)} - 1\right] d\alpha_A$$

$$= \frac{1}{k_1 C} \left[-2\cdot 5 \ln(1 - \alpha_A) - \alpha_A\right]_0^{\alpha_{Af}} = \frac{1}{k_1 C}\left[2\cdot 5 \ln \frac{1}{1 - \alpha_{Af}} - \alpha_{Af}\right]$$

Introducing the numerical values:

$$\frac{V_t}{F_A} = \frac{1}{12\cdot 8 \times 0\cdot 0143}\left[2\cdot 303 \times 2\cdot 5 \log_{10} \frac{1}{(1 - 0\cdot 6)} - 0\cdot 6\right]$$

$$= 9\cdot 27 \text{ m}^3 \text{ s/kmol}$$

The feed rate of ethane (molecular wt. 30) is to be 20 tonne/h which is equivalent to $F_A = 0\cdot 185$ kmol/s. Therefore $V_t = 9\cdot 27 \times 0\cdot 185 = \underline{1\cdot 72 \text{ m}^3}$, which is the volume of the reactor required.

The pyrolysis reaction is strongly endothermic so that one of the main problems in designing the reactor is to provide for sufficiently high rates of heat transfer. The volume calculated above would be made up of a series of tubes, probably in the range 50–150 mm diameter, arranged in a furnace similar to Fig. 1.5c. (For further details of ethylene plants see MILLER[14].)

Calculation with reversible reaction. At 900°C the equilibrium constant K_p for ethane decomposition $P_{C_2H_4} P_{H_2}/P_{C_2H_6}$ is 3·2 bar; using the method described on page 13 the equilibrium conversion of ethane under the conditions above (i.e. 1·4 bar, 0·5 kmol steam added) is 0·86. This shows that the influence of the reverse reaction is appreciable.

The rate equation for a reversible reaction $\mathbf{A} \rightleftharpoons \mathbf{M} + \mathbf{N}$, which is first order in the forward direction and first order with respect to each of $\mathbf{M}$ and $\mathbf{N}$ in the reverse direction, may be written in terms of the equilibrium constant K_c:

$$\mathscr{R}_A = k_1\left[C_A - \frac{C_M C_N}{K_c}\right]$$

Expressing this relation in terms of K_p $(= K_c P/C)$ and mol fractions y_A etc. of the various species, $(C_A = C y_A$ etc):

$$\mathscr{R}_A = k_1 C \left[y_A - \frac{y_M y_N}{K_p} P \right]$$

where P is the total pressure. From the above stoichiometric balance $y_M = y_N = \alpha_A/(1·5 + \alpha_A)$. Hence, substituting, equation 1.21 becomes:

$$\frac{V_t}{F_A} = \int_0^{\alpha_{Af}} \frac{d\alpha_A}{k_1 C \left[\dfrac{(1 - \alpha_A)}{(1·5 + \alpha_A)} - \dfrac{\alpha_A^2}{(1·5 + \alpha_A)^2} \dfrac{P}{K_p} \right]}$$

Substituting numerical values for k_1, C, P and K_p and evaluating this integral for $a_{Af} = 0·6$ gives the result $V_t/F_A = 10·0$ m³ s/kmol: hence $V_t = \underline{1·86}$ m³ which may be compared with the previous result above.

Residence time and space velocity

So far in dealing with tubular reactors we have considered a spatial coordinate as the variable, i.e. an element of volume δV_t situated at a distance z from the reactor inlet (Fig. 1.14), although z has not appeared explicitly in the equations. For a continuous flow reactor operating in a steady state, the spatial coordinate is indeed the most satisfactory variable to describe the situation, because the compositions do not vary with time, but only with position in the reactor.

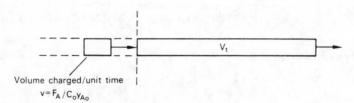

Volume charged/unit time
$v = F_A / C_o y_{Ao}$

FIG. 1.15. Tubular reactor: residence time $\tau = V_t/v$ if volumetric flow rate constant throughout reactor

There is, however, another way of looking at a tubular reactor in which plug flow occurs (Fig. 1.15). If we imagine that a small volume of reaction mixture is encapsulated by a membrane in which it is free to expand or contract at constant pressure, it will behave as a miniature batch reactor, spending a time τ, said to be the *residence time*, in the reactor, and emerging with the conversion α_{Af}. If there is *no expansion or contraction* of the element, i.e. the volumetric rate of flow is constant and equal to v throughout the reactor, the residence time or contact time is related to the volume of the reactor simply by $\tau = V_t/v$ (e.g. if V_t is 10 m³ and v is 0·5 m³/s, i.e. the volume charged to the reactor per second is 0·5 m³, then obviously the time spent in the reactor is $10/0·5 = 20$ s, i.e. V_t/v). To determine the volume of a tubular reactor we may therefore calculate the reaction time t_r for batchwise operation from equation 1.16 and set this equal to the residence time:

$$\frac{V_t}{v} = \tau = t_r = \int\limits_{0}^{\chi_f} \frac{d\chi}{\mathcal{R}_A}$$

which is exactly equivalent to equation 1.22. For comparison with equation 1.21 it may be noted that $v = F_A/(C_0 y_{A0})$ where y_{A0} is the mol fraction of **A** at the inlet to the reactor and C_0 is the molar density of the reaction mixture at inlet.

The situation is more complicated if expansion or contraction of a volume element does occur and the volumetric flowrate is not constant throughout the reactor. The ratio V_t/v, where v is the volume flow into the reactor, no longer gives the true residence time or contact time. However, the ratio V_t/v may still be quoted but is called the *space time* and its reciprocal v/V_t the *space velocity*. The space velocity is not in fact a velocity at all; it has dimensions of $(time)^{-1}$ and is therefore really a reactor volume displacement frequency. When a space velocity is quoted in the literature, its definition needs to be examined carefully; sometimes a ratio v_l/V_t is used, where v_l is a liquid volume rate of flow of a reactant which is metered as a liquid but subsequently vaporised before feeding to the reactor.

The space velocity for a given conversion is often used as a ready measure of the performance of a reactor. The use of equation 1.16 to calculate reaction time, as if for a batch reactor, is not to be recommended as normal practice; it can be equated to V_t/v only if there is no change in volume. Further, the method of using reaction time is a blind alley in the sense that it has to be abandoned when the theory of tubular reactors is extended to take into account longitudinal and radial dispersion and other departures from the plug flow hypothesis which are important in the design of catalytic tubular reactors (pages 152 *et seq.*).

Tubular reactors – non-isothermal operation

In designing and operating a tubular reactor when the heat of reaction is appreciable, strictly isothermal operation is rarely achieved and usually is not economically justifiable, although the aim may be to maintain the local temperatures within fairly narrow limits. On the assumption of plug flow, the rate of temperature rise or fall along the reactor dT/dz is determined by a heat balance (equation 1.13) around an element of reactor volume $\delta V_t = A_c \delta z$, where A_c is the area of cross-section of the reactor:

$$- Gc\frac{dT}{dz}\delta z - \frac{dQ}{dz}\delta z + (-\Delta H_A)\mathcal{R}_A A_c \delta z = \quad 0 \qquad \ldots .(1.23)$$

| Difference between heat inflow and outflow with fluid stream | Heat transferred from element | Heat released by reaction | (in steady state) |

where G is the mass flowrate, and

dQ/dz is the rate of heat transfer per unit length of reactor.

To obtain the temperature and concentration profiles along the reactor, equation 1.23 must be solved by numerical methods simultaneously with equation 1.21 for the material balance.

The design engineer can arrange for the heat transferred per unit length dQ/dz to vary with position in the reactor according to the requirements of the reaction. Consider, for example, the pyrolysis of ethane for which a reactor similar to that shown in Fig. 1.5c might be used. The cool feed enters the convection section, the duty of which is to heat the reactant stream to the reaction temperature. As the required reaction temperature is approached, the reaction rate increases, and a high rate of heat transfer to the fluid stream is required to offset the large endothermic heat of reaction. This high heat flux at a high temperature is effected in the radiant section of the furnace. Detailed numerical computations are usually made by splitting up the reactor in the furnace into a convenient number of sections.

From a general point of view, three types of expression for the heat transfer term can be distinguished:

(a) *Adiabatic operation*: $dQ/dz = 0$. The heat released in the reaction is retained in the reaction mixture so that the temperature rise along the reactor parallels the extent of the conversion α. The material balance and heat balance equations can be solved in a manner similar to that used in the example on an adiabatic batch reactor (page 35). Adiabatic operation is important in heterogeneous tubular reactors and is considered further under that heading in Chapter 2.

(b) *Constant heat transfer coefficient*: This case is again similar to the one for batch reactors and is also considered further in Chapter 2, under heterogeneous reactors.

(c) *Constant heat flux*: If part of the tubular reactor is situated in the radiant section of a furnace, as in Fig. 1.5c, and the reaction mixture is at a temperature considerably lower than that of the furnace walls, heat transfer to the reactor occurs mainly by radiation and the rate will be virtually independent of the actual temperature of the reaction mixture. For this part of the reactor dQ/dz may be virtually constant.

Pressure drop in tubular reactors

For a homogeneous tubular reactor, the pressure drop corresponding to the desired flowrate is often relatively small and does not usually impose any serious limitation on the conditions of operation. The pressure drop must, of course, be calculated as part of the design so that ancillary equipment may be specified. Only for gases at low pressures or, liquids of high viscosity, e.g. polymers, is the pressure drop likely to have a major influence on the design.

In heterogeneous systems, however, the question of pressure drop may be more serious. If the reaction system is a two-phase mixture of liquid and gas, or if the gas flows through a deep bed of small particles, the pressure drop should be checked at an early stage in the design so that its influence can be assessed. The methods of calculating such pressure drops are much the same as those for flow without reaction (e.g. for packed beds see Volume 2, Chapter 4).

Kinetic data from tubular reactors

In the laboratory, tubular reactors are very convenient for gas-phase reactions, and for any reaction which is so fast that it is impractical to follow it batchwise. Measurements are usually made when the reactor is operating in a steady state, so that the conversion at the outlet or at any intermediate point does not change with time. For fast reactions particularly, a physical method of determining the conversion, such as ultra-violet or infra-red absorption, is preferred to avoid disturbing the reaction. The conversion obtained at the outlet is regulated by changing either the flowrate or the volume of the reactor.

The reactor may be set up either as a differential reactor, in which case concentrations are measured over a segment of the reactor with only a small change in conversion, or as an integral reactor with an appreciable change in conversion. When the integral method is used for gas-phase reactions in particular, the pressure drop should be small; when there is a change in the number of mols between reactants and products, integrated forms of equation 1.21 which allow for constant pressure expansion or contraction must be used for interpretation of the results. Thus, for an irreversible first order reaction of the kind $A \rightarrow v_M M$, using a feed of pure A the integrated form of equation 1.21, assuming plug flow, is:

$$\frac{V_t}{F_A} = \frac{1}{k_1 C_{A0}} \left[v_M \ln \frac{1}{(1 - \alpha_A)} - (v_M - 1) \, \alpha_A \right]$$

In order to obtain basic kinetic data, laboratory tubular reactors are usually operated as closely as possible to isothermal conditions. If however a full scale tubular reactor is to be operated adiabatically, it may be desirable to obtain data on the small scale adiabatically. If the small reactor has the same length as the full scale version but a reduced cross-section, it may be regarded as a longitudinal element of the large reactor and, assuming plug flow applies to both, scaling up to the large reactor is simply a matter of increasing the cross-sectional area in proportion to the feed rate. One of the problems in operating a small reactor adiabatically, especially at high temperatures, is to prevent heat loss since the surface to volume ratio is large. This difficulty may be overcome by using electrical heating elements wound in several sections along the tube, each controlled by a servo-system with thermocouples which sense the temperature difference between the reaction mixture and the outside of the tube (Fig. 1.16). In this way, the temperature of the heating jacket follows exactly the adiabatic temperature path of the reaction, and no heat is lost or gained by the reaction mixture itself. This type of reactor is particularly valuable in developing heterogeneous packed bed reactors for which it is still reasonable to assume the plug flow model even for large diameters.

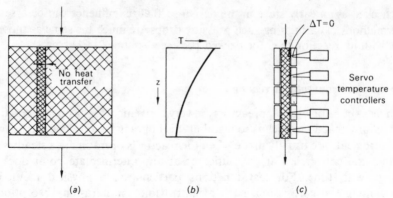

FIG. 1.16. Laboratory-scale reproduction of adiabatic tubular reactor
temperature profile
(a) Large-scale reactor
(b) Adiabatic temperature profile, endothermic reaction
(c) Laboratory-scale reactor

CONTINUOUS STIRRED TANK REACTORS

The stirred tank reactor in the form of either a single tank, or more often a series of tanks (Fig. 1.1d), is particularly suitable for liquid-phase reactions, and is widely used in the organic chemicals industry for medium and large scale production. It can form a unit in a continuous process, giving consistent product quality, ease of automatic control, and low manpower requirements. Although, as we shall see below, the volume of a stirred tank reactor must be larger than that of a plug flow tubular reactor for the same production rate, this is little disadvantage because large volume tanks are relatively cheap to construct. If the reactor has to be cleaned periodically, as happens sometimes in polymerisations or in plant used for manufacturing a variety of products, the open structure of a tank is an advantage.

In a stirred tank reactor, the reactants are diluted immediately on entering the tank; in many cases this favours the desired reaction and suppresses the formation of byproducts. Because fresh reactants are rapidly mixed into a large volume, the temperature of the tank is readily controlled, and hot spots are much less likely to occur than in tubular reactors. Moreover, if a series of stirred tanks is used, it is relatively easy to hold each tank at a different temperature so that an optimum temperature sequence can be attained.

Assumption of ideal mixing. Residence time

In the theory of continuous stirred tank reactors, an important basic assumption is that the contents of each tank are *well mixed*. This means that the compositions in the tank are everywhere uniform and that the product stream leaving the tank has the same composition as the mixture within the tank. This assumption is reasonably well borne out in practice unless the tank is exceptionally large, the stirrer inadequate, or the reaction mixture very viscous.

In the treatment which follows, it will be assumed that the mass density of the reaction mixture is constant throughout a series of stirred tanks. Thus, if the volumetric feed rate is v, then in the steady state, the rate of outflow from each tank will also be v. Material balances may then be written on a volume basis, and this considerably simplifies the treatment. In practice, the constancy of the density of the mixture is a reasonable assumption for liquids, and any correction which may need to be applied is likely to be small.

The mean residence time for a continuous stirred tank reactor of volume V_c may be defined as V_c/v in just the same way as for a tubular reactor. However in a homogeneous reaction mixture, it is not possible to identify particular elements of fluid as having any particular residence time, because there is complete mixing on a molecular scale. If the feed consists of a suspension of particles, it may be shown that, although there is a distribution of residence times among the individual particles, the mean residence time does correspond to V_c/v if the system is ideally mixed.

Flow of tracer through a system of stirred tanks

Several useful ideas and conclusions about the fundamental characteristics of stirred tanks can be deduced from studying the flow of an unreactive tracer

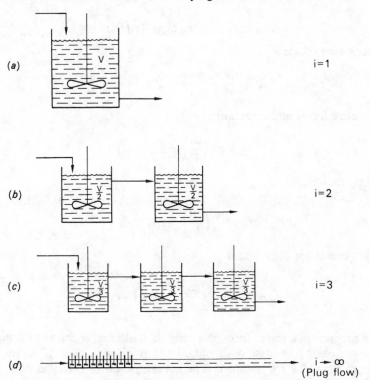

FIG. 1.17. Sets of equal stirred tanks in series. each set having the same total volume V

substance such as a coloured dye through a series of tanks. So that we may readily compare the effects of increasing the number of tanks, let us consider a volume V which can be made up of one tank only; or two tanks each of volume $V/2$; or three tanks each of volume $V/3$; or in general i equally sized tanks each of volume V/i (Fig. 1.17). Through this series of tanks, a pure liquid flows at a volumetric flow rate v. Suppose that at time $t = 0$ when all the tanks are full and the whole system is in a state of steady flow, a shot of n_0 mols of tracer is injected into the feed. The tracer is assumed to be completely miscible with the liquid in the tanks and to be in the form of a concentrated solution whose volume is negligible compared with V. The aim is to calculate the number of mols which have left the system completely, as a function of time and the total number of tanks to be employed, i.

At any time t, let tank 1 contain n_1 mols of tracer, whence the concentration in the tank $C_1 = n_1/(V/i)$.

Therefore rate of outflow of tracer from tank 1 $= C_1 v = (n_1 i/V) v$.

Applying a material balance to the tracer over tank 1, and noting that as far as tracer concentration is concerned, the system is not in a steady state:

Inflow	$-$	Outflow	$-$	Reaction	$=$	Accumulation
0	$-$	$\dfrac{n_1 iv}{V}$	$-$	0	$=$	$\dfrac{dn_1}{dt}$

$$\therefore \quad n_1 = n_0 \exp\left(-vit/V\right)$$

Similarly for tank 2:

$$\frac{n_1 iv}{V} - \frac{n_2 iv}{V} - 0 = \frac{dn_2}{dt}$$

Substituting for n_1 and integrating:

$$n_2 = n_0 \left(\frac{vit}{V}\right) \exp\left(-vit/V\right)$$

Similarly for tank 3:

$$n_3 = \frac{n_0}{2!} \left(\frac{vit}{V}\right)^2 \exp\left(-vit/V\right)$$

and, in general, for the rth tank:

$$n_r = \frac{n_0}{(r-1)!} \left(\frac{vit}{V}\right)^{r-1} \exp\left(-vit/V\right)$$

The progress of a tracer through a series of tanks can be followed using these equations. Taking the case of three tanks as an example, Fig. 1.17c, the curves shown in Fig. 1.18 are obtained. The concentration of tracer in tank 1 after its injection ($t = 0$) is given by $3n_0/V$ and thereafter it diminishes steadily. On the other hand the concentrations in tanks 2 and 3 rise to a peak as the wave of tracer passes through, the peak of the wave becoming attenuated and the wave more drawn out as it proceeds from one tank to the next.

As the basis for further comparison, let us return to the case of a fixed volume V which may be a single tank (Fig. 1.17a) or two tanks (Fig. 1.17b) or three tanks and so on. In the general case there will be i tanks and the concentration of the tracer leaving the last tank will be C_i. If we now plot C_i/C'_0, where $C'_0 = n_0/V$, against the reduced time (vt/V), the family of curves shown in Fig. 1.19a is obtained. The curves are known as C (or outlet concentration) curves.

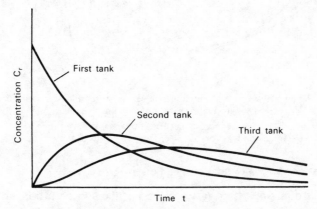

FIG. 1.18. Progress of a tracer through a series of three stirred tanks

Alternatively, we may plot F, the fraction of tracer which has escaped completely after a time t, against (vt/V). F may be obtained by noting that at any time t the number of mols remaining in the series of i tanks is:

$$n_t = n_1 + n_2 + n_3 + \ldots + n_i$$

therefore:

$$F = \frac{n_0 - n_t}{n_0} = 1 - \frac{1}{n_0}(n_1 + n_2 + n_3 + \ldots + n_i)$$

or

$$F = 1 - \exp(-vit/V)\left\{1 + \frac{vit}{V} + \frac{1}{2!}\left(\frac{vit}{V}\right)^2 + \ldots + \frac{1}{(i-1)!}\left(\frac{vit}{V}\right)^{i-1}\right\}$$

As $i \to \infty$ the term in brackets above tends to $\exp(vit/V)$ for $t < (V/v)$, and F becomes zero. Thus, for an infinite number of tanks the fraction of tracer that has escaped is zero for all times less than the residence time V/v. This is exactly the same as for the case of an ideal tubular reactor with plug flow.

We see therefore from the curves in Fig. 1.19 that for a single stirred tank a high proportion of the tracer, 0·632, has escaped from the tank within a time equal to the residence time V/v. This constitutes a bypassing effect which is an inherent disadvantage of a stirred tank. However, as the number of tanks in the system is increased, the proportion of tracer which escapes within the residence time is diminished, showing that the bypassing effect is reduced by having a larger number of tanks. In the limit as $i \to \infty$, the system becomes identical to an ideal tubular reactor with plug flow.

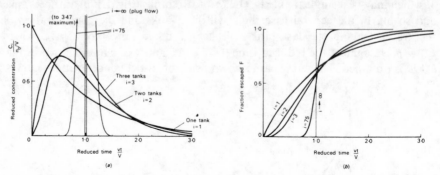

FIG. 1.19. Curves for tracer leaving different sets of stirred tanks in series, each set having the same total volume V

(a) C_θ-curves: reduced concentration versus reduced time
(b) F-curves: fraction of tracer escaped versus reduced time

Design equations for continuous stirred tank reactors

In the case of a single shot of tracer above, the concentrations in the tanks vary with time, and there is no reaction to take into account. However, when a series of stirred tanks is used as a chemical reactor, and the reactants are fed at a constant rate, eventually the system reaches a steady state such that the concentrations in the individual tanks, although different, do not vary with time. When the general material balance of equation 1.12 is applied, the accumulation term is therefore zero. Considering first of all the most general case in which the mass density of the mixture is not necessarily constant, the material balance on the reactant **A** is made on the basis of F_A mols of **A** per unit time fed to the first tank. Then a material balance for the rth tank of volume V_{cr} (Fig. 1.20) is, in the steady state:

$$F_A(1 - \alpha_{Ar-1}) - F_A(1 - \alpha_{Ar}) - \mathscr{R}_A V_{cr} = 0$$

Inflow   Outflow   Reaction

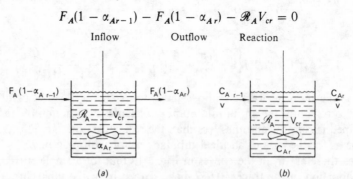

FIG. 1.20. Continuous stirred tank reactor: material balance over rth tank in steady state

(a) General case: feed to first tank F_A
(b) Constant volume flow rate v

where α_{Ar-1} is the fractional conversion of **A** in the mixture leaving tank $r-1$ and entering tank r, and

α_{Ar} is the fractional conversion of **A** in the mixture leaving tank r.

Assuming that the contents of the tank are well mixed, α_{Ar} is also the fractional conversion of the reactant **A** in tank r. Therefore:

$$\frac{V_{cr}}{F_A} = \frac{\alpha_{Ar} - \alpha_{Ar-1}}{\mathscr{R}_A} \qquad \dots (1.24)$$

which is the counterpart of equation 1.21 for a tubular reactor.

Stirred tanks are usually employed for reactions in liquids and, in most cases, the mass density of the reaction mixture may be assumed constant. Material balances may then be taken on the basis of the volume rate of flow v which is constant throughout the system of tanks. The material balance on **A** over tank r may thus be written, in the steady state:

$$vC_{Ar-1} - vC_{Ar} - V_{cr}\mathscr{R}_A = 0$$
$$\text{Inflow} \qquad \text{Outflow} \quad \text{Reaction}$$

where C_{Ar-1} is the concentration of **A** in the liquid entering tank r from tank $r-1$, and C_{Ar} is the concentration of **A** in the liquid leaving tank r. Therefore:

$$\frac{V_{cr}}{v} = \frac{C_{Ar-1} - C_{Ar}}{\mathscr{R}_A} \qquad \dots (1.25)$$

Equation 1.25 may be written in terms of $\chi_r = (C_{A0} - C_{Ar})$; C_{A0} is here the concentration of **A** in the feed to the first tank. Thus:

$$\frac{V_{cr}}{v} = \frac{\chi_r - \chi_{r-1}}{\mathscr{R}_A} \qquad \dots (1.26)$$

In these equations $V_{cr}/v = \tau_r$, the residence time in tank r. Equation 1.26 may be compared with equation 1.22 for a tubular reactor. The difference between them is that, whereas 1.22 is an integral equation, 1.26 is a simple algebraic equation. If the reactor system consists of only one or two tanks the equations are fairly simple to solve. If a large number of tanks is employed, the equations whose general form is given by 1.26 constitute a set of finite difference equations and must be solved accordingly. If there is more than one reactant involved, in general a set of material balance equations must be written for each reactant.

In order to proceed with a solution, a rate equation is required for $\mathscr{R}_A$. Allowance must be made for the fact that the rate constant will be a function of temperature, and may therefore be different for each tank. The temperature distribution will depend on the heat balance for each tank (equation 1.13), and this will be affected by the amount of heating or cooling that is carried out. The example which follows concerns an isothermal system of two tanks with two reactants, one of which is in considerable excess.

Example: *A two-stage continuous stirred tank reactor*

A solution of an ester $R \cdot COOR'$ is to be hydrolysed with an excess of caustic soda solution. Two stirred tanks of equal size will be used. The ester and caustic soda solutions flow separately into the first tank at rates of 0·004 and 0·001 m^3/s and with concentrations of 0·02 and 1·0 kmol/m^3 respectively. The reaction:

$$R \cdot COOR' + NaOH \rightarrow R \cdot COO\,Na + R'OH$$

is second order with a velocity constant of 0·033 m^3/kmol s at the temperature at which both tanks operate. Determine the volume of the tanks required to effect 95 per cent conversion of the ester ($\alpha_{A_2} = 0.95$).

Solution

Although the solutions are fed separately to the first tank, we may for the purpose of argument consider them to be mixed together just prior to entering the tank as shown in Fig. 1.21. If the ester

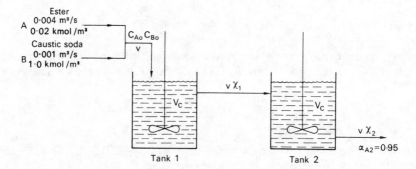

FIG. 1.21. Continuous stirred tank reactor: worked example

is denoted by A and the caustic soda by B. and χ is the number of mols of ester (or caustic soda) which have reacted per litre of the combined solutions, a material balance on the ester over the first tank gives, for the steady state:

$$\underset{\text{Inflow}}{vC_{A0}} \quad - \quad \underset{\text{Outflow}}{v(C_{A0} - \chi_1)} \quad - \quad \underset{\text{Reaction}}{V_{c1}k_2(C_{A0} - \chi_1)(C_{B0} - \chi_1)} \quad = \quad 0$$

Similarly for tank 2:

$$v(C_{A0} - \chi_1) - v(C_{A0} - \chi_2) - V_{c2}k_2(C_{A0} - \chi_2)(C_{B0} - \chi_2) = 0$$

Because we have used the variable χ, a further set of material balance equations for reactant B is not required; they are implicit in the equations for A above. Rearranging these, and putting $V_{c1} = V_{c2}$ because the tanks are of equal size:

$$v\chi_1 = V_{c1}k_2(C_{A0} - \chi_1)(C_{B0} - \chi_1) \qquad \qquad \dots (1.27)$$

and

$$v(\chi_2 - \chi_1) = V_{c1}k_2(C_{A0} - \chi_2)(C_{B0} - \chi_2) \qquad \qquad \dots (1.28)$$

In these equations we know $\chi_2 = \alpha_{A2}C_{A0}$ and all the other quantities excepting χ_1 and V_{c1}; hence by eliminating χ_1 we may find V_{c1}.

Considering numerical values, the total volume flow rate $v = 0.005$ m³/s

$$C_{A0} = 0.02 \times \frac{0.004}{0.005} = 0.016 \text{ kmol/m}^3$$

$$C_{B0} = 1.0 \times \frac{0.001}{0.005} = 0.20 \text{ kmol/m}^3$$

$$\chi_2 = 0.95 \times 0.016 = 0.0152 \text{ kmol/m}^3$$

$$(C_{A0} - \chi_2) = 0.0008 \text{ kmol/m}^3$$

$$(C_{B0} - \chi_2) = 0.1848 \text{ kmol/m}^3$$

At this stage we note that because a substantial excess of caustic soda is used, as a first approximation χ_1 may be neglected in comparison with C_{B0}. We thus avoid the necessity of solving a cubic equation. At a later stage the value of χ_1 obtained using this approximation can be substituted in the term $(C_{B0} - \chi_1)$ and the calculation repeated in an iterative fashion. Setting $(C_{B0} - \chi_1) = 0.20$, equations 1.27 and 1.28 become:

$$0.005 \, \chi_1 = V_{c1} \times 0.033 \times (0.016 - \chi_1) \times 0.20 \qquad \dots (1.29)$$

$$0.005 \, (0.0152 - \chi_1) = V_{c1} \times 0.033 \times 0.0008 \times 0.1848 \qquad \dots (1.30)$$

Eliminating χ_1, and rearranging:

$$V_{c1}^2 + 1.56 \, V_{c1} - 11.6 = 0$$

Hence

$$V_{c1} = 2.71 \text{ m}^3$$

Thus as a first estimate, the two tanks should each be of this capacity. An improved estimate may be obtained by using the value of V_{c1} calculated above and substituting in equation 1.30, giving $\chi_1 = 0.0125$ kmol/m³. Thus the second approximation to the term $(C_{B0} - \chi_1)$ is 0.1875. Repeating the calculation,

$$\underline{\underline{V_{c1} = 2.80 \text{ m}^3}}$$

Graphical methods

For second order reactions, graphs showing the fractional conversion for various residence times and reactant feed ratios have been drawn up by ELDRIDGE and PIRET[18]. These graphs, which were prepared from numerical calculations based on equation 1.26, provide a convenient method for dealing with sets of equal sized tanks of up to five in number, all at the same temperature.

A wholly graphical method arising from equation 1.25 or 1.26 may be used providing that the rate of reaction $\mathscr{R}_A$ is a function of a single variable only, either C_A or χ. The tanks must therefore all be at the same temperature. Experimental rate data may be used directly in graphical form without the necessity of fitting a rate equation. In order to establish the method, equation 1.25 is firstly rearranged to give:

$$\mathscr{R}_{Ar} = -\frac{v}{V_{cr}}(C_{Ar} - C_{Ar-1}) \qquad \dots (1.31)$$

where the subscript r has been added to stress that $\mathcal{R}_{Ar}$ is the value of $\mathcal{R}_A$ for the tank r. Consider now a graph of $\mathcal{R}_A$ versus C_A as shown in Fig. 1.22a. From a point on the C_A axis, $\mathcal{R}_A = 0$, $C_A = C_{Ar-1}$ we construct a line of slope:

$$- \frac{v}{V_{cr}}$$

which therefore has the equation:

$$\mathcal{R}_A = - \frac{v}{V_{cr}} (C_A - C_{Ar-1})$$

Equation 1.31 shows that the point $(\mathcal{R}_{Ar}, C_{Ar})$ lies on this line; however $\mathcal{R}_{Ar}$ must also lie on the rate of reaction versus C_A curve, so that $(\mathcal{R}_{Ar}, C_{Ar})$ is the point of intersection of the two as shown. Thus starting with the first tank, we can draw the first line from the point $O_0(0, C_{A0})$ on Fig. 1.22b and locate C_{A1}, the concentration of the reactant leaving the first tank. Then from the point

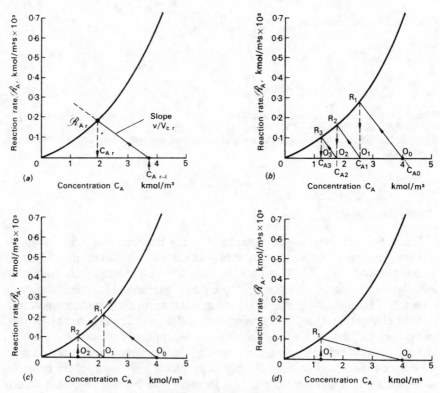

Fig. 1.22. Graphical construction for continuous stirred tank reactors

    (a) General method
    (b) Three equal tanks, outlet concentration C_{A3} unknown
    (c) Two equal tanks, volume unknown
    (d) One tank, volume unknown

$O_1(0, C_{A1})$ we can draw a second line to locate C_{A2} for the second tank, and so on for the whole series as the following example shows.

Example: *Graphical construction for a three-stage continuous stirred tank reactor*

(a) A system of three stirred tanks is to be designed to treat a solution containing $4.0 \, \text{kmol/m}^3$ of a reactant **A**. Experiments with a small reactor in the laboratory gave the kinetic data shown as a graph of rate of reaction versus C_A in Fig. 1.22b. If the feed rate to the reactor system is $1.2 \times 10^{-4} \, \text{m}^3/\text{s}$ what fractional conversion will be obtained if each of the three tanks has a volume of $0.60 \, \text{m}^3$?

(b) Calculate the volumes of the tanks required for the same overall conversion if two equal tanks are used, and if only one tank is used.

Solution

(a) Referring to Fig. 1.22b, from the point O_0 representing the feed composition of $4.0 \, \text{kmol/m}^3$, a line $O_0 R_1$ is drawn of slope $-1.2 \times 10^{-4}/0.60 = -2 \times 10^{-4} \, \text{s}^{-1}$ to intersect the rate curve at R_1. This point of intersection gives the concentration of **A** in the first tank C_{A1}. A perpendicular $R_1 O_1$ is dropped from R_1 to the C_A axis, and from the point O_1 a second line also of slope $-2 \times 10^{-4} \, \text{s}^{-1}$ is drawn. The construction is continued until O_3 is reached which gives the concentration of **A** leaving the last tank C_{A3}. Reading from the figure $C_{A3} = 1.23 \, \text{kmol/m}^3$. The fractional conversion is therefore $(4.0 - 1.23)/4.0 = \underline{0.69}$.

(b) When the volumes of the identical tanks are unknown the graphical construction must be carried out on a trial and error basis. The procedure for the case of two tanks is shown in Fig. 1.22c; the points O_0 and O_2 are known, but the position of R_1 has to be adjusted to make $O_0 R_1$ and $O_1 R_2$ parallel because these lines must have the same slope if the tanks are of equal size. From the figure this slope is $1.13 \times 10^{-4} \, \text{s}^{-1}$. The volume of each tank must therefore be $1.2 \times 10^{-4}/1.13 \times 10^{-4} = 1.06 \, \text{m}^3$.

For a single tank the construction is straightforward as shown in Fig. 1.22d and the volume obtained is $3.16 \, \text{m}^3$.

It is interesting to compare the total volume required for the same duty in the three cases.

$$\text{Total volume for 3 tanks is } 3 \times 0.60 = 1.80 \, \text{m}^3$$
$$\text{for 2 tanks is } 2 \times 1.06 = 2.12 \, \text{m}^3$$
$$\text{for 1 tank is } 1 \times 3.16 = 3.16 \, \text{m}^3$$

These results illustrate the general conclusion that, as the number of tanks is increased, the total volume required diminishes and tends in the limit to the volume of the equivalent plug flow reactor. The only exception is in the case of a zero order reaction for which the total volume is constant and equal to that of the plug flow reactor for all configurations.

Autothermal operation

One of the advantages of the continuous stirred tank reactor is the fact that it is ideally suited to autothermal operation. Feed-back of the reaction heat from products to reactants is indeed a feature inherent in the operation of a continuous stirred tank reactor consisting of a single tank only, because fresh reactants are mixed directly into the products. An important, but less obvious, point about autothermal operation is the existence of two possible stable operating conditions.

To understand how this can occur, consider a heat balance over a single tank operating in a steady state. The tank is equipped with a cooling coil of area A_t through which flows a cooling medium at a temperature T_C. Because conditions are steady, the accumulation term in the general heat balance, equation 1.13, is zero. The remaining terms in the heat balance may then be arranged as follows (cf. equations 1.19 for a batch reactor and 1.23 for a tubular reactor):

$$(-\Delta H_A)\mathcal{R}_A V_c \quad = \quad Gc(T - T_0) \quad + \quad UA_t(T - T_c)$$

Rate of heat production by reaction | Rate of heat removal by outflow of products | Rate of heat removal by heat transfer

Here, the enthalpy of the products of mass flowrate G and specific heat c is measured relative to T_0, the inlet temperature of the reactants. The rate of heat production term on the left-hand side of this equation varies with the temperature of operation, T, as shown in diagram (a) of Fig. 1.23; as T increases,

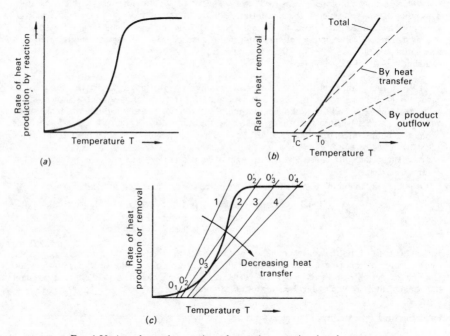

(a)

(b)

(c)

FIG. 1.23. Autothermal operation of a continuous stirred tank reactor
(a) and (b) show rates of heat production and removal
(c) shows the effects of different amounts of heat transfer on possible stable operating states

$\mathcal{R}_A$ increases rapidly at first but then tends to an upper limit as the reactant concentration in the tank approaches zero, corresponding to almost complete conversion. On the other hand, the rate of heat removal by both product outflow and heat transfer is virtually linear, as shown in diagram (b). To satisfy the heat balance equation above, the point representing the actual operating temperature must lie on both the rate of heat production curve and the rate of heat removal line, i.e. at the point of intersection as shown in (c).

In Fig. 1.23c, it may be seen how more than one stable operating temperature can sometimes occur. If the rate of heat removal is high (line 1), due either to rapid outflow or to a high rate of heat transfer, there is only one point of inter-section O_1, corresponding to a low operating temperature close to the reactant inlet temperature T_0 or the cooling medium temperature T_C. With a somewhat smaller flowrate or heat transfer rate (line 2) there are two points of intersection, O_2 at a low temperature and conversion, and O_2' at a considerably higher temperature and conversion. If the reactor is started up from cold, it will settle down in the lower operating state O_2. However, if a disturbance causes the temperature to rise above the intermediate point of intersection beyond which the rate of heat production exceeds the rate of loss, the system will pass into the upper operating state O_2'. Line 3 represents a heat removal rate for which the lower operating state O_3 can only just be realised, and for line 4, only an upper operating temperature O_4' is possible.

Obviously, in designing and operating a stirred tank reactor it is necessary to be aware of these different operating conditions. Further discussion of the dynamic response and control of an autothermal continuous stirred tank reactor is given by KRAMERS and WESTERTERP[6].

Kinetic data from continuous stirred tank reactors

For fast or moderately fast liquid phase reactions, the stirred tank reactor can be very useful for establishing kinetic data in the laboratory. When a steady state has been reached, the composition of the reaction mixture may be deter-mined by a physical method using a flow cell attached to the reactor outlet, as in the case of a tubular reactor. The stirred tank reactor, however, has a number of further advantages in comparison with a tubular reactor. With an appropriate ratio of reactor volume to feed rate and with good mixing, the difference in reactant composition between feed and òutflow can be large, without changing the basic situation whereby all the reaction takes place at a single uniform concentration level in the reactor. Comparatively large differences between inlet and outlet compositions are required in order to determine the rate of the reaction with reasonable accuracy. With a tubular reactor, on the other hand, if a large difference in reactant composition is set up across the reactor, the integral method of interpreting the results must be employed and this may give rise to problems when dealing with complex reactions.

There is one further point of comparison. Interpretation of results from a stirred tank reactor depends on the assumption that the contents of the tank are well mixed. Interpretation of results from a tubular reactor rests on the assump-tion of plug flow unless the flow is laminar and is treated as such. Which of these two assumptions can be met most satisfactorily in practical experiments? Unless the viscosity of the reaction mixture is high or the reaction extremely fast, a high speed stirrer is very effective in maintaining the contents of a stirred tank uniform. On the other hand, a tubular reactor may have to be very carefully designed if back-mixing is to be completely eliminated, and in most practical situations there is an element of uncertainty about whether the plug flow assumption is valid.

COMPARISON OF BATCH, TUBULAR AND STIRRED
TANK REACTORS FOR A SINGLE REACTION. REACTOR OUTPUT

There are two criteria which can be used to compare the performances of different types of reactor. The first, which is a measure of reactor productivity, is the output of product in relation to reactor size. The second, which relates to reactor selectivity, is the extent to which formation of unwanted byproducts can be suppressed. When comparing reactions on the basis of output as in the present section, only one reaction need be considered, but when in the next section the question of byproduct formation is taken up, more complex schemes of two or more reactions must necessarily be introduced.

In defining precisely the criterion of reactor output, it is convenient to use one particular reactant as a reference rather than the product formed. The distinction is unimportant if there is only one reaction, but is necessary if more than one reaction is involved. The *unit output* W_A of a reactor system may thus be defined as the mols of reactant A converted, per unit time, per unit volume of reaction space; in calculating this quantity it should be understood that the total mols converted by the whole reactor in unit time is to be divided by the total volume of the system. The unit output is therefore an average rate of reaction for the reactor as a whole, and is thus distinct from the specific rate $\mathscr{R}_A$ which is the local rate of reaction.

We shall now proceed to compare the three basic types of reactor —batch tubular and stirred tank —in terms of their performance in carrying out a single first order irreversible reaction:

$$\text{A} \rightarrow \text{Products}$$

it will be assumed that there is no change in mass density and that the temperature is uniform throughout. However, it has already been shown that the conversion in a tubular reactor with plug flow is identical to that in a batch reactor irrespective of the order of the reaction, if the residence time of the tubular reactor τ is equal to reaction time of the batch reactor t_r. Thus, the comparison rests between batch and plug flow tubular reactors on the one hand, and stirred tank reactors consisting of one, two or several tanks on the other.

Batch reactor and tubular plug flow reactor

In terms of χ, mols reacted in unit volume after time t, the material balance for a first order reaction is simply:

$$\frac{d\chi}{dt} = k_1(C_{A0} - \chi)$$

where C_{A0} is the initial concentration of the reactant A. (The subscript A will from now on be omitted from C, W and α because there is only the one reactant throughout). Integrating:

$$t_r = \frac{1}{k_1} \ln \frac{C_0}{C_0 - \chi}$$

as shown in Table 1.1. Since in unit volume a total of χ mols of **A** have reacted after a time t_r, the unit output W_b from a batch reactor is given by:

$$W_b = \chi/t_r$$

or in terms of the fractional conversion α ($\chi = \alpha C_0$):

$$W_b = \frac{\alpha C_0}{t_r} = \frac{\alpha C_0 k_1}{\ln\left[1/(1-\alpha)\right]}$$

therefore:

$$\frac{W_b}{k_1 C_0} = \frac{\alpha}{\ln\left[1/(1-\alpha)\right]} \qquad \ldots (1.32)$$

Continuous stirred tank reactor

One tank

A material balance on the reactant gives (Fig. 1.24a), for the steady state:

$$\underset{\text{Inflow}}{vC_0} - \underset{\text{Outflow}}{v(C_0 - \chi)} - \underset{\text{Reaction}}{V_c k_1(C_0 - \chi)} = 0$$

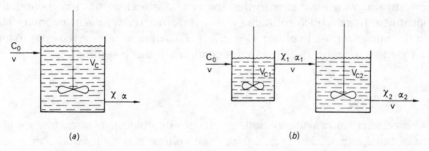

(a)   (b)

FIG. 1.24. Continuous stirred tank reactors: calculation of unit output

Considering the outflow, it may be seen that in unit time, $v\chi$ mols of reactant are converted in the reactor of volume V_c. Thus:

$$W_{c1} = \frac{v\chi}{V_c}$$

Hence:

$$\frac{v\chi}{V_c} = k_1(C_0 - \chi) = k_1 C_0(1 - \alpha)$$

therefore:

$$\frac{W_{c1}}{k_1 C_0} = (1 - \alpha) \qquad \ldots (1.33)$$

Two tanks

Let us first of all consider the general case in which the tanks are not necessarily of equal size (Fig. 1.24b).

Taking a material balance on the reactant **A** over each tank in succession:

Tank 1:

$$vC_0 - v(C_0 - \chi_1) - V_{c1}k_1(C_0 - \chi_1) = 0$$

Tank 2:

$$v(C_0 - \chi_1) - v(C_0 - \chi_2) - V_{c2}k_1(C_0 - \chi_2) = 0$$

i.e.

$$v\chi_1 = V_{c1}k_1(C_0 - \chi_1)$$

$$v(\chi_2 - \chi_1) = V_{c2}k_1(C_0 - \chi_2).$$

i.e.

$$\alpha_1 = \frac{V_{c1}}{v} k_1(1 - \alpha_1)$$

$$\alpha_2 - \alpha_1 = \frac{V_{c2}}{v} k_1(1 - \alpha_2)$$

At this point there arises the question of whether the two tanks should be the same size or of different sizes. Mathematically, this question needs to be investigated with some care so that the desired objective function is correctly identified. If we wished to design a two-stage reactor we might be interested in the minimum total volume $V_c = V_{c1} + V_{c2}$ required for a given conversion α_2. Writing the ratio $V_{c2}/V_{c1} = s$, this condition is met by setting:

$$\left(\frac{\partial V_c}{\partial s}\right)_{\alpha_2} = 0$$

whence, after some manipulation, $s = 1$, showing that the two reactors should be of equal size. (If, however, a fixed total volume is considered and the ratio s is varied to find the effect on α_2, i.e.:

$$\left(\frac{\partial \alpha_2}{\partial s}\right)_{V_c} = 0$$

the maximum does not occur at $s = 1$, but this is not a likely situation in practice). In general, it may be shown that the optimum value of the ratio s depends on the order of reaction and is unity only for a first order reaction. However, the convenience and reduction in costs associated with having all tanks the same size will in practice always outweigh any small increase in total volume that this may entail.

We will assume henceforth that the tanks are of equal size, i.e.:

$$V_{c1} = V_{c2} = \tfrac{1}{2}V_c$$

Eliminating α_1 from the equations above, we may proceed to determine the unit output W_{c2} of a series of two tanks, where:

$$W_{c2} = \frac{v\chi_2}{V_c} = \frac{vC_0\alpha_2}{V_c}$$

After fairly lengthy manipulation, we obtain

$$\frac{W_{c2}}{k_1 C_0} = \frac{\alpha_2}{2}\left[\frac{(1-\alpha_2)^{\frac{1}{2}}}{1-(1-\alpha_2)^{\frac{1}{2}}}\right] \qquad \dots (1.34)$$

Comparison of reactors

It may be seen from equations 1.32, 1.33 and 1.34 that unit output is a function of conversion. Some numerical values of the dimensionless quantity $W/k_1 C_0$ representing the unit output are shown in Table 1.3. Shown also in Table 1.3 are values of the ratio:

$$\frac{\text{Unit output batch reactor}}{\text{Unit output stirred tank reactor}} = \frac{\text{Volume stirred tank reactor}}{\text{Volume batch reactor}}$$

for various values of the conversion. These show that a single continuous stirred tank reactor must always be larger than a batch or tubular plug flow reactor for the same duty, and for high conversions the stirred tank must be very much larger indeed. If two tanks are used, however, the total volume is

TABLE 1.3. *Comparison of continuous stirred tank reactors and batch reactors with respect to unit output $W/k_1 C_0$ and reactor volume. First order reaction*

Reactor type		Conversion			
		0·50	0·90	0·95	0·99
Batch or tubular plug flow	Unit output	0·722	0·391	0·317	0·215
C.S.T.R. one tank	Unit output	0·50	0·10	0·05	0·01
	Vol. ratio C.S.T.R./Batch	1·44	3·91	6·34	21·5
C.S.T.R. two tanks	Unit output	0·604	0·208	0·137	0·055
	Vol. ratio C.S.T.R./Batch	1·19	1·88	2·31	3·91

less than that of a single tank. Although the detailed calculations for systems of three or more tanks are not given here, it can be seen in Fig. 1.25, which is based on charts prepared by LEVENSPIEL [5], that the total volume is progressively reduced as the number of tanks is increased. This principle is evident also in the example on stirred tank reactors solved by the graphical method on page 55, which does not refer to a first order reaction. Calculations such as those in Table 1.3 can be extended to give results for orders of reaction both greater than and less than one. As the order of the reaction increases, so the comparison becomes even less favourable to the stirred tank reactor.

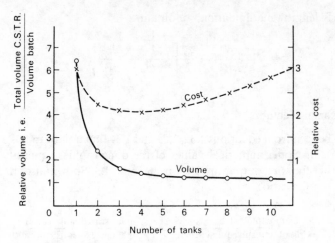

FIG. 1.25. Comparison of size and cost of continuous stirred
tank reactors with a batch or a tubular plug flow reactor:
first order reaction, conversion 0·95

As the number of stirred tanks in a series is increased, so is the total volume of the system reduced. In the limit with an infinite number of tanks, we can expect the volume to approach that of the equivalent batch or tubular reactor because, as we have seen from tracer injection (page 47), in the limiting case plug flow is obtained. However, although the total volume of a series of tanks progressively decreases with increasing number, this does not mean that the total cost will continue to fall. The cost of a tank and its associated mixing and heat transfer equipment will be proportional to approximately the 0·6 power of its volume. When total cost is plotted against the number of tanks, as in the second curve of Fig. 1.25, the curve passes through a minimum. This usually occurs in the region of 3 to 6 tanks and it is most likely that a number in this range will be employed in practice.

COMPARISON OF BATCH, TUBULAR AND STIRRED TANK REACTORS FOR MULTIPLE REACTIONS. REACTOR YIELD

If more than one chemical reaction can take place in a reaction mixture, the type of reactor used may have a quite considerable effect on the products formed. The choice of operating conditions is also important, especially the temperature and the degree of conversion of the reactants. The economic importance of choosing the type of reactor which will suppress any unwanted byproducts to the greatest extent has already been stressed (page 2).

In this section, our aim will be to take certain model reaction schemes and work out in detail the product distribution which would be obtained from each of the basic types of reactor. It is fair to say that in practice there are often difficulties in attempting to design reactors from fundamental principles when multiple reactions are involved. Information on the kinetics of the individual reactions is often incomplete, and in many instances an expensive and time-consuming laboratory investigation would be needed to fill in all the gaps. Nevertheless, the model reaction schemes examined below are valuable in indicating firstly how such limited information as may be available can be used to the best advantage, and secondly what key experiments should be undertaken in any research and development programme.

Types of multiple reactions

Multiple reactions are of two basic kinds. Taking the case of one reactant only, these are:

(a) Reactions in parallel or competing reactions of the type:

$$\mathbf{A} \to \mathbf{P} \text{ (desired product)}$$
$$\searrow \mathbf{Q} \text{ (unwanted product)}$$

(b) Reactions in series or consecutive reactions of the type:

$$\mathbf{A} \to \mathbf{P} \to \mathbf{Q}$$

where again $\mathbf{P}$ is the desired product and $\mathbf{Q}$ the unwanted byproduct.

When a second reactant $\mathbf{B}$ is involved, the situation is basically unchanged in the case of parallel reactions:

$$\mathbf{A} + \mathbf{B} \to \mathbf{P}$$
$$\mathbf{A} + \mathbf{B} \to \mathbf{Q}$$

The reactions are thus in parallel with respect to both $\mathbf{A}$ and $\mathbf{B}$. For reactions in series, however, if the second reactant $\mathbf{B}$ participates in the reaction with the product $\mathbf{P}$ as well as with $\mathbf{A}$, i.e. if:

$$\mathbf{A} + \mathbf{B} \to \mathbf{P}$$
$$\mathbf{P} + \mathbf{B} \to \mathbf{Q}$$

then although the reactions are in series with respect to $\mathbf{A}$ they are in parallel with respect to $\mathbf{B}$. In these circumstances, we have:

As we shall see, however, the series character of these reactions is the more important, because $\mathbf{B}$ cannot react to give $\mathbf{Q}$ until a significant amount of $\mathbf{P}$ has been formed.

More complex reaction schemes can be regarded as combinations of these basic types of individual reaction steps.

Yield and selectivity

When a mixture of reactants undergoes treatment in a reactor and more than one product is formed, part of each reactant is converted into the desired product, part is converted into undesired products and the remainder escapes unreacted. The amount of the desired product actually obtained is therefore smaller than the amount expected had all the reactant been transformed into the desired product alone. The reaction is then said to give a certain yield of the desired product. Unfortunately, the term yield has been used by different authors for two somewhat different quantities and care must be taken to avoid confusion. Here these two usages will be distinguished by employing the terms *relative yield* and *operational yield*; in each case the amount of product formed will be expressed in terms of the stoichiometrically equivalent amount of the reactant **A** from which it was produced.

The *relative yield* Φ_A is defined by:

$$\Phi_A = \frac{\text{Mols of } \mathbf{A} \text{ transformed into desired product}}{\text{Total mols of } \mathbf{A} \text{ which have reacted}}$$

The relative yield is therefore a net yield based on the amount of **A** actually consumed.

The *operational yield* Θ_A is defined by:

$$\Theta_A = \frac{\text{Mols of } \mathbf{A} \text{ transformed into desired product}}{\text{Total mols of } \mathbf{A} \text{ fed to the reactor}}$$

It is based on the total amount of reactant **A** entering the reactor, irrespective of whether it is consumed in the reaction or passes through unchanged.

Both these quantities are fractions and it follows from the definitions above that Φ_A always exceeds Θ_A, unless all the reactant is consumed, when they are equal.

If unreacted **A** can be recovered from the product mixture at low cost and then recycled, the relative yield is the more significant, and the reactor can probably be operated economically at quite a low conversion per pass. If it cannot be recovered and no credit can be allotted to it, the operational yield is the more relevant, and the reactor will probably have to operate at a high conversion per pass.

Another way of expressing product distribution is the *selectivity* of the desired reaction. Once more expressing the amount of product formed in terms of the amount of **A** reacted, the selectivity is defined as:

$$\frac{\text{Mols of } \mathbf{A} \text{ transformed into the desired product}}{\text{Mols of } \mathbf{A} \text{ transformed into unwanted products}}$$

It is thus a product ratio and can have any value, the higher the better. It is often used to describe catalyst performance in heterogeneous reactions.

Reactor type and backmixing

When more than one reaction can occur, the extent of any backmixing of products with reactants is one of the most important factors in determining the yield of the desired product. In a well stirred batch reactor, or in an ideal tubular reactor with plug flow, there is no backmixing, whereas in a single continuous stirred tank reactor there is complete backmixing. Intermediate between these two extremes are systems of two or more continuous stirred tanks in series, and non-ideal tubular reactors in which some degree of backmixing occurs (often termed longitudinal dispersion).

Backmixing in a reactor affects the yield for two reasons. The first and most obvious reason is that the products are mixed into the reactants; this is undesirable if the required product is capable of reacting further with the reactants to give an unwanted product, as in some series reactions. The second reason is that backmixing affects the level of reactant concentration at which the reaction is carried out. If there is no backmixing, the concentration level is high at the start of the reaction and has a low value only towards the end of the reaction. With backmixing, as in a single continuous stirred tank reactor, the concentration of reactant is low throughout. As we shall see, for some reactions high reactant concentrations favour high yields, whereas for other reactions low concentrations are more favourable.

If two reactants are involved in a reaction, high concentrations of both, at least initially, may be obtained in a batch or tubular plug flow reactor. and low concentrations of both in a single continuous stirred tank reactor. In some circumstances, however, a high concentration of reactant **A** coupled with a low concentration of reactant **B** may be desirable. This may be achieved in a number of ways:

(a) *Without recycle*: For continuous operation a cross-flow type of reactor may be used as illustrated in Fig. 1.2. The reactor can consist of, either a tubular reactor with multiple injection of **B** (Fig. 1.2*b*), or a series of several stirred tanks with the feed of **B** divided between them (Fig. 1.2*c*). If a batch type of reactor were preferred, the semi-batch mode of operation would be used (Fig. 1.2*a*). Reaction without recycle is the normal choice where the cost of separating unreacted **A** from the reaction mixture is high.

(b) *With recycle*: If the cost of separating **A** is low, then a large excess of **A** can be maintained in the reactor. A single continuous stirred tank reactor will provide a low concentration of **B**, while the large excess of **A** ensures a high concentration of **A**. Unreacted **A** is separated and recycled as shown in Fig. 1.26*f* (page 69).

Reactions in parallel

Let us consider the case of one reactant only but different orders of reaction for the two reaction paths, i.e.

$$A \xrightarrow{k_P} P \text{ desired product}$$

$$A \xrightarrow{k_Q} Q \text{ unwanted product}$$

with the corresponding rate equations:

$$\mathcal{R}_{AP} = k_P C_A^p$$

$$\mathcal{R}_{AQ} = k_Q C_A^q$$

In these equations it is understood that C_A may be (a) the concentration of **A** at a particular time in a batch reactor, (b) the local concentration in a tubular reactor operating in a steady state, or (c) the concentration in a stirred tank reactor, possibly one of a series, also in a steady state. Let δt be an interval of time which is sufficiently short for the concentration of **A** not to change appreciably in the case of the batch reactor; the length of the time interval is not important for the flow reactors because they are each in a steady state. Per unit volume of reaction mixture, the mols of **A** transformed into **P** is thus $\mathcal{R}_{AP}\delta t$, and the total amount reacted $(\mathcal{R}_{AP} + \mathcal{R}_{AQ})\,\delta t$. The relative yield under the circumstances may be called the instantaneous or point yield ϕ_A, because C_A will change (a) with time in the batch reactor, or (b) with position in the tubular reactor. Thus:

$$\phi_A = \frac{\mathcal{R}_{AP}\delta t}{(\mathcal{R}_{AP} + \mathcal{R}_{AQ})\,\delta t} = \frac{k_P C_A^p}{k_P C_A^p + k_Q C_A^q}$$

or

$$\phi_A = \left[1 + \frac{k_Q}{k_P} C_A^{(q-p)}\right]^{-1} \qquad \dots (1.35)$$

Similarly the local selectivity is given by:

$$\frac{\mathcal{R}_{AP}\delta t}{\mathcal{R}_{AQ}\delta t} = \frac{k_P}{k_Q} C_A^{(p-q)}$$

In order to find the overall relative yield Φ_A, i.e. the yield obtained at the end of a batch reaction or at the outlet of a tubular reactor, consider an element of unit volume of the reaction mixture. If the concentration of **A** decreases by δC_A either (a) with time in a batch reactor or (b) as the element progresses downstream in a tubular reactor, the amount of **A** transformed into **P** is $-\phi_A \delta C_A$. The total amount of **A** transformed into **P** during the whole reaction is therefore $\int_{C_{A0}}^{C_{Af}} - \phi_A \, dC_A$. The total amount of **A** reacted in the element is $(C_{A0} - C_{Af})$.

The overall relative yield is therefore:

$$\Phi_A = -\frac{1}{(C_{A0} - C_{Af})} \int_{C_{A0}}^{C_{Af}} \phi_A \, dC_A$$

and thus represents an average of the instantaneous value ϕ_A over the whole concentration range. Thus, from equation 1.35:

$$\Phi_A = -\frac{1}{(C_{A0} - C_{Af})} \int_{C_{A0}}^{C_{Af}} \frac{dC_A}{1 + \dfrac{k_Q}{k_P} C_A^{(q-p)}} \qquad \ldots\ldots(1.36)$$

For a stirred tank reactor consisting of a single tank in a steady state, the overall yield is the same as the instantaneous yield given by equation 1.35 because concentrations do not vary with either time or position. If more than one stirred tank is used, however, an appropriate average must be taken.

Requirements for high yield

Reactant concentration and reactor type: Although equation 1.36 gives the exact value of the final yield obtainable from a batch or tubular reactor, the nature of the conditions required for a high yield can be seen more readily from equation 1.35. If $p > q$, i.e. the order of the desired reaction is higher than that of the undesired reaction, a high yield ϕ_A will be obtained when C_A is high. A batch reactor, or a tubular reactor, gives a high reactant concentration at least initially and should therefore be chosen in preference to a single stirred tank reactor in which reactant concentration is low. If the stirred tank type of reactor is chosen on other grounds, it should consist of several tanks in series. In operating the reactor, any recycle streams which might dilute the reactants should be avoided. Conversely, if $p < q$, a high yield is favoured by a low reactant concentration and a single stirred tank reactor is the most suitable. If a batch or tubular reactor were nevertheless chosen, dilution of the reactant by a recycle stream would be an advantage. Finally, if $p = q$, the yield will be unaffected by reactant concentration.

Pressure in gas phase reactions: If a high reactant concentration is required, i.e. $p > q$, the reaction should be carried out at high pressure and the presence of inert gases in the reactant stream should be avoided. Conversely, if $p < q$ and low concentrations are required, low pressures should be used.

Temperature of operation: Adjusting the temperature affords a means of altering the ratio k_P/k_Q, provided that the activation energies of the two reactions are different. Thus

$$\frac{k_P}{k_Q} = \frac{\mathscr{A}_P}{\mathscr{A}_Q} \exp\left[-\frac{(E_P - E_Q)}{RT}\right]$$

Choice of catalyst: If a catalyst can be found which will enable the desired reaction to proceed at a satisfactory rate at a temperature which is sufficiently low for the rate of the undesired reaction to be negligible, this will usually be the best solution of all to the problem.

Yield and reactor output

The concentration at which the reaction is carried out affects not only the yield but also the reactor output. If a high yield is favoured by a high reactant concentration, there is no conflict, because the average rate of reaction and therefore the reactor output will be high also. However, if a high yield requires a low reactant concentration, the reactor output will be low. An economic optimum must be sought, balancing the cost of reactant wasted in undesired byproducts against the initial cost of a larger reactor. In most cases the product distribution is the most important factor, especially (a) when raw material costs are high, and (b) when the cost of equipment for the separation, purification and recycle of the reactor products greatly exceeds the cost of the reactor.

Reactions in parallel—two reactants

If a second reactant **B** is involved in a system of parallel reactions, then the same principles apply to **B** as to **A**. The rate equations are examined to see whether the order of the desired reaction with respect to **B** is higher or lower than that of the undesired reaction, and to decide whether high or low concentrations of **B** favour a high yield of desired product.

There are three possible types of combination between the concentration levels of **A** and **B** that may be required for a high yield:

- (a) C_A, C_B both high. In this case a batch or a tubular plug flow reactor is the most suitable.
- (b) C_A, C_B both low. A single continuous stirred tank reactor is the most suitable.
- (c) C_A high, C_B low (or C_A low and C_B high). A cross-flow reactor is the most suitable for continuous operation without recycle, and a semi-batch reactor for batchwise operation. If the reactant required in high concentration can be easily recycled, a single continuous stirred tank reactor can be used.

These ways of matching reactor characteristics to the concentration levels required are illustrated in Fig. 1.26.

Example

Two reactants undergo parallel reactions as follows:

$$\mathbf{A} + \mathbf{B} \rightarrow \mathbf{P} \text{ (desired product)}$$

$$2\,\mathbf{B} \rightarrow \mathbf{Q} \text{ (unwanted product)}$$

with the corresponding rate equations

$$\mathscr{R}_{AP} = \mathscr{R}_{BP} = k_P C_A C_B,$$

$$\mathscr{R}_{BQ} = k_Q C_B^2.$$

Suggest suitable continuous contacting schemes which will give high yields of **P** (a) if the cost of separating **A** is high and recycling is not feasible, (b) if the cost of separating **A** is low and recycling can be employed. For the purpose of quantitative treatment set $k_P = k_Q$. The desired conversion of reactant **B** is 0·95.

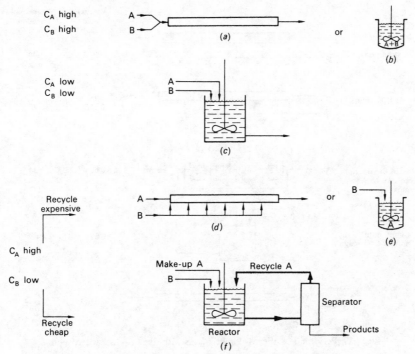

FIG. 1.26. Contacting schemes to match possible concentration levels required for high relative yields in parallel reactions

(a) Plug flow tubular reactor
(b) Batch reactor
(c) Single continuous stirred tank reactor
(d) Cross-flow tubular reactor
(e) Semi-batch reactor
(f) Single continuous stirred tank reactor with recycle of A

Alternatively a series of several continuous stirred tanks could be used in place of the tubular reactors in (a) and (d)

Solution

Inspection of the rate equations shows that, with respect to A, the order of the desired reaction is unity, and the order of the undesired reaction is effectively zero because A does not participate in it. The desired reaction is therefore favoured by high values of C_A. With respect to B, the order of the desired reaction is unity, and the order of the unwanted reaction is two. The desired reaction is therefore favoured by low values of C_B.

(a) If recycling is not feasible, a cross-flow type of reactor will be the most suitable, the feed of B being distributed between several points along the reactor. Cross–flow reactors for this particular reaction system have been studied in considerable detail. Figure 1.27 shows some results obtained with reactors employing five equidistant feed positions, together with the performances of a single continuous stirred tank reactor and of a straight tubular reactor for the purpose of comparison. In case (d) the amounts of B fed at each point in the tubular reactor have been calculated to give the maximum yield of desired product. It may be seen from (b), however, that there is little disadvantage in having the feed of B distributed in equal parts. Furthermore, the five sections of the tubular reactor can be replaced by five stirred tanks, as in (c), without appreciably diminishing the yield although the total reactor volume is somewhat greater. By way of contrast, a simple tubular reactor (e) gives a substantially lower yield.

(b) If A can be recycled, a high concentration of A together with a low concentration of B can be maintained in a single continuous stirred tank reactor, as shown in Fig. 1.26f. By suitable adjustment

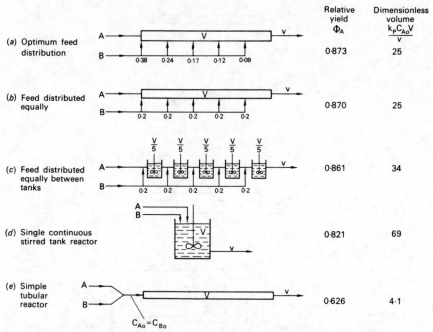

FIG. 1.27. Performance of cross-flow reactors with five equidistant feed points

Parallel reactions: $A + B \rightarrow P$ $\mathscr{R}_{AP} = \mathscr{R}_{BP} = k_P C_A C_B$
                              $2B \rightarrow Q$ $\mathscr{R}_{BQ} = k_Q C_B^2$

with $k_P = k_Q$. Equal molar feed rates of A and B. Final conversion of B = 0·95

of the rate of recycle of A and the corresponding rate of outflow, the ratio of concentrations in the reactor $C_A/C_B = r'$ may be set to any desired value. The relative yield based on B, Φ_B, will then be given by:

$$\Phi_B = \frac{k_P C_A C_B}{k_P C_A C_B + k_Q C_B^2}$$

If $k_P = k_Q$:

$$\Phi_B = \frac{C_A}{C_A + C_B} = \frac{r'}{1 + r'}$$

Even if A can be separated from the product mixture relatively easily as the recycle rate and hence r' is increased, so the operating costs will increase and the volumes of the reactor and separator must also be increased. These costs have to be set against the cash value of the increased yield of desired product as r' is increased. Thus, the optimum setting of the recycle rate will be determined by an economic balance.

Reactions in series

When reactions in series are considered, it is not possible to draw any very satisfactory conclusions without working out the product distribution completely for each of the basic reactor types. The general case in which the reactions are of arbitrary order is more complex than for parallel reactions. Only the case of two first order reactions will therefore be considered:

$$A \xrightarrow{k_{11}} P \xrightarrow{k_{12}} Q$$

where **P** is the desired product and **Q** is the unwanted product.

Batch reactor or tubular plug flow reactor

Let us consider unit volume of the reaction mixture in which concentrations are changing with time: this unit volume may be situated in a batch reactor or moving in plug flow in a tubular reactor. Material balances on this volume give the following equations:

$$-\frac{dC_A}{dt} = k_{11}C_A; \quad \frac{dC_P}{dt} = k_{11}C_A - k_{12}C_P; \quad \frac{dC_Q}{dt} = k_{12}C_P$$

If $C_P = 0$ when $t = 0$, the concentration of **P** at any time t is given by the solution to these equations which is:

$$C_P = C_{A0} \frac{k_{11}}{k_{12} - k_{11}} \left[\exp\left(-k_{11}t\right) - \exp\left(-k_{12}t\right) \right] \qquad \ldots (1.37)$$

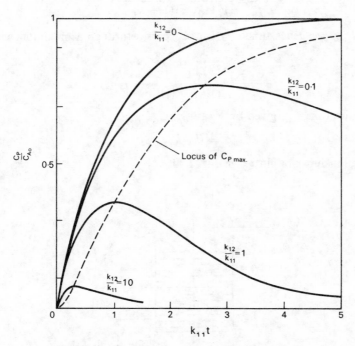

FIG. 1.28. Reaction in series −batch or tubular plug flow reactor. Concentration C_P of intermediate product **P** for consecutive first order reactions. $A \rightarrow P \rightarrow Q$

Differentiation and setting $dC_P/dt = 0$ shows that C_P passes through a maximum given by:

$$\frac{C_{Pmax}}{C_{A0}} = \left(\frac{k_{11}}{k_{12}}\right)^{\frac{k_{12}}{k_{12}-k_{11}}} \qquad \dots (1.38)$$

which occurs at a time:

$$t_{max} = \frac{\ln(k_{12}/k_{11})}{(k_{12} - k_{11})} \qquad \dots (1.39)$$

The relationships 1.37, 1.38 and 1.39 are plotted in Fig. 1.28 for various values of the ratio k_{12}/k_{11}.

Continuous stirred tank reactor—one tank

Taking material balances in the steady state as shown in Fig. 1.29:

(a) on **A**: $\qquad vC_{A0} - vC_A - V_c k_{11} C_A = 0$

(b) on **P**: $\qquad 0 - vC_P - V_c(k_{12}C_P - k_{11}C_A) = 0$

whence

$$\frac{C_P}{C_{A0}} = \frac{k_{11}\tau}{(1 + k_{11}\tau)(1 + k_{12}\tau)} \qquad \dots (1.40)$$

$$\frac{C_Q}{C_{A0}} = \frac{k_{11}k_{12}\tau^2}{(1 + k_{11}\tau)(1 + k_{12}\tau)}$$

where τ is the residence time $\tau = V/v$. C_P passes through a maximum in this case also:

$$\frac{C_{Pmax}}{C_{A0}} = \frac{1}{[(k_{12}/k_{11})^{\frac{1}{2}} + 1]^2} \qquad \dots (1.41)$$

at a residence time τ_{max} given by

$$\tau_{max} = (k_{11}k_{12})^{-\frac{1}{2}} \qquad \dots (1.42)$$

These relationships are plotted in Fig. 1.30.

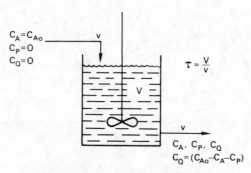

FIG. 1.29. Continuous stirred tank reactor: single tank, reactions in series, $A \to P \to Q$

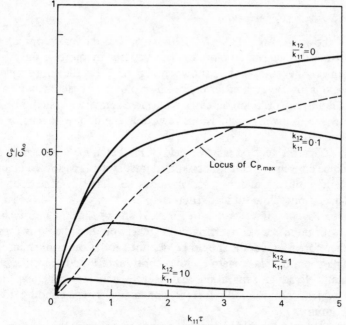

FIG. 1.30. Reactions in series—single continuous stirred tank reactor. Concentration C_P of intermediate product **P** for consecutive first order reactions, $\mathbf{A} \rightarrow \mathbf{P} \rightarrow \mathbf{Q}$

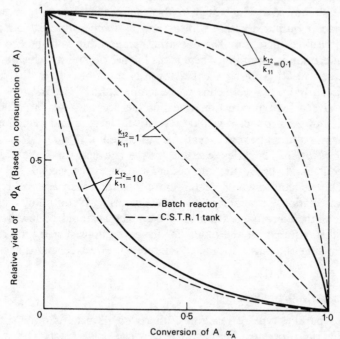

FIG. 1.31. Reactions in series—comparison between batch or tubular plug flow reactor and a single continuous stirred tank reactor. Consecutive first order reactions, $\mathbf{A} \rightarrow \mathbf{P} \rightarrow \mathbf{Q}$

Reactor comparison and conclusions

The curves shown in Figs. 1.28 and 1.30, which are curves of operational yield versus reduced time, can be more easily compared by plotting the relative yield of **P** against conversion of **A** as shown in Fig. 1.31. It is then apparent that the relative yield is always greater for the batch or plug flow reactor than for the single stirred tank reactor, and decreases with increasing conversion. We may therefore draw the following conclusions regarding the choice of reactor type and mode of operation:

Reactor type: For the highest relative yield of **P** a batch or tubular plug flow reactor should be chosen. If a continuous stirred tank system is adopted on other grounds, several tanks should be used in series so that the behaviour may approach that of a plug flow tubular reactor.

Conversion in reactor: If $k_{12}/k_{11} \gg 1$, Fig. 1.31 shows that the relative yield falls sharply with increasing conversion, i.e. **P** reacts rapidly once it is formed. If possible, therefore, the reactor should be designed for a low conversion of **A** per batch or pass in a tubular reactor, with separation of **P** and recycling of the unused reactant. Product separation and recycle may be quite expensive, however, in which case we look for the conversion corresponding to the economic optimum.

Temperature: We may be able to exercise some control over the ratio k_{12}/k_{11} by sensible choice of the operating temperature. If $E_1 > E_2$ a high temperature should be chosen, and conversely a low temperature if $E_1 < E_2$. When a low temperature is required for the best yield, there arises the problem that reaction rates and reactor output decrease with decreasing temperature, i.e. the size of the reactor required increases. The operating temperature will thus be determined by an economic optimum. There is the further possibility of establishing a temperature variation along the reactor. For example, if for the case of $E_1 < E_2$ two stirred tanks were chosen, the temperature of the first tank could be high to give a high production rate but the second tank, in which the concentration of the product **P** would be relatively large, could be maintained at a lower temperature to avoid excessive degradation of **P** to **Q**.

General conclusions: In series reactions, as the concentration of the desired intermediate **P** builds up, so the rate of degradation to the second product **Q** increases. The best course would be to remove **P** continuously as soon as it was formed by distillation, extraction or a similar operation. If continuous removal is not feasible, the conversion attained in the reactor should be low if a high relative yield is required. As the results for the continuous stirred tank reactor show, backmixing of a partially reacted mixture with fresh reactants should be avoided.

Reactions in series—two reactants

The series–parallel type of reaction outlined on page 63 is quite common among industrial processes. For example, ethylene oxide reacts with water to give monoethylene glycol, which may then react with more ethylene oxide to give diethylene glycol.

$$H_2O + C_2H_4O \rightarrow HO \cdot C_2H_4 \cdot OH$$

$$HO \cdot C_2H_4 \cdot OH + C_2H_4O \rightarrow HO \cdot (C_2H_4O)_2 \cdot H$$

i.e.
$$A + B \rightarrow P$$
$$P + B \rightarrow Q$$

In such cases the order with respect to **B** is usually the same for both the first and the second reaction. Under these circumstances, the level of concentration of **B** at which the reactions are carried out has no effect on the relative rates of the two reactions, as may be seen by writing these as parallel reactions with respect to **B**:

These reactions will therefore behave very similarly to the reactions in series above where only one reactant was involved i.e.:

$$A \xrightarrow{+B} P \xrightarrow{+B} Q$$

The same general conclusions apply; since backmixing of products with reactants should be avoided, a tubular plug flow reactor or a batch reactor is preferred. However, there is one respect in which a series reaction involving a second reactant **B** does differ from simple series reaction with one reactant, even when the orders are the same. This is in the stoichiometry of the reaction; the reaction cannot proceed completely to the product **Q**, even in infinite time, if less than two mols of **B** per mol of **A** are supplied. Some control over the maximum extent of the reaction can therefore be achieved by choosing the appropriate ratio of **B** to **A** in the feed. For reactions which are first order in **A**, **B** and **P**, charts[5] are available showing yields and end points reached for various feed ratios.

GAS–LIQUID REACTORS

In a number of important industrial processes, it is necessary to carry out a reaction between a gas and a liquid. Usually the object is to make a particular product, for example, a chlorinated hydrocarbon such as chlorobenzene by the reaction of gaseous chlorine with liquid benzene. Sometimes the liquid is simply the reaction medium, perhaps containing a catalyst, and all the reactants and products are gaseous. In other cases the main aim is to separate a constituent such as CO_2 from a gas mixture; although pure water could be used to remove CO_2, a solution of caustic soda, potassium carbonate or ethanolamine has the advantages of increasing both the absorption capacity of the liquid and the rate of absorption. The subject of gas–liquid reactor design thus really includes absorption with chemical reaction which is discussed in Volume 2, Chapter 12.

The type of equipment used for gas–liquid reactions is shown in Fig. 1.32. Firstly in *a* there is the conventional packed column which is often used when the purpose is to absorb a constituent from a gas. The liquid is distributed as a series of films over the packing, and the gas is the continuous phase. The pressure drop for the gas is relatively low and the packed column is therefore very suitable for treating large volume flows of gas mixtures. Fig. 1.32*b* shows a spray column; as in the packed column the liquid hold-up is comparatively low. and the gas is the continuous phase. In the sieve tray column shown in *c*, however, and in bubble columns, gas is dispersed in the liquid passing over the tray.

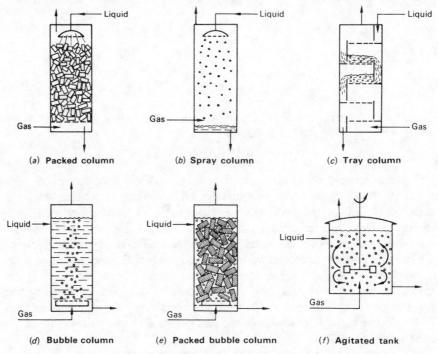

(*a*) **Packed column**   (*b*) **Spray column**   (*c*) **Tray column**

(*d*) **Bubble column**   (*e*) **Packed bubble column**   (*f*) **Agitated tank**

Fig. 1.32. Equipment used for gas–liquid reactions: in (*a*), (*b*) and (*c*) liquid hold-up is low; in (*d*), (*e*) and (*f*) liquid hold-up is high

Because the tray is relatively shallow, the pressure drop in the gas phase is fairly low, and the liquid hold-up, although a little larger than for a packed column, is still relatively small. The tray column is useful when stagewise operation is required and a relatively large volume flow of gas is to be treated.

When a high liquid hold-up is required in the reactor, one of the types shown in Fig. 1.32*d*, *e* and *f* may be used. The bubble column *d* is simply a vessel filled with liquid, with a sparger ring at the base for dispersing the gas. In some cases a draught tube is used to direct recirculation of the liquid and to influence the bubble motion. One of the disadvantages of the simple bubble column is that coalescence of the bubbles tends to occur with the formation of large slugs whose upper surfaces are in the form of spherical caps. By packing the vessel with Raschig rings, for example, as in e, the formation of very large bubbles is avoided. The reactor thus becomes an ordinary packed column operated in a flooded condition and with a sparger to disperse the gas; naturally the maximum superficial gas velocity is much less than in an unflooded packed column. Finally in f an agitator is used to disperse the gas into the liquid contained in a tank. A vaned-disc type of impeller is normally used (see Volume 2, Chapter 12). An agitated tank provides small bubbles and thus a high interfacial area of contact between gas and liquid phases, but its greater mechanical complication compared with a simple bubble column is a disadvantage with corrosive materials and at high pressures and temperatures. If necessary, stagewise operation can be achieved by having several compartments in a vertical column, with impellers mounted on a common shaft.

Equations for mass transfer with chemical reaction

In designing a gas–liquid reactor, there is the need not only to provide for the required temperature and pressure for the reaction, but also to ensure adequate interfacial area of contact between the two phases. Although the reactor as such is classed as heterogeneous, the chemical reaction itself is really homogeneous, occurring in either the liquid phase, which is the most common, or in the gas phase, and only rarely in both.

The absorption of a gas by a liquid with simultaneous reaction in the liquid phase is the most important case. There are several theories of mass transfer between two fluid phases (see Volume 1, Chapter 8; Volume 2, Chapter 12), but for the purpose of illustration the film theory will be used here. Results from the possibly more realistic penetration theory are similar numerically, although more complicated in their mathematical form[22].

Consider a *second order* reaction in the liquid phase between a substance **A** which is transferred from the gas phase and reactant **B** which is in the liquid phase only. The gas will be taken as consisting of pure **A** so that complications arising from gas film resistance are avoided. The stoichiometry of the reaction is represented by:

$$v_A\mathbf{A} + v_B\mathbf{B} \rightarrow \text{Products}$$

with the rate equation:

$$\mathscr{R}_A = k_2 C_A C_B \qquad \dots\dots(1.43)$$

Note that:

$$\mathscr{R}_B = \frac{v_B}{v_A}\mathscr{R}_A$$

In the film theory, steady state conditions are assumed in the film such that in any volume element the difference between the rate of mass transfer into and out of the element is just balanced by the rate of reaction within the element.

Carrying out such a material balance on reactant **A** the following differential equation results:

$$D_A \frac{d^2 C_A}{dx^2} - \mathscr{R}_A = 0$$

Similarly a material balance on **B** gives:

$$D_B \frac{d^2 C_B}{dx^2} - \mathscr{R}_B = 0$$

Introducing the rate equation (1.43) above, the differential equations become:

$$D_A \frac{d^2 C_A}{dx^2} - k_2 C_A C_B = 0 \qquad \qquad \dots (1.44)$$

$$D_B \frac{d^2 C_B}{dx^2} - k_2 C_A C_B \frac{v_B}{v_A} = 0 \qquad \qquad \dots (1.45)$$

Typical conditions within the film may be seen in Fig. 1.33a. Boundary conditions at the gas–liquid interface are:

for **A**:

$$C_A = C_{Ai}, \quad x = 0 \qquad \qquad \dots (1.46)$$

for **B**:

$$\frac{dC_B}{dx} = 0, \quad x = 0 \qquad \qquad \dots (1.47)$$

On the liquid side of the film where $x = \delta$, the boundary condition for **B** is simply:

$$C_B = C_{BL}, \quad x = \delta \qquad \qquad \dots (1.48)$$

To obtain the boundary condition for **A** we note that, except for the amount of **A** which reacts in the film, **A** is transferred across this boundary and reacts in the bulk of the liquid. For unit area of interface, the volume of this bulk liquid may be written as $[(\varepsilon/a) - \delta]$, where ε is the volume of liquid per unit volume of reactor space (i.e. the liquid hold-up), a is the gas–liquid interfacial area per unit volume of reactor space (i.e. specific area), and δ is the thickness of the film. The boundary condition for **A** is thus:

$$- D_A \left(\frac{dC_A}{dx}\right)_{x=\delta} = k_2 C_{AL} C_{BL} \left(\frac{\varepsilon}{a} - \delta\right) \qquad \qquad \dots (1.49)$$

$$\underbrace{\phantom{- D_A \left(\frac{dC_A}{dx}\right)_{x=\delta}}}_{\substack{\text{Mass transfer} \\ \text{across boundary}}} \qquad \underbrace{\phantom{k_2 C_{AL} C_{BL}}}_{\substack{\text{Reaction in} \\ \text{bulk liquid}}}$$

Although a complete analytical solution of the set of equations 1.44 to 1.49 is not possible, analytical solutions are obtainable for part of the range of variables[19, 22] and a numerical solution has been obtained for the remainder[15]

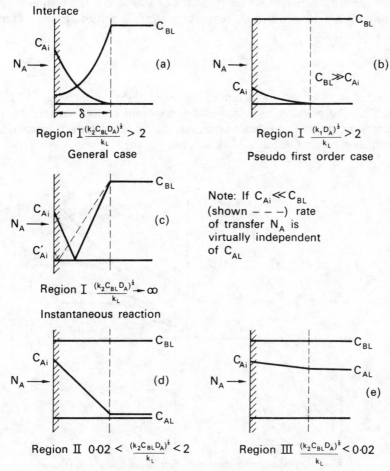

FIG. 1.33. Liquid phase concentration profiles for mass transfer with chemical reaction –film theory

The results of these solutions will be discussed in terms of a reaction factor f_A which is defined by the expression:

$$N_A = k_L C_{Ai} f_A \qquad \qquad \dots (1.50)$$

which may be compared with equation 1.51 which applies when there is no reaction:

$$N_A = k_L (C_{Ai} - C_{AL}) \qquad \qquad \dots (1.51)$$

where N_A is the molar flux of **A** at the interface, and k_L the mass transfer coefficient for physical absorption of **A**. When greater than 1, f_A represents the enhancement of the rate of transfer of **A** caused by the chemical reaction as compared with pure physical absorption with a zero concentration of **A** in the bulk liquid.

The complete solution of the set of equations 1.44 to 1.49 is shown in graphical form in Fig. 1.34, in which the reaction factor f_A is plotted against a dimensionless parameter:

$$\frac{\sqrt{(k_2 C_{BL} D_A)}}{k_L}.$$

(The right-hand side of this diagram, region I, is the same as Fig. 12.12 of Volume 2.) The physical significance of the various regions covered by this diagram is important and is best appreciated by considering the corresponding concentration profiles of Fig. 1.33 as follows.

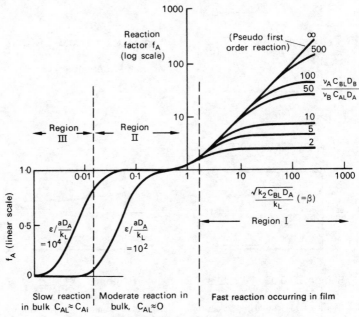

FIG. 1.34. Reaction factor f_A for a second order reaction (numerical solution) and pseudo-first order reaction (analytical solution)

Rate of transformation of A *per unit volume reactor*

The value of the parameter

$$\frac{\sqrt{(k_2 C_{BL} D_A)}}{k_L}$$

provides an important indication of whether a large specific interfacial area a or a large liquid hold-up ε is required in a reactor to be designed for a particular reaction of rate constant k_2. Let the rate of transformation of A per unit volume of reactor be J_A. Three regions may be distinguished as shown in Fig. 1.34.

Region I:

$$\frac{\sqrt{(k_2 C_{BL} D_A)}}{k_L} > 2$$

The reaction is fast and occurs mainly in the film as **A** is being transported (Fig. 1.33a). C_{AL}, the concentration of **A** in the bulk of the liquid, is virtually zero. Thus:

$$J_A = k_L C_{Ai} f_A a \qquad \dots (1.52)$$

Therefore, in this set of circumstances, J_A will be large if a is large; a large interfacial area a is required in the reactor but the liquid hold-up is not important. A packed column, for example, would be suitable.

If the concentration of **B** in the bulk liquid, C_{BL}, is much greater than C_{Ai}, a common case being where **A** reacts with a pure liquid **B**, the kinetics of the reaction become pseudo-first order, and the above equations can be solved analytically to give:

$$f_A = \frac{\beta}{\tanh \beta} \quad \text{where } \beta = \frac{\sqrt{(k_1 D_A)}}{k_L}$$

k_1 is the pseudo-first order rate constant. The concentration profile is sketched in Fig. 1.33b. Note that for large β, $f_A \to \beta$, hence:

$$N_A = \sqrt{(k_1 D_A)} \, C_{Ai}$$

N_A is then independent of mass transfer coefficient k_L.

On the other hand, if the reaction between **A** and **B** is virtually instantaneous, as shown in Fig 1.33c, and $C_{Ai} \ll C_{BL}$, the rate of transformation is nearly independent of C_{Ai}, being determined mainly by the rate of mass transfer of **B** to the reaction zone.

Region II:
$$0 \cdot 02 < \frac{\sqrt{(k_2 C_{BL} D_A)}}{k_L} < 2$$

This is an intermediate region in which the reaction is sufficiently fast to hold C_{AL} close to zero (Fig. 1.33d), but, although C_{AL} is small, nearly all the reaction occurs in the bulk of the liquid. The hold-up ε is important because unless:

$$\varepsilon \sqrt{\frac{a D_A}{k_L}} > 10^2$$

f_A will be substantially less than 1 in the part of the region (see Fig. 1.34) where:

$$\frac{\sqrt{(k_2 C_{BL} D_A)}}{k_L} < 0 \cdot 1$$

When $f_A = 1$:

$$J_A = k_L C_{Ai} a \qquad \dots (1.53)$$

Thus, both interfacial area and hold-up should be high. For example, an agitated tank will give a high value of J_A.

Region III:
$$\frac{\sqrt{(k_2 C_{BL} D_A)}}{k_L} < 0 \cdot 02$$

The reaction is slow and occurs in the bulk of the liquid. Mass transfer serves to

D

keep the bulk concentration of A, C_{AL}, close to the saturation value C_{Ai} (Fig. 1.33e) and sufficient interfacial area should be provided for this purpose. However, a high liquid hold-up is the more important requirement, and a bubble column is likely to be suitable. If $C_{AL} \approx C_{Ai}$, then:

$$J_A = k_2 C_{Ai} C_{BL} \varepsilon \qquad \qquad \dots (1.54)$$

Near the boundary between the regions II and III, where virtually none of the reaction occurs in the film, the concentration of A in the bulk of the liquid, C_{AL}, is determined by the simple relation:

$$J_A = N_A a = k_L a (C_{Ai} - C_{AL}) = k_2 C_{AL} C_{BL} \varepsilon \qquad \dots (1.55)$$

$$\underbrace{\phantom{J_A = N_A a = k_L a (C_{Ai} - C_{AL})}}_{\substack{\text{Mass transfer} \\ \text{through film}}} \qquad \underbrace{\phantom{k_2 C_{AL} C_{BL} \varepsilon}}_{\substack{\text{Reaction in} \\ \text{bulk}}}$$

i.e.

$$C_{AL} = C_{Ai} \left(1 + \frac{k_2 C_{BL} \varepsilon}{k_L a} \right)^{-1} \qquad \dots (1.56)$$

Choice of a suitable reactor

Choosing a suitable reactor for a gas–liquid reaction is a question of matching the characteristics of the reaction system, especially the reaction kinetics, with the characteristics of the reactors under consideration. As we have seen above, two of the most important characteristics of gas–liquid reactors are the specific interfacial area a, and the liquid hold-up ε. Table 1.4 shows some representative values of these quantities for various gas–liquid reactors.

TABLE 1.4. *Comparison of specific interfacial area a and liquid hold-up ε for various types of reactor*

Type of contactor	Specific area a m²/m³	Liquid hold-up ε (fraction)
Spray column	60	0·05
Packed column (2.5 cm Raschig rings)	220	0·08
Plate column	150	0·15
Bubble contactor	200	0·85
Agitated tank	500	0·80

Example

You are asked to recommend, with reasons, the most suitable type of equipment for carrying out a gas–liquid reaction between a gas A and a solution of a reactant B. Particulars of the system are as follows:

Concentration of B in solution = 5 kmol/m³
Diffusivity of A in the solution = 1·5 × 10⁻⁹ m²/s
Rate constant of the reaction (A + B → Products; second order overall) = 0·03 m³/kmol s.

For bubble dispersions (plate columns, bubble columns, agitated vessels) take k_L as having a range from 2 × 10⁻⁴ to 4 × 10⁻⁴ m/s. For a packed column, take k_L as having a range 0·5 × 10⁻⁴ to 1·0 × 10⁻⁴ m/s.

Solution

We first calculate the value of the parameter:

$$\frac{\sqrt{(k_2 C_{BL} D_A)}}{k_L}$$

From the data above:

$$\sqrt{(k_2 C_{BL} D_A)} = \sqrt{(0.03 \times 5 \times 1.5 \times 10^{-9})} = 1.5 \times 10^{-5} \text{ m/s}$$

For the bubble dispersions:

$$\frac{\sqrt{(k_2 C_{BL} D_A)}}{k_L}$$

will therefore lie in the range:

$$\frac{1.5 \times 10^{-5}}{2 \times 10^{-4}} = 0.075 \quad \text{to} \quad \frac{1.5 \times 10^{-5}}{4 \times 10^{-4}} - 0.038$$

For a packed column, k_L will have the range:

$$\frac{1.5 \times 10^{-5}}{0.5 \times 10^{-4}} = 0.3 \quad \text{to} \quad \frac{1.5 \times 10^{-5}}{1.0 \times 10^{-4}} = 0.15$$

Referring to Fig. 1.34 these values lie in region II indicating that the reaction is only moderately fast and that a relatively high liquid hold-up is required. A packed column would in any case therefore be unsuitable. We therefore conclude from the above considerations that an agitated tank, a simple bubble column or a packed bubble column should be chosen. The final choice between these will depend on such factors as operating temperature and pressure, corrosiveness of the system, allowable pressure drop in the gas, and the possibility of fouling.

Notice that to continue further with the design of a reactor by taking a value of f_A from Fig. 1.34, or using equations 1.53 or 1.55, requires a knowledge of ε and a for the bubble dispersion. One of the problems still outstanding in gas–liquid reactor design is to establish reliable values for these quantities[13]. One of the difficulties is that, not only do they depend on the design of the device used to disperse the gas, and the gas flowrate, but also on how the surface properties of the liquid affect the coalescence between bubbles. In practice, laboratory or pilot plant experiments are usually required. For packed columns, design procedures are more satisfactory, and for details the works of DANCKWERTS[15] and of DANCKWERTS and SHARMA[21] should be consulted.

In most practical cases laboratory experiments are also required to provide knowledge about kinetics of the reaction. For fast reactions, a laminar liquid jet or a wetted wall column can enable a thorough investigation of the kinetics to be made[15], although this procedure is time consuming. For moderately fast or slow reactions, less ambitious experiments with an agitated vessel may provide valuable information. Generally, as the speed of the agitator increases so the area a and the gas hold-up $(1 - \varepsilon)$ increase, f_A thereby decreasing (Fig. 1.34). As f_A approaches zero, it is found that a further increase in agitator speed produces no further increase in the overall rate of reaction, i.e. the effect of mass transfer has been eliminated and equation 1.54, for example, will apply.

Although we can distinguish the principles of gas–liquid reactor design by examining the cases of second order and pseudo-first order reactions above, in practice many reactions have more complex kinetics. Concentration profiles for other model reaction schemes have been worked out, but generally a great deal of information, not usually available, is required to proceed strictly from fundamental principles. In such cases, successful design must proceed on a more empirical basis, with general principles serving as a guide, for example, in deciding whether the reaction occurs mainly in a zone close to the surface or in the bulk of the liquid phase.

NON-IDEAL FLOW AND MIXING IN CHEMICAL REACTORS

So far we have developed calculation methods only for the ideal cases of plug flow in tubular reactors (page 36), and complete mixing in stirred tank reactors (page 46). In reality, the flow of fluids in reactors is rarely ideal, and although for some reactors design equations based on the assumption of ideal flow give acceptable results, in other cases the departures from the ideal flow state need to

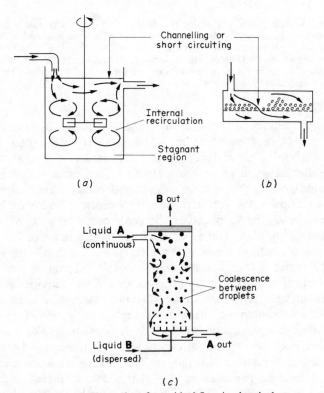

FIG. 1.35. Examples of non-ideal flow in chemical reactors
(a) Continuous stirred tank.
(b) Gas–solid reaction with maldistribution of solid on a shallow tray.
(c) Liquid–liquid reaction: Note the circulation patterns in the continuous phase **A** induced by the rising droplets of **B**; coalescence may occur between phase **B** droplets giving a range of sizes and upwards velocities.

be taken into account. Following the development by DANCKWERTS[24] of the basic ideas, one of the leading contributors to the subject of non-ideal flow has been LEVENSPIEL whose papers and books[5,25,26], especially *Chemical Reaction Engineering*, Chapters 9 and 10, should be consulted.

It is possible to distinguish various types of non-ideal flow patterns in reactors (and process vessels generally) the most important being *channelling*, *internal recirculation* and the presence of *stagnant regions*. These are illustrated in the examples shown in Fig. 1.35. In two-phase (and three-phase) reactors the flow patterns in one phase interact with the flow patterns in the other phase as in the case of the liquid–liquid reactor of Fig. 1.35c. One of the major problems in the scale-up of reactors is that the flow patterns often change with a change of scale, especially in reactors involving two or more phases where flow interactions may occur. A gas–solid fluidised bed is another example of an important class of reactor whose characteristics on scale-up are difficult to predict[25] (see Volume 2, Chapter 6).

Experimental tracer methods

It has already been seen how, in principle, experiments with a non-reactive tracer can be very useful in establishing the flow characteristics of a series of well-

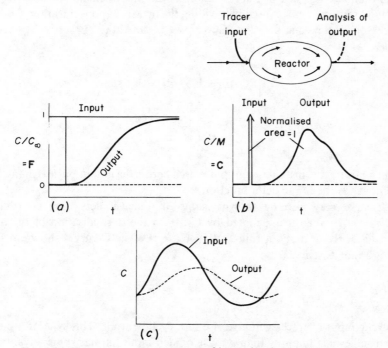

FIG. 1.36. Tracer measurements; types of input signals and output responses
(a) Step input—F-curve
(b) Pulse input—C-curve
(c) Sinusoidal input

mixed continuous stirred tanks (pages 47–50). When the flow through a reactor or any other type of process vessel is non-ideal, experiments with tracers can provide most valuable information on the nature of the flow.

The injection of a tracer and the subsequent analysis of the exit stream is an example of the general stimulus–response methods described under Process Control in Chapter 3. In tracer experiments various input signals can be used, especially the following (Fig. 1.36):

(a) Step input—F-curves

The inlet concentration is increased suddenly from zero to C_∞ and maintained thereafter at this value. The concentration–time curve at the outlet expressed as the fraction C/C_∞ vs. t is known as an "F-curve". The time scale may also be expressed in a dimensionless form as $\theta = t/\tau$ where τ is the *holding time* (or *residence time*), i.e. $\tau = V/v$ where V is the volume of the reactor and v the volumetric rate of flow.

(b) Pulse input—C-curves

An instantaneous pulse of tracer is injected into the stream entering the vessel. The outlet response, normalised by dividing the measured concentration C by M, the area under the concentration–time curve, is called the "C-curve".

Thus:

$$\int_0^\infty \mathbf{C}\, dt = \int_0^\infty \frac{C}{M}\, dt = 1 \qquad \dots (1.57)$$

where

$$M = \int_0^\infty C\, dt \qquad \dots (1.58)$$

(Mathematically the instantaneous pulse used here is the unit impulse (Chapter 3) also known as the Dirac delta function.)

A C-curve may also be shown against a dimensionless time coordinate $\theta = t/\tau = vt/V$ when it is denoted by $\mathbf{C}_\theta$; from the normalisation of the area under this curve to unity, it follows that $\mathbf{C}_\theta = \tau\mathbf{C}$. The C-curves shown in Fig. 1.19a are thus $\mathbf{C}_\theta$ curves.

(c) Cyclic, sinusoidal and other inputs

A cyclic input of tracer will give rise to a cyclic output. This type of input is more troublesome to apply than a pulse or step input but has some advantages for frequency response analysis. A truly instantaneous pulse input is also difficult to apply in practice, but an idealised input pulse is not essential[5,26] for obtaining the desired information.

Age distribution of a stream leaving a vessel—E-curves

The distribution of residence times for a stream of fluid leaving a vessel is called the *exit age distribution* **E** (synonymous with *residence time distribution* or

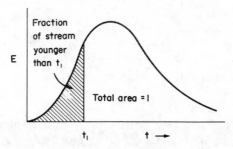

FIG. 1.37. Exit age distribution or E-curve; also known as the residence time distribution

RTD). By definition, the fraction of the exit stream of age between t and $(t + \delta t)$ is $\mathbf{E} \, \delta t$ (Fig. 1.37). Integrating over all ages:

$$\int_0^\infty \mathbf{E} \, dt = 1.$$

The fraction of material younger than t_1 is:

$$\int_0^{t_1} \mathbf{E} \, dt$$

Relation between **F**-, **C**- *and* **E**-*curves*

For *steady-state flow* in a *closed vessel* (i.e. one in which fluid enters and leaves solely by plug flow) the residence time distribution for any batch of fluid entering must be the same as that leaving (otherwise accumulation would occur).

Hence the **C**-curve generated by injecting a pulse of fluid at the entrance must be identical with the **E**-curve, i.e.

$$\mathbf{C} \equiv \mathbf{E}$$

Consider an **F**-curve generated by switching, say, to red fluid at $t = 0$. At any time $t > 0$ red fluid and only red fluid in the exit stream is younger than age t.
Thus

(Fraction of red fluid   (Fraction of exit stream
    in exit stream)   =   younger than age t)

i.e.
$$\mathbf{F} = \int_0^t \mathbf{E} \, dt \qquad \dots (1.59)$$

or
$$\frac{dF}{dt} \equiv E \equiv C \qquad \dots (1.60)$$

Note that it may be shown also that the holding time τ is equal to $\bar{t}_E$, the mean of the residence time distribution, i.e.

$$\tau = V/v = \bar{t}_E \qquad \dots (1.61)$$

Application of tracer information to reactors

Direct application of exit age distribution

Information on residence time distribution obtained from tracer measurements can be used directly to predict the performance of a reactor in which non-ideal flow occurs. A good example is the conversion obtained when solid particles undergo a reaction in a rotary kiln (Fig. 1.38). Using labelled particles

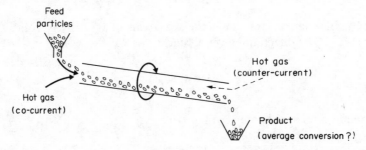

FIG. 1.38. Average conversion from residence time distribution—example of rotary kiln

as tracers, a **C**-curve and hence an exit age distribution can be determined; the fraction of particles in the exit stream which have stayed in the reactor for times between t and $(t + \Delta t)$ may thus be expressed as $E \Delta t$. From other experiments α_t the fractional conversion of reactant in those particles staying in the reactor for a time t can be found. Using α_t for the range of times t to $t + \Delta t$ providing Δt is small, the average conversion $\bar{\alpha}$ for the mixture of particles which constitute the product is thus $\bar{\alpha} = \sum \alpha_t E \Delta t$ where the summation is taken over the whole range of residence times, if necessary 0 to ∞.

Macromixing and micromixing

In the above example of the rotary kiln, although the particles intermingle, each particle remains as a distinct and separate aggregate of molecules which retains its identity as it passes through the process. This type of intermingling of aggregates is called *macromixing*, i.e. mixing but only on a *macroscopic scale*. Perfect macromixing means that the composition averaged over a volume sufficiently great to contain a relatively large number of aggregates does not vary with location in the vessel.

In contrast to this, when a solution of a substance is diluted, the molecules of the solute mix on a *molecular scale* with fresh solvent and essentially lose their identity in the process. This is called *micromixing*. In the same way solutions of two different solutes can undergo micromixing.

If, however, the two solutions to be mixed are very viscous, and diffusion is slow, complete micromixing may not occur; instead we could have a mixture which was intermediate in its *degree of segregation* between macromixing at one extreme and micromixing at the other. As we shall see below, it transpires that in attempting to use tracer measurements directly to predict, say, the average conversion from a reactor, further information on the degree of segregation is required, unless the process (here the reaction) occurring is a *first order* or *linear* process, in this case, a first order reaction.

Significance of linear and non-linear processes

As an example, consider the various cases of mixing and reaction shown in Fig. 1.39. In Case I we have a solution of a reactant concentration C_0 and pure

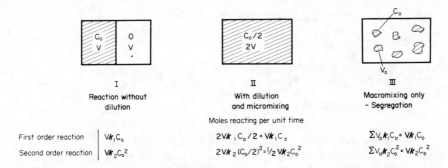

FIG. 1.39. Influence of state of mixing on first and second order reaction rates

solvent in separate compartments each of volume V. In Case II, the partition is removed and there is dilution of the reactant solution with micromixing so that the reactant concentration is halved. In Case III the partition is removed but there is macromixing only and the reactant solution remains segregated. It is seen that the number of moles reacting per unit time in all three cases is the same for a first order reaction which is a linear process. However, for a second order reaction, which is an example of a non-linear process, the rate of reaction depends on the degree of segregation resulting from the mixing: if micromixing occurs, the rate is only half the rate with complete segregation.

In general, it may be concluded:

Reaction order > 1, reaction rate greatest with complete segregation.
Reaction order = 1, reaction rate unaffected by segregation.
Reaction order < 1, reaction rate greatest with complete micromixing.

Occurrence of micromixing in flow reactors

In flow reactors two extremes of micromixing can be visualised:

(a) Completely segregated flow until mixing occurs at the outlet or as the fluid is sampled for analysis, i.e. segregation throughout the entire system to the outlet, or *late* mixing.

(b) A condition of *maximum mixedness* meaning intimate micromixing at the earliest possible stage, i.e. *early* mixing.

The significance of early or late mixing may be seen from Kramers' example[27] (Fig. 1.40) of a plug flow reactor in series with an ideally mixed stirred tank in

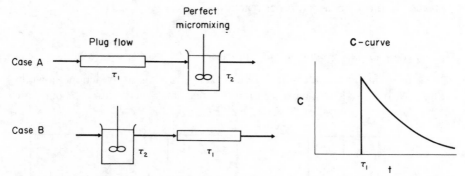

FIG. 1.40. Kramers' example of a plug flow tubular reactor (residence time τ_1) and an ideal stirred tank (residence time τ_2) in series. Note that the C-curve is the same for both configurations

which macromixing occurs. The two units may be connected either with the tubular reactor first (Case A) or with the stirred tank first (Case B). As far as the response to a pulse input of tracer is concerned, the plug flow section merely delays the tracer by a time τ_1, the plug flow residence time, irrespective of the configuration, and the C-curve is the same for both cases (for the equation of the part due to the stirred tank, see page 48). If we suppose that a *first order* reaction with rate constant k_1 occurs in this system, the reader may like to show as an exercise that again in both Case A and Case B:

$$\frac{C_e}{C_0} = \frac{e^{-k_1\tau_1}}{1 + k_1\tau_2} \qquad \dots (1.62)$$

where C_0 is the inlet concentration of reactant, C_e the outlet concentration, and τ_2 is the residence time in the stirred tank.

However, for a *second order* reaction the conversions in the two cases A and B are *different*. With late mixing as in Case A, the conversion is higher than with early mixing as in Case B. Again as a general rule it is concluded that if the reaction order is >1 late mixing, i.e. segregation, gives the highest conversion: if the reaction order is <1 early mixing gives the highest conversion; and for a first order reaction the conversion is independent of the type of mixing.

A further important conclusion is that for a given **C**-curve or residence time distribution obtained from tracer studies, a *unique* value of the conversion in a chemical reaction is not necessarily obtainable unless the reaction is first order. Tracer measurements can certainly tell us about departures from good macro-mixing. However, *tracer measurements cannot give any further information about the extent of micromixing* because the tracer stimulus-response is a first order (linear) process as is a first order reaction.

Non-ideal flow modelling

Returning to the example of Fig. 1.40, if a reactor on which tracer measure-ments were made gave a **C**-curve similar to that shown, an ideal plug-flow tubular section in series with an ideal stirred-tank section, it could be used as a model of the reactor (within the limitations outlined above). These ideal reactor sections can be regarded as building blocks for the model.

Dispersed plug flow model

The dispersed plug flow model is a type of flow which may be used either by itself to represent departures from plug flow by means of a single parameter D_L, the longitudinal dispersion coefficient, or in combination with other flow regions. The basic equation of the dispersed plug flow model is:

$$\frac{\partial C}{\partial t} + u \frac{\partial C}{\partial z} = D_L \frac{\partial^2 C}{\partial z^2} \qquad \dots (1.63)$$

where C is the concentration of non-reacting tracer and u is the velocity along the pipe in the z-direction (Fig. 1.41) (see Volume 2, Chapter 4). In dimensionless form, using L, the distance between inlet and outlet positions as a characteristic length, this becomes:

$$\frac{\partial C}{\partial \theta} + \frac{\partial C}{\partial z'} = \frac{D_L}{uL} \frac{\partial^2 C}{\partial z'^2} \qquad \dots (1.64)$$

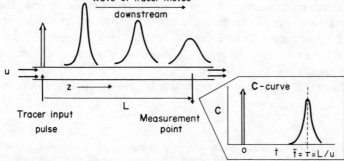

FIG. 1.41. Response of the dispersed plug flow model to a pulse input of tracer. If longitudinal dispersion occurs to only a small extent (small D_L/uL) the C-curve is almost symmetrical

where $z' = z/L$ and $\theta = ut/L$. The dimensionless group D_L/uL is called the vessel dispersion number; if $D_L/uL \to 0$ we have ideal plug flow with no dispersion; if $D_L/uL \to \infty$ we have complete backmixing as in a stirred tank.

The response of the dispersed plug flow model to a pulse input of tracer is shown in Fig. 1.41. If D_L/uL is small, an almost symmetrical C-curve is obtained; in dimensionless form its analytical equation is:

$$\mathbf{C}_\theta = \frac{1}{2\sqrt{[\pi(D_L/uL)]}} \exp \frac{(1 - \theta)^2}{4(D_L/uL)} \qquad \dots (1.65)$$

which represents a family of Gaussian (i.e. normal or error) curves with mean $\bar{\theta}$ and variance σ_θ^2:

$$\bar{\theta} = \frac{\bar{t}}{\tau} = 1 \quad \text{and} \quad \sigma_\theta^2 = \frac{\sigma^2}{t^2} = 2\left(\frac{D_L}{uL}\right)$$

Hence an estimate of the dispersion coefficient D_L may be calculated. An extensive literature[28, 29] exists on the theory, application and best means of determining D_L. As a further extension of the model a radial dispersion coefficient D_R may be introduced (Volume 2, Chapter 4).

Tanks-in-series model

Another one-parameter model of a reactor is provided by a set of i equal ideally stirred tanks in series. The equations for this model have already been examined and the $\mathbf{C}_\theta$-curves are shown in Fig. 1.19a. By comparing these C-curves with the experimentally determined C-curve for the reactor, the appropriate value of the parameter i can be found.

Other types of region used in modelling

As well as regions corresponding to ideal plug flow, dispersed plug flow, an ideal stirred tank, and a series of i stirred tanks, a further type of region is a stagnant or deadwater zone. In the construction of mixed models we may also need to introduce by-pass flow, internal recirculation, and cross-flow (which involves interchange but no net flow between regions). Of course, the more complicated the model, the larger is the number of parameters which have to be determined by matching C- or F-curves. A wide variety of possibilities exists; for example, over twenty somewhat different models have been proposed for representing a gas–solid fluidised bed[25].

In conclusion the following example shows how a real stirred tank might be modelled[30].

Example

As a model of a certain poorly agitated continuous stirred tank reactor of total volume V, it is supposed that only a fraction w is well mixed, the remainder being a deadwater zone. Furthermore, a fraction f of the total volume flowrate v fed to the tank by-passes the well-mixed zone completely (Fig. 1.42a).

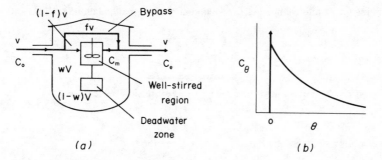

FIG. 1.42. Model of a poorly agitated continuous stirred tank reactor
(a) Flow model: fraction f of flow v bypasses; only a fraction w of tank volume V is well stirred
(b) Equivalent C_θ-curves for pulse input

(i) If the model is correct, sketch the C-curve expected if a pulse input of tracer is applied to the reactor.

(ii) On the basis of the model, determine the fractional conversion α_f which will be obtained under steady conditions if the feed contains a reactant which undergoes a homogeneous first order reaction with rate constant k_1.

Solution

(i) Because of the by-pass which is assumed to have no holding capacity, part of the input pulse appears instantly in the outlet stream. If the input pulse is normalised to unity, the magnitude of this immediate output pulse is f. The remainder of the C_θ-curve (Fig. 1.42b) is a simple exponential decay similar to the curve for one tank shown in Fig. 1.19a on page 50. However, in the present case, because only a fraction $(1 - f)$ of the unit pulse enters the well–mixed region volume wV, the equation of the curve with $\theta = vt/V$ is:

$$C_\theta = (1 - f)^2 \frac{1}{w} \exp\left\{ -\frac{(1 - f)}{w} \theta \right\}$$

as may be shown by an analysis similar to that on page 48. Notice that values for f and w could be determined by comparing experimentally obtained data with this expression.

(ii) Consider a material balance over the well-mixed region only, in the steady state. Let the reactant inlet concentration by C_0 and C_m be the intermediate concentration at the outlet of this region before mixing with the by-pass stream.

$$\underset{\text{Inflow}}{v(1 - f)C_0} - \underset{\text{Outflow}}{v(1 - f)C_m} - \underset{\text{Reaction}}{k_1 C_m wV} = 0$$

Hence:

$$C_m = \frac{C_0}{1 + \dfrac{k_1 wV}{v(1 - f)}}$$

Next consider mixing of this stream with the by-pass giving a final exit concentration C_e.

$$v(1 - f)C_m + fvC_0 = vC_e$$

$$\therefore \frac{C_e}{C_0} = (1 - f)\frac{C_m}{C_0} + f$$

The fractional conversion at the exit

$$\alpha_f = \frac{C_0 - C_e}{C_0} = 1 - \frac{C_e}{C_0}$$

Substituting for C_e and eliminating C_m we find after some manipulation:

$$\alpha_f = \frac{1}{\dfrac{v}{k_1 w V} + \dfrac{1}{(1-f)}}$$

Note that if $f = 0$, $w = 1$ and writing $V/v = \tau$ this expression reduces to:

$$\alpha_f = \frac{k_1 \tau}{1 + k_1 \tau}$$

as expected for a single ideally mixed stirred-tank reactor.

GAS–SOLID NON-CATALYTIC REACTORS

There is a large class of industrially important heterogeneous reactions in which a gas or a liquid is brought into contact with a solid and reacts with the solid transforming it into a product. Among the most important are: the reduction of iron oxide to metallic iron in a blast furnace; the combustion of coal particles in a pulverised fuel boiler; and the incineration of solid wastes; but these examples also happen to be some of the most complex chemically. Further simple examples are the roasting of sulphide ores such as zinc blende:

$$ZnS + \tfrac{3}{2}O_2 = ZnO + SO_2$$

and two reactions which are used in the carbonyl route for the extraction and purification of nickel:

$$NiO + H_2 = Ni + H_2O$$
$$Ni + 4CO = Ni(CO)_4$$

In the first of these reactions the product, impure metallic nickel, is in the form of a porous solid, whereas in the second, the nickel carbonyl is volatile and only the impurities remain as a solid residue.

It can be seen from even these few examples that a variety of circumstances can exist. As an initial approach to the subject, however, it is useful to distinguish two extreme ways in which reaction within a particle can develop.

(a) *Shrinking core reaction mode.* If the reactant solid is non-porous, the reaction begins on the outside of the particle and the reaction zone moves towards the centre. The core of unreacted material shrinks until it is entirely consumed. The product formed may be a coherent but porous solid like the zinc oxide from the oxidation of zinc sulphide above, in which case the size of the particle may be virtually unchanged (Fig. 1.43a(i)). On the other hand, if the product is a gas or if the solid product formed is a friable ash, as in the combustion of some types of coal, the particle decreases in size during the reaction until it disappears (Fig. 1.43a(ii)). When the reactant solid is porous a *fast* reaction may nevertheless still proceed via a shrinking core reaction mode because diffusion of the gaseous reactant into the interior of the solid will be a relatively slow process.

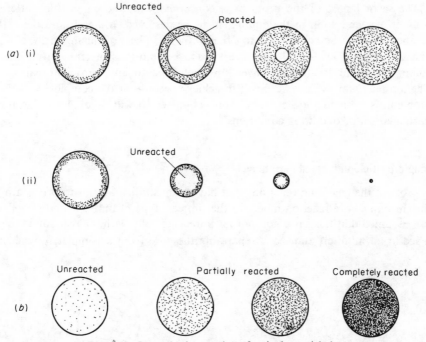

FIG. 1.43. Stages in the reaction of a single particle by
(a) Shrinking core reaction mode with the formation of either (i) a coherent porous solid or
(ii) friable ash or gaseous product
(b) Progressive conversion reaction mode

(b) *Progressive conversion reaction mode.* In a porous reactant solid an in-herently *slower* reaction can, however, proceed differently. If the rate of diffusion of the gaseous reactant into the interior of the particle can effectively keep pace with the reaction, the whole particle may be progressively and uniformly con-verted into the product (Fig. 1.43b). If the product is itself a coherent solid, the particle size may be virtually unchanged, but if the product is volatile or forms only a weak solid structure, the particle will collapse or disintegrate towards the end of the reaction.

Modelling and design of gas–solid reactors

The basic approach is to consider the problem in two parts. Firstly, the reaction of a single particle with a plentiful excess of the gaseous reactant is studied. A common technique is to suspend the particle from the arm of a thermobalance in a stream of gas at a carefully controlled temperature; the course of the reaction is followed through the change in weight with time. From the results a suitable kinetic model may be developed for the progress of the reaction within a single particle. Included in this model will be a description of any mass transfer resistances associated with the reaction and of how the reaction is affected by concentration of the reactant present in the gas phase.

The second part of the problem is concerned with the contacting pattern between gas and solid in the equipment to be designed. Material and thermal balances have to be considered, together with the effects of mixing in the solid and gas phases. These will influence the local conditions of temperature, gas composition and fractional conversion applicable to any particular particle. The ultimate aim, which is difficult to achieve because of the complexity of the problem, is to estimate the overall conversion which will be obtained for any particular set of operating conditions.

Single particle unreacted core models

Most of the gas–solid reactions that have been studied appear to proceed by the shrinking core reaction mode. In the simplest type of unreacted core model it is assumed that there is a non-porous unreacted solid with the reaction taking place in an infinitely thin zone separating the core from a completely reacted

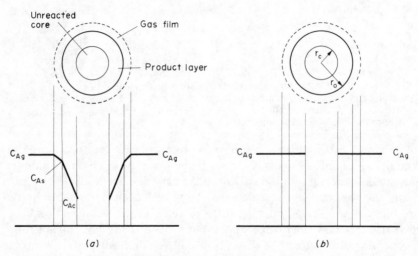

FIG. 1.44. Unreacted core model, impermeable solid, showing gas-phase reactant concentration
(a) With significant gas-film and product-layer resistances
(b) Negligible gas-film and product-layer resistances

product as shown in Fig. 1.44 for a spherical particle. Considering a reaction between a gaseous reactant **A** and a solid **B** and assuming that a coherent porous solid product is formed, five consecutive steps may be distinguished in the overall process:

(1) Mass transfer of the gaseous reactant **A** through the gas film surrounding the particle to the particle surface.
(2) Penetration of **A** by diffusion through pores and cracks in the layer of product to the surface of the unreacted core.
(3) Chemical reaction at the surface of the core.

(4) Diffusion of any gaseous products back through the product layer.

(5) Mass transfer of gaseous products through the gas film.

If no gaseous products are formed, steps (4) and (5) cannot contribute to the resistance of the overall process. Similarly, when gaseous products are formed, their counterdiffusion away from the reaction zone may influence the effective diffusivity of the reactants, but gaseous products will not otherwise affect the course of the reaction unless it is a reversible one. Of the three remaining steps (1) to (3), if the resistance associated with one of these steps is much greater than the other resistances then, as with catalytic reactors (Chapter 2), that step becomes *rate-determining* for the overall reaction process.

The shrinking core reaction mode is not necessarily limited to non-porous unreacted solids. With a fast reaction in a porous solid, diffusion into the core and chemical reaction occur in parallel. The mechanism of the process is very similar to the mechanism of the catalysed gas–solid reactions where the Thiele modulus is large, the effectiveness factor is small and the reaction is confined to a thin zone (Chapter 2, page 121). This combination of reaction with core diffusion gives rise to a reaction zone which, although not infinitely thin but diffuse, nevertheless advances into the core at a steady rate.

Before a simple mathematical analysis is possible a further restriction needs to be applied; it is assumed that the rate of advance of the reaction zone is small compared with the diffusional velocity of **A** through the product layer, i.e. that a pseudo-steady state exists.

Although general models in which external film mass transfer, diffusion through the product layer, diffusion into the core and chemical reaction are all taken into account have been developed[31], it is convenient to consider the special cases which arise when one of these stages is rate-determining. Some of these correspond to the shrinking core mode of reaction, while others lead to the progressive conversion mode. Owing to limitations of space, only the shrinking core reaction mode in which chemical reaction or a combination of chemical reaction and core diffusion is rate-determining will be considered further. A more extended treatment may be found in the book by SZEKELY *et al.* (See section on Further Reading.)

Unreacted core model—Fast chemical reaction

Consider a reaction of stoichiometry

$$\mathbf{A}(g) + b\mathbf{B}(s) = \text{Products}$$

which is first order with respect to the gaseous reactant **A**. If neither the gas-film nor the solid product-layer presents any significant resistance to mass transfer, the concentration of **A** at the reaction surface at the radial position r_c will be C_{Ag}, the same as in the bulk of the gas (Fig. 1.44b).

If the reacting core is impermeable, reaction will take place at the surface of the core, whereas if the core has some degree of porosity the combination of chemical reaction and limited core diffusivity will give rise to a more extended reaction zone. In either case, the overall rate of reaction will be proportional to the area of the reaction front.

Taking therefore unit area of the core surface as the basis for the reaction rate, and writing the first order rate constant as k_s, then the rate at which moles of **A** are consumed in the reaction is given by:

$$-\frac{dM_A}{dt} = 4\pi r_c^2 k_s C_{Ag} \qquad \ldots (1.66)$$

If the core is porous, k_s is not a simple rate constant but incorporates the core diffusivity as well.

From the stoichiometry of the reaction:

$$-\frac{dM_A}{dt} = -\frac{1}{b}\frac{dM_B}{dt}$$

where M_B is the moles of **B** in the core. $M_B = C_B\frac{4}{3}\pi r_c^3$, C_B being the molar density of the solid **B**. Thus:

$$-\frac{1}{b}\frac{dM_B}{dt} = -\frac{1}{b} C_B 4\pi r_c^2 \frac{dr_c}{dt} \qquad \ldots (1.67)$$

Hence equating (1.66) and (1.67):

$$-C_B \frac{dr_c}{dt} = bk_s C_{Ag}$$

$$\text{Integrating:} -C_B \int_{r_0}^{r_c} dr_c = bk_s C_{Ag} \int_0^t dt$$

i.e.
$$t = \frac{C_B}{bk_s C_{Ag}}(r_0 - r_c) \qquad \ldots (1.68)$$

The time t_f for complete conversion corresponds to $r_c = 0$,

i.e.
$$t_f = \frac{C_B r_0}{bk_s C_{Ag}} \qquad \ldots (1.69)$$

Thus in terms of t_f, the time t at which the core has a radius r_c and the fractional conversion for the particle as a whole is α_B is given by

$$t/t_f = (1 - r_c/r_0) = 1 - (1 - \alpha_B)^{\frac{1}{3}} \qquad \ldots (1.70)$$

Example

Spherical particles of a sulphide ore 2 mm in diameter are roasted in an air stream at a steady temperature. Periodically small samples of the ore are removed, crushed and analysed with the following results:

Time (min)	15	30	60
Fractional conversion	0·334	0·584	0·880

Are these measurements consistent with a shrinking core and chemical reaction rate proportional to the area of the reaction zone? If so, estimate the time for complete reaction of the 2-mm particles, and the time for complete reaction of similar 500-μm particles.

Solution

Using the above data, time t can be plotted against $[1 - (1 - \alpha_B)^{\frac{1}{3}}]$ according to equation 1.70; a straight line confirms the assumed model and the slope gives t_f, the time to complete conversion.

Alternatively, calculate $t_f = t/[1 - (1 - \alpha_B)^{\frac{1}{3}}]$ from each data point:

Time (min)	15	30	60
Fractional conversion	0·334	0·584	0·880
Calculated t_f (min)	118	118	118

The constancy of t_f confirms the model; estimated time to complete conversion of 2-mm particles is thus 118 min.

Because $t_f \propto r_0$, the estimated time to complete conversion of 0·5-mm diam. particles is $118·4 \times (0·5/2) = 29·6$ min.

Limitations of simple models—solids structure

In practice, reaction in a particle may occur within a diffuse front instead of a thin reaction zone, indicating a mechanism intermediate between the shrinking core and progressive conversion modes of reaction. Often it is not realistic to single out a particular step as being rate-determining because there may be several factors, each of which affects the reaction to a similar extent. As a further complication there may be significant temperature gradients around and within the particle with fast chemical reactions; for example, in the oxidation of zinc sulphide temperature differences of up to 55 degK between the reaction zone and the surrounding furnace atmosphere have been measured[32].

Recent progress in understanding and modelling uncatalysed gas–solid reactions has been based on the now well-established theory of catalysed gas–solid reactions (Chapter 2). The importance of characterising the structure of the solids, i.e. porosity, pore size and shape, internal surface area and adsorption behaviour, is now recognised. The problem with uncatalysed reactions is that structural changes necessarily take place during the course of the reaction; pores in the reacting solid are enlarged as reaction proceeds; if a solid product is formed there is a process of nucleation of the second solid phase. Furthermore, at the reaction temperature, some sintering of the product or reactant phases may occur and, if a highly exothermic reaction takes place in a thin reaction zone, drastic alterations in structure may occur near the reaction front.

Some reactions are brought about by the action of heat alone, for example the thermal decomposition of carbonates, and baking bread and other materials. These constitute a special class of solid reactions somewhat akin to the progressive conversion type of reaction models but with the rate limited by the rate of heat penetration from the exterior.

Types of equipment and contacting patterns

There is a wide choice of contacting methods and equipment for gas–solid reactions. As with other solids-handling problems, the solution finally adopted may depend very much on the physical condition of the reactants and products, i.e. particle size, any tendency of the particles to fuse together, ease of gas–solid separation, etc. One type of equipment, the rotary kiln, has already been mentioned (Fig. 1.38) and some further types of equipment suitable for continuous operation are shown in Fig. 1.45. The concepts of macromixing in the solid phase and dispersion in the gas phase as discussed in the previous section will be involved in the quantitative treatment of such equipment.

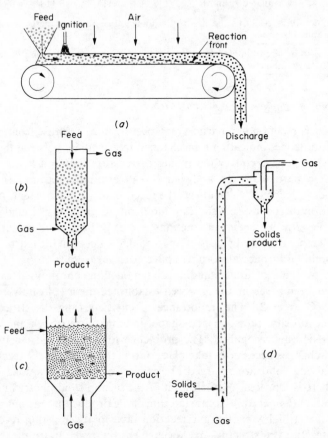

FIG. 1.45. Types of reactor and contacting patterns for gas–solid reactions
(a) Moving bed in cross flow
(b) Hopper type of reactor, particles moving downwards in counter-current flow
(c) Fluidised bed reactor, particles well mixed
(d) Transfer line reactor, solids transported as a dilute phase in co-current flow

The principle of the moving bed of Fig. 1.45a in cross-flow to the air supply is used for roasting zinc blende and for the combustion of large coal on chain grate stokers. Another kind of moving bed is found in hopper-type reactors in which particles are fed at the top and continuously move downwards to be discharged at the bottom; an example is the decomposer used for nickel carbonyl where pure nickel is deposited on metallic nickel balls which grow in size as they pass down the reactor (Fig. 1.45b).

Rotary kilns (Fig. 1.38) have advantages where the solid particles tend to stick together as in cement manufacture, and in the reduction and carbonylation steps in the purification of nickel. In rotary kilns, the flow of gas may be co- or counter-current to the solids.

If the solid particles can be maintained in the fluidised state without problems of agglomeration or attrition, the fluidised bed reactor Fig. 1.45c is likely to be preferred. For short contact times at high temperatures the dilute phase transfer line reactor (Fig. 1.45d) has advantages.

Raked hearth reactors were once extensively used in the metals extraction industries but are now being superseded.

Semi-batch reactors where the gas passes through a fixed bed of solids which are charged and removed batchwise are common for small-scale operations.

Fluidised bed reactor

As the following example shows, the fluidised bed reactor is one of the few types that can be analysed in a relatively simple manner, providing (a) complete mixing is assumed in the solid phase, (b) a sufficient excess of gas is used so that the gas-phase composition may be taken as uniform throughout.

Example

Particles of a sulphide ore are to be roasted in a fluidised bed using excess air: the particles may be assumed spherical and uniform in size. Laboratory experiments indicate that the oxidation proceeds by the unreacted core mechanism with the reaction rate proportional to the core area, the time for complete reaction of a single particle being 16 min at the temperature at which the bed will operate. The particles will be fed and withdrawn continuously from the bed at a steady rate of 6 tonnes of product per hour (1·67 kg/s). The solids hold-up in the bed at any time is 10 tonnes.
   (a) Estimate what proportion of the particles leaving the bed still contain some unreacted material in the core.
   (b) Calculate the corresponding average conversion for the product as a whole.
   (c) If the amount of unreacted material so determined is unacceptably large, what plant modifications would you suggest to reduce it?

Solution

The fluidised bed will be considered as a continuous stirred tank reactor in which ideal macromixing of the particles occurs. As shown in the section on mixing, page 87, in the steady state the required *exit age distribution* is the same as the C-curve obtained using a single shot of tracer. In fact the desired C-curve is identical with that derived on page 48 for a tank containing a liquid with ideal micromixing but now the argument is applied to particles as follows:

In Fig. 1.46 let M_F be the mass hold-up of particles in the reactor, m the mass rate of outflow in the steady state (i.e. the holding time or residence time $\tau = M_F/m$), and c' the number of tracer particles per unit mass of all particles. Consider a shot of n_{m0} tracer particles input at $t = 0$, giving an initial value $c'_0 = n_{m0}/M_F$ in the bed. Applying a material balance:

$$\text{Inflow} - \text{Outflow} = \text{Accumulation}$$

$$0 \quad - \quad c'm \quad = \quad M_F \frac{dc'}{dt}$$

$$\text{i.e. } \frac{dc'}{dt} = -\frac{m}{M_F}c' = -\frac{c'}{\tau}; \text{ whence } c' = c'_0\, e^{-t/\tau}$$

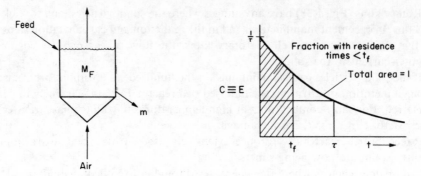

FIG. 1.46. Flow of particles through a fluidised bed showing a normalised C-curve which is identical with the exit age distribution function E

For a normalised C-curve we require

$$\mathbf{C} = \frac{c'}{\displaystyle\int_0^\infty c' \, dt} = \frac{c'_0 \, e^{-t/\tau}}{c'_0 \displaystyle\int_0^\infty e^{-t/\tau} \, dt} = \frac{1}{\tau} e^{-t/\tau} = \mathbf{E}$$

By definition the exit age distribution function $\mathbf{E}$ is such that the fraction of the exit stream with residence times between t and $t + \delta t$ is given by $\mathbf{E} \, \delta t$.

(a) To determine what proportion of the particles leaving the bed still contain some unreacted material in the core we calculate $\mathbf{F}$, that fraction of particles which leave the bed with residence times less than t_f, the time for complete reaction of a particle

$$\mathbf{F} = \int_0^{t_f} \frac{1}{\tau} e^{-t/\tau} \, dt = 1 - e^{-t_f/\tau}$$

For the fluidised bed $\tau = \dfrac{10}{6/60} = 100$ min.

For a particle $t_f = 16$ min.

$$\therefore \text{ Fraction with some unreacted material} = 1 - e^{-16/100} = \underline{0\cdot148}.$$

(b) The conversion α of reactant in a single particle depends on its length of stay in the bed. Working in terms of the fraction unconverted $(1 - \alpha)$ which from equation 1.70 is given by:

$$(1 - \alpha) = (1 - t/t_f)^3$$

the mean value of the fraction unconverted

$$(1 - \bar{\alpha}) = \sum (1 - \alpha)\mathbf{E} \, \delta t$$

where $\mathbf{E} \, \delta t$ is the fraction of the exit stream which has stayed in the reactor for times between t and $t + \delta t$ and in which $(1 - \alpha)$ is still unconverted,

i.e.
$$(1 - \bar{\alpha}) = \int_0^{t_f} (1 - \alpha) \frac{1}{\tau} e^{-t/\tau} \, dt$$

where t_f is taken as the upper limit rather than ∞ because even if the particle stays in the reactor longer than the time for complete reaction, the conversion cannot exceed 100 per cent and for such particles $(1 - \alpha)$ is zero. Substituting for $(1 - \alpha)$:

$$(1 - \bar{\alpha}) = \int_0^{t_f} (1 - t/t_f)^3 \frac{1}{\tau} e^{-t/\tau} \, dt$$

Using repeated integration by parts:

$$(1 - \bar{\alpha}) = 1 + 3\left(\frac{\tau}{t_f}\right) - 6\left(\frac{\tau}{t_f}\right)^2 + 6\left(\frac{\tau}{t_f}\right)^3 (1 - e^{-t_f/\tau})$$

If t_f/τ is small, it is useful to expand the exponential when several of the above terms cancel,
(c) To reduce the amount of unreacted material the following modifications could be

$$(1 - \bar{\alpha}) = \frac{1}{4}\left(\frac{t_f}{\tau}\right) - \frac{1}{20}\left(\frac{t_f}{\tau}\right)^2 + \frac{1}{120}\left(\frac{t_f}{\tau}\right)^3 - \cdots$$

In the present problem $t_f/\tau = 16/100$, i.e.

$$(1 - \bar{\alpha}) = 0.04 - 0.0013 + \cdots = 0.039$$

Hence the average conversion for the product $\bar{\alpha} = 0.961$.
(c) To reduce the amount of unreacted material the following modifications could be suggested:

   (i) reduce the solids feed rate,
   (ii) increase the solids hold-up in the bed,
   (iii) construct a second fluidised bed in series,
   (iv) partition the proposed fluidised bed.

FURTHER READING

CARBERRY, J. J.: *Chemical and Catalytic Reaction Engineering* (McGraw-Hill 1976).
LAPIDUS, L. and AMUNDSON, N. R. (eds.): *Chemical Reactor Theory—A Review* (Prentice-Hall 1977).
LEVENSPIEL, O.: *Chemical Reaction Engineering*, 2nd ed. (Wiley 1972).
RASE, H. F.: *Chemical Reactor Design for Process Plants*, Vols. 1 and 2 (Wiley 1977).
SMITH, J. M.: *Chemical Engineering Kinetics*, 2nd ed. (McGraw-Hill 1970).
SZEKELY, J., EVANS, J. M. and SOHN, H. Y.: *Gas–Solid Reactions* (Academic Press 1976).

REFERENCES TO CHAPTER 1

[1] SMITH, J. M.: *Chemical Engineering Kinetics*, 2nd ed. (McGraw-Hill 1970).
[2] GROGGINS, P. H.: *Unit Process in Organic Synthesis*, 5th ed. (McGraw-Hill 1958).
[3] WALAS, S. M.: *Reaction Kinetics for Chemical Engineers* (McGraw-Hill 1959).
[4] FROST, A. A. and PEARSON, R. G.: *Kinetics and Mechanism*, 2nd ed. (Wiley 1961).
[5] LEVENSPIEL, O.: *Chemical Reaction Engineering*, 2nd ed. (Wiley 1972).
[6] KRAMERS, H. and WESTERTERP, K. R.: *Chemical Reactor Design and Operation* (Chapman & Hall 1963).
[7] ARIS, R.: *Elementary Chemical Reactor Analysis* (Prentice-Hall 1969).
[8] BRÖTZ, W.: *Fundamentals of Chemical Reaction Engineering* (Addison-Wesley 1965).
[9] LAIDLER, K. J.: *Chemical Kinetics*, 2nd ed. (McGraw-Hill 1965).
[10] DENBIGH, K. G.: *Principles of Chemical Equilibria*, 2nd ed. (Cambridge 1966).
[11] DENBIGH, K. G. and TURNER, J. C. R.: *Chemical Reactor Theory*, 3rd ed. (Cambridge 1971).
[12] ASTARITA, G.: *Mass Transfer with Chemical Reaction* (Elsevier 1967).
[13] VALENTIN, F. H. H.: *Absorption in Gas-Liquid Dispersions* (Spon 1967).
[14] MILLER, S. A. (ed.): *Ethylene and its Industrial Derivatives* (Benn 1969).
[15] DANCKWERTS, P. V.: *Gas-Liquid Reactions* (McGraw-Hill 1970).
[16] MITCHELL, J. E.: *Trans. Am. Inst. Chem. Eng.* **42** (1946) 293. The Dow process for styrene production.
[17] WENNER, R. R. and DYBDAL, E. C.: *Chem. Eng. Prog.* **44** (1948) 275. Catalytic dehydrogenation of ethylbenzene.
[18] ELDRIDGE, J. W. and PIRET, E. L.: *Chem. Eng. Prog.* **46** (1950) 290. Continuous flow stirred tank reactor systems.

(19) LIGHTFOOT, E. N.: *A.I.Ch.E.Jl.* **4** (1958) 499. Steady state absorption of a sparingly soluble gas in an agitated tank with simultaneous irreversible first order reaction.

(20) SCHUTT, H. C.: *Chem. Eng. Prog.* **55** (1) (1959) 68. Light hydrocarbon pyrolysis.

(21) DANCKWERTS, P. V. and SHARMA, M. M.: *Chem. Engr, London* (1966) CE 244. The absorption of carbon dioxide into solutions of alkalis and amines.

(22) RESNICK, W. and GAL-OR, B.: *Advances in Chemical Engineering*, Vol. 7 (Academic Press 1968) 295. Gas–liquid dispersions.

(23) SHEEL, J. G. P. and CROWE, C. M.: *Can. J. Chem. Eng.* **47** (1969) 183. Simulation and optimization of an existing ethylbenzene dehydrogenation reactor.

(24) DANCKWERTS, P. V.: *Chem. Eng. Sci.* **2** (1953) 1. Continuous flow systems. Distribution of residence times.

(25) KUNII, D. and LEVENSPIEL, O.: *Fluidization Engineering* (Wiley 1969).

(26) LEVENSPIEL, O. and BISCHOFF, K. G.: *Advances in Chemical Engineering*, Vol. 4 (Academic Press 1963). Patterns of flow in chemical process vessels.

(27) KRAMERS, H.: *Chem. Eng. Sci.* **8** (1958) 45. Physical factors in chemical reaction engineering.

(28) GUNN, D. J.: *Chem. Engr.* London (1968) CE153. Mixing in packed and fluidised beds.

(29) SHERWOOD, T. K., PIGFORD, R. L. and WILKE, C. R.: *Mass Transfer* (McGraw-Hill 1975).

(30) CHOLETTE, A. and CLOUTIER, L.: *Can. J. Chem. Eng.* **37** (1959) 105. Mixing efficiency determinations for continuous flow systems.

(31) WEN, C. Y.: *Ind. Eng. Chem.* **60** (9) (1968) 34. Noncatalytic heterogeneous solid–fluid reaction models.

(32) DENBIGH, K. G. and BEVERIDGE, G. S. G.: *Trans. Inst. Chem. Eng.* **40** (1962) 23. The oxidation of zinc sulphide spheres in an air stream.

LIST OF SYMBOLS USED IN CHAPTER 1

Asterisk (*) indicates that these dimensions are dependent on order of reaction

$\mathscr{A}$	Frequency factor in rate equation		*
A_t	Area of cooling coils		L^2
a	Interfacial area per unit volume in gas–liquid contacting		L^{-1}
A_c	Area of cross-section of a tubular reactor		L^2
b	Moles of **B** reacting per mole of **A**		—
C	Molar density of reaction mixture (at time t)		ML^{-3}
C_A, C_B	Molar concentration of **A**; molar concentration of **B**, etc.		ML^{-3}
C_{Ae}	Concentration of **A** at equilibrium		ML^{-3}
C_{Af}	Concentration of **A** at end of reaction		ML^{-3}
C_{Ag}	Concentration of **A** in bulk of gas		ML^{-3}
C_{Ai}	Value of C_A at interface		ML^{-3}
C_{AL}, C_{BL}	Value of C_A, C_B in bulk of liquid		ML^{-3}
C_{A0}	Concentration of **A** initially or in feed		ML^{-3}
C_e	Final exit concentration		ML^{-3}
C_i	Concentration of tracer leaving tank i		ML^{-3}
C_m	Intermediate concentration at end of well-mixed region		ML^{-3}
C_0	Molar density of reaction mixture at inlet		ML^{-3}
C'_0	n_0/V		ML^{-3}
C_∞	Value of C in step input		ML^{-3}
C	Outlet response of concentration divided by area under C–t curve		T^{-1}
$\mathbf{C}_\theta$	Value of **C** when t is replaced by t/τ		—
c	Specific heat at constant pressure	$(HM^{-1}\theta^{-1})$	$L^2T^{-2}\theta^{-1}$
c_j	Specific heat at constant pressure of component j	$(HM^{-1}\theta^{-1})$	$L^2T^{-2}\theta^{-1}$
c'	Number of tracer particles per unit mass		M^{-1}
c'_0	Initial value of c'		M^{-1}
D_A	Liquid phase diffusivity of reactant **A**		L^2T^{-1}
D_L	Longitudinal dispersion coefficient		L^2T^{-1}
D_R	Radial dispersion coefficient		L^2T^{-1}
E	Activation energy per mol	(HM^{-1})	L^2T^{-2}
E	Exit age distribution		T^{-1}
F	Fraction of tracer escaped at time t		—
F_A	Molar feed rate of **A** into reactor		MT^{-1}
F	C/C_∞		—

Symbol	Description		Dimensions
f	Fraction of feed by-passing reactor		—
G	Mass flow rate		MT^{-1}
ΔH	Enthalpy change per mol (Heat of reaction $= -\Delta H$)	(HM^{-1})	L^2T^{-2}
ΔH_A	Enthalpy change in reaction per mol of A	(HM^{-1})	L^2T^{-2}
i	Number of tanks in a series		—
J_A	Rate of transformation of A per unit volume of reactor		$ML^{-3}T^{-1}$
j	Component of reaction mixture		—
K_c	Equilibrium constant in terms of concentrations		*
K_p	Equilibrium constant in terms of partial pressure		*
k_L	Liquid film transfer coefficient		LT^{-1}
k	Rate constant		*
k_f	Rate constant of forward reaction in a reversible reaction		*
k_P, k_Q	Rate constants for reactions giving products P, Q, etc.		*
k_p	Rate constant with concentrations expressed as partial pressures		*
k_r	Rate constant of reverse reaction in a reversible reaction		*
k_1	First order or pseudo first order rate constant		T^{-1}
k_2	Second order rate constant		$M^{-1}L^3T^{-1}$
k_{11}, k_{12}	Rate constants of first and second first order reactions in a series		T^{-1}
k_s	First order rate constant at solid surface		LT^{-1}
L	Distance between inlet and outlet		L
M	Area under concentration–time curve		$ML^{-3}T$
M_A, M_B	Moles of A, B in core		M
M_F	Mass of particles in bed		M
m	Mass outflow rate of particles		MT^{-1}
m_j	Mass of component j present in a batch reactor		M
N_A	Mol flux across a gas–liquid interface		$ML^{-2}T^{-1}$
n_A	Mols of A in a given volume of reaction mixture		M
n_{A0}	Mols of A in a given volume of reaction mixture initially		M
n_{m0}	Number of tracer particles injected		—
n_r	Number of mols in rth tank		M
n_t	Mols remaining in a series of tanks after time t		M
n_0	Mols of tracer injected into feed		M
P	Total pressure		$ML^{-1}T^{-2}$
P_A	Partial pressure of A		$ML^{-1}T^{-2}$
p	Order of reaction		—
Q	Heat flow from batch or tubular reactor	(HT^{-1})	ML^2T^{-3}
q	Order or reaction		—
R	Gas constant	$(HM^{-1}\theta^{-1})$	$L^2T^{-2}\theta^{-1}$
$\mathscr{R}_A, \mathscr{R}_B$	Rate of reaction per unit volume of reactor with respect to reactant A, reactant P		$ML^{-3}T^{-1}$
$\mathscr{R}_{+A}$	Rate of forward reaction with respect to A in a reversible reaction		$ML^{-3}T^{-1}$
$\mathscr{R}_{-A}$	Rate of reverse reaction with respect to A in a reversible reaction		$ML^{-3}T^{-1}$
r	Order of reaction		—
r'	Concentration ratio C_A/C_B		—
r_c	Radius of reaction surface		L
r_0	Initial value of r_c		L
s	V_{c2}/V_{c1}		—
T	Temperature absolute		θ
T_C	Temperature of cooling water		θ
T_{mx}	Maximum safe operating temperature		θ
T_0	Initial temperature		θ
t	Time		T
$\bar{t}$	Mean time		T
$\bar{t}_E$	Mean of residence time distribution		T
t_f	Time for complete conversion		T
t_r	Reaction time, batch reactor		T
t_s	Shut-down time, batch reactor		T
$t_{\frac{1}{2}}$	Half-life of reaction, i.e. time for half reactant to be consumed		T
U	Heat transfer coefficient, overall	$(HL^{-2}T^{-1}\theta^{-1})$	$MT^{-3}\theta^{-1}$

u	Velocity in pipe	LT^{-1}
V	Volume of reaction mixture	L^3
V_a	Volume of dispersed element of reactant	L^3
V_b	Volume of batch reactor	L^3
V_c	Volume of a continuous stirred tank reactor; total volume if more than one tank	L^3
V_{cr}	Volume of the rth tank in a series of continuous stirred tank reactors	L^3
V_t	Volume of a tubular reactor	L^3
v	Volume rate of flow into reactor	L^3T^{-1}
W_A	Unit output of a reactor with respect to reactor A	$ML^{-3}T^{-1}$
W_b	Unit output for batch reactor	$ML^{-3}T^{-1}$
W_{c1}, W_{c2}	Unit output for a continuous stirred tank reactor comprising one tank; two tanks	$ML^{-3}T^{-1}$
w	Fraction of contents of reactor which are well mixed	—
x	Distance from interface in direction of transfer	L
y_A	Mol fraction of A	—
y_{A0}	Mol fraction of A at inlet to reactor	—
z	Distance along tubular reactor	L
z'	Dimensionless derivative of z (z/L)	—
α	Fractional conversion	—
α_A	Fractional conversion of reactant A, i.e. the fraction of A which has reacted	—
α_{Af}	Fractional conversion of reactant A; final value on discharge from reactor	—
α_B	Fractional conversion for particle	—
α_e	Fractional conversion at equilibrium in a reversible reaction	—
α_f	Fractional conversion under steady state conditions	—
α_t	Fractional conversion in material staying in reactor for time t	—
$\bar{\alpha}$	Average conversion	—
β	$\sqrt{(k_1 D_A)}/k_L$ or $\sqrt{(k_2 C_{BL} D_A)}/k_L$	—
δ	Film thickness	L
ε	Volume fraction liquid phase	—
Θ	Operational yield of desired product	—
θ	Dimensionless time derivative	—
$\bar{\theta}$	Mean value of θ	—
v_A, v_B	Stoichiometric coefficients of A, B	—
σ^2	Variance of t	T^2
σ_θ^2	Variance of θ	—
τ	Residence time: V_t/v or V_c/v	T
τ_1	Plug flow residence time	T
τ_2	Stirred-tank residence time	T
Φ	Relative yield of desired product, overall	—
ϕ	Relative yield of desired product, instantaneous	—
χ	Mols of reactant transformed in unit volume of reaction mixture	ML^{-3}
χ_e	Value of χ at equilibrium	ML^{-3}
χ_f	Mols of reactant transformed in unit volume of reaction mixture; final value on discharge from reactor	ML^{-3}

CHAPTER 2

The design of catalytic reactors

OUTLINE OF THE DESIGN PROBLEM

The design of heterogeneous chemical reactors falls into a special category because an additional complexity intrudes into the problem. We must now concern ourselves with the transfer of matter between phases, as well as considering the fluid dynamics and chemistry of the system. Thus, in addition to an equation describing the rate at which the chemical reaction proceeds, one must also provide a relationship or algorithm to account for the various physical processes which occur. For this purpose it is convenient to classify the reactions as gas–liquid, gas–solid and liquid–gas–solid processes.

In gas–liquid systems the factors which should be recognised as contributing to the overall process are mass transfer of the gas to the liquid interface (Chapter 1) and the subsequent chemical reaction between the gas and liquid. Typical of many such processes are the absorption of carbon dioxide in alkaline solutions, the removal of ammonia from tail gases by scrubbing with acids and the solution of hydrogen sulphide in alkaline aqueous media. It is, of course, essential that there should be efficient contact between the gas and liquid phases, and the various types of countercurrent contacting devices which are used in industrial plant include spray towers, plate towers and packed towers. The problem with which one is faced in designing such equipment is to choose the type of contacting device required and subsequently to assess the height of tower necessary to reduce the gas concentration to within a specified limit.

The upgrading of petroleum by catalytic cracking and reforming reactions, the catalytic synthesis of ammonia and the gasification of coal are all examples of gas–solid reactions in which the reactant gas must first be transported through the fluid phase to the surface of the solid and, if the solid is a porous type of material, diffuse along the honeycomb of pores towards the interior of the particle. The reactant will then adsorb at the active surface and be chemically transformed into a product which subsequently desorbs, diffuses to the exterior of the particle, and is then transported to the bulk fluid phase. For such reactions the designer usually has to assess the amount of solid material to pack into a reactor to achieve a specified conversion. The rate of transport of gas to and from the surface of the solid and within the pore structure of the particle will affect both the dimensions of the reactor unit required and the selection of particle size. Compared with a reaction uninfluenced by transport processes, the complexity of the design problem is exacerbated since, in the steady state, the rate of each of the transport processes is coupled with the rates of chemical reaction, of adsorption and of desorption.

An example of a liquid–gas–solid reaction is the hydrosulphurisation of petroleum feedstocks. A trickle bed reactor, in which the petroleum feed is allowed to flow downwards through a catalyst bed and hydrogen is passed countercurrent to the liquid, is the basis of an effective commercial process for this reaction. The gaseous reactant must now be absorbed by the liquid and transported across a liquid film to the catalyst surface before chemical reaction occurs. Here we have a combination of the problems encountered in gas–liquid and gas–solid systems.

Whatever the nature of the reaction, and whether the vessel chosen for the operation be a packed tubular, fluidised bed or trickle bed reactor, the essence of the design problem is to estimate the size of reactor required. This is achieved by solving the transport and chemical rate equations appropriate to the system. Prior to this however, the operating conditions, such as initial temperature, pressure and reactant concentrations, must be chosen and a decision made concerning the type of reactor to be used. For example, one might have selected a packed tubular reactor operated adiabatically, or perhaps a fluidised bed reactor fitted with an internal cooling arrangement. Such operating variables constitute the design conditions and must be chosen before a mathematical model is constructed.

Rate limiting step

From what has been said it is clear that in a heterogeneous system the conversion of a reactant into a product depends on a number of processes occurring consecutively. A moderately complicated system which we can consider, involving a large number of distinct physical and chemical steps, is a heterogeneous gas–solid reaction such as the gasification of coal. We visualise a model in which reaction first occurs at the exterior surface of the solid particles. The reaction zone then moves towards the interior of each particle, leaving behind the reacted and spent inert solid which we may refer to as ash. The reactant gas, which in this particular case is oxygen, is first transported through the bulk air to the exterior surface of the particle, penetrates the ash surrounding the unreacted core by a diffusion process, and chemisorbs at the reactive surface. Chemical reaction then ensues and the gaseous products desorb. In order to reach the bulk fluid phase the gaseous products necessarily have to diffuse outwards through the ash and thence through the stagnant gas film adjacent to the surface. We may recognise that a heterogeneous catalytic gas reaction involves similar steps except for the formation of ash. The reactant and product gases in this case diffuse through the internal porous structure of the catalyst enabling the maximum amount of the reactive surface to be accessible.

Any one, or a combination, of these physical and chemical rate processes may determine the rate of the overall process. It is important to appreciate that, in the steady state, each of the steps occurs at exactly the same rate; otherwise an accumulation of material would result at some stage and this would be contrary to the definition of a steady state. The rate limiting step is that step in the sequence of rate processes whose rate coefficient determines the rate of overall conversion. For heuristic purposes let us consider the absorption of carbon dioxide in alkaline solution. The rate at which the carbon dioxide is transported to the gas–liquid interface may be written in terms of the driving force existing between the bulk gas and the interface. If C_i represents the concentration of carbon dioxide at the interface and C represents the concentration in the bulk gas, the flux of carbon dioxide reaching the interface per unit volume of particle is:

$$Na = h_D a(C - C_i) \qquad \ldots .(2.1)$$

where h_D represents the mass transfer coefficient, and a is the interfacial area per unit volume of particle. Chemical reaction between the carbon dioxide and alkali may be represented by a pseudo first order kinetic equation, provided alkali is in excess. The rate $\mathscr{R}'_v$ of chemical reaction per unit volume of particle will therefore be:

$$\mathscr{R}'_v = kC_i \qquad \ldots .(2.2)$$

The rate is proportional to the concentration C_i of carbon dioxide available at the interface. No more carbon dioxide will react (and therefore be absorbed) than will be transported to the interface. Thus, in the steady state, the flux of carbon dioxide will balance the rate of chemical reaction. From equations 2.1 and 2.2 we find the unknown concentration, since:

$$Na = \mathscr{R}'_v$$

and hence:

$$C_i = \frac{h_D a C}{h_D a + k} \qquad \ldots .(2.3)$$

The rate at which the overall process of mass transfer and chemical reaction occurs may be found by substituting for C_i in equation 2.2. This overall rate is:

$$\mathscr{R}'_v = \left\{ \frac{1}{1/k + 1/h_D a} \right\} C \qquad \ldots .(2.4)$$

If, under the operating conditions, $h_D a \gg k$ the overall rate reduces to kC. In this case the chemical reaction is said to be rate determining. If, on the other hand, $h_D a \ll k$ the rate is $h_D a C$ and it is the transport process which is rate determining.

The concept of resistance can be usefully employed when a process involving both external mass transfer and chemical reaction steps is being analysed. If the chemical reaction can be represented by a first order equation such as 2.2 then an overall rate coefficient may be defined such that:

$$\frac{1}{\bar{k}} = \frac{1}{k} + \frac{1}{h_D a} \qquad \ldots .(2.5)$$

In writing such an equation we imply that there is a resistance to chemical reaction and a resistance to external mass transfer. If the operating conditions were such that the process occurred in a regime intermediate between the chemically controlled region and the mass transfer controlled region, the overall rate would be a composite of the two steps and equation 2.4 should be used rather than either one of the two limiting cases. Although the concept of resistance can also be applied in the case of a more complex process (such as a heterogeneous catalytic reaction) the overall rate coefficient cannot, in this case, be written in the form of equation 2.5 for the problem would involve C_i as a boundary condition and C as a variable dependent on penetration within the porous structure. Nevertheless, we can still speak of internal diffusion as rate determining for, under such circumstances, internal diffusion influences the kinetic behaviour of the system. Difficulties also arise if the rate of reaction is not a linear function of concentration. If the chemical reaction were anything but first order the rate coefficients for chemical reaction and mass transfer steps would have different dimensions and one could only refer to a step as rate limiting within a certain concentration range.

Diffusion and mass transfer limited processes tend to occur in relatively high temperature regions, whereas chemically controlled processes occur at lower temperatures. This is because the rate of a chemical reaction (also chemisorption and desorption) depends, according to the Arrhenius equation, on the factor $e^{-E/RT}$ while rates of diffusion and mass transfer are not so strongly influenced by temperature. Thus, in high temperature regions, the chemical reaction rate coefficient would be much greater than the mass transfer coefficient or the diffusion coefficient, and so the process would be limited by mass transfer in the interior of a particle or exterior to the particle. Conversely, in low temperature regions the chemical reaction would tend to be rate limiting, for then the chemical rate constant would tend to be the smaller of the coefficients. When reaction occurs in the chemically controlled regime the effect of temperature is typical of chemical reaction whereas, if reaction occurs in a regime where external mass transfer is rate limiting, the effect of temperature is characteristic of mass transfer processes. If intraparticle diffusion is rate limiting the effect of temperature is, as we shall see on page 124, a composite of chemical reaction and diffusion.

MASS TRANSFER WITHIN POROUS SOLIDS

For gas–solid heterogeneous reactions particle size and average pore diameter will influence the reaction rate per unit mass of solid if internal diffusion happens to be rate limiting. The actual mode of transport within the porous structure will depend largely on the average pore radius and the conditions of pressure within the reactor. Before developing equations which will enable us to predict reaction rates in porous solids, a brief consideration of transport in pores is pertinent.

The effective diffusivity

The diffusion of gases through the tortuous narrow channels of a porous solid generally occurs by one or more of three mechanisms. When the mean free path of the gas molecules is considerably greater than the pore diameter, collisions between molecules in the gas are much less numerous than those between molecules and pore walls. Under these conditions the mode of transport is Knudsen diffusion. When the mean free path of the gas molecules is much smaller than the pore diameter, gaseous collisions will be more frequent than collisions of molecules with pore walls, and under these circumstances ordinary bulk diffusion occurs. A third mechanism of transport which is possible when a gas is adsorbed on the inner surface of a porous solid is surface diffusion. Transport occurs by the movement of molecules over the surface in the direction of decreasing surface concentration. Although there is not much evidence on this point, it is unlikely that surface diffusion is of any importance in catalysis at elevated temperatures. Nevertheless, surface diffusion may contribute to the overall transport process in low temperature reactions of some vapours. Finally, it should be borne in mind that when a total pressure difference is maintained across a pore, as is reputed to be the case for some catalytic cracking reactions, forced flow in pores is likely to occur, transport being due to a total concentration gradient.

Both Knudsen diffusion and bulk flow can be described adequately for homogeneous media. However, a porous mass of solid usually contains pores of non-uniform cross-section which pursue a very tortuous path through the particle and which may intersect with many other pores. Thus the flux predicted by an equation for normal bulk diffusion (or for Knudsen diffusion) should be multiplied by a geometric factor which takes into account the tortuosity and the fact that the flow will be impeded by that fraction of the total pellet volume which is solid. It is therefore expedient to define an effective diffusivity D_e in such a way that the flux of material may be thought of as flowing through an equivalent homogeneous medium. We may then write:

$$D_e = D \; \frac{\psi}{\tau} \qquad\qquad \dots.(2.6)$$

where D is the normal (molecular) bulk diffusion coefficient,
   ψ is the porosity of the particles, and
   τ is a tortuosity factor.

We thus imply that the effective diffusion coefficient is calculated on the basis of a flux resulting from a concentration gradient in a homogeneous medium which has been made equivalent to the heterogeneous porous mass by invoking the geometric factor ψ/τ. Experimental techniques for estimating the effective diffusivity include diffusion and flow through pelletised particles. A common procedure is to expose the two faces of a porous disc of the material to the pure components at the same total pressure. An interesting method due to BARRER and GROVE[14] relies on the measurement of the time lag required to reach a steady pressure gradient, while a promising new technique employed by GUNN and PRYCE[51] depends on measuring the dispersion of a sinusoidal pulse of tracer in a bed packed with the porous material. Gas chromatographic methods for evaluating dispersion coefficients have also been employed[54].

Just as one considers two regions of flow for homogeneous media, so one may have molecular or Knudsen transport for heterogeneous media.

The molecular flow region

HIRSCHFELDER, CURTISS and BIRD[3] obtained a theoretical expression for the molecular diffusion coefficient for two interdiffusing gases, modifying the kinetic theory of gases by taking into account the nature of attraction and repulsion forces between gas molecules. Their expression for the diffusion coefficient has been successfully applied to many gaseous binary mixtures and represents one of the best methods for estimating unknown values. On the other hand, Maxwell's formula, modified by GILLILAND[8] and discussed in Volume 1, Chapter 8, also gives satisfactory results. Experimental methods for estimating diffusion coefficients rely on the measurement of flux per unit concentration gradient. An extensive tabulation of experimental diffusion coefficients for binary gas mixtures is to be found in a report by EERKENS and GROSSMAN[23].

To calculate the effective diffusivity in the region of molecular flow, the estimated value of D must be multiplied by the geometric factor ψ/τ which is descriptive of the heterogeneous nature of the porous medium through which diffusion occurs.

The porosity ψ of the porous mass is included in the geometric factor to account for the fact that the flux per unit total cross-section is ψ times the flux if there were no solid present. The porosity may conveniently be measured by finding the particle density ρ_p in a pyknometer using an inert non-penetrating liquid. The true density ρ_s of the solid should also be found by observing the pressure of a gas (which is not adsorbed) before and after expansion into a vessel containing a known weight of the material. The ratio ρ_p/ρ_s then gives the fraction of solid present in the particles and $(1 - \rho_p/\rho_s)$ is the porosity.

The tortuosity τ is also included in the geometric factor to account for the tortuous nature of the pores. It is the ratio of the path length which must be traversed by molecules in diffusing between two points within a pellet to the direct linear separation between those points. Theoretical predictions of τ rely on somewhat inadequate models of the porous structure, but experimental values may be obtained from measurements of D_e, D and ψ.

The Knudsen flow region

In the region of flow where collisions of molecules with the container walls are more frequent than intermolecular gaseous collisions, KNUDSEN[7] demonstrated that the net flow of molecules in the direction of gas flow is proportional to the gradient of the molecular flux. From geometrical considerations it may be shown[1] that, for the case of a capillary of circular cross-section and radius r, the proportionality factor is $8\pi r^3/3$. This results in a Knudsen diffusion coefficient:

$$D_K = \frac{8r}{3}\sqrt{\frac{RT}{2\pi M}} \qquad \ldots\ldots(2.7)$$

This equation, however, cannot be directly applied to the majority of porous solids since they are not well represented by a collection of straight cylindrical capillaries. EVERETT[25] showed that pore radius is related to the specific surface area S_g per unit mass and to the specific pore volume V_g per unit mass by the equation:

$$r = \frac{2}{\alpha}\frac{V_g}{S_g} \qquad \ldots\ldots(2.8)$$

where α is a factor characteristic of the particular pore geometry. Values of α depend on the pore structure: for uniform non-intersecting cylindrical capillaries $\alpha = 1$, but for non-intersecting close packed cylindrical rods $\alpha = 0{\cdot}104$. Although an estimation of precise values of V_g and S_g from experimental data is obtained by the somewhat arbitrary selection of points on an adsorption isotherm representing complete pore filling and the completion of a monolayer, some significance may be given to an average pore dimension derived from pore volume and surface area measurements. Thus if, for the purposes of calculating a Knudsen diffusion coefficient, the pore model adopted consists of non-intersecting cylindrical capillaries and the radius computed from equation 2.8 is a radius r_e equivalent to the radius of a cylinder having the same surface to volume ratio as the pore, then equation 2.7 may be applied. In terms of the porosity ψ, specific surface area S_g and particle density ρ_p (mass per unit total particle volume, including the volume occupied by pore space):

$$D_K = \frac{16}{3}\frac{\psi}{\rho_p S_g}\sqrt{\frac{RT}{2\pi M}} \qquad \ldots\ldots(2.9)$$

In the region of Knudsen flow the effective diffusivity D_{eK} for the porous solid may be computed in a similar way to the effective diffusivity in the region of molecular flow, i.e. D_K is simply multiplied by the geometric factor.

The transition region

Under conditions where Knudsen or molecular diffusion does not predominate, SCOTT and DULLIEN[35] obtained a relation for the effective diffusivity. The formula they obtained for a binary mixture of gases is:

$$D_e = \cfrac{1}{\cfrac{1}{D_{eM}} + \cfrac{1}{D_{eK}} - \cfrac{x_A(1 + N_B/N_A)}{D_{eM}}} \qquad \ldots\ldots(2.10)$$

E

where D_{eM} and D_{eK} are the effective diffusivities in the molecular and Knudsen regions,

N_A and N_B are the molar fluxes of species **A** and **B**, and

x_A is the mol fraction of **A**.

If we set $N_A = -N_B$ and hold the total pressure constant, we obtain from equation 2.10 an expression for the effective self-diffusion coefficient in the transition region:

$$\frac{1}{D_e} = \frac{1}{D_{eM}} + \frac{1}{D_{eK}} \qquad \qquad \ldots . (2.11)$$

This result has also been obtained independently by other workers[10, 12].

Forced flow in pores

Many heterogeneous reactions give rise to an increase or decrease in the total number of mols present in the porous solid due to the reaction stoichiometry. In such cases there will be a pressure difference between the interior and exterior of the particle and forced flow occurs. When the mean free path of the reacting molecules is large compared with the pore diameter, forced flow is indistinguishable from Knudsen flow and is not affected by pressure differentials. When, however, the mean free path is small compared with the pore diameter and a pressure difference exists across the pore, forced flow (Poiseuille flow: see Volume 1, Chapter 3) resulting from this pressure difference will be superimposed on molecular flow. The diffusion coefficient D_P for forced flow depends on the square of the pore radius and on the total pressure difference ΔP:

$$D_P = \frac{-\Delta P r^2}{8\mu} \qquad \qquad \ldots . (2.12)$$

The viscosity of most gases at atmospheric pressure is of the order of 10^{-7} Ns m^{-2}, so for pores of about 1 μm radius D_P is approximately 10^{-5} m^2 s^{-1}. Molecular diffusion coefficients are of similar magnitude so that in small pores forced flow will compete with molecular diffusion. For fast reactions accompanied by an increase in the number of mols an excess pressure is developed in the interior recesses of the porous particle which results in the forced flow of excess product and reactant molecules to the particle exterior. Conversely, for pores greater than about 100 μm radius, D_P is as high as 10^{-3} m^2 s^{-1} and the coefficient of diffusion which will determine the rate of intraparticle transport will be the coefficient of molecular diffusion.

Except in the case of reactions at high pressure, the pressure drop which must be maintained to cause flow through a packed bed of particles is usually insufficient to produce forced flow in the capillaries of the solid, and the gas flow is diverted around the exterior periphery of the pellets. Reactants then reach the interior of the porous solid by Knudsen or molecular diffusion.

CHEMICAL REACTION IN POROUS CATALYST PELLETS

A porous solid catalyst, whose behaviour is usually specific to a particular system, enhances the approach to equilibrium of a gas phase chemical reaction. Employment of such a material therefore enables thermodynamic equilibrium to be achieved at moderate temperatures in a comparatively short time interval.

When designing a heterogeneous catalytic reactor it is important to know, for the purposes of calculating throughputs, the rate of formation of desired product on the basis of unit reactor volume and under the hydrodynamic conditions obtaining in the reactor. Whether the volume is defined with respect to the total reactor volume, including that occupied by solid, or with respect to void volume is really a matter of convenience. What is important however is either that rate data for the reaction be known for identical physical conditions prevailing within the reactor (this usually means obtaining rate data *in situ*) or, alternatively, that rate data obtained in the absence of mass transport effects be available. If pilot plant experiments are performed, conditions can usually be arranged to match those within a larger reactor, and then the rate data obtained can be immediately applied to a reactor design problem since the measurements will have taken into account mass transfer effects. If rate data from laboratory experiments are utilised, it is essential to ensure that they are obtained in the absence of mass and heat transfer effects. This being so, the rate should be multiplied by two factors to transpose the experimental rate to the basis of reaction rate per unit volume of a reactor packed with catalyst particles. If we wish to calculate throughput on the basis of total reactor volume the bed voidage e should be taken into account and the rate multiplied by $(1 - e)$ which is the fraction of reactor volume occupied by solid. Account must also be taken of mass transfer effects and so the rate is multiplied by an effectiveness factor, η, which is defined as the ratio of the rate of reaction in a pellet to the rate at which reaction would occur if the concentration and temperature within the pellet were the same as the respective values external to the pellet. The factor η therefore accounts for the influence which concentration and temperature gradients (which exist within the porous solid and result in mass and heat transfer effects) have on the chemical reaction rate. The reaction rate per unit volume of reactor is thus written:

$$\mathscr{R}_v = (1 - e)k\mathrm{f}(C)\eta \qquad \qquad \ldots (2.13)$$

where $\mathrm{f}(C)$ is that function of concentration which describes the specific rate per unit volume in the absence of any mass and heat transfer effects within the particle and k is the specific rate constant per unit volume of reactor. An experimental determination of η merely involves comparing the observed rate of reaction with the rate of reaction on catalyst pellets sufficiently small not to cause diffusion effects.

Isothermal reactions in porous catalyst pellets

THIELE[9], who predicted how in-pore diffusion would influence chemical reaction rates, employed a geometric model with isotropic properties. Both the effective diffusivity and the effective thermal conductivity are independent of position for such a model. Although idealised geometric shapes are used to depict the situation within a particle such models, as we shall see later, are quite good approximations to practical catalyst pellets.

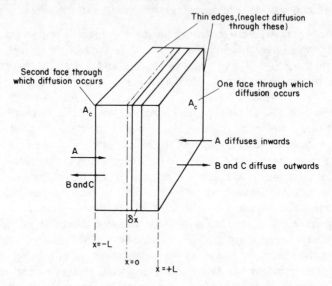

FIG. 2.1. Geometry of slab model for catalyst pellets

The simplest case we shall consider is that of a first order chemical reaction occurring within a rectangular slab of porous catalyst, the edges of which are sealed so that diffusion occurs in one dimension only. Figure 2.1 illustrates the geometry of the slab. Consider that the first order irreversible reaction:

$$A \rightarrow B$$

occurs within the volume of the particle and suppose its specific velocity constant on the basis of unit surface area is k_a. For heterogeneous reactions uninfluenced by mass transfer effects experimental values for rate constants are usually based on unit surface area. The corresponding value in terms of unit total volume of particle would be $\rho_p S_g k_a$ where ρ_p is the apparent density of the catalyst pellet and S_g is the specific surface area per unit mass of the solid including the internal pore surface area. We shall designate the specific rate constant based on unit volume of particle as k. A component material balance for A across the element δx gives:

$$-A_c D_e \left(\frac{dC_A}{dx}\right)_{x+\delta x} = -A_c D_e \left(\frac{dC_A}{dx}\right)_x - k\, C_A A_c \delta x \qquad \ldots .(2.14)$$

since, in the steady state, the flux of A into the element at $(x + \delta x)$ must be balanced by the flux out of the element at x minus the amount lost by reaction within the volume element $A_c \delta x$. Note that for the coordinate system considered, since the concentration of the reacting component **A** decreases in the direction of decreasing x, the concentration gradient is positive, and hence the flux is negative. If the concentration gradient term at the point $(x + \delta x)$ in equation 2.14 is expanded in a Taylor series about the point x and differential coefficients of order greater than two are ignored, the equation simplifies to:

$$\frac{d^2 C_A}{dx^2} - \frac{k C_A}{D_e} = 0 \qquad \qquad \dots.(2.15)$$

An analogous equation may be written for component **B**. By reference to Fig. 2.1, it will be seen that, because the product **B** diffuses outward, its flux is positive. Reaction produces **B** within the slab of material and hence makes the term depicting the rate of formation of **B** in the material balance equation positive, resulting in an equation similar in form to equation 2.15.

The boundary conditions for the problem may be written by referring to Fig. 2.1. At the exterior surface of the slab the concentration will be that corresponding to the conditions in the bulk gas phase, provided there is no resistance to mass transfer in the gas phase. Hence:

$$C_A = C_{A\infty} \quad \text{at} \quad x = \pm L \qquad \qquad \dots.(2.16)$$

At the centre of the slab considerations of symmetry demand that:

$$\frac{dC_A}{dx} = 0 \quad \text{at} \quad x = 0 \qquad \qquad \dots.(2.17)$$

so that the net flux through the plane at $x = 0$ is zero, diffusion across this boundary being just as likely in the direction of increasing x as in the direction of decreasing x. The solution of equation 2.14 with the boundary conditions given by equations 2.16 and 2.17 is:

$$C_A = C_{A\infty} \frac{\cosh \lambda x}{\cosh \lambda L} \qquad \qquad \dots.(2.18)$$

where λ denotes the quantity $\sqrt{(k/D_e)}$. Equation 2.18 describes the concentration profile of **A** within the catalyst slab. In the steady state the total rate of consumption of **A** must be equal to the total flux of **A** at the external surfaces. From symmetry, the total transfer across the external surfaces will be twice that at $x = L$. Thus:

$$-A_c k \int_{-L}^{+L} C_A(x)\,dx = -2A_c D_e \left(\frac{dC_A}{dx}\right)_{x = \pm L}$$

If the whole of the catalyst surface area were available to the exterior concentration $C_{A\infty}$ there would be no diffusional resistance and the rate would then be $-2A_c L k C_{A\infty}$. The ratio of these two rates is the effectiveness factor:

$$\eta = \frac{-A_c k \displaystyle\int_{-L}^{+L} C_A(x)\,dx}{-2A_c L k C_{A\infty}} = \frac{-2 A_c D_e \left(\dfrac{dC_A}{dx}\right)_{x=L}}{-2A_c L k C_{A\infty}} \qquad \dots.(2.19)$$

By evaluating either the integral or the differential in the numerator of equation 2.19 the effectiveness factor may be calculated. In either case, by substitution from equation 2.18:

$$\eta = \frac{\tanh \lambda L}{\lambda L} = \frac{\tanh \phi}{\phi} \qquad \ldots . (2.20)$$

where $\phi \, [= \lambda L = L\sqrt{(k/D_e)}]$ is a dimensionless quantity known as the Thiele modulus. If the function η is plotted from equation 2.20, corresponding to the case of a first order irreversible reaction in a slab with sealed edges, it may be seen from Fig. 2.2 that when $\phi < 0\cdot2$, η is close to unity. Under these conditions there would be no diffusional resistance, for the rate of chemical reaction is not limited by diffusion. On the other hand when $\phi > 5\cdot0$, $\eta = 1/\phi$ is a good approximation and for such conditions internal diffusion is the rate determining process. Between these two limiting values of ϕ the effectiveness factor is calculated from equation 2.20 and the rate process is in a region where neither in-pore mass transport nor chemical reaction is overwhelmingly rate determining.

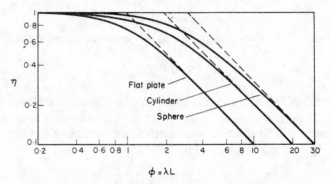

$\phi = \lambda L$

FIG. 2.2. Effectiveness factors for flat plate, cylinder and sphere

Only a very limited number of manufactured catalysts could be approximately described by the slab model but there appear to be many which conform to the shape of a cylinder or sphere. Utilising the same principles as for the slab, it may be shown (see the next example) that for a cylinder of radius r sealed at the flat ends the effectiveness factor is:

$$\eta = \frac{2}{\lambda r} \frac{I_1(\lambda r)}{I_0(\lambda r)} \qquad \ldots . (2.21)$$

where I_0 and I_1 denote zero and first order modified Bessel functions of the first kind.

For a sphere of radius r (see the example on page 120):

$$\eta = \frac{3}{\lambda r} \left\{ \coth (\lambda r) - \frac{1}{\lambda r} \right\} \qquad \ldots . (2.22)$$

$\text{Gunn}^{(47)}$ has recently discussed the case of the hollow cylindrical catalyst particle. Such catalyst particles reduce the difficulties caused by excessive pressure drops.

Example

Derive an expression for the effectiveness factor of a cylindrical catalyst pellet, sealed at both ends, in which a first order chemical reaction occurs.

Solution

The pellet has cylindrical symmetry about its central axis. Construct an annulus with radii $(r + \delta r)$ and r and consider the diffusive flux of material into and out of the cylindrical annulus, length L.

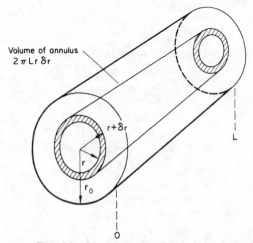

Volume of annulus
$2\pi L r \, \delta r$

$r + \delta r$

r

L

r_0

O

FIG. 2.3. Geometry of cylindrical catalyst pellet

A material balance for the reactant gives (see Fig. 2.3):

Diffusive flux in at $(r + \delta r)$
$-$ Diffusive flux at r = Amount reacted in volume $2\pi L r \delta r$

i.e.

$$\left\{-2\pi D_e L \left(r \frac{dC}{dr}\right)_{r+\delta r}\right\} - \left\{-2\pi D_e L \left(r \frac{dC}{dr}\right)_r\right\} = 2\pi L r \delta r k C$$

Expanding the first term and ignoring terms higher than δr^2:

$$\frac{d^2 C}{dr^2} + \frac{1}{r}\frac{dC}{dr} - \lambda^2 C = 0 \quad \text{where} \quad \lambda = \sqrt{(k/D_e)}$$

This is a standard modified Bessel equation of zero order whose solution is:

$$C = A I_0(\lambda r) + B K_0(\lambda r)$$

where I_0 and K_0 represent zero order modified Bessel functions of the first and second kind respectively.

The boundary conditions for the problem are $r = r_0$, $C = C_0$; $r = 0$, C is finite. Since C remains finite at $r = 0$ and $K_0(0) = \infty$ then we must put $B = 0$ to satisfy the physical conditions. Substituting the boundary conditions therefore gives the solution:

$$\frac{C}{C_0} = \frac{I_0(\lambda r)}{I_0(\lambda r_0)}$$

For the cylinder:

$$\eta = \frac{2\pi r D_e L (dC/dr)_{r_0}}{\pi r_0^2 L k C_0}$$

From the relation between C and r:
$$\left(\frac{dC}{dr}\right)_{r_0} = \lambda C_0 \frac{I_1(\lambda r)}{I_0(\lambda r)}$$

since:
$$\frac{d}{dr}\{I_0(\lambda r)\} = I_1(\lambda r)$$

so:
$$\eta = \frac{2}{\lambda r_0} \frac{I_1(\lambda r_0)}{I_0(\lambda r_0)}$$

Example

Derive an expression for the effectiveness factor of a spherical catalyst pellet in which a first order isothermal reaction occurs.

Solution

Take the origin of coordinates at the centre of the pellet, radius r_0, and construct an infinitesimally

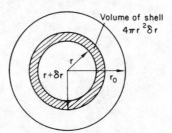

Fig. 2.4. Geometry of spherical catalyst pellet

thin shell of radii $(r + \delta r)$ and r (see Fig. 2.4). A material balance for the reactant across the shell gives:

$$\begin{array}{c}\text{Diffusive flux in at } (r + \delta r) \\ - \text{ Diffusive flux at } r\end{array} = \text{Amount reacted in volume } 4\pi r^2 \delta r$$

i.e.
$$\left\{ 4\pi D_e \left(r^2 \frac{dC}{dr} \right)_{r+\delta r} \right\} - \left\{ 4\pi D_e \left(r^2 \frac{dC}{dr} \right)_r \right\} = 4\pi r^2 \delta r k C$$

Expanding the first term and ignoring terms higher than δr^2:

$$\frac{d^2C}{dr^2} + \frac{2}{r}\frac{dC}{dr} = \frac{k}{D_e} C$$

or
$$\frac{1}{r^2}\frac{d}{dr}\left\{ r^2 \frac{dC}{dr} \right\} - \lambda^2 C = 0 \quad \text{where} \quad \lambda = \sqrt{(k/D_e)}$$

Substituting
$$C = f(r)/r :$$
$$f''(r) - \lambda^2 f(r) = 0$$

therefore
$$f(r) = Ae^{\lambda r} + Be^{-\lambda r}$$

The boundary conditions for the problem are $r = r_0$, $C = C_0$; $r = 0$, C is finite. Now if, at $r = 0$, C is to remain finite then $f(0) = 0$. At r_0 we have $f(r_0) = C_0 r_0$. Substituting these boundary conditions:

$$f(r) = Cr = \frac{C_0 r_0 \sinh(\lambda r)}{\sinh(\lambda r_0)}$$

Now for a sphere:

$$\eta = \frac{4\pi r_0^2 D_e (dC/dr)_{r_0}}{\frac{4}{3}\pi r_0^3 k C_0}$$

From the relation between C and r:

$$\left(\frac{dC}{dr}\right)_{r_0} = \frac{1}{r_0}\{\lambda r_0 \coth(\lambda r_0) - 1\}$$

Hence:

$$\eta = \frac{3}{\lambda r_0}\left\{\coth(\lambda r_0) - \frac{1}{\lambda r_0}\right\}$$

The Thiele moduli for the cylinder and sphere differ from that for the slab. In the case of the slab we recall that $\phi = \lambda L$, whereas for the cylinder it is conveniently defined as $\phi = \lambda r/2$ and for the sphere as $\phi = \lambda r/3$. In each case the reciprocal of this corresponds to the respective asymptote for the curve representing the slab, cylinder or sphere. We may note here that the ratio of the geometric volume V_p of each of the models to the external surface area S_x is L for the slab, $r/2$ for the cylinder and $r/3$ for the sphere. Thus, if the Thiele modulus is defined as:

$$\phi = \lambda \frac{V_p}{S_x} = \frac{V_p}{S_x}\sqrt{\frac{k}{D_e}} \qquad \ldots\ldots(2.23)$$

the asymptotes become coincident. The asymptotes for large ϕ correspond to $\eta = 1/\phi$ for any shape of particle because, as ARIS[4] points out, diffusion is rate determining under these conditions and reaction occurs, therefore, in only a very thin region of the particle adjacent to the exterior surface. The curvature of the surface is thus unimportant.

The effectiveness factor for the slab model may also be calculated for reactions other than first order. It turns out that when the Thiele modulus is large the asymptotic value of η for all reactions is inversely proportional to the Thiele modulus, and when the latter approaches zero the effectiveness factor tends to unity. However, just as we found that the asymptotes for a first order reaction in particles of different geometry do not coincide unless we choose a definition for the Thiele modulus which forces them to become superimposed, so we find that the asymptotes for reaction orders $n = 0$, 1 and 2 do not coincide unless we define a generalised Thiele modulus:

$$\bar{\phi} = \frac{V_p}{S_x}\sqrt{\left(\frac{n+1}{2}\frac{kC_\infty^{(n-1)}}{D_e}\right)} \qquad \ldots\ldots(2.24)$$

The modulus $\bar{\phi}$ defined by equation 2.24 has the advantage that the asymptotes to η are approximately coincident for all particle shapes and for all reaction orders except $n = 0$; for this latter case[32] $\eta = 1$ for $\bar{\phi} < 2$ and $\eta = 1/\bar{\phi}$ for $\bar{\phi} > 2$. Thus η may be calculated from the simple slab model, using equation 2.24 to define the Thiele modulus. The curve of η as a function of $\bar{\phi}$ is therefore quite general for practical catalyst pellets. For $\bar{\phi} > 3$ it is found that $\eta = 1/\bar{\phi}$

to an accuracy within 0·5 per cent, while the approximation is within 3·5 per cent for $\bar{\phi} > 2$. It is best to use this generalised curve (i.e. η as a function of $\bar{\phi}$) because the asymptotes for different cases can then almost be made to coincide. The errors involved in using the generalised curve are probably no greater than errors perpetrated by estimating values of parameters in the Thiele modulus.

Effect of intraparticle diffusion on experimental parameters

When intraparticle diffusion is rate determining, the kinetic behaviour of the system is different from that which prevails when chemical reaction is rate determining. For conditions of diffusion control ϕ will be large, and then the effectiveness factor η ($= 1/\phi$ tanh ϕ, from equation 2.20) becomes $\bar{\phi}^{-1}$. From equation 2.23, it is seen therefore that η is proportional to $k^{-\frac{1}{2}}$. The chemical reaction rate on the other hand is directly proportional to k so that, from equation 2.13 at the beginning of this section, the overall reaction rate is proportional to $k^{\frac{1}{2}}$. Since the specific rate constant is directly proportional to $e^{-E/RT}$, where E is the activation energy for the chemical reaction in the absence of diffusion effects, we are led to the important result that for a diffusion limited reaction the rate is proportional to $e^{-E/2RT}$. Hence the apparent activation energy E_D, measured when reaction occurs in the diffusion controlled region, is only half the true value:

$$E_D = E/2 \qquad\qquad ....(2.25)$$

A further important result which arises because of the functional form of $\bar{\phi}$ is that the apparent order of reaction in the diffusion controlled region differs from that which is observed when chemical reaction is rate determining. Recalling that the reaction order is defined as the exponent n to which the concentration C_∞ is raised in the equation depicting the chemical reaction rate, we replace $f(C)$ in equation 2.13 by C_∞^n. Hence the overall reaction rate per unit volume is $(1 - e)\eta k C_\infty^n$. When diffusion is rate determining, η is (as already mentioned) equal to ϕ^{-1}; from equation 2.24 it is therefore proportional to $C_\infty^{-(n-1)/2}$. Thus the overall reaction rate depends on $C_\infty^n C_\infty^{-(n-1)/2} = C_\infty^{(n+1)/2}$. The apparent order of reaction n_D as measured when reaction is dominated by intraparticle diffusion effects is thus related to the true order as follows:

$$n_D = (n + 1)/2 \qquad\qquad ....(2.26)$$

A zero order reaction thus becomes a half-order reaction, a first order reaction remains first order, whereas a second order reaction has an apparent order of 3/2 when strongly influenced by diffusional effects. Because k and n are modified in the diffusion controlled region then, if the rate of the overall process is estimated by multiplying the chemical reaction rate by the effectiveness factor (as in equation 2.13), it is imperative to know the true rate of chemical reaction uninfluenced by diffusion effects.

The functional dependence of other parameters on the reaction rate also becomes modified when diffusion determines the overall rate. If we write the rate of reaction for an nth order reaction in terms of equation 2.13 and substitute the general expression obtained for the effectiveness factor at high values of $\bar{\phi}$, where η is proportional to $1/\bar{\phi}$ and $\bar{\phi}$ is defined by equation 2.24, we obtain:

$$\mathscr{R}_V = (1 - e)kC_\infty^n \eta = (1 - e)kC_\infty^n \frac{S_x}{V_p} \sqrt{\left(\frac{2D_e}{(n + 1)kC_\infty^{n-1}}\right)} \quad \ldots (2.27)$$

The way in which experimental parameters are affected when intraparticle diffusion is important may be deduced by inspection of equation 2.27. Referring the specific rate constant to unit surface area, rather than unit reactor volume, the term $(1 - e)\,k$ is equivalent to $\rho_b S_g k_a$ where ρ_b is the bulk density of the catalyst and S_g is its surface area per unit mass. On the other hand the rate constant k appearing in the denominator under the square root sign in equation 2.27 is based on unit particle volume and is therefore equal to $\rho_p S_g k_a$ where ρ_p is the particle density. Thus, if bulk diffusion controls the reaction, the rate becomes dependent on the square root of the specific surface area, rather than being directly proportional to surface area, in the absence of transport effects. We do not include the external surface area S_x in this reckoning since the ratio V_p/S_x is, for a given particle shape, an independent parameter characteristic of the particle size. On the other hand, if Knudsen diffusion determines the rate, then because the effective diffusivity for Knudsen flow is inversely proportional to the specific surface area (equation 2.9) the reaction rate becomes independent of surface area.

The pore volume per unit mass V_g (a measure of the porosity) is also a parameter which is important and is implicitly contained in equation 2.27. Since the product of the particle density ρ_p and specific pore volume V_g represents the porosity, then ρ_p is inversely proportional to V_g. Therefore, when the rate is controlled by bulk diffusion, it is proportional not simply to the square root of the specific surface area but to the product of $\sqrt{S_g}$ and $\sqrt{V_g}$. If Knudsen diffusion controls the reaction then the overall rate is directly proportional to V_g since the effective Knudsen diffusivity contained in the quantity $\sqrt{(D_e/\rho_p)}$ is, from equation 2.9, proportional to the ratio of the porosity ψ and the particle density ρ_p.

Table 2.1 summarises the effect which intraparticle mass transfer effects have on parameters involved explicitly or implicitly in the expression for the overall rate of reaction.

TABLE 2.1. *Effect of intraparticle diffusion on various parameters*

Rate determining step	Order	Activation energy	Surface area	Pore volume
Chemical reaction	n	E	S_g	independent
Bulk diffusion	$(n + 1)/2$	$E/2$	$\sqrt{S_g}$	$\sqrt{V_g}$
Knudsen diffusion	$(n + 1)/2$	$E/2$	independent	V_g

Non-isothermal reactions in porous catalyst pellets

So far the effect of temperature gradients within the particle has been ignored. Strongly exothermic reactions generate a considerable amount of heat which, if conditions are to remain stable, must be transported through the particle to the exterior surface where it may then be dissipated. Similarly an endothermic reaction requires a source of heat and in this case the heat must permeate the particle from the exterior to the interior. In any event a temperature gradient within the particle is established and the chemical reaction rate will vary with position.

We may consider the problem by writing a material and heat balance for the slab of catalyst depicted in Fig. 2.1. For an irreversible first order exothermic reaction the material balance is:

$$\frac{d^2 C_A}{dx^2} - \frac{k C_A}{D_e} = 0 \qquad\qquad \text{(equation 2.15)}$$

A heat balance over the element δx gives:

$$\frac{d^2 T}{dx^2} + \frac{(-\Delta H) k C_A}{k_e} = 0 \qquad\qquad \ldots .(2.28)$$

where ΔH is the enthalpy change resulting from reaction, and k_e is the effective thermal conductivity of the particle defined by analogy with the discussion on effective diffusivity on page 111 *et seq.* In writing these two equations it should be remembered that the specific rate constant k is a function of temperature, usually of the Arrhenius form ($k = \mathscr{A} \, e^{-E/RT}$, where $\mathscr{A}$ is the frequency factor for reaction). These two simultaneous differential equations are to be solved together with the boundary conditions:

$$C_A = C_{A\infty} \quad \text{at} \quad x = \pm L \qquad\qquad \ldots .(2.29)$$

$$T = T_\infty \quad \text{at} \quad x = \pm L \qquad\qquad \ldots .(2.30)$$

$$\frac{dC_A}{dx} = \frac{dT}{dx} = 0 \quad \text{at} \quad x = 0 \qquad\qquad \ldots .(2.31)$$

Because of the non-linearity of the equations the problem can only be solved in this form by numerical techniques[31, 32]. However, an approximation may be made which gives an asymptotically exact solution[36], or, alternatively, the exponential function of temperature may be expanded to give equations which can be solved analytically[34, 42]. A convenient solution to the problem may be presented in the form of families of curves for the effectiveness factor as a function of the Thiele modulus. Figure 2.5 shows these curves for the case of a first order irreversible reaction occurring in spherical catalyst particles. Two additional independent dimensionless parameters are introduced into the problem and these are defined as:

$$\beta = \frac{(-\Delta H) D_e C_\infty}{k_e T_\infty} \qquad\qquad \ldots .(2.32)$$

$$\varepsilon = E/\mathbf{R}T \qquad\qquad \ldots .(2.33)$$

The parameter β represents the maximum temperature difference that could exist in the particle relative to the temperature at the exterior surface, for if we recognise that in the steady state the heat flux within an elementary thickness of the particle is balanced by the heat generated by chemical reaction then:

$$k_e \frac{dT}{dx} = \Delta H D_e \frac{dC}{dx} \qquad \ldots\ldots(2.34)$$

If equation 2.34 is then integrated from the exterior surface where $T = T_\infty$ and $C = C_\infty$ to the centre of the particle where (say) $T = T_M$ and $C = C_M$ we obtain

$$\frac{T_M - T_\infty}{T_\infty} = \frac{(-\Delta H)D_e}{k_e T_\infty}(C_\infty - C_M) \qquad \ldots\ldots(2.35)$$

When the Thiele modulus is large C_M is effectively zero and the maximum difference in temperature between the centre and exterior of the particle is $(-\Delta H)D_e C_\infty / k_e$. Relative to the temperature outside the particle this maximum temperature difference is therefore β. For exothermic reactions β is positive while for endothermic reactions it is negative. The curve in Fig. 2.5 for $\beta = 0$ represents isothermal conditions within the pellet. It is interesting to note that for a reaction in which $-\Delta H = 10^5$ kJ/kmol, $k_e = 1$ W/mK, $D_e = 10^{-5}$ m^2 s^{-1} and $C_\infty = 10^{-1}$ kmol/m^3, the value of $T_M - T_\infty$ is 100°C. In practice much lower values than this are observed but it does serve to show that serious errors may be introduced into calculations if conditions within the pellet are arbitrarily assumed to be isothermal.

On the other hand, it has been argued that the resistance to heat transfer is effectively within a thin gas film enveloping the catalyst particle[53]. Thus, for the whole practical range of heat transfer coefficients and thermal conductivities, the catalyst particle may be considered to be at a uniform temperature. Any temperature increase arising from the exothermic nature of a reaction would therefore be across the fluid film rather than in the pellet interior.

Figure 2.5 shows that, for exothermic reactions ($\beta > 0$), the effectiveness factor may exceed unity. This is because the increase in rate caused by the temperature rise inside the particle more than compensates for the decrease in rate caused by the negative concentration gradient which effects a decrease in concentration towards the centre of the particle. A further point of interest is that, for reactions which are highly exothermic and at low values of the Thiele modulus, the value of η is not uniquely defined by the Thiele modulus and the parameters β and ε. The shape of the curves in this region indicates that the effectiveness factor may correspond to any one of three values for a given value of the Thiele modulus. In effect there are three different conditions for which the rate of heat generation within the particle is equal to the rate of heat removal. One condition represents a metastable state and the remaining two conditions correspond to a region in which the rate is limited by chemical reaction (relatively low temperatures) and a region where there is diffusion limitation (relatively high temperatures). The region of multiple solutions in Fig. 2.5, however, corresponds to large values of β and ε seldom encountered in practice.

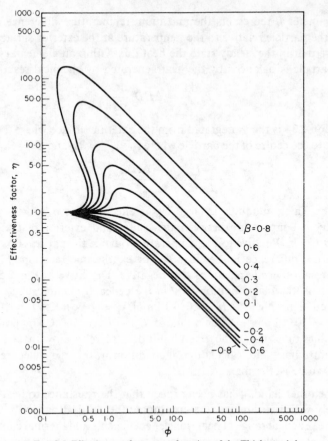

FIG. 2.5. Effectiveness factor as a function of the Thiele modulus
for non-isothermal conditions

McGreavy and Thornton[53] have developed an alternative approach to
the problem of identifying such regions of unique and multiple solutions.
Recognising that the resistance to heat transfer is probably due to a thin gas
film surrounding the particle, but that the resistance to mass transfer is within
the porous solid, they solved the mass and heat balance equations for a pellet
with modified boundary conditions. Thus the heat balance for the pellet
represented by equation 2.28 was replaced by:

$$h(T - T_\infty) = h_D(-\Delta H)(C_\infty - C) \qquad \dots \text{(2.36)}$$

and solved simultaneously with the mass balance represented by equation 2.15.
Boundary conditions represented by equations 2.29 and 2.30 were replaced by:

$$D_e \frac{dC}{dx} = h_D(C_\infty - C) \quad \text{at} \quad x = L \qquad \dots \text{(2.37)}$$

and

$$k_e \frac{dT}{dx} = h(T - T_\infty) \quad \text{at} \quad x = L \qquad \dots \text{(2.38)}$$

respectively.

A modified Thiele modulus may be defined by rewriting $\phi = L\sqrt{(k/D_e)}$ (see page 118) in the form:

$$\phi' = L\sqrt{(\mathscr{A}/D_e)} \qquad \ldots (2.39)$$

where $\mathscr{A}$ is the frequency factor in the classical Arrhenius equation. The numerical solution is then depicted in Fig. 2.6, which resembles a plot of effectiveness factor as a function of Thiele modulus (cf. Fig. 2.5). Whereas an effectiveness factor chart describes the situation for a given single particle in a packed bed (and is therefore of limited value in reactor design), Fig. 2.6 may be used to

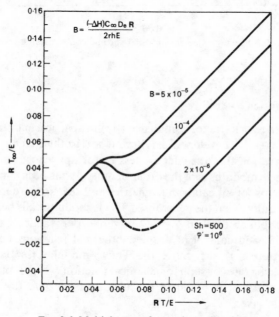

FIG. 2.6. Multiple states for catalyst pellets in reactor

identify the region of multiple solutions for the whole reactor. If local extrema are calculated from Fig. 2.6 by finding conditions for which $dT_\infty/dT = 0$, the bounds of T_∞ may be located. It is then possible to predict the region of multiple solutions corresponding to unstable operating conditions. Figure 2.7, for example, shows two reactor trajectories which would intersect the region of metastable conditions within the reactor. Such a method can predict regions of instability for the packed reactor rather than for a single particle, because use has been made of the modified Thiele modulus employing the kinetic Arrhenius factor $\mathscr{A}$ which is independent of position along the bed.

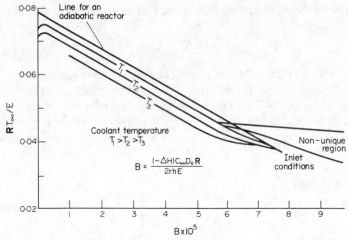

FIG. 2.7. Reactor trajectories in adiabatic and cooled catalyst beds

Criteria for diffusion control

In assessing whether a reactor is influenced by intraparticle mass transfer effects WEISZ and PRATER[18] developed a criterion for isothermal reactions based upon the observation that the effectiveness factor approaches unity when the generalised Thiele modulus is of the order of unity. It has been shown[4] that the effectiveness factor for all catalyst geometries and reaction orders (except zero order) tends to unity when the generalised Thiele modulus falls below a value of one. Since η is about unity when $\phi < \sqrt{2}$ for zero order reactions, a quite general criterion for diffusion control of simple isothermal reactions not affected by product inhibition is $\bar{\phi} < 1$. Since the Thiele modulus (see equation 2.23) contains the specific rate constant for chemical reaction, which is often unknown, a more useful criterion is obtained by substituting $\mathcal{R}'_V/C_\infty$ (for a first order reaction) for k to give:

$$\left\{\frac{V_p}{S_x}\right\}^2 \frac{\mathcal{R}'_V}{D_e C_\infty} < 1 \qquad \qquad \dots (2.40)$$

where $\mathcal{R}'_V$ is the measured rate of reaction per unit volume of catalyst particle.

PETERSEN[41] points out that this criterion is invalid for more complex chemical reactions whose rate is retarded by products. In such cases the observed kinetic rate expression should be substituted into the material balance equation for the particular geometry of particle concerned. An asymptotic solution to the material balance equation then gives the correct form of the effectiveness factor. The results indicate that the inequality 2.40 is applicable only at high partial pressures of product. For low partial pressures of product (often the condition in an experimental differential tubular reactor) the criterion will depend on the magnitude of the constants in the kinetic rate equation.

The usual experimental criterion for diffusion control involves an evaluation of the rate of reaction as a function of particle size. At a sufficiently small particle size the measured rate of reaction will become independent of particle size and the rate of reaction can be safely assumed to be independent of intraparticle mass transfer effects. At the other extreme, if the observed rate is inversely proportional to particle size the reaction is strongly influenced by intraparticle diffusion. For a reaction whose rate is inhibited by the presence of products there is an attendant danger of misinterpreting experimental results obtained for different particle sizes when a differential reactor is used, for, under these conditions, the effectiveness factor is sensitive to changes in the partial pressure of product.

When reaction conditions within the particle are non-isothermal WEISZ and HICKS[34] showed that a suitable criterion defining conditions under which a reaction is not controlled by mass and heat transfer effects in the solid is:

$$\left\{\frac{V_p}{S_x}\right\}^2 \frac{\mathscr{R}'_v}{D_e C_\infty} \exp\left\{\frac{\varepsilon\beta}{1+\beta}\right\} < 1 \qquad \ldots\ldots(2.41)$$

Selectivity in catalytic reactions influenced by mass and heat transfer effects

It is rare that a catalyst can be chosen for a reaction such that it is entirely specific or unique in its behaviour. More often than not products additional to the main desired product are generated concomitantly. The ratio of the specific chemical rate constant of a desired reaction to that for an undesired reaction is termed the kinetic selectivity factor (which we shall designate S) and is of central importance in catalysis. Its magnitude is determined by the relative rates at which adsorption, surface reaction and desorption occur in the overall process and, for consecutive reactions, whether or not the intermediate product forms a localised or mobile adsorbed complex with the surface. In the case of two parallel competing catalytic reactions a second factor, the thermodynamic factor, is also of importance. This latter factor depends exponentially on the difference in free energy changes associated with the adsorption–desorption equilibria of the two competing reactants. The thermodynamic factor also influences the course of a consecutive reaction where it is enhanced by the ability of the intermediate product to desorb rapidly and also the reluctance of the catalyst to re-adsorb the intermediate product after it has vacated the surface.

The kinetic and thermodynamic selectivity factors are quantities which are functions of the chemistry of the system. When an active catalyst has been selected for a particular reaction (often by a judicious combination of theory and experiment) we ensure that the kinetic and thermodynamic factors are such that they favour the formation of desired product. Many commercial processes, however, employ porous catalysts since this is the best means of increasing the extent of surface at which the reaction occurs. We, as chemical engineers, are therefore interested in the effect which the porous nature of the catalyst has on the selectivity of the chemical process.

WHEELER[13] considered the problem of chemical selectivity in porous catalysts. Although he employed a cylindrical pore model and restricted his conclusions to the effect of pore size on selectivity, the following discussion will be based on the simple geometrical model of the catalyst pellet introduced earlier (see Fig. 2.1, and p. 116 et seq.).

Isothermal conditions

Sometimes it may be necessary to convert into a desired product only one component in a mixture. For example, it may be required to dehydrogenate a six-membered cycloparaffin in the presence of a five-membered cycloparaffin without affecting the latter. In this case it is desirable to select a catalyst which favours the reaction:

$$A \xrightarrow{k_1} B$$

when it might be possible for the reaction:

$$X \xrightarrow{k_2} Y$$

to occur simultaneously.

Suppose the desired product is **B** and also suppose that the reactions occur in a packed tubular reactor in which we may neglect both longitudinal and radial dispersion effects. If both reactions are first order the ratio of the rates of the respective reactions is:

$$\frac{\mathscr{R}'_{VA}}{\mathscr{R}'_{VX}} = \frac{k_1 C_A}{k_2 C_X} \qquad \qquad ....(2.42)$$

If the reactions were not influenced by in-pore diffusion effects the intrinsic kinetic selectivity would be $k_1/k_2 (=S)$. When mass transfer is important, the rate of reaction of both **A** and **X** must be calculated with this in mind. We recall from page 117 that the rate of reaction for the slab model is:

$$-A_c D_e \left(\frac{dC_A}{dx}\right)_{x=\pm L}$$

The concentration profile of **A** through the slab is given by:

$$C_A = C_{A\infty} \frac{\cosh(\lambda x)}{\cosh(\lambda L)} \qquad \qquad \text{(equation 2.18)}$$

so that by differentiation of C_A we obtain the rate of decomposition of **A**:

$$\mathscr{R}'_{VA} = \frac{-A_c D_e}{V_p} \left(\frac{dC_A}{dx}\right)_{x=L} = \frac{-A_c D_e C_{A\infty}}{L V_p} \phi_1 \tanh \phi_1 \qquad ....(2.43)$$

where $\phi_1 = L \sqrt{\dfrac{k_1}{D_e}}$ is the Thiele modulus pertaining to the decomposition of **A**.

A similar equation may be written for the decomposition of **X**:

$$\mathscr{R}'_{VX} = \frac{-A_c D_e C_{X\infty}}{L V_p} \phi_2 \tanh \phi_2 \qquad \ldots.(2.44)$$

where ϕ_2 corresponds to the Thiele modulus for the decomposition of **X**. If we were dealing with a general type of catalyst pellet, then because of the properties of the general Thiele parameter $\bar{\phi}$ we need only replace ϕ_1 and ϕ_2 by $\bar{\phi}_1$ and $\bar{\phi}_2$ respectively and substitute (V_p/S_x) for the characteristic dimension L. The ratio of the rates of decomposition of **A** and **X** then becomes:

$$\frac{\mathscr{R}'_{VA}}{\mathscr{R}'_{VX}} = \frac{C_{A\infty}}{C_{X\infty}} \frac{\bar{\phi}_1 \tanh \bar{\phi}_1}{\bar{\phi}_2 \tanh \bar{\phi}_2} \qquad \ldots.(2.45)$$

Although equation 2.45 is only applicable to competing first order reactions in catalyst particles, at large values of $\bar{\phi}$ where diffusion is rate controlling the equation is equivalent at the asymptotes to equations obtained for reaction orders other than one.

Since $\bar{\phi}$ is proportional to $\sqrt{k}$ we may conclude that for competing simultaneous reactions strongly influenced by diffusion effects (where $\bar{\phi}$ is large and $\tanh \bar{\phi} \eqsim 1$) the selectivity depends on $\sqrt{S}$ (where $S = k_1/k_2$), the square root of the ratio of the respective rate constants. The corollary is that, for such reactions, maximum selectivity is displayed by small sized particles and, in the limit, if the particle size is sufficiently small (small $\bar{\phi}$ so that $\tanh \bar{\phi} \eqsim \bar{\phi}$) the selectivity is the same as for a non-porous particle, i.e. S itself.

We should add a note of caution here, however, for in the Knudsen flow region D_e is proportional to the pore radius. When the pores are sufficiently small for Knudsen diffusion to occur then the selectivity will also be influenced by pore size. Maximum selectivity would be obtained for small particles which contain large diameter pores.

When two simultaneous reaction paths are involved in a process the routes may be represented:

$$A \begin{array}{c} \overset{k_1}{\nearrow} B \\ \underset{k_2}{\searrow} C \end{array}$$

A classical example of this type of competitive reaction is the conversion of ethanol by a copper catalyst at about 300°C. The principal product is acetaldehyde but ethylene is also evolved in smaller quantities. If, however, an alumina catalyst is used, ethylene is the preferred product. If, in the above reaction scheme, **B** is the desired product then the selectivity may be found by comparing the respective rates of formation of **B** and **C**. Adopting the slab model for simplicity and remembering that, in the steady state, the rates of formation of **B** and **C** must be equal to the flux of **B** and **C** at the exterior surface of the particle, assuming that the effective diffusivities of **B** and **C** are equal:

$$\frac{\mathscr{R}'_{VB}}{\mathscr{R}'_{VC}} = \frac{\left(\dfrac{dC_B}{dx}\right)_{x=L}}{\left(\dfrac{dC_C}{dx}\right)_{x=L}} \qquad \ldots.(2.46)$$

The respective fluxes may be evaluated by writing the material balance equations for each component and solving the resulting simultaneous equations. If the two reactions are of the same kinetic order, then it is obvious from the form of equation 2.46 that the selectivity is unaffected by mass transfer in pores. If, however, the kinetic orders of the reactions differ then, as the example below shows, the reaction of the lowest kinetic order is favoured. The rate of formation of **B** with respect to **C** would therefore be impeded and the selectivity reduced if the order of the reaction producing **B** were less than that for the reaction producing **C**. In such cases the highest selectivity would be obtained by the use of small diameter particles or particles in which the effective diffusivity is high.

Example

Two gas-phase concurrent irreversible reactions

occur isothermally in a flat slab-shaped porous catalyst pellet. The desirable product **B** is formed by a first order chemical reaction and the wasteful product **C** is formed by a zero order reaction. Deduce an expression for the catalyst selectivity.

Solution

Taking a material balance across the element δx in Fig. 2.1, the flux in at $(x + \delta x)$ minus the flux out at x is equal to the amount reacted in volume $2A_c\delta x$ where A_c represents the area of each of the faces. If the slab is thin, diffusion through the edges may be neglected. For the three components the material balance equations therefore become:

$$\frac{d^2C_A}{dx^2} - f^2 C_A = g^2$$

$$\frac{d^2C_B}{dx^2} + f^2 C_A = 0$$

$$\frac{d^2C_C}{dx^2} = -g^2$$

where $f^2 = k_1/D_e$ and $g^2 = k_2/D_e$.

The boundary conditions are:

$$x = \pm L, \quad C_A = C_{A0}, \ C_B = C_C = 0$$

and

$$x = 0, \quad \frac{dC_A}{dx} = \frac{dC_B}{dx} = \frac{dC_C}{dx} = 0$$

The solution satisfying the above set of differential equations is $C_A = A\,e^{fx} - \dfrac{g^2}{f^2}$ where the term g^2/f^2 represents the particular integral. On inserting the boundary conditions the complete solution is:

$$C_A = \left(C_{A0} + \frac{g^2}{f^2}\right)\frac{\cosh(fx)}{\cosh(fL)} - \frac{g^2}{f^2}$$

The concentration C_B may now be found, but since the selectivity will be given by:

$$S = \frac{(dC_B/dx)_{x=L}}{(dC_C/dx)_{x=L}}$$

the material balance equation need only be integrated once. We obtain:

$$\left(\frac{dC_B}{dx}\right)_{x=L} = -f^2 \int_0^L \left\{\left(C_{A0} + \frac{g^2}{f^2}\right)\frac{\cosh(fx)}{\cosh(fL)} - \frac{g^2}{f^2}\right\}dx = -f\left(C_{A0} + \frac{g^2}{f^2}\right)\tanh(fL) + g^2L$$

and

$$\left(\frac{dC_C}{dx}\right)_{x=L} = -g^2L$$

Hence:

$$S = \left(\frac{k_1}{k_2}C_{A0} + 1\right)\frac{\tanh\phi}{\phi} - 1$$

where

$$\phi = L\sqrt{\frac{k_1}{D_e}}.$$

Another important class of reactions, which is common in petroleum reforming reactions, may be represented by the scheme:

$$\mathbf{A} \xrightarrow{k_1} \mathbf{B} \xrightarrow{k_2} \mathbf{C}$$

and exemplified by the dehydrogenation of six-membered cycloparaffins to aromatics (e.g. cyclohexane converted to cyclohex-1-ene and ultimately benzene) catalysed by transition metals and metal oxides. Again we suppose **B** to be the desired product while **C** is a waste product. (If **C** were the desired product we would require a low selectivity for the formation of **B**). On the basis of first order kinetics and using the flat plate model, the material balance equation for component **B** is:

$$D_e\frac{d^2C_B}{dx^2} = k_2C_B - k_1C_A \qquad \qquad \ldots(2.47)$$

For component **A** the material balance equation is, as previously obtained:

$$D_e\frac{d^2C_A}{dx^2} = k_1C_A \qquad \qquad \text{(equation 2.15)}$$

The boundary conditions for the problem are:

$$C_A = C_{A\infty} \quad \text{and} \quad C_B = C_{B\infty} \quad \text{at} \quad x = L \qquad \ldots(2.48)$$

$$dC_A/dx = dC_B/dx = 0 \qquad \qquad \text{at} \quad x = 0 \qquad \ldots(2.49)$$

Solving the two simultaneous linear differential equations with the above boundary conditions leads to:

$$C_A = C_{A\infty}\frac{\cosh(\lambda_1 x)}{\cosh(\lambda_1 L)} \qquad \qquad \text{(equation 2.18)}$$

and

$$C_B = C_{A\infty}\left\{\frac{k_1}{k_1 - k_2}\right\}\left\{\frac{\cosh(\lambda_2 x)}{\cosh(\lambda_2 L)} - \frac{\cosh(\lambda_1 x)}{\cosh(\lambda_1 L)}\right\} + C_{B\infty}\frac{\cosh(\lambda_2 x)}{\cosh(\lambda_2 L)} \qquad \ldots(2.50)$$

where λ_1 and λ_2 are $\sqrt{(k_1/D_e)}$ and $\sqrt{(k_2/D_e)}$ respectively. The selectivity of the reaction will be the rate of formation of **B** with respect to **A** which, in the steady state, will be equal to the ratio of the fluxes of **B** and **A** at the exterior surface of the particle. Thus:

$$-\frac{(dC_B/dx)_{x=L}}{(dC_A/dx)_{x=L}} = \left\{\frac{k_1}{k_1 - k_2}\right\}\left\{1 - \frac{\phi_2 \tanh \phi_2}{\phi_1 \tanh \phi_1}\right\} - \frac{C_{B\infty}}{C_{A\infty}}\frac{\phi_2 \tanh \phi_2}{\phi_1 \tanh \phi_1} \quad \ldots(2.51)$$

Although the ratio of the fluxes of **B** and **A** at the exterior surface of the particle is really a point value, we may conveniently regard it as representing the rate of change of the concentration of **B** with respect to **A** at a position in the reactor corresponding to the location of the particle. The left side of equation 2.51 may thus be replaced by $-\dfrac{dC_B}{dC_A}$ where C_B and C_A are now gas phase concentrations. With this substitution and for large values of ϕ_1 and ϕ_2 one of the limiting forms of equation 2.51 is obtained:

$$-\frac{dC_B}{dC_A} = \frac{\sqrt{S}}{1 + \sqrt{S}} - \frac{C_B}{C_A}\frac{1}{\sqrt{S}} \quad \ldots(2.52)$$

where S is the kinetic selectivity ($= k_1/k_2$). Integrating equation 2.52 from the reactor inlet (where the concentration of **A** is, say, C_{A0} and that of **B** is taken as zero) to any point along the reactor:

$$\frac{C_B}{C_A} = \left\{\frac{S}{S - 1}\right\}\left\{\left(\frac{C_A}{C_{A0}}\right)^{\frac{1 - \sqrt{S}}{\sqrt{S}}} - 1\right\} \quad \ldots(2.53)$$

The other limiting form of equation 2.51 is obtained when ϕ_1 and ϕ_2 are small:

$$-\frac{dC_B}{dC_A} = 1 - \frac{1}{S}\frac{C_B}{C_A} \quad \ldots(2.54)$$

which, on integration, gives:

$$\frac{C_B}{C_A} = \left(\frac{S}{S - 1}\right)\left\{\left(\frac{C_A}{C_{A0}}\right)^{\frac{1 - S}{S}} - 1\right\} \quad \ldots(2.55)$$

Comparison of equations 2.52 and 2.54 shows that at low effectiveness factors (when ϕ_1 and ϕ_2 are large) the selectivity is less than it would be if the effectiveness factor for each reaction were near unity (small ϕ_1 and ϕ_2). The consequence of this is seen by comparing equations 2.53 and 2.55, the respective integrated forms of equations 2.52 and 2.54. The yield of **B** is comparatively low when in-pore diffusion is a rate limiting process and, for a given fraction of **A** reacted, the conversion to **B** is impeded. A corollary to this is that it should be possible to increase the yield of a desired intermediate by using smaller catalyst particles or, alternatively, by altering the pore structure in such a way as to increase the effective diffusivity. If, however, the effectiveness factor is below about 0·3 (at which value the yield of **B** becomes independent of diffusion effects) a large reduction in pellet size (or large increase in D_e) is required to achieve any significant improvement in selectivity. WHEELER[13] suggests that when such a drastic reduction in pellet size is necessary, a fluidised bed reactor may be used to improve the yield of an intermediate.

Non-isothermal conditions

The influence which the simultaneous transfer of heat and mass in porous catalysts has on the selectivity of first order concurrent catalytic reactions has recently been investigated by ØSTERGAARD[38]. As shown previously, selectivity is not affected by any limitations due to mass transfer when the process corresponds to two concurrent first order reactions:

However, with heat transfer between the interior and exterior of the pellet (made possible by temperature gradients resulting from an exothermic diffusion limited reaction) the selectivity may be substantially altered. For the flat-plate model the material and heat balance equations to solve are:

$$\frac{d^2 C_A}{dx^2} - \left(\frac{k_1 + k_2}{D_e}\right) C_A = 0 \qquad \ldots\ldots(2.56)$$

$$\frac{d^2 C_B}{dx^2} + \frac{k_1}{D_e} C_A = 0 \qquad \ldots\ldots(2.57)$$

and

$$\frac{d^2 T}{dx^2} + \frac{(-\Delta H_1) k_1 + (-\Delta H_2) k_2}{k_e} C_A = 0 \qquad \ldots\ldots(2.58)$$

where

$$k_i = \mathscr{A}_i \exp(-E_i/RT), \quad i = 1, 2 \qquad \ldots\ldots(2.59)$$

and ΔH_1 and ΔH_2 correspond to the respective enthalpy changes. The boundary conditions are:

$$\frac{dC_A}{dx} = \frac{dC_B}{dx} = \frac{dT}{dx} = 0 \quad \text{at} \quad x = 0 \qquad \ldots\ldots(2.60)$$

$$C_A = C_{A\infty}, \quad C_B = C_{B\infty}, \quad T = T_\infty \quad \text{at} \quad x = L \qquad \ldots\ldots(2.61)$$

This is a two point boundary-value problem and, because of the non-linearity of the equations, cannot be solved analytically. If, however, $E_1 = E_2$ the selectivity is the same as if there were no resistance to either heat or mass transfer (see the example overleaf). For the case where $\Delta H_1 = \Delta H_2$ but $E_1 \neq E_2$, the selectivity is determined by the effects of simultaneous heat and mass transfer. Figure 2.8 shows that if the activation energy of the desired reaction is lower than that of the reaction leading to the wasteful product ($E_2/E_1 > 1$) the best selectivity is obtained for high values of the Thiele modulus. When $E_2/E_1 < 1$ a decrease in selectivity results. In the former case the selectivity approaches an upper limit asymptotically, and it is not worth while increasing the Thiele modulus beyond a value where there would be a significant decrease in the efficiency of conversion.

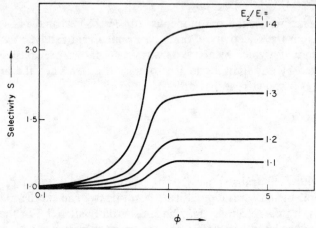

FIG. 2.8. Selectivity as a function of the Thiele modulus for non-isothermal conditions

Example

Show that the selectivity of two concurrent first order reactions occurring in flat-shaped porous catalyst pellets is independent of the effect of either heat or mass transfer if the activation energies of both reactions are equal.

Solution

For the concurrent first order reactions:

$$A\underset{k_2}{\overset{k_1}{<}}\begin{matrix}B\\C\end{matrix}$$

the mass and heat transfer equations are:

$$\frac{d^2C_A}{dx^2} - \frac{k_1 + k_2}{D_e}C_A = 0$$

$$\frac{d^2C_B}{dx^2} + \frac{k_1}{D_e}C_A = 0$$

$$\frac{d^2T}{dx^2} + \frac{(-\Delta H_1)k_1 + (-\Delta H_2)k_2}{k_e}C_A = 0$$

where $k_i = \mathscr{A}_i \exp(-E_i/\mathbf{R}T)$, $i = 1, 2$.

The boundary conditions are given by:

$$\frac{dC_A}{dx} = \frac{dC_B}{dx} = \frac{dT}{dx} = 0 \quad \text{at} \quad x = 0$$

$$C_A = C_{A\infty}, C_B = C_{B\infty}, T = T_\infty \quad \text{at} \quad x = L$$

When $E_1 = E_2$ we see that $k_1/k_2 = \mathscr{A}_1/\mathscr{A}_2$ – a ratio independent of temperature. Hence, if the mass-transfer equations are divided they may, for the case $E_1 = E_2$, be integrated directly and the gradients evaluated at the slab surface, $x = L$. Thus:

$$\left(\frac{\mathrm{d}C_A}{\mathrm{d}x}\right)_{x=L} = -\frac{\left(1 + \dfrac{\mathscr{A}_2}{\mathscr{A}_1}\right)}{\left(1 + \dfrac{(-\Delta H_2)\mathscr{A}_2}{(-\Delta H_1)\mathscr{A}_1}\right)} \frac{k_e}{D_e(-\Delta H_1)} \left(\frac{\mathrm{d}T}{\mathrm{d}x}\right)_{x=L}$$

$\left(\dfrac{\mathrm{d}C_B}{\mathrm{d}x}\right)_{x=L}$ may be found similarly. The selectivity is determined by the ratio of the reaction rates at the surface. In the steady state, this is equal to the ratio of the fluxes of **C** and **B** at the slab surface. Hence we obtain for the selectivity:

$$\frac{\left(\dfrac{\mathrm{d}C_C}{\mathrm{d}x}\right)_{x=L}}{\left(\dfrac{\mathrm{d}C_B}{\mathrm{d}x}\right)_{x=L}} = \frac{\left(\dfrac{\mathrm{d}C_A}{\mathrm{d}x}\right)_{x=L}}{\left(\dfrac{\mathrm{d}C_B}{\mathrm{d}x}\right)_{x=L}} - 1 = 1 + \frac{\mathscr{A}_2}{\mathscr{A}_1} - 1 = \underline{\underline{\frac{\mathscr{A}_2}{\mathscr{A}_1}}}$$

and this is the same result as would have been obtained if there were no resistance to either mass or heat transfer within the pellets.

Selectivity of bifunctional catalysts

Certain heterogeneous catalytic processes require the presence of more than one catalyst to achieve a significant yield of desired product. The conversion of n-heptane to i-heptane, for example, requires the presence of a dehydrogenation catalyst, such as platinum, together with an isomerisation catalyst such as silica-alumina. In this particular case the n-heptane would be dehydrogenated by the platinum catalyst and the product isomerised to i-heptene which, in turn, would be hydrogenated to give finally i-heptane. When the hydrogenation and the isomerisation are carried out simultaneously the catalyst is said to act as a bifunctional catalyst.

Many organic reactions involving the upgrading of petroleum feedstocks are enhanced if a bifunctional catalyst is used. Some of the reactions that take place may be typified by one or more of the following:

1:
$$\mathbf{A} \xrightarrow{X} \mathbf{B} \xrightarrow{Y} \mathbf{C}$$

2:
$$\mathbf{A} \underset{}{\overset{X}{\rightleftharpoons}} \mathbf{B} \underset{}{\overset{Y}{\rightleftharpoons}} \mathbf{C}$$

3:
$$\mathbf{A} \overset{X}{\rightleftharpoons} \mathbf{B} \overset{\overset{\textstyle \mathbf{C}}{{\scriptstyle Y}\nearrow}}{\underset{{\scriptstyle X}\searrow}{\underset{\textstyle \mathbf{D}}{}}}$$

in which **A** represents the initial reactant, **B** and **D** unwanted products and **C** the desired product. **X** and **Y** represent hydrogenation and isomerisation catalysts respectively. These reaction schemes implicitly assume the participation of hydrogen and pseudo first order reaction kinetics. To a first approximation, the assumption of first order chemical kinetics is not unrealistic, for SINFELT[39] has shown that under some conditions many reactions involving upgrading of petroleum may be represented by first order kinetics. GUNN and THOMAS[40] examined mass transfer effects accompanying such reactions occurring iso-thermally in spherical catalyst particles containing the catalyst components **X** and **Y**. They demonstrated that it is possible to choose the volume fraction of **X** in such a way that the formation of **C** may be maximised. Curves 1, 2 and 3 in Fig. 2.9 indicate that a tubular reactor may be packed with discrete spherical particles of **X** and **Y** in such a way that the throughput of **C** is maximised. For given values of the kinetic constants of each step, the effective diffusivity, and the particle size, the amount of **C** formed is maximised by choosing the ratio of **X** to **Y** correctly. Curve 1 corresponds to the irreversible reaction 1 and requires more of component **X** than either reactions 2 or 3 for the same value of chosen parameters. On the other hand the mere fact that reaction 2 has a reversible step means that more of **Y** is required to produce the maximum throughput of **C**. When a second end-product **D** results, as in reaction 3, even more of catalyst component **Y** is needed if the formation of **D** is catalysed by **X**.

Even better throughputs of **C** result if components **X** and **Y** are incorporated in the same catalyst particle, rather than if they exist as separate particles. In effect, the intermediate product **B** no longer has to be desorbed from particles of the **X** type catalyst, transported through the gas phase and thence readsorbed on **Y** type particles prior to reaction. Resistance to intraparticle mass transfer is therefore reduced or eliminated by bringing **X** type catalyst sites into close proximity to **Y** type catalyst sites. Curve 4 in Fig. 2.9 illustrates this point and shows that for such a composite catalyst, containing both **X** and **Y** in the same particle, the throughput of **C** for reaction 3 is higher than it would have been had discrete particles of **X** and **Y** been used (curve 3).

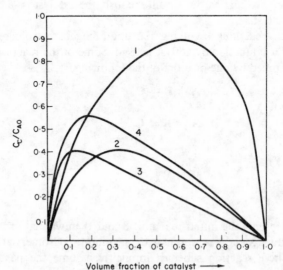

Volume fraction of catalyst ⟶

FIG. 2.9. Optimum yields of desired product in bifunctional catalyst systems

The yields of **C** found for reactions 1, 2 and 3 were evaluated with the restriction that the catalyst composition should be uniform along the reactor length. However, there seems to be no point in **Y** being present at the reactor inlet where there is no **B** to convert. Similarly, there is little point in much **X** being present at the reactor outlet where there may be only small amounts of **A** remaining unconverted to **B**. THOMAS and WOOD[48] examined this question for reaction schemes represented by reactions 1 and 2. For an irreversible consecutive reaction, such as 1, they showed analytically that the optimal profile consists of two catalyst zones, one containing pure **X** and the other containing pure **Y**. When one of the steps is reversible, as in 2, numerical optimisation techniques showed that a catalyst concentration profile diminishing along the reactor length is superior to either a constant catalyst composition or a two-zone reactor. However, it was also shown that, in most cases, a bifunctional catalyst of constant composition will give a yield which is close to the optimum, thus obviating the practical difficulty of packing a reactor with a catalyst having a variable composition along the reactor length. The exception noted was for reactions in which there may be a possibility of the desired product undergoing further unwanted side reactions. Under such circumstances the holding time should be restricted and a two-zone reactor, containing zones of pure **X** and pure **Y**, is to be preferred. Similar results were obtained independently by GUNN[46] who calculated, using a numerical method, the maximum yield of desired product for discrete numbers of reactor stages containing different proportions of the two catalyst components. JACKSON[49] has produced an analytical solution to the problem for the case of reaction 2. JENKINS and THOMAS[52] have demonstrated that, for the reforming of methylcyclopentane at 500°C (773 K) and 1 bar pressure, theoretical predictions of optimum catalyst compositions based on a kinetic model are close to the results obtained employing an experimental hill-climbing procedure.

Catalyst poisoning

Catalysts become poisoned when the feed stream to a reactor contains impurities which are deleterious to the activity of the catalyst. Particularly strong poisons are substances whose molecular structure contains lone electron pairs capable of forming covalent bonds with catalyst surfaces. Examples are ammonia, phosphine, arsine, hydrogen sulphide, sulphur dioxide and carbon monoxide. Other poisons include hydrogen, oxygen, halogens and mercury. The surface of a catalyst becomes poisoned by virtue of the foreign impurity adsorbed within the porous structure of the catalyst and covering a fraction of its surface, thus reducing the overall activity. The reactants participating in the desired reaction must now be transported to the unpoisoned part of the surface before any further reaction ensues, and so poisoning increases the average distance over which the reactants must diffuse prior to reaction at the surface. We may distinguish between two types of poisoning:

(i) homogeneous poisoning, in which the impurity is distributed evenly over the active surface;

Poisoned fraction ζ

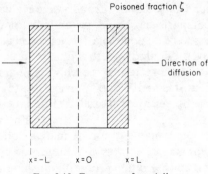

Direction of diffusion

x = –L   x = 0   x = L

FIG. 2.10. Geometry of partially
poisoned (selective) slab

(ii) selective poisoning in which an extremely active surface first becomes poisoned at the exterior surface of the particle, and then progressively becomes poisoned towards the centre of the particle.

When homogeneous poisoning occurs, since no reaction will be possible on the poisoned fraction (ζ, say) of active surface it is reasonable to suppose that the intrinsic activity of the catalyst decreases in proportion to the fraction of active surface remaining unpoisoned. To find the ratio of activity of the poisoned catalyst to the activity of an unpoisoned catalyst one would compare the stationary flux of reactant to the particle surface in each case. For a first order reaction occurring in a flat plate (slab) of catalyst one finds (see the example below) this ratio to be:

$$F = \frac{\sqrt{(1 - \zeta)}\tanh\{\phi\sqrt{(1 - \zeta)}\}}{\tanh \phi} \qquad \dots\dots(2.62)$$

where ϕ is the Thiele modulus for a first order reaction occurring in a flat plate of catalyst and ζ is the fraction of active surface poisoned. The two limiting cases of equation 2.62 correspond to extreme values of ϕ. Where ϕ is small F becomes equal to $(1 - \zeta)$ and the activity decreases linearly with the amount of poison added, as shown by curve 1 in Fig. 2.11. On the other hand when ϕ is large $F = \sqrt{(1 - \zeta)}$ and the activity decreases less rapidly than linearly due to the reactants penetrating the interior of the particle to a lesser extent for large values of the Thiele modulus (curve 2, Fig. 2.11).

Example

A fraction ζ of the active surface of some porous slab-shaped catalyst pellets becomes poisoned. The pellets are used to catalyse a first order isothermal chemical reaction. Find an expression for the ratio of the activity of the poisoned catalyst to the original activity of the unpoisoned catalyst when (a) homogeneous poisoning occurs, (b) selective poisoning occurs.

Solution

(a) If homogeneous poisoning occurs the activity decreases in proportion to the fraction $(1 - \zeta)$ of surface remaining unpoisoned. In the steady state the rate of reaction is equal to the flux of reactant to the surface. The ratio of activity F of the poisoned slab to the unpoisoned slab will be equal to the ratio of the reactant fluxes under the respective conditions. Hence:

$$F = \frac{\left(\dfrac{dC}{dx}\right)'_{x=L}}{\left(\dfrac{dC}{dx}\right)_{x=L}}$$

where the prime denotes conditions in the poisoned slab.

Now from equation 2.18 the concentration of reactant is a function of the distance the reactant has penetrated the slab. Thus:

$$C = C_\infty \frac{\cosh(\lambda x)}{\cosh(\lambda L)} \quad \text{where} \quad \lambda = \sqrt{\frac{k}{D_e}}.$$

If the slab were poisoned the activity would be $k(1 - \zeta)$ rather than k and then:

$$C = C_\infty \frac{\cosh(\lambda' x)}{\cosh(\lambda' L)} \quad \text{where} \quad \lambda' = \sqrt{\frac{k(1 - \zeta)}{D_e}}$$

Evaluating the respective fluxes at $x = L$:

$$F = \frac{\sqrt{(1 - \zeta)}\,\tanh\{\phi\sqrt{(1 - \zeta)}\}}{\tanh\phi} \quad \text{where} \quad \phi = L\sqrt{\frac{k}{D_e}}$$

(b) When selective poisoning occurs the exterior surface of the porous pellet becomes poisoned initially and the reactants must then be transported to the unaffected interior of the catalyst before reaction may ensue. When the reaction rate in the unpoisoned portion is chemically controlled the activity merely falls off in proportion to the fraction of surface poisoned. However if the reaction is diffusion limited, in the steady state the flux of reactant past the boundary between poisoned and unpoisoned surfaces is equal to the chemical reaction rate (see Fig. 2.14). Thus:

$$\text{Flux of reactant at the boundary between poisoned and unpoisoned portion of slab} = D_e \frac{C_\infty - C_L}{\zeta L}$$

$$\text{Reaction rate in unpoisoned length } (1 - \zeta) L = D_e \left(\frac{dC}{dx}\right)_{x=(1-\zeta)L}$$

The concentration profile in the unpoisoned length is, by analogy with equation 2.18:

$$C = C_L \frac{\cosh(\lambda x)}{\cosh\{\lambda(1 - \zeta) L\}}$$

therefore:

$$D_e \left(\frac{dC}{dx}\right)_{x=(1-\zeta)L} = \frac{D_e}{L} C_L \phi \tanh\{\phi(1 - \zeta)\}$$

where

$$\phi = \lambda L = L\sqrt{\frac{k}{D_e}}.$$

In the steady state then:

$$\frac{D_e(C_\infty - C_L)}{\zeta L} = \frac{D_e}{L} C_L \phi \tanh\{\phi(1 - \zeta)\}$$

Solving the above equation explicitly for C_L:

$$C_L = \frac{C_\infty}{1 + \phi\zeta \tanh\{\phi(1 - \zeta)\}}$$

Hence the rate of reaction in the partially poisoned slab is:

$$\frac{D_e(C_\infty - C_L)}{L} = \frac{C_\infty D_e}{L} \frac{\phi \tanh\{\phi(1 - \zeta)\}}{1 + \phi\zeta \tanh\{\phi(1 - \zeta)\}}$$

In an unpoisoned slab the reaction rate is $\left(\dfrac{C_\infty D_e}{L}\right)\phi \tanh \phi$ and so:

$$F = \frac{\tanh\{\phi(1 - \zeta)\}}{1 + \phi\zeta \tanh\{\phi(1 - \zeta)\}} \frac{1}{\tanh \phi}$$

For an active catalyst, $\tanh\{\phi(1 - \zeta)\} \to 1$ and then $F \to (1 + \phi\zeta)^{-1}$.

Selective poisoning occurs with very active catalysts. Initially, the exterior surface is poisoned and then, as more poison is added, an increasing depth of the interior surface becomes poisoned and inaccessible to reactant. If the reaction rate in the unpoisoned portion of catalyst happens to be chemically controlled, the reaction rate will fall off directly in proportion to the fraction of surface poisoned and the activity decreases linearly with the amount of poison added (curve 1, Fig. 2.11). When the reaction is diffusion influenced, in the steady state the flux of reactant past the boundary between the poisoned and unpoisoned parts of the surface will equal the reaction rate in the unpoisoned portion. For the slab model, as the foregoing example shows, the ratio of activity in a poisoned catalyst to that in an unpoisoned catalyst is

$$F = \frac{\tanh\{\phi(1 - \zeta)\}}{1 + \phi\zeta \tanh\{\phi(1 - \zeta)\}} \frac{1}{\tanh \phi} \qquad \dots(2.63)$$

For large values of the Thiele modulus the fraction $\phi(1 - \zeta)$ will usually be sufficiently large that $F = (1 + \phi\zeta)^{-1}$. Curve 3 in Fig. 2.11 depicts selective poisoning of active catalysts near the particle exterior and is the function represented by equation 2.63. Curve 4 describes the effect of selective poisoning for large values of the Thiele modulus. For the latter case the activity decreases drastically, after only a small amount of poison has been added.

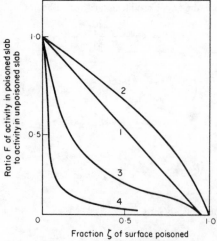

Fraction ζ of surface poisoned

FIG. 2.11. Catalyst poisoning

It is apparent from the above discussion that ζ is time dependent, for an increasing amount of poison is being added quite involuntarily as the impure feed continually flows to the catalyst contained in the reactor. The general problem of obtaining $\zeta(t)$ is complex but, in principle, can be treated by solving the unsteady state conservation equation for the poison. When the poison is strongly and rapidly adsorbed, the fraction of surface poisoned is dependent on the square root of the time for which the feed has been flowing. In general, if $\zeta(t)$ is known, a judgement can be made concerning the optimum time for which the catalyst may be used before its activity falls to a value which produces an uneconomical throughput or yield of product[30, 44, 45, 50].

MASS TRANSFER FROM THE FLUID STREAM TO A SOLID SURFACE

Under some circumstances there will be a resistance to the transport of material from the bulk fluid stream to the exterior surface of a catalyst particle. When such a resistance to mass transfer exists, the concentration C of a reactant in the bulk fluid will differ from its concentration C_i at the solid-gas interface. Because C_i is usually unknown it is necessary to eliminate it from the rate equation describing the external mass transfer process. Since, in the steady state, the rates of all of the steps in the process are equal, it is possible to obtain an overall rate expression in which C_i does not appear explicitly. The algebraic manipulation of equations 2.1, 2.2 and 2.3 leading to the overall rate given by equation 2.4 is a simple example of how an interface concentration can be eliminated from the rate expression.

Based on such analyses, which of course do imply a film model in which the resistance to mass transfer is supposed to be confined to a film of finite thickness (see Volume 1, Chapter 8), it is possible to estimate the effect which mass transport external to the solid surface has on the overall reaction rate. For equimolar counterdiffusion of a component A in the gas phase, the rate of transfer of A from the bulk gas to the interface can be expressed as:

$$N_A = k_G(P_A - P_{Ai}) = h_D(C_A - C_{Ai}) \qquad \ldots .(2.64)$$

where k_G is the gas film mass transfer coefficient per unit external surface area, and P_A and P_{Ai} are the partial pressure of the component A in the bulk gas and at the interface respectively.

The right-hand side of equation 2.64 contains the mass transfer coefficient h_D which is used if the driving force is expressed in terms of gas concentrations. Because of the stoichiometric demands imposed by chemical reaction, equimolar counterdiffusion of components may not necessarily occur. In the latter case the driving force must be divided by the drift factor[2] which is the logarithmic mean value of $(P + \delta_A P_A)$ and of $(P + \delta_A P_{Ai})$,

where P is the total pressure,
$\quad$ P_A is the partial pressure of component A of the gas mixture,

P_{Ai} is the value of P_A at the interface, and

δ_A is the net difference in number of mols of product and reactant per mol of component A.

The transfer coefficient can be correlated in the form of a dimensionless Sherwood number $Sh. (= h_D d_p/D)$. The particle diameter d_p is often taken to be the diameter of the sphere having the same area as the (irregular shaped) pellet. THALLER and THODOS[28] correlated the mass transfer coefficient in terms of the gas velocity u and the Schmidt number $Sc. (= \mu/\rho D)$:

$$j_d = \frac{Sh.}{Re.Sc.^{\frac{1}{3}}} = \frac{h_D}{u} Sc.^{\frac{1}{3}} \qquad \qquad \ldots.(2.65)$$

where j_d is the mass transfer factor (see Volume 1, Chapter 8), and

Re. is the Reynolds number $u\rho d_p/\mu$ based on the particle diameter.

The mass transfer factor has also been correlated as a function of the Reynolds number only and therefore takes account of hydrodynamic conditions. If e is the voidage of the packed bed and the total volume occupied by all of the catalyst pellets is V_p, then the total reactor volume is $V_p/(1 - e)$. Hence the rate of mass transfer of component A per unit volume of reactor is $N_A S_x(1 - e)/V_p$. If we now consider a case in which only external mass transfer controls the overall reaction rate we have:

$$\mathscr{R}_V = \frac{N_A S_x(1 - e)}{V_p} = (1 - e)\frac{S_x}{V_p} h_D(C_A - C_{Ai}) \qquad \ldots.(2.66)$$

Alternatively, equation 2.66 may be written in terms of the j-factor. The unknown interface concentration C_{Ai} can now be eliminated in the usual way, by equating the rate of mass transfer to the rate of chemical reaction.

We see that, in principle, the overall reaction rate can be expressed in terms of coefficients such as the reaction rate constant and the mass transfer coefficient. To be of any use for design purposes, however, we must have knowledge of these parameters. By measuring the kinetic constant in the absence of mass transfer effects and using correlations to estimate the mass transfer coefficient we are really implying that these estimated parameters are independent of one another. This would only be true if each element of external surface behaved kinetically as all other surface elements. Such conditions are only fulfilled if the surface is uniformly accessible. It is fortuitous, however, that predictions of overall rates based on such assumptions are often within the accuracy of the kinetic information, and for this reason values of k and h_D obtained independently are frequently employed for substitution into overall rate expressions.

It is important to realise that the resistance to mass transfer within a porous catalyst is greater than the resistance to mass transfer from the bulk phase to the solid-gas interface. This may be demonstrated by considering a first order reaction in a slab of catalyst material and supposing that there is no convective transport of mass to the catalyst surface. Clearly this will represent a limiting case since, if there were convective transport to the surface, transfer would be enhanced and there would be a smaller gas phase resistance. The rate of reaction per unit external surface area is $-D_e \left(\dfrac{dC}{dx}\right)_{x=L}$. This may be found by differentiating the equation representing the concentration profile within the slab, and finding the value at $x = L$ corresponding to the slab surface. Since the concentration profile within the slab is $C = C_i \cosh(\lambda x)/\cosh(\lambda L)$ (see equation 2.18), we have from equation 2.20 that:

$$-D_e \left(\frac{dC}{dx}\right)_{x=L} = C_i D_e \lambda \tanh(\lambda L) = C_i \frac{D_e}{L} \phi^2 \eta \qquad \ldots (2.67)$$

Because we are assuming that there is resistance to mass transfer in the gas phase we must now distinguish between the concentration C_i at the slab surface and that in the bulk gas. The quantity $D_e \phi^2 \eta / L$ is really equivalent to a rate constant per unit surface area k_a so that the rate per unit area is $k_a C_i$. Since the interface concentration C_i is unknown we have to express it in terms of the bulk gas concentration C_∞ before this result can be of any use. To do this we consider the unidirectional transfer of the component in question from the bulk gas to the interface. If no reaction occurs in the gas phase a steady state material balance over an element of gas volume gives the Laplace equation:

$$D \frac{d^2 C}{dx^2} = 0 \qquad \ldots (2.68)$$

where D is the diffusion coefficient of the component through the gas phase. The physical boundary conditions are best dealt with by a change of variables. The point $x = 0$ is conveniently taken as that point along the abscissa of coordinates beyond which the gas concentration is that of the bulk gas C_∞. We suppose that the interface is a distance δ (the film thickness) away from this plane. The boundary conditions are then:

$$x = 0, \qquad\qquad C = C_\infty \qquad \ldots (2.69)$$

$$x = \delta, \; -D\left(\frac{dC}{dx}\right)_{x=\delta} = k_a C_i \qquad \ldots (2.70)$$

The second boundary condition merely asserts that, in the steady state, the flux to the surface equals the chemical reaction rate at the interface. The solution to equation 2.68 with these boundary conditions is:

$$C - C_\infty = \frac{-k_a C_i}{D} x \qquad \ldots (2.71)$$

At the interface ($x = \delta$) therefore, where the gas concentration is C_i:

$$\frac{C_i}{C_\infty} = \frac{1}{1 + k_a \delta / D} \qquad \ldots (2.72)$$

F

Equation 2.72 now expresses C_i in terms of the bulk concentration C_∞. The rate of reaction per unit external area is thus:

$$k_a C_i = \frac{C_\infty}{\dfrac{1}{k_a} + \dfrac{\delta}{D}} \qquad \qquad \ldots\ldots(2.73)$$

Note that equation 2.73 is similar in form to equation 2.4. Thus $1/k_a$ may be regarded as the resistance to conversion by chemical reaction and δ/D as the resistance to transport through the gas phase. From the discussion following equation 2.67 we recall that $\dfrac{D_e \phi^2 \eta}{L}\cdot$ is equivalent to k_a; the term $k_a \, \delta/D$ may therefore be rewritten $(\delta/L)(D_e/D)\phi^2\eta$ and so equation 2.72 may be expressed in the alternative form:

$$\frac{C_i}{C_\infty} = Tr. = \frac{1}{(\delta/L)(D_e/D)\phi^2\eta + 1} \qquad \ldots\ldots(2.74)$$

where the ratio of C_i/C_∞ may be referred to as the transport factor $Tr.$. Equation 2.74 enables us to compare the transport factor with the effectiveness factor η. Since the ratio of the porosity to the tortuosity is less than unity, equation 2.6 implies that $D_e < D$. For small values of ϕ the effectiveness factor η becomes unity and equation 2.74 shows that $Tr. \to 1$ when ϕ becomes sufficiently small. On the other hand at large values of ϕ the effectiveness factor $\eta \to 1/\phi$ and $Tr. \to (LD/\delta D_e)\eta$. By substituting $\eta = \tanh \phi/\phi$ intermediate values can be estimated. The conclusion is that $Tr. \geqslant \eta$. Because $Tr.$ really represents the ratio of the reaction rate when there is a resistance to mass transfer in the gas phase to the rate when there is no resistance, $Tr. \geqslant \eta$ implies that resistance to diffusion through the gas phase is less than resistance to diffusion within the porous slab except when ϕ is small for which condition both η and $Tr.$ approach unity. This may be generalised to include convective transport through the gas phase by substituting h_D for D/δ in equation 2.72. If h_D is then expressed in terms of the Sherwood number $Sh.$ $(= h_D L/D)$ and $D_e \phi^2 \eta/L$ substituted for k_a from equation 2.67 then the transport factor becomes:

$$Tr. = \frac{1}{1 + k_a/h_D} = \frac{1}{1 + \left(\dfrac{D_e \phi^2 \eta}{D Sh.}\right)} \qquad \ldots\ldots(2.75)$$

With increase in flux past the particle the value of $Sh.$ increases since it is a function of the Reynolds number. Hence, as convective transport of mass to the particle becomes more important with increased flowrate, the transport resistance becomes less significant.

CHEMICAL KINETICS OF HETEROGENEOUS CATALYTIC REACTIONS

To complete the discussion of factors involved in the design of gas–solid heterogeneous catalytic reactors we will examine several aspects of the kinetics of chemical reactions occurring in the presence of a catalyst surface. We consider, for heuristic purposes, the equilibrium reaction:

$$\mathbf{A} + \mathbf{B} \underset{\mathbf{k}}{\overset{\mathbf{k}}{\rightleftharpoons}} \mathbf{P}$$

occurring at an active catalyst surface. The net rate* of surface chemical reaction will be the difference between the rates of the forward and reverse reactions. If the forward rate is determined by the simultaneous presence of chemisorbed **A** and **B** (i.e. a Langmuir–Hinshelwood mechanism applies) then it will be proportional to the number of pairs of adjacent catalyst sites occupied by **A** and **B**. This is given by the product of the number n_A of adsorbed reactant molecules and the ratio of occupied to unoccupied sites. This latter quantity is expressed as $\theta_B/(1 - \theta_A - \theta_B)$ where θ_A and θ_B represent the fractions of active surface covered by **A** and **B** respectively. The reverse rate, on the other hand, is proportional to the product of the number n_P of adsorbed product molecules and the fraction of active sites left vacant $(1 - \sum_j \theta_j)$ where $\sum_j \theta_j$ is the total fraction of sites occupied by any species. The net rate of the surface reaction per unit surface is thus,

$$\mathscr{R}_S'' = \mathbf{k} n_A \left(\frac{\theta_B}{1 - \theta_A - \theta_B} \right) - \mathbf{k}' n_P (1 - \sum_j \theta_j) \qquad \ldots (2.76)$$

We may conveniently define n_A/S', where S' is the surface area of the catalyst, as the surface concentration C_A. The corresponding surface concentration, C_B, of **B** is defined as n_B/S' and the total concentration, C_S, of catalyst sites as n_S/S' where n_S is the number of available active sites. The fraction θ_A of surface covered by **A** will be n_A/n_S, which, in terms of surface concentrations becomes C_A/C_S. Similar substitutions may be made for any other species so the term $(1 - \sum_j \theta_j)$ representing the fraction of vacant sites can be written C_V/C_S where C_V is $(C_S - \sum_j C_j)$. Substituting these quantities into equation 2.76† gives:

$$\mathscr{R}_S'' = \mathbf{k}S' \frac{C_A C_B}{C_S} - \mathbf{k}'S' \frac{C_P C_V}{C_S} = \mathbf{k}_S \frac{C_A C_B}{C_S} - \mathbf{k}_S' \frac{C_P C_V}{C_S} \qquad \ldots (2.77)$$

provided the surface coverage is sufficiently small that $1 \gg (\theta_A + \theta_B)$, which condition is usually fulfilled during catalysis. If the forward rate of reaction is determined by gaseous or physically adsorbed **B** interacting with chemisorbed **A** (an Eley–Rideal mechanism) the expression corresponding to equation 2.77 would be:

$$\mathscr{R}_S'' = \mathbf{k}S' C_A a_B - \mathbf{k}'S' \frac{C_P C_V}{C_S} = \mathbf{k}_S C_A a_B - \mathbf{k}_S' \frac{C_P C_V}{C_S} \qquad \ldots (2.78)$$

where a_B is the (thermodynamic) activity of gaseous or physically adsorbed **B**.

---

* If the rate is measured as the number of molecules transformed per unit area per unit time, **k** and **k'** have units $\mathbf{L}^{-2}\mathbf{T}^{-1}$.

† The rate constants $\mathbf{k}_S$ and $\mathbf{k}_S'$ in equations 2.77 and 2.78 have been substituted in place of $\mathbf{k}S'$ and $\mathbf{k}'S'$ respectively and, since the rate of surface reaction $\mathscr{R}_S''$ has units $\mathbf{L}^{-2}\mathbf{T}^{-1}$, $\mathbf{k}_S$ and $\mathbf{k}_S'$ have units $\mathbf{T}^{-1}$.

As well as the net rate of surface reaction we must also consider the net rate of each reactant at the surface. We may think of this as the difference between the rates of adsorption and desorption of both **A** and **B**. Since the rate of adsorption is usually expressed in terms of the product of the prevailing partial pressure of adsorbate and the extent of free surface, while the rate of desorption is proportional to the fraction of surface covered, the net rate of adsorption of **A** in molecules per unit area will be:

$$\mathscr{R}''_{Aa} = k_{Aa} P_A (1 - \sum_j \theta_j) - k_{Ad} \theta_A \qquad \ldots (2.79)$$

which in terms of surface concentrations becomes:

$$\mathscr{R}''_{Aa} = k_{Aa} P_A \frac{C_V}{C_S} - k_{Ad} \frac{C_A}{C_S} \qquad \ldots (2.80)$$

A similar equation can be written for the reactant **B**.

Since **P** is the final product of reaction it is also necessary to consider the net rate of desorption. In an analogous way to the net rate of adsorption of **A**, the net rate of desorption of **P** may be written:

$$\mathscr{R}''_{Pd} = k_{Pd} \frac{C_P}{C_S} - k_{Pa} P_P \frac{C_V}{C_S} \qquad \ldots (2.81)$$

It should be noted that the partial pressures in equations 2.80 and 2.81 are those corresponding to the values at the interface between solid and gas. If there were no resistance to transport through the gas phase then, as discussed in an earlier section, the partial pressures will correspond to those in the bulk gas.

The overall chemical rate may be written in terms of the partial pressures of **A**, **B** and **P** by equating the rates $\mathscr{R}''_S$, $\mathscr{R}''_{Aa}$ and $\mathscr{R}''_{Pd}$ and eliminating the surface concentrations C_A, C_B and C_P from equations 2.77 (or 2.78 as the case may be), 2.80 and 2.81. The final equation so obtained is cumbersome and unwieldy and contains several constants which, for practical reasons, cannot be determined independently. For this reason it is convenient to consider limiting cases in which either surface reaction, adsorption or desorption is rate determining.

Adsorption of a reactant as the rate determining step

If the adsorption of **A** is the rate determining step in the sequence of adsorption, surface reaction and desorption processes, then equation 2.80 will be the appropriate equation to use for expressing the overall chemical rate. To be of use, however, it is first necessary to express C_A, C_V and C_S in terms of the partial pressures of reactants and products. To do this an approximation is made; it is assumed that all processes except the adsorption of **A** are at equilibrium. Thus the processes involving **B** and **P** are in a state of pseudo-equilibrium. The surface concentration of **B** can therefore be expressed in terms of an equilibrium constant K_B for the adsorption–desorption equilibrium of **B**:

$$K_B = \frac{C_B}{P_B C_V} \qquad \ldots (2.82)$$

and similarly for P:

$$K_P = \frac{C_P}{P_P C_V} \qquad \qquad ....(2.83)$$

The equation for the surface reaction between **A** and **B** may be written in terms of the equilibrium constant for the surface reaction:

$$K_S = \frac{C_P C_V}{C_A C_B} \qquad \qquad ....(2.84)$$

Substituting equations 2.82, 2.83 and 2.84 into equation 2.80:

$$\mathscr{R}''_{Aa} = \frac{k_{Aa} C_V}{C_S} \left\{ P_A - \frac{K_P}{K_A K_B K_S} \frac{P_P}{P_B} \right\} \qquad \qquad ....(2.85)$$

where K_A is the ratio k_{Aa}/k_{Pd}. The ratio C_V/C_S can be written in terms of partial pressures and equilibrium constants since the concentration of vacant sites is the difference between the total concentration of sites and the sum of the surface concentrations of **A**, **B** and **P**:

$$C_V = C_S - (C_A + C_B + C_P) \qquad \qquad ....(2.86)$$

and so, from equations 2.82, 2.83 and 2.84:

$$\frac{C_V}{C_S} = \frac{1}{1 + K_B P_B + K_P P_P + \dfrac{K_P}{K_B K_S} \dfrac{P_P}{P_B}} \qquad \qquad ....(2.87)$$

If we define the equilibrium constant for the adsorption of **A** as:

$$K_A = \frac{C_A}{P_A C_V} \qquad \qquad ....(2.88)$$

and recognise that the thermodynamic equilibrium constant for the overall equilibrium is:

$$K = \frac{P_P}{P_A P_B} \qquad \qquad ....(2.89)$$

equation 2.85 gives, for the rate of chemical reaction:*

$$\mathscr{R}''_A = \frac{k_{Aa} \left\{ P_A - \dfrac{1}{K} \dfrac{P_P}{P_B} \right\}}{1 + K_B P_B + K_P P_P + \dfrac{K_A}{K} \dfrac{P_P}{P_B}} \qquad \qquad ....(2.90)$$

It should be noted that equation 2.90 contains a driving force term in the numerator. This is the driving force tending to drive the chemical reaction toward the equilibrium state. The collection of terms in the denominator is usually referred to as the adsorption term, since terms such as $K_B P_B$ represent the retarding effect of the adsorption of species **B** on the rate of disappearance of **A**.

---

* The rate $\mathscr{R}''_A$ expressed in units $L^{-2} T^{-1}$ must be transposed to units of mols per unit time per unit catalyst mass for reactor design purposes. This is accomplished by multiplying $\mathscr{R}''_A$ by S_g/N, where N is Avogadro's number.

New experimental techniques enable the constants K_B etc. to be determined separately during the course of a chemical reaction[27] and hence, if it were found that the adsorption of **A** controls the overall rate of conversion, equation 2.90 could be used directly as the rate equation for design purposes. If, however, external mass transfer were important the partial pressures in equation 2.90 would be values at the interface and an equation (such as equation 2.66) for each component would be required to express interfacial partial pressures in terms of bulk partial pressures. If internal diffusion were also important, the overall rate equation would be multiplied by an effectiveness factor either estimated experimentally, or alternatively obtained by theoretical considerations similar to those discussed earlier.

Surface reaction as the rate determining step

If, on the other hand, surface reaction determined the overall chemical rate, equation 2.77 (or 2.78 if an Eley–Rideal mechanism operates) would represent the rate. If it is assumed that a pseudo-equilibrium state is reached for each of the adsorption–desorption processes then, by a similar method to that already discussed for reactions where adsorption is rate determining, it can be shown that the rate of chemical reaction is (for a Langmuir–Hinshelwood mechanism):

$$\mathscr{R}''_{AS} = \frac{C_S k_S K_A K_B \left\{ P_A P_B - \frac{1}{K} P_P \right\}}{(1 + K_A P_A + K_B P_B + K_P P_P)^2} \qquad \ldots (2.91)$$

This equation also contains a driving force term and an adsorption term. A similar equation may be derived for the case of an Eley–Rideal mechanism and its form is interpreted in Table 2.2.

Desorption of a product as the rate determining step

In this case pseudo-equilibrium is assumed for the surface reaction and for the adsorption–desorption processes involving **A** and **B**. By similar methods to those employed in deriving equation 2.90, it can be shown that the rate of chemical reaction is:

$$\mathscr{R}''_{Pd} = \frac{k_{Pa} K \left\{ P_A P_B - \frac{1}{K} P_P \right\}}{1 + K_A P_A + K_B P_B + K K_P P_A P_B} \qquad \ldots (2.92)$$

when the desorption of **P** controls the rate. This equation, it should be noted, also contains a driving force term and an adsorption term.

General remarks in relation to rate determining steps for other mechanisms

In principle it is possible to write down the rate equation for any rate determining chemical step assuming any particular mechanism. To take a specific example, the overall rate may be controlled by the adsorption of **A** and the reaction may involve the dissociative adsorption of **A**, only half of which then reacts with adsorbed **B** by a Langmuir–Hinshelwood mechanism. The basic rate equation which represents such a process can be transposed into an equivalent expression in terms of partial pressures and equilibrium constants by methods similar to those employed to obtain the rate equations 2.90, 2.91 and

TABLE 2.2. *Structure of reactor design equations*

Reaction	Mechanism	Driving force	Adsorption term
1. $A \rightleftharpoons P$	(i) Adsorption of A controls rate	$P_A - \dfrac{P_P}{K}$	$1 + \dfrac{K_A}{K}P_P + K_P P_P$
(Equilibrium constant K is dimensionless)	(ii) Surface reaction controls rate, single site mechanism	$P_A - \dfrac{P_P}{K}$	$1 + K_A P_A + K_P P_P$
	(iii) Surface reaction controls rate, adsorbed A reacts with adjacent vacant site	$P_A - \dfrac{P_P}{K}$	$(1 + K_A P_A + K_P P_P)^2$
	(iv) Desorption of P controls rate	$P_A - \dfrac{P_P}{K}$	$1 + KK_P P_A + K_A P_A$
	(v) A dissociates when adsorbed and adsorption controls rate	$P_A - \dfrac{P_P}{K}$	$\left\{1 + \left(\dfrac{K_A}{K}P_P\right)^{\frac{1}{2}} + K_P P_P\right\}^2$
	(vi) A dissociates when adsorbed and surface reaction controls rate	$P_A - \dfrac{P_P}{K}$	$\{1 + (K_A P_A)^{\frac{1}{2}} + K_P P_P\}^2$
2. $A + B \rightleftharpoons P$	Langmuir-Hinshelwood mechanism (adsorbed A reacts with adsorbed B)		
(Equilibrium constant K has dimensions $M^{-1}LT^2$)	(i) Adsorption of A controls rate	$P_A - \dfrac{P_P}{KP_B}$	$1 + \dfrac{K_A P_P}{KP_B} + K_B P_B + K_P P_P$
	(ii) Surface reaction controls rate	$P_A P_B - \dfrac{P_P}{K}$	$\{1 + K_A P_A + K_B P_B + K_P P_P\}^2$
	(iii) Desorption of P controls rate	$P_A P_B - \dfrac{P_P}{K}$	$1 + K_A P_A + K_B P_B + KK_P P_A P_B$

Footnote:
1. The expression for the rate of reaction in terms of partial pressures is proportional to the driving force divided by the adsorption term.
2. To derive the corresponding kinetic expressions for a bimolecular-unimolecular reversible reaction proceeding via an Eley–Rideal mechanism (adsorbed **A** reacts with gaseous or physically adsorbed **B**), the term $K_B P_B$ should be omitted from the adsorption term. When the surface reaction controls the rate the adsorption term is not squared and the term $K_B P_B$ is omitted.
3. To derive the kinetic expression for irreversible reactions simply put K large, thus omitting the second term in the driving force.
4. If two products are formed (**P** and **Q**) the equations are modified by (a) multiplying the second term of the driving force by P_Q, and (b) adding $K_Q P_Q$ within the bracket of the adsorption term.
5. If **A** or **B** dissociates during the process of chemisorption, both the driving force and the adsorption term should be modified (see for example cases 1(v) and 1(vi) above). For a full discussion of such situations see HOUGEN and WATSON[2].

2.92. Table 2.2 contains a number of selected mechanisms for each of which the basic rate equation, the driving force term and the adsorption term are given.

A useful empirical approach to the design of heterogeneous chemical reactors often consists of selecting a suitable equation, such as one in Table 2.2, which with numerical values substituted for the kinetic and equilibrium constants represents the chemical reaction in the absence of mass transfer effects. Graphical methods are often employed to aid the selection of an appropriate equation[43] and the constants determined by a least squares approach[2]. It is important to stress, however, that while the equation selected may well represent the experimental data, it does not necessarily represent the true mechanism of reaction; as is evident from Table 2.2, many of the equations are similar in form. Nevertheless the equation will be sufficient to represent the behaviour of the reaction for the conditions investigated and can be used for design purposes. A fuller and more detailed account of a wide selection of mechanisms has been given by HOUGEN and WATSON[2].

DESIGN CALCULATIONS

The problems which a chemical engineer has to solve when contemplating the design of a chemical reactor packed with a catalyst or reacting solid are, in principle, similar to those encountered during the design of an empty reactor, except that the presence of the solid somewhat complicates the material and heat balance equations. The situation is further exacerbated by the designer having to predict and avoid those conditions which might lead to instability within the reactor.

We shall consider, in turn, the various problems which have to be faced when designing isothermal, adiabatic and other non-isothermal tubular reactors, and we shall also briefly discuss fluidised bed reactors. Problems of instability arise when inappropriate operating conditions are chosen and when reactors are started up. A detailed discussion of this latter topic is outside the scope of this chapter but, since reactor instability is undesirable, we shall briefly inspect the problems involved.

Packed tubular reactors

The effect which the solid packing has on the flow pattern within a tubular reactor can sometimes be of sufficient magnitude to cause significant departures from plug flow conditions. The presence of solid particles in a tube causes elements of flowing gas to become displaced randomly and therefore produces a mixing effect. An eddy diffusion coefficient can be ascribed to this mixing effect and becomes superimposed on the transport processes which normally occur in unpacked tubes—either a molecular diffusion process at fairly low

Reynolds numbers or eddy motion due to turbulence at high Reynolds numbers. Both transverse and longitudinal components of the flux attributed to this dispersion effect are of importance but operate in opposite ways. Transverse dispersion tends to bring the performance of the reactor closer to that which would be predicted by the simple design equation based upon plug flow. On the other hand, longitudinal dispersion is inclined to invalidate the plug flow assumptions so that the conversion would be less than would be expected if plug flow conditions obtained. The reason for this is that transverse mixing of the fluid elements helps to smooth out the parabolic velocity profile which normally develops in an unpacked tube, whereas longitudinal dispersion in the direction of flow causes some fluid elements to spend less time in the reactor than they would if this additional component of flux due to eddy motion were not superimposed.

The magnitude of the dispersion effect due to transverse or radial mixing can be assessed by relying on theoretical predictions[19] and experimental observations[22] which assert that the value of the Peclet number $Pe.$ ($= ud_p/D$, where d_p is the particle diameter) for transverse dispersion in packed tubes is approximately 10. At Reynolds numbers of around 100 the diffusion coefficient to be ascribed to radial dispersion effects is about four times greater than the value for molecular diffusion. At higher Reynolds numbers the radial dispersion effect is correspondingly larger.

Longitudinal dispersion in packed reactors is thought to be caused by interstices between particles acting as mixing chambers. Theoretical analysis of a model[21] based on this assumption shows that the Peclet number $Pe._l$ for longitudinal dispersion is about 2 and this has been confirmed by experiment[16, 22]. Thus the diffusion coefficient for longitudinal dispersion is approximately five times that for transverse dispersion for the same flow conditions. The flux which results from the longitudinal dispersion effect is, however, usually much smaller than the flux resulting from transverse dispersion because axial concentration gradients are very much less steep than radial concentration gradients if the ratio of the tube length to diameter is large. Whether or not longitudinal dispersion is important depends on the ratio of the reactor length to particle size. If the ratio is less than about 100 then the flux resulting from longitudinal dispersion should be considered in addition to the flux due to bulk flow when designing the reactor.

Isothermal conditions

The design equation for the isothermal fixed bed tubular reactor with no longitudinal dispersion effects represents the simplest form of reactor to analyse. No net exchange of mass or energy occurs in the radial direction, so transverse dispersion effects can be neglected. If we also suppose that the ratio of the tube length to particle size is large then we can safely ignore longitudinal dispersion effects compared with the effect of bulk flow. Hence, in writing the conservation equation over an element dz of the length of the reactor (Fig. 2.12), we may consider that the fluid velocity u is independent of radial position; this implies a flat velocity profile (plug flow conditions) and ignores dispersion effects in the direction of flow.

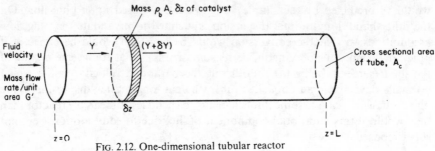

FIG. 2.12. One-dimensional tubular reactor

Suppose that a mass of catalyst, whose bulk density is ρ_b, is contained within an elementary length δz of a reactor of uniform cross-section A_c. The mass of catalyst occupying the elementary volume will therefore by $\rho_b A_c \delta z$. Let G' be the steady-state mass flow rate per unit area to the reactor. If the number of mols of product per unit mass of fluid entering the elementary section is Y and the amount emerging is $(Y + \delta Y)$, the net difference in mass flow per unit area across the element will be $G' \delta Y$. In the steady state this will be balanced by the amount of product formed by chemical reaction within the element so:

$$G' \delta Y = \mathscr{R}_Y^M \rho_b \delta z \qquad \qquad \dots (2.93)$$

where the isothermal reaction rate $\mathscr{R}_Y^M$ is expressed in units of mols per unit time per unit mass of catalyst*. To calculate the length of reactor required to achieve a conversion corresponding to an exit concentration Y_L equation 2.93 is integrated from Y_0 (say) to Y_L, where the subscripts 0 and L refer to inlet and exit conditions respectively. Thus:

$$L = \frac{G'}{\rho_b} \int_{Y_0}^{Y_L} \frac{\mathrm{d}Y}{\mathscr{R}_Y^M} \qquad \qquad \dots (2.94)$$

The mass of catalyst W contained within this length will be $\rho_b L A_c$ while the residence time, clearly, will be $e\rho L/G'$ where ρ is the fluid density under the conditions of reaction. Replacing L by $W/\rho_b A_c$, we see that the residence time is $e\rho W/\rho_b G' A_c$, i.e. the void volume eW/ρ_b divided by the volumetric flow rate $G' A_c/\rho$. It is equivalent to the quantity V/v for an unpacked tubular reactor, where V is the reactor volume and v the volumetric flow rate. The reciprocal of the residence time $\rho_b G' A_c/e\rho W$ is the space velocity, useful for comparing reactor performance; the physical significance of this is simply the rate of replacement of fluid within the void volume of the reactor.

* If intraparticle diffusion effects are important and an effectiveness factor η is employed (as in equation 2.13) to correct the chemical kinetics observed in the absence of transport effects, then it is necessary to adopt a stepwise procedure for solution. First the pellet equations (such as 2.15) are solved in order to calculate η for the entrance to the reactor and then the reactor equation (2.93) may be solved in difference form, thus providing a new value of Y at the next increment along the reactor. The whole procedure may then be repeated at successive increments along the reactor.

If the reaction rate is a function of total pressure as well as concentration, and there is a pressure drop along the reactor due to the solid packing, the conversion within the reactor will be affected by the drop in total pressure along the tube. In most cases it is perfectly reasonable to neglect the change in momentum flux due to the expansion of the fluid in comparison with the pressure force and the component of the drag force in the direction of flow created by the surface of the solid particles. The pressure drop dP along an elementary length dz of the packed bed may thus be written in terms of R_1, the component of the drag force per unit surface area of particles in the direction of flow:

$$-\frac{dP}{dz} = \frac{R_1}{d'_m} = \left(\frac{R_1}{\rho u_1^2}\right)\frac{\rho u_1^2}{d'_m} \qquad \ldots\ldots(2.95)$$

in which u_1 is the mean velocity in the voids ($= u/e$ where u is the average velocity as measured over the whole cross-sectional area of the bed and e is the bed voidage) and d'_m ($=$ volume of voids/total surface area) is one quarter of the hydraulic mean diameter. ERGUN[15] has correlated the friction factor $(R_1/\rho u_1^2)$ in terms of the modified Reynolds number (see equation 4.19 in Volume 2). In principle, therefore, a relation between P and z may be obtained by integration of equation 2.95 and it is therefore possible to allow for the effect of total pressure on the reaction rate by substituting $P = f(z)$ in the expression for the reaction rate. Equation 2.94 may then be integrated directly.

When $\mathcal{R}_Y$ is not a known function of Y an experimental programme to determine the chemical reaction rate is necessary. If the experimental reactor is operated in such a way that the conversion is sufficiently small—small enough that the flow of product can be considered to be an insignificant fraction of the total mass flow—the reactor is said to be operating differentially. Provided the small change in composition can be detected quantitatively, the reaction rate may be directly determined as a function of the mol fraction of reactants. On this basis a differential reactor only provides initial rate data. It is therefore important to carry out experiments in which products are added at the inlet, thereby determining the effect of any retardation by products. An investigation of this kind over a sufficiently wide range of conditions will yield the functional form of $\mathcal{R}_Y$ and, by substitution in equation 2.94, the reactor size and catalyst mass can be estimated. If it is not possible to operate the reactor differentially conditions are chosen so that relatively high conversions are obtained and the reactor is now said to be an integral reactor. By operating the reactor in this way Y may be found as a function of W/G' and $\mathcal{R}_Y$ determined, for any given Y, by evaluating the slope of the curve at various points.

Example

Carbon disulphide is produced in an isothermally operated packed tubular reactor at a pressure of 1 bar by the catalytic reaction between methane and sulphur.

$$CH_4 + 2S_2 = CS_2 + 2H_2S$$

At 600°C ($=873$ K) the reaction is chemically controlled the rate being given by:

$$\mathscr{R}_Y^M = \frac{8 \times 10^{-2} \exp(-115{,}000/\mathbf{R}T)P_{CH_4}P_{S_2}}{1 + 0{\cdot}6{\cdot}P_{S_2} + 2{\cdot}0\,P_{CS_2} + 1{\cdot}7\,P_{H_2S}} \text{ kmol/s(kg catalyst)}$$

where the partial pressures are expressed in bar and the temperature in K and $\mathbf{R}$ as kJ/kmol.

Estimate the mass of catalyst required to produce 1 tonne per day of CS_2 at a conversion level of 90 per cent.

Solution

Assuming plug flow conditions within the reactor, the following equation applies:

$$L = \frac{G'}{\rho_b} \int_0^{Y_L} \frac{dY}{\mathscr{R}_Y^M}$$

where Y is the concentration of CS_2 in mols per unit mass, G' is the mass flow per unit area and ρ_b the catalyst bulk density. In terms of catalyst mass this becomes:

$$W = G \int_0^{Y_L} \frac{dY}{\mathscr{R}_Y^M}$$

where G is now the molar flow rate through the tube.

The rate expression is given in terms of partial pressures and this is now rewritten in terms of the number of mols γ of methane converted to CS_2. Consider 1 mol of gas entering the reactor, then at any cross-section (distance z along the tube from the inlet) the number of mols of reactant and product may be written in terms of γ as follows:

	Mols at inlet	Mols at position z
CH_4	1/3	$1/3 - \gamma$
S_2	2/3	$2/3 - 2\gamma$
CS_2	—	γ
H_2S	—	2γ
Total	1·00	1·00

For a pressure of 1 bar it follows that the partial pressures in the rate expression become:

$$P_{CH_4} = (1/3 - \gamma); \; P_{S_2} = (2/3 - 2\gamma); \; P_{CS_2} = \gamma \text{ and } P_{H_2S} = 2\gamma$$

Now if M is the mean molecular weight of the gas at the reactor inlet then γ kmol of CS_2 are produced for every M kg of gas entering the reactor. Hence:

$$Y = \frac{\gamma}{M}$$

and the required catalyst mass is now:

$$W = \frac{G}{M} \int_0^{\gamma_L} \frac{d\gamma}{\mathscr{R}_\gamma^M}$$

For an expected conversion level of 0·9, the upper limit of the integral will be given by $\dfrac{\gamma_L}{\frac{1}{3} - \gamma_L} = 0{\cdot}9$, i.e. by $\gamma_L = 0{\cdot}158$. Thus 1 tonne per day ($=1{\cdot}48$ kmol/s CS_2) is produced by using an inlet flow of $G = \dfrac{1{\cdot}48 \times 10^{-4} \times M}{0{\cdot}158}$ kg/s. Hence:

$$W = \frac{533}{0{\cdot}158} \int_0^{0{\cdot}158} \frac{d\gamma}{\mathscr{R}_\gamma^M}$$

Substituting the expressions for partial pressures into the rate equation the mass of catalyst required may be determined by direct numerical (or graphical) integration. Thus:

$$\underline{W = 23 \text{ kg}}$$

If axial dispersion were an important effect, the reactor performance would tend to fall below that of a plug flow reactor. In this event an extra term must be added to equation 2.93 to account for the net dispersion occurring within the elementary length δz. A longitudinal dispersion coefficient may be defined by analogy with BOUSSINESQ's concept of eddy viscosity[6]. Thus both molecular diffusion in the direction of flow and eddy diffusion due to local turbulence contribute to the overall dispersion coefficient, which is the effective diffusivity for the bed of solid. The mass of fluid entering the element δz by longitudinal diffusion will be $-D'_L \left(\dfrac{dY}{dz}\right)_z$, where D'_L is now the dispersion coefficient in the axial direction and has units $ML^{-1}T^{-1}$ (since the concentration gradient has units $\text{MOLS}\, M^{-1}L^{-1}$). The amount leaving the element will be $-D'_L \left(\dfrac{dY}{dz}\right)_{z+\delta z}$. The material balance equation will therefore be:

$$-D'_L \left(\frac{dY}{dz}\right)_z + G'Y_z = -D'_L \left(\frac{dY}{dz}\right)_{z+\delta z} + G'Y_{z+\delta z} + \rho_b \mathscr{R}^M_Y \delta z \ldots (2.96)$$

which on expansion becomes:

$$D'_L \frac{d^2 Y}{dz^2} - G' \frac{dY}{dz} = \rho_b \mathscr{R}^M_Y \qquad \ldots (2.97)$$

The boundary condition at $z = 0$ may be written down by invoking the conservation of mass for an element bounded by the plane $z = 0$ and any plane upstream of the inlet where $Y = Y_0$, say:

$$G'Y_0 = G'(Y)_{z=0} - D'_L \left(\frac{dY}{dz}\right)_{z=0} \qquad \ldots (2.98)$$

from which we obtain the condition:

$$z = 0, \quad G'Y = G'Y_0 + D'_L \frac{dY}{dz} \qquad \ldots (2.99)$$

Similarly, at the reactor exit we obtain:

$$z = L, \quad \frac{dY}{dz} = 0 \qquad \ldots (2.100)$$

since there is no further possibility of Y changing from Y_L after the reactor exit. Equation 2.96 may be solved analytically for zero or first order reactions, but for other cases resort to numerical methods is generally necessary. In either event Y is obtained as a function of z and so the length L of catalyst bed required to achieve a conversion corresponding to an exit concentration Y_L may be calculated. The mass of catalyst corresponding to this length L is $\rho_b A_c L$.

Having pointed out the modifications to be made to a design based upon the plug flow approach, it is salutary to note that axial dispersion is seldom of importance in fixed bed tubular reactors.

Adiabatic conditions

Adiabatic reactors are more frequently encountered in practice than iso-thermal reactors. Because there is no exchange of heat with the surroundings, radial temperature gradients are absent. All of the heat generated or absorbed by the chemical reaction manifests itself by a change in enthalpy of the fluid stream. It is therefore necessary to write a heat balance equation for the reactor in addition to the material balance equation 2.93. Generally heat transfer between solid and fluid is sufficiently rapid for it to be justifiable to assume that all the heat generated or absorbed at any point in the reactor is transmitted instantaneously to or from the solid. It is therefore only necessary to take a heat balance for the fluid entering and emerging from an elementary section δz. Referring to Fig. 2.12 and neglecting the effect of longitudinal heat con-duction:

$$G'\bar{c}_p\delta T = \rho_b(-\Delta H)\mathscr{R}^M_{YT}\,\delta z \qquad \text{....(2.101)}$$

where $\bar{c}_p$ is the mean heat capacity of the fluid,

$\quad\Delta H$ is the enthalpy change on reaction, and

$\quad\mathscr{R}^M_{YT}$ is the reaction rate (mols per unit time and unit mass of catalyst)—now a function of Y and T.

Simultaneous solution of the mass balance equation 2.93 and the heat balance equation 2.101 with the appropriate boundary conditions gives z as a function of Y.

A simplified procedure is to assume $-\Delta H/\bar{c}_p$ as constant. If equation 2.101 (the heat balance equation) is divided by equation 2.93 (the mass balance equation) and integrated, we immediately obtain:

$$T = T_0 + \frac{(-\Delta H)Y}{\bar{c}_p} \qquad \text{....(2.102)}$$

where T_0 is the inlet temperature. This relation implies that the adiabatic reaction path is linear. If equation 2.102 is substituted into the mass balance equation 2.93:

$$\frac{dY}{dz} = \frac{\rho_b}{G'}\mathscr{R}^M_{YT_0} \qquad \text{....(2.103)}$$

where the reaction rate $\mathscr{R}^M_{YT_0}$ along the adiabatic reaction path is now expressed as a function of Y only. Integration then gives z as a function of Y directly.

Because the adiabatic reaction path is linear a graphical solution, also applicable to multi-bed reactors, is particularly apposite. (See the next example as an illustration.) If the design data are available in the form of rate data $\mathscr{R}^M_{YT}$ for various temperatures and conversions they may be displayed as contours of equal reaction rate in the (T, Y) plane. Figure 2.13 shows such contours upon which is superimposed an adiabatic reaction path of slope

$\bar{c}_p/(-\Delta H)$ and intercept T_0 on the abscissa. The reactor size may be evaluated by computing:

$$\frac{\rho_b L}{G'} = \int_{Y_0}^{Y_L} \frac{\mathrm{d}Y}{\mathscr{R}_{YT_0}^M} \qquad \ldots\ldots(2.104)$$

from a plot of $1/\mathscr{R}_{YT_0}^M$ as a function of Y. The various values of $\mathscr{R}_{YT_0}^M$ are simply those points at which the adiabatic reaction path intersects the contours.

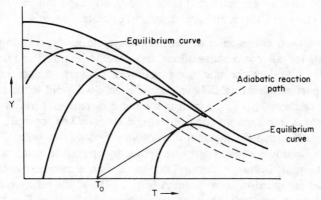

FIG. 2.13. Graphical solution for an adiabatic reactor

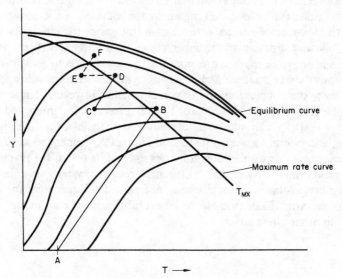

FIG. 2.14. Optimum design of a two-stage and three-stage adiabatic reactor

It is often necessary to employ more than one adiabatic reactor to achieve a desired conversion. In the first place chemical equilibrium may have been established in the first reactor and it would be necessary to cool and/or remove the product before entering the second reactor. This, of course, is one good reason for choosing a catalyst which will function at the lowest possible temperature. Secondly, for an exothermic reaction, the temperature may rise to a point at which it is deleterious to the catalyst activity. At this point the products from the first reactor are cooled prior to entering a second adiabatic reactor. To design such a system it is only necessary to superimpose on the rate contours the adiabatic temperature paths for each of the reactors. The volume requirements for each reactor can then be computed from the rate contours in the same way as for a single reactor. It is necessary, however, to consider carefully how many reactors in series it is economic to operate.

Should we wish to minimise the size of the system it would be important to ensure that, for all conversions along the reactor length, the rate is at its maximum[11,24]. Since the rate is a function of conversion and temperature, setting the partial differential $\partial \mathcal{R}_{YT}/\partial T$ equal to zero will yield, for an exothermic reaction, a relation $T_{MX}(Y_{MX})$ which is the locus of temperatures at which the reaction rate is a maximum for a given conversion. The locus T_{MX} of these points passes through the maxima of curves of Y as a function of T shown in Fig. 2.14 as contours of constant rate. Thus, to operate a series of adiabatic reactors along an optimum temperature path, hence minimising the reactor size, the feed is heated to some point A (Fig. 2.14) and the reaction allowed to continue along an adiabatic reaction path until a point such as B, in the vicinity of the optimum temperature curve, is reached. The products are then cooled to C before entering a second adiabatic reactor in which reaction proceeds to an extent indicated by point D, again in the vicinity of the curve T_{MX}. The greater the number of adiabatic reactors in the series the closer the optimum path is followed. For a given number of reactors, three say, there will be six design decisions to be made corresponding to the points A to F inclusive. These six decisions may be made in such a way as to minimise the capital and running costs of the system of reactors and heat exchangers. However, such an optimisation problem is outside the scope of this chapter and the interested reader is referred to ARIS[4]. It should be pointed out, nevertheless, that the high cost of installing and operating heat transfer and control equipment so as to maintain the optimum temperature profile militates against its use. If the reaction is not highly exothermic an optimal isothermal reactor system may be a more economic proposition and its size may not be much larger than the adiabatic system of reactors. Each case has to be examined on its own merits and compared with other alternatives.

Example

SO$_3$ is produced by the catalytic oxidation of SO$_2$ in two packed adiabatic tubular reactors arranged in series with intercooling between stages. The molar composition of the mixture entering the first reactor is 7 per cent SO$_2$, 11 per cent O$_2$ and 82 per cent N$_2$. The inlet temperature of the reactor is controlled at 688 K and the inlet flow is 0·17 kmol/s. Calculate the mass of catalyst required for each stage so that 78 per cent of the SO$_2$ is converted in the first stage and a further 20 per cent in the second stage.

Thermodynamic data for the system are as follows:

	Mean specific heat in range 415°C to 600°C (688 to 873 K)
Component	kJ/kmol K
SO_2	51·0
SO_3	75·5
O_2	33·0
N_2	30·5

For the reaction:

$$SO_2 + \tfrac{1}{2}O_2 \rightleftharpoons SO_3$$

the standard enthalpy change at 415°C ($=688$ K) is $\Delta H^0_{688} = -97{,}500$ kJ/kmol

Solution

The chemical reaction which ensues is:

$$SO_2 + \tfrac{1}{2}O_2 \rightleftharpoons SO_3$$

and if χ kmol of SO_2 are converted to χ kmol of SO_3 a material balance for the first reactor may be drawn up:

Component	kmol at inlet	kmol at outlet
SO_2	$0·17 \times 0·07$	$0·17 \times 0·07 - \chi$
O_2	$0·17 \times 0·11$	$0·17 \times 0·11 - \chi/2$
SO_3	—	χ
N_2	$0·17 \times 0·82$	$0·17 \times 0·82$

From the last column of the above table and the specific heat data the difference in enthalpy between reactants and products is:

$$H_P - H_R = (T - 688)\{[(0·17 \times 0·07) - \chi] \times 51·0 + [(0·17 \times 0·11) - \chi/2] \times 33·0 + [75·5\chi] + [0·17 \times 0·82 \times 30·5]\}$$

where T (K) is the temperature at the exit from the first reactor. The percentage conversion x may be written:

$$x = \frac{\chi}{0·17 \times 0·07} \times 100$$

and on substitution and simplification:

$$H_P - H_R = (T - 688)(5·482 + 8·0x)$$

An enthalpy balance (in kJ) over the first reactor gives:

$$H_P - H_R = \Delta H^0_{688} = 97{,}500\,\chi = 11·7\,x$$

where $\Delta H^0_{688}(= -97{,}500$ kJ/kmol is the standard enthalpy of reaction at 688K).

From the two expressions obtained:

$$T = 688 + \frac{11·7\,x}{5·482 + 0·001\,x} = 688 + 2·11\,x$$

This provides a linear relation between temperature and conversion in the first reactor.

Figure 2.15 is a conversion chart for the reactant mixture and shows rate curves in the conversion–temperature plane. The line AA' is plotted on the chart from the linear $T - x$ relation. The rate of reaction at any conversion level within the reactor may therefore be obtained by reading the rate corresponding to the intersection of the line with various conversion ordinates.

Now for the first reactor:

$$W_1 = 170 \times 0·07 \int_0^{0·78} \frac{dx}{\mathscr{R}^M}$$

and this is easily solved by graphical integration as depicted in Figure 2.16, by plotting $1/\mathscr{R}$ versus x and finding the area underneath the curve corresponding to the first reactor. The area corresponds to 3.36×10^5 kg s (kmol)$^{-1}$ and $W_1 = 4$ kg.

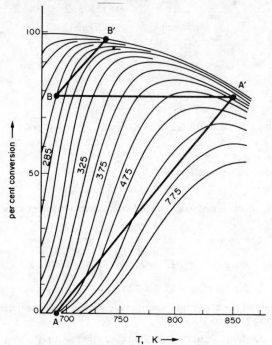

FIG. 2.15. Graphical solution for design of SO_2 reactor. Figures on curves represent the reaction rate in units of kmol s^{-1} kg^{-1}

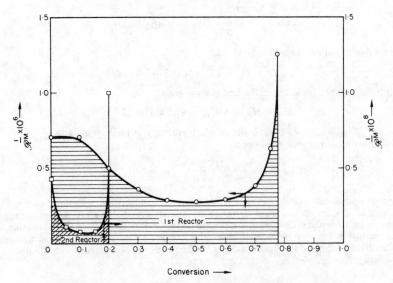

FIG. 2.16. Graphical integration to determine catalyst weights

A similar calculation is now undertaken for the second reactor but, in this particular case, a useful approximation may be made. Because the reactant mixture is highly diluted with N_2, the same heat conservation equation may be used for the second reactor as for the first reactor. The line BB′ starting at the point (0·78, 688) and parallel to the first line AA′ may therefore be drawn. For the second reactor:

$$W_2 = 0.17 \times 0.07 \times (1 - 0.78) \int_0^{} \frac{dx}{\mathscr{R}^M}$$

and this is also solved by the same procedure as before. $1/\mathscr{R}^M$ is plotted as a function of x by noting the rate corresponding to the intersection of the line BB′ with the selected conversion ordinate. The area under the curve $1/\mathscr{R}^M$ versus x for the second reactor is 49×10^5 kg s/kmol. Hence $\underline{\underline{W_2 = 12 \text{ kg.}}}$

Non-isothermal conditions

When the reactor exchanges heat with the surroundings radial temperature gradients exist and this causes transverse diffusion of the reactant. For an exothermic reaction, the reaction rate will be highest along the tube axis since the temperature there will be greater than at any other radial position. Reactants, therefore, will be rapidly consumed at the tube centre resulting in a steep transverse concentration gradient causing an inward flux of reactant and a corresponding outward flux of products. The existence of radial temperature and concentration gradients, of course, renders the simple plug flow approach to design inadequate.

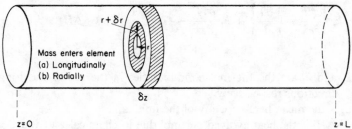

FIG. 2.17. Two-dimensional tubular reactor

It is now essential to write the mass and energy balance equations for the two dimensions z and r. For the sake of completeness, we will include the effect of longitudinal dispersion and heat conduction and deduce the material and energy balances for one component in an elementary annulus of radius δr and length δz. We assume that equimolar counterdiffusion occurs and by reference to Fig. 2.17 write down, in turn, the components of mass which are entering the element in unit time longitudinally and radially:

Mass entering by longitudinal bulk flow $\quad = 2\pi r \delta r G'(Y)_z$

Mass entering by transverse diffusion $\quad = -D'_r 2\pi r \delta z \left(\dfrac{\partial Y}{\partial r}\right)_r$

Mass entering by longitudinal diffusion $\quad = -D'_L 2\pi r \delta r \left(\dfrac{\partial Y}{\partial z}\right)_z$

In general the longitudinal and radial dispersion coefficients D'_r and D'_L will, as discussed on page 157, differ. The mass leaving the element in unit time can be written similarly as a series of components:

Mass leaving by longitudinal bulk flow $= 2\pi r \delta r G'(Y)_{z+\delta z}$

Mass leaving by transverse diffusion $= -D'_r 2\pi (r + \delta r) \delta z \left(\dfrac{\partial Y}{\partial r}\right)_{r+\delta r}$

Mass leaving by longitudinal diffusion $= -D'_L 2\pi r \delta r \left(\dfrac{\partial Y}{\partial z}\right)_{z+\delta z}$

Mass of component produced by chemical reaction $= 2\pi r \delta r \delta z \rho_b \mathscr{R}^M_{YT}$

In the steady state the algebraic sum of the components of mass entering and leaving the element will be zero. By expanding terms evaluated at $(z + \delta z)$ and $(r + \delta r)$ in a Taylor series about the points z and r respectively and neglecting second order differences, the material balance equation becomes:

$$D'_r \left\{\frac{\partial^2 Y}{\partial r^2} + \frac{1}{r}\frac{\partial Y}{\partial r}\right\} + D'_L \frac{\partial^2 Y}{\partial z^2} - G'\frac{\partial Y}{\partial z} = \rho_b \mathscr{R}^M_{YT} \qquad \ldots \ldots (2.105)$$

A heat balance equation may be deduced analogously:

$$k_r \left\{\frac{\partial^2 T}{\partial r^2} + \frac{1}{r}\frac{\partial T}{\partial r}\right\} + k_l \frac{\partial^2 T}{\partial z^2} - G'\bar{c}_p \frac{\partial T}{\partial z} = \rho_b(-\Delta H)\, \mathscr{R}^M_{YT}$$

$$\ldots \ldots (2.106)$$

where k_r and k_l are the thermal conductivities in the radial and longitudinal directions respectively,

$\bar{c}_p$ is the mean heat capacity of the fluid, and

$(-\Delta H)$ is the heat evolved per mol due to chemical reaction.

When writing the boundary conditions for the above pair of simultaneous equations the heat transferred to the surroundings from the reactor may be accounted for by ensuring that the tube wall temperature correctly reflects the total heat flux through the reactor wall. If the reaction rate is a function of pressure then the momentum balance equation must also be invoked, but if the rate is insensitive or independent of total pressure then it may be neglected.

It is useful at this stage to note the assumptions which are implicit in the derivation of equations 2.105 and 2.106. These are as follows: laminar flow, constant dispersion and thermal conductivity coefficients, instantaneous heat transfer between solid catalyst and the reacting ideal gas mixture, and the neglect of potential energy. It should also be noted that the two simultaneous equations are coupled and highly non-linear because of the effect of temperature on the reaction rate. Numerical methods of solution are therefore generally adopted and those employed are based upon the use of finite differences. The neglect of longitudinal diffusion and conduction simplifies the equations considerably. If we also suppose that the system is isotropic we can write a single effective diffusivity D'_e for the bed in place of D'_r and D'_L and a single effective thermal conductivity k_e.

The results of calculations typical of an exothermic catalytic reaction are shown in Figs. 2.18 and 2.19. It is clear that the conversion is higher along the tube axis than at other radial positions and that the temperature first increases to a maximum before decreasing at points further along the reactor length. The exothermic nature of the reaction, of course, leads to an initial increase in temperature, but because in the later stages of reaction the radial heat transfer to the wall and surroundings becomes larger than the heat evolved by chemical reaction, the temperature steadily decreases. The mean temperature and conversion over the tube radius for any given position along the reactor length may

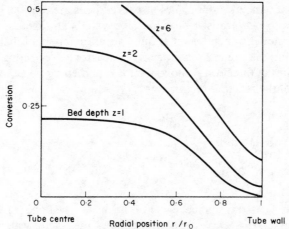

Fig. 2.18. Conversion profiles as a function of tube length and radius

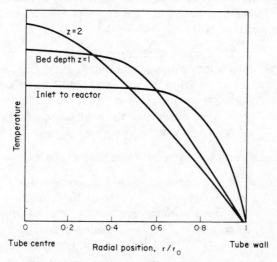

Fig. 2.19. Temperature profiles as a function of tube length and radius

easily be computed from such results and consequently the bed depth required for a given specified conversion readily found.

Further advancements in the theory of fixed bed reactor design have been made[29, 33] but it is unusual for experimental data to be of sufficient precision and extent to justify the application of sophisticated methods of calculation. Uncertainties in the knowledge of effective thermal conductivities and heat transfer between gas and solid make the calculation of temperature distribution in the bed susceptible to inaccuracies, particularly in view of the pronounced effect of temperature on the reaction rate. A useful approach to the preliminary design of a non-isothermal fixed bed reactor is to assume that all the resistance to heat transfer is in a thin layer near the tube wall. This is a fair approximation because radial temperature profiles in packed beds are parabolic with most of the resistance to heat transfer near the tube wall. With this assumption a one-dimensional model, which becomes quite accurate for small diameter tubes, is satisfactory for the approximate design of reactors. Neglecting diffusion and conduction in the direction of flow, the mass and energy balances for a single component of the reacting mixture are:

$$G' \frac{dY}{dz} = \rho_b \mathcal{R}_{YT}^M \qquad\qquad \text{(equation 2.93)}$$

$$G' \bar{c}_p \frac{dT}{dz} = \rho_b (-\Delta H)\ \mathcal{R}_{YT}^M - \frac{h}{a'}(T - T_w) \qquad \ldots .(2.107)$$

where h is the heat transfer coefficient (dimensions $\mathbf{MT^{-3}\theta^{-1}}$) expressing the resistance to the transfer of heat between the reactor wall and the reactor contents,

a' is the surface area for heat transfer per unit hydraulic radius (equal to the area of cross-section divided by the perimeter), and

T_w is the wall temperature.

If the wall temperature is constant, inspection of the above equations shows that, for a given inlet temperature, a maximum temperature is attained somewhere along the reactor length if the reaction is exothermic. It is desirable that this should not exceed the temperature at which the catalyst activity declines. In Fig. 2.20 the curve ABC shows a non-isothermal reaction path for an inlet temperature T_0 corresponding to A. Provided $T_0 > T_w$, it is obvious that $\frac{dT}{dY} < \frac{-\Delta H}{\bar{c}_p}$ and the rate of temperature increase will be less than in the adiabatic case. The point B, in fact, corresponds to the temperature at which the reaction rate is at a maximum and the locus of such points is the curve T_{MX} described previously. The maximum temperature attained from any given inlet temperature may be calculated by solving, using an iterative method[5], the pair of simultaneous equations 2.93 and 2.107 and finding the temperature at which $\frac{d\mathcal{R}_{YT}^M}{dT} = 0$, or, equivalently, $\frac{dY}{dT} = 0$. We will see later that a packed tubular reactor is very sensitive to change in wall temperature. It is therefore important to estimate the maximum attainable temperature, for a given inlet temperature, from the point of view of maintaining both catalyst activity and reactor stability.

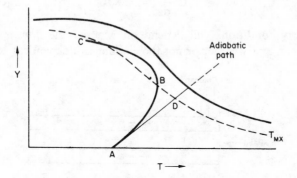

FIG. 2.20. Reaction path of a cooled tubular reactor

An important class of reactors is that for which the wall temperature is not constant but varies along the reactor length. Such would be the case when the cooling tubes and reactor tubes form an integral part of a composite heat exchanger. Figures 2.21a and 2.21b show, respectively, cocurrent and counter-current flow of coolant and reactant mixture, the coolant fluid being entirely independent and separate from the reactants and products. However, the reactant feed itself may be used as coolant prior to entering the reactor tubes and again may flow cocurrent or countercurrent to the reactant mixture (Fig. 2.21c and 2.21d respectively). In each case heat is exchanged between the reactant mixture and the cooling fluid. A heat balance for a component of the reactant mixture leads to:

$$G'\bar{c}_p \frac{dT}{dz} = \rho_b(-\Delta H)\,\mathcal{R}_{YT}^M - \frac{U}{a'}\,(T - T_C) \qquad \ldots\ldots(2.108)$$

an equation analogous to 2.107 but in which T_C is a function of z and U is an overall heat transfer coefficient for the transfer of heat between the fluid streams. The variation in T_C may be described by taking a heat balance for an infinitesimal section of the cooling tube:

$$G'_C\bar{c}_{pC} \frac{dT_C}{dz} \pm \frac{U}{a'}\,(T - T_C) = 0 \qquad \ldots\ldots(2.109)$$

where G'_C is the mass flow rate of coolant per unit cross-section of the reactor
  tube, and
  $\bar{c}_{pC}$ is the mean heat capacity of the coolant.
If flow is cocurrent the upper sign is used; if countercurrent the lower sign is used. Since the mass flow rate of the cooling fluid is based upon the cross-sectional area of the reactor tube the ratio $G'\bar{c}_p/G'_C\bar{c}_{pC}\,(=\Gamma)$ is a measure of the capacities of the two streams to exchange heat. In terms of the limitations

imposed by the one-dimensional model, the system is fully described by equations 2.108 and 2.109 together with the mass balance equation:

$$G' \frac{dY}{dz} = \rho_b \mathscr{R}_{YT}^M \qquad \text{(equation 2.93)}$$

The boundary conditions will depend on whether the flow is cocurrent or countercurrent, and whether or not the coolant is independent of the reactant mixture.

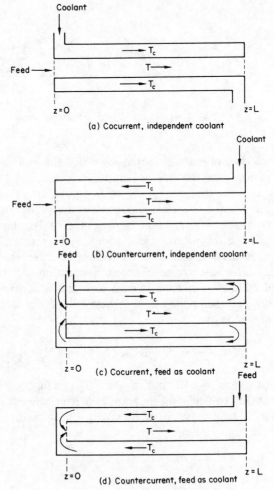

FIG. 2.21. Cocurrent and countercurrent
cooled tubular reactors

The reaction path in the T, Y plane could be plotted by solving the above set of equations with the appropriate boundary conditions. A reaction path similar to the curve ABC in Fig. 2.20 would be obtained. The size of reactor necessary to achieve a specified conversion could be assessed by tabulating

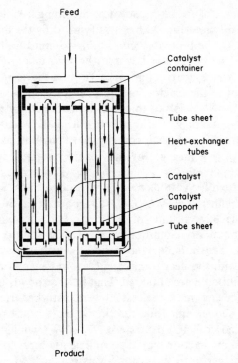

FIG. 2.22. Ammonia converter

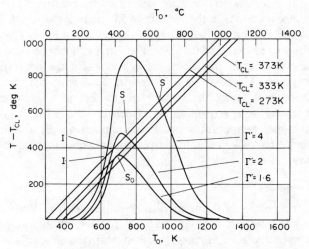

FIG. 2.23. Heat transfer characteristics of countercurrent cooled tubular reactor

points at which the reaction path crosses the constant rate contours, hence giving values of $\mathscr{R}_{YT}$ which could be used to integrate the mass balance equation 2.93. The reaction path would be suitable provided the maximum temperature attained was not deleterious to the catalyst activity.

In the case of a reactor cooled by incoming feed flowing countercurrent to the reaction mixture (Fig. 2.21d), and typified by the ammonia converter sketched in Fig. 2.22, the boundary conditions complicate the integrations of the equations. At the point $z = L$ where the feed enters the cooling tubes, $T_C = T_{CL}$, whereas at the reactor entrance ($z = 0$) the feed temperature is, of course, equivalent to the reactor inlet temperature, so here $T = T_0 = T_C$ and $Y = Y_0$. VAN HEERDEN[17] solved this two-point boundary value problem by assuming various values for T_0 and then integrating the equations to find the reactor exit temperature T_L and the incoming feed temperature T_{CL} at $z = L$. The quantity $(T - T_{CL})$ is a measure of the heat transferred between reactor and cooling tubes so a plot of $(T - T_{CL})$ versus T_0 will describe the heat exchanging capacity of the system for various reactor inlet conditions. Such curves are displayed in Fig. 2.23 with $\Gamma' (= UaL/G'_C \bar{c}_{pC})$ as a parameter. Superimposed on the diagram are parallel straight lines of unit slope, each line corresponding to the value of T_{CL}, the boundary condition at $z = L$. The solution to the three simultaneous equations and boundary conditions will be given by the intersections of the line T_{CL} with the heat transfer curves. In general an unstable solution I and a stable solution S are obtained. That S corresponds to a stable condition may be seen by the fact that if the system were disturbed from the semi-stable condition at I by a sudden small increase in T_0, more heat would be transferred between reactor and cooling tubes, since $(T - T_{CL})$ would now be larger and the heat generated by reaction would be sufficiently great to cause the inlet temperature to rise further until the point S was reached, whereupon the heat exchanging capacity would again match the heat generated by reaction. The region in which it is best to operate is in the vicinity of S_0, the point at which I and S are coincident. As the catalyst activity declines the curves of $(T - T_{CL})$ versus T_0 are displaced downwards and so the system will immediately become quenched. For continuing operation, therefore, more heat exchange capacity must be added to the system. This may be achieved by decreasing the mass flowrate. The temperature level along the reactor length therefore increases and the system now operates along a reaction path (such as ABC in Fig. 2.20) displaced to regions of higher temperature but lower conversion. Production is thus maintained but at the cost of reduced conversion concomitant with the decaying catalyst activity.

Fluidised bed reactors

The general properties of fluidised beds were discussed in Volume 2. Their application to gas-solid and catalytic reactions has certain advantages over the use of tubular type reactors. High wall-to-bed heat transfer coefficients enable heat to be abstracted from, or absorbed by, the reactor with considerable efficiency. A mechanical advantage is also gained by the relative ease with which solids may be conveyed and, because of solids mixing, the whole of the gas in the reactor is at substantially the same temperature. Extremely valuable is the large external surface area exposed by the solid to the gas; because of this, reactions limited by intraparticle diffusion will give a higher conversion in a fluidised bed than in a packed bed tubular reactor.

MAY[26] considered a model for describing reactions in a fluidised bed. He wrote the mass conservation equations for the bubble phase (identified with rising bubbles containing little or no solid particles) and the dense phase (in which the solid particles are thoroughly mixed). He tacitly assumed that mass transferred by cross-flow between the two phases was not a function of bed height, and took an average concentration gradient for the whole height of the bed. Conversions within the bed were calculated by solving the conservation equations for particular cross-flow ratios. May compared the predicted conversion in beds in which there was no mixing of solids with beds in which there was complete back-mixing. It was found that conversion was smaller for complete back-mixing and relatively low cross-flow ratios than for no mixing and high cross-flow ratios. Thus, if back-mixing is appreciable, a larger amount of catalyst has to be used in the fluidised bed to achieve a given conversion than if a plug flow reactor, in which there is no back-mixing, is used. It was also shown that, for two concurrent first order reactions, the extent of gas-solid contact affects the selectivity. Although the same total conversion may be realised, more catalyst is required when the contact efficiency is poor than when the contact efficiency is good, and proportionately more of the less reactive component in the feed would be converted at the expense of the more reactive component.

The difficulty with an approach in which flow and diffusive parameters are assigned to a model is that the assumptions do not conform strictly to the pattern of behaviour in the bed. Furthermore, it is doubtful whether either solid dispersion or gas mixing can be looked upon as a diffusive flux. ROWE[37] calculated, from a knowledge of the aerodynamics in the bed, a mean residence time for particles at the surface of which a slow first order gas–solid reaction occurred. Unless the particle size of the solid material was chosen correctly most of the reactant gas passed through the bed as bubbles and had insufficient time to react. The most effective way of increasing the contact efficiency was to increase the particle size, for this caused more of the reactant gas to pass through the dense phase. Doubling the particle size almost halved the contact time required for the reactants to be completely converted. Provided that the gas flow is sufficiently fast to cause bubble formation in the bed, the heat transfer characteristics are good.

Thermal characteristics of packed reactors

There are several aspects of thermal sensitivity and instability which are important to consider in relation to reactor design. When an exothermic catalytic reaction occurs in a non-isothermal reactor, for example, a small change in coolant temperature may, under certain circumstances, produce undesirable hotspots or regions of high temperature within the reactor. Similarly, it is of central importance to determine whether or not there is likely to be any set of operating conditions which may cause thermal instability in the sense that the reaction may either become extinguished or continue at a higher temperature level as a result of fluctuations in the feed condition. We will briefly examine these problems.

Sensitivity of countercurrent cooled reactors

To illustrate the problem of thermal sensitivity we will analyse the simple one-dimensional model of the counter-current cooled packed tubular reactor described earlier and illustrated in Fig. 2.17. We have already seen that the mass and heat balance equations for the system may be written:

$$G' \frac{dY}{dz} = \rho_b \mathscr{R}^M \qquad\qquad \text{(equation 2.93)}$$

$$G' \bar{c}_p \frac{dT}{dz} = -\Delta H \rho_b \mathscr{R}^M - \frac{h}{a'}(T - T_w) \qquad \text{(equation 2.107)}$$

where T_w represents a constant wall temperature. Solution of these simultaneous equations with initial conditions $z = 0$, $T = T_0$ and $Y = Y_0$ enables us to plot the reaction path in the (T,Y) plane. The curve ABC, portrayed in Fig. 2.20, is typical of the reaction path that might be obtained. There will be one such curve for any given value T_0 of the reactor inlet temperature. By locating the points at which $\frac{dT}{dY} = 0$ the loci of maximum temperature may be plotted, each locus corresponding to a chosen value T_w for the constant coolant temperature along the reactor length. Thus, a family of curves such as that sketched in Fig. 2.24 would be obtained, representing loci of maximum temperatures for constant wall temperatures T_w, increasing sequentially from curve 1 to curve 5. Since the maximum temperature along the non-isothermal reaction path (point B in Fig. 2.20) must be less than the temperature given by the intersection of the adiabatic line with the locus of possible maxima (point D in Fig. 2.20), the highest temperature achieved in the reactor must be bounded by this latter point. If the adiabatic line—and therefore the non-isothermal reaction path—lies entirely below the locus of possible maxima, then $\frac{dT}{dY} > 0$, and the temperature will increase along the reactor length because $\frac{dT}{dz} > 0$ for all points along the reactor; if, however, the adiabatic line lies above the locus, $\frac{dT}{dY} < 0$ and the temperature will therefore decrease along the reactor length. For a particular T_w corresponding to the loci 1 or 2 in Fig. 2.24, T_0, the inlet temperature to the reactor, must be the maximum temperature in the reactor because the adiabatic line lies entirely above these two curves making $\frac{dT}{dz}$ always negative. If T_w corresponds to curve 3, however, we might expect the temperature to increase from T_0 to a maximum and thence to decrease along the remaining reactor length since the adiabatic line now intersects this particular locus at point A. In the case of curve 4 the adiabatic line is tangent to the maximum temperature locus at B, and then intersects it at point C where the temperature would be much higher. The adiabatic line intersects curve 5 at point D. From this analysis, therefore, we are led to expect that, if the steady wall temperature of the reactor is increased through a sequence corresponding to curves 1 to 4, the maximum temperature in the reactor gradually increases from T_0 to B as T_W increases from T_{W1} to T_{W4}. If the wall temperature were to be increased to T_{W5} (the maximum temperature locus corresponding to curve 5), there would be a discontinuity and the maximum temperature in the reactor would suddenly jump to point D. This type of sensitivity was predicted by BILOUS and AMUNDSON[20] who calculated temperature maxima within a non-isothermal reactor for various constant wall temperatures. The results of their computations are shown in Fig. 2.25 and it is seen that, if the wall temperature increases from 300 K to 335 K, a temperature maximum appears in the reactor. Further increase in wall temperature causes the maximum to increase sharply to higher temperature. Bilous and Amundson also found that sensitivity to heat transfer can produce temperature maxima in much the same way as sensitivity to change in wall temperature. In view of the nature of equations 2.93 and 2.107 this is to be expected.

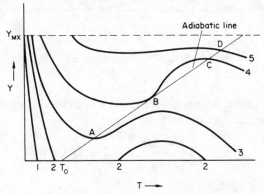

FIG. 2.24. Loci of maximum wall temperatures

The autothermal region

Clearly it is desirable to utilise economically the heat generated by an exothermic reaction. If the heat dissipated by reaction can be used in such a way that the cold incoming gases are heated to a temperature sufficient to initiate a fast reaction, then by judicious choice of operating conditions the process may be rendered thermally self-sustaining. This may be accomplished by transferring heat from the hot exit gases to the incoming feed. The differential equations already discussed for the countercurrent self-cooled reactor are applicable in this instance. Combining the three simultaneous equations 2.108, 2.109 and 2.93 (utilising the upper sign in equation 2.109 for the countercurrent case

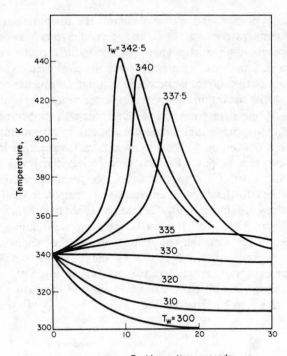

FIG. 2.25. Sensitivity of reactor to change in wall
temperature

and remembering that for a self-cooled—or in this particular case self-heated—
reactor $G'_C \bar{c}_{pC} = G' \bar{c}_p$) we obtain:

$$G' \frac{d}{dz} \left\{ \frac{(-\Delta H)}{\bar{c}_p} Y - T + T_C \right\} = 0 \qquad \ldots (2.110)$$

Integration of this equation yields:

$$T_C = T - \frac{(-\Delta H)}{\bar{c}_p}(Y - Y_0) \qquad \ldots (2.111)$$

Substituting equation 2.111 into the heat balance equation 2.108 for the reactor:

$$G' \frac{dT}{dz} = \frac{(-\Delta H)}{\bar{c}_p} \rho_b \mathscr{R}^M - \frac{G'(-\Delta H)}{\bar{c}_p} \frac{\Gamma}{L}(Y - Y_0) \qquad \ldots (2.112)$$

where the heat exchanging capacity of the system has been written in terms of
$\Gamma (= UaL/G'\bar{c}_p)$. The mass balance equation:

$$G' \frac{dY}{dz} = \overset{\circ}{\rho}_b \mathscr{R}^M \qquad \text{(equation 2.93)}$$

must be integrated simultaneously with equation 2.112 from $z = 0$ to $z = L$
with the initial conditions (at $z = 0$), $Y = Y_0$ and $T = T_0$ to obtain T and Y

as functions of z. In this event the exit temperature T_L and concentration Y_L can, in principle, be computed. Numerical integration of the equations is generally necessary. To estimate whether the feed stream has been warmed sufficiently for the reaction to give a high conversion and be thermally self-supporting, we seek a condition for which the heat generated by reaction equals the heat gained by the cold feed stream. The heat generated per unit mass of feed by chemical reaction within the reactor may be written in terms of the con-centration of product:

$$Q_1^M = (-\Delta H)(Y_L - Y_0) \qquad \dots(2.113)$$

whereas the net heat gained by the feed would be:

$$Q_2^M = \bar{c}_p(T_L - T_{CL}) \qquad \dots(2.114)$$

The system is thermally self-supporting if $Q_1^M = Q_2^M$, i.e. if:

$$T_L - T_{CL} = \frac{(-\Delta H)}{\bar{c}_p}(Y_L - Y_0) \qquad \dots(2.115)$$

The temperature of the non-reacting heating fluid at the reactor inlet is T_0 and at the exit T_{CL} so the difference, say ΔT, is

$$\Delta T = T_0 - T_{CL} = T_0 - T_L + \frac{(-\Delta H)}{\bar{c}_p}(Y_L - Y_0) \qquad \dots(2.116)$$

Since the numerical values of T_L and Y_L are dependent on the heat exchanging capacity (as shown by equation 2.112), the quantity on the right-hand side of equation 2.116 may be displayed as a function of the inlet temperature to the bed, T_0, with Γ' as a parameter. The three bell-shaped curves in Fig. 2.26 are for

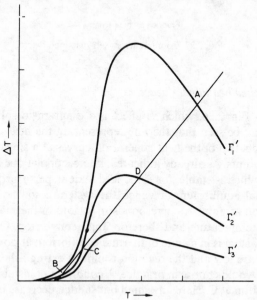

FIG. 2.26. Temperature increase of feed as function of inlet temperature

different values of Γ' and each represents the locus of values given by the right-hand side of this equation. The left-hand side of the equation may be represented by a straight line of unit slope through the point $(T_{CL}, 0)$. The points at which the line intersects the curve represent solutions to equation 2.116. However, we seek only a stable solution which coincides with a high yield. Such a solution would be represented by the point A in Fig. 2.26, for the straight line intersects the curve only once and at a bed temperature corresponding to a high yield of product. The condition for autothermal reaction would therefore be represented by the operating condition T_{CL}, Γ'_3, i.e. the cold feed temperature, should be $T_0 = T_{CL}$ and the heat exchange capacity (which determines the ratio L/G, the length of solid packing divided by the mass flow rate) should correspond to Γ'_3. The corresponding temperature profiles for the reactor and exchanger are sketched in Fig. 2.27.

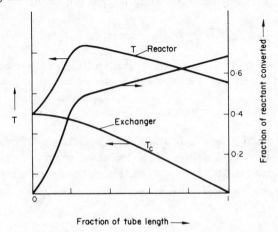

Fig. 2.27. Countercurrent self-sustained reactor

Stability of packed-bed tubular reactors

Referring to Fig. 2.26, which in effect is a diagram to solve a heat balance equation, it may be seen that the line representing the heat gained by the feed stream intersects two of the heat generation curves at a single point and one of them at three points. As already indicated, intersection at one point corresponds to a solution which is stable, for if the feed state is perturbed there is no other state of thermal equilibrium at which the system could continue to operate. The intersection at point A corresponds to a state of thermal equilibrium at a high reaction temperature and therefore high conversion. On the other hand, although the system is in a state of thermal equilibrium at point B, the temperature and yield are low and the reaction is almost extinguished—certainly not a condition one would choose in practice. A condition of instability is represented by the intersection at C. Here any small but sudden decrease in feed temperature would cause the system to become quenched, for the equilibrium state would revert to B. A sudden increase in feed temperature would displace the system to D. Although D represents a relatively stable condition, such a choice should be avoided for a large perturbation might result in the system restabilising at B where the reaction is quenched. A reactor should therefore be designed such that thermal equilibrium can be established at only one point (such as A in Fig. 2.26) where the conversion is high.

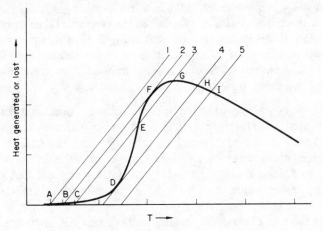

FIG. 2.28. Stability states in reactors

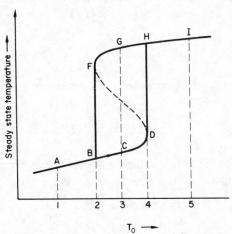

FIG. 2.29. Hysteresis effect in packed-bed reactor

It is interesting to note from Fig. 2.28 that if the feed to an independently cooled reactor (in which an exothermic reaction is occurring) is gradually varied through a sequence of increasing temperatures such as 1 to 5, the system will have undergone some fairly violent changes in thermal equilibrium before finally settling at point I. Corresponding to a sequential increase in feed temperature from 1 to 4 the reactor will pass through states A, B, and C to D, at which last point any slight increase in feed temperature causes the thermal

G

equilibrium condition to jump to a point near H. If the feed temperature is then raised to 5 the system gradually changes along the smooth curve from H to I. On reducing the feed temperature from 5 to 2 on the other hand, the sequence of equilibrium states will alter along the path IHGF. Any further small decrease in feed temperature would displace the system to a point near B and subsequently along the curve BA as the feed temperature gradually changes from 2 to 1. Thus, there is a hysteresis effect when the system is taken through such a thermal cycle and this is clearly illustrated in Fig. 2.29, which traces the path of equilibrium states as the feed temperature is first increased from 1 to 5 and subsequently returned to 1. These effects are important to consider when start-up or shut-down times are estimated.

Another type of stability problem arises in reactors containing reactive solid or catalyst particles. During chemical reaction the particles themselves pass through various states of thermal equilibrium, and regions of instability will exist along the reactor bed. Consider, for example, a first order catalytic reaction in an adiabatic tubular reactor and further suppose that the reactor operates in a region where there is no diffusion limitation within the particles. The steady state condition for reaction in the particle may then be expressed by equating the rate of chemical reaction to the rate of mass transfer. The rate of chemical reaction per unit reactor volume will be $(1 - e)kC_i$ since the effectiveness factor η is considered to be unity. From equation 2.66 the rate of mass transfer per unit volume is $(1 - e) (S_x/V_p)h_D(C - C_i)$ so the steady state condition is:

$$kC_i = h_D \frac{S_x}{V_p}(C - C_i) \qquad \qquad ....(2.117)$$

Similarly, equating the heat generated by reaction to the heat flux from the particle:

$$(- \Delta H)kC_i = h \frac{S_x}{V_p}(T_i - T) \qquad \qquad ....(2.118)$$

where h is the heat transfer coefficient between the particle and the gas, and T_i is the temperature at the particle surface.

Eliminating C_i from equations 2.117 and 2.118:

$$T_i - T = (-\Delta H) \frac{V_p}{S_x} \frac{1}{h} \frac{kC}{1 + \dfrac{V_p k}{S_x h_D}} \qquad \qquad ....(2.119)$$

Now the maximum temperature T_{MX} achieved for a given conversion in an adiabatic reactor may be adduced from the heat balance:

$$T_{MX} = T + \frac{(-\Delta H)C}{\rho \bar{c}_p} \qquad \qquad ....(2.120)$$

which is analogous to equation 2.102, except that concentration has been expressed in terms of mols per unit volume, C ($=\rho Y$), with the gas density ρ

regarded as virtually constant. Combining equations 2.119 and 2.120, the temperature difference between the particle surface and the bulk gas may be expressed in dimensionless form:

$$\frac{T_i - T}{T_{MX} - T} = \frac{ak}{1 + bk} \qquad \qquad \dots(2.121)$$

where k is the usual Arrhenius' function of temperature,

a is the constant $\dfrac{\rho \bar{c}_p V_p}{S_x h}$, and

b is the constant $\dfrac{V_p}{S_x h_D}$.

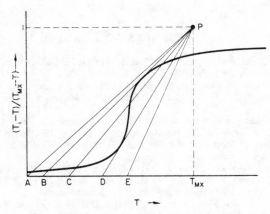

FIG. 2.30. Stability states of particles in adiabatic bed

The condition of thermal equilibrium for a particle at some point along the reactor is thus established by equation 2.121. The function on the right-hand side of this equation is represented by a sigmoid-shaped curve as shown in Fig. 2.30, and is as a function of bed temperature $T(z)$ for some position z along the reactor. The left-hand side of equation 2.121 (as a function of T) is a family of straight lines all of which pass through the common pole point $(T_{MX}, 1)$ designated P. It is apparent from Fig. 2.30 that as the temperature increases along the bed it is possible to have a sequence of states corresponding to lines such as PA, PB, PC, PD and PE. As soon as a position has been reached along the reactor where the temperature is B there is the possibility of two stable temperatures for a particle. Such a situation would persist until a position along the reactor has been reached where the temperature is D, after which only one stable state is possible. Thus, there will be an infinite family of steady state profiles within this region of the bed. Which profile actually obtains depends on start-up and bed entrance conditions. However, for steady state operating conditions, the work of McGreavy and Thornton[53] offers some hope to designers in that it may be possible to predict and thus avoid regions of instability in cooled packed-bed catalytic reactors.

FURTHER READING

ARIS, R.: *Elementary Chemical Reactor Analysis* (Prentice-Hall 1969).
ARIS, R.: *The Mathematical Theory of Diffusion and Reaction in Permeable Catalysts*, Vols. I and II (Oxford University Press 1975).
CARBERRY, J. J.: *Chemical and Catalytic Reaction Engineering* (McGraw-Hill 1976).
I.C.I. LTD.: *Catalyst Handbook* (Wolfe Scientific Books 1970).
LAPIDUS, L. and AMUNDSON, N. R. (eds.): *Chemical Reactor Theory—A Review* (Prentice-Hall 1977).
PETERSEN, E. E.: *Chemical Reaction Analysis* (Prentice-Hall 1965).
RASE, H. F.: *Chemical Reactor Design for Process Plants*, Vols. 1 and 2 (Wiley 1977).
SATTERFIELD, C. N.: *Mass Transfer in Heterogeneous Catalysis* (M.I.T. Press 1970).
SMITH, J. M.: *Chemical Engineering Kinetics*, 2nd ed. (McGraw-Hill 1970).
SZEKELY, J., EVANS, J. M. and SOHN, H. Y.: *Gas–Solid Reactions* (Academic Press 1976).
THOMAS, J. M. and THOMAS, W. J.: *Introduction to the Principles of Heterogeneous Catalysis* (Academic Press 1967).

REFERENCES TO CHAPTER 2

[1] HERZFELD, K. and SMALLWOOD, M.: in *Treatise on Physical Chemistry* (edited by H. S. Taylor) **1**, 169 (Princeton Univ. Press 1931).
[2] HOUGEN, O. A. and WATSON, K. M.: *Chemical Process Principles*, Part 3, 938 (Wiley 1947).
[3] HIRSCHFELDER, J. O., CURTISS, C. F. and BIRD, R. B.: *Molecular Theory of Gases and Liquids* (Wiley 1954).
[4] ARIS, R.: *Introduction to the Analysis of Chemical Reactors*, 236 (Prentice-Hall 1965).
[5] PETERSEN, E. E.: *Chemical Reaction Analysis*, 186 (Prentice-Hall 1965).
[6] BOUSSINESQ, M. J.: *Mém. prés. div. Sav. Acad. Sci. Inst. Fr.* **23** (1877) 1–680 (see 46). Essai sur la théorie des eaux courantes.
[7] KNUDSEN, M.: *Ann. Phys.* **28** (1909) 75. Die Gesetze der Molekularströmung und der inneren Reibungströmung der Gase durch Röhren. *Ann. Phys.* **35** (1911) 389. Molekularströmung des Wasserstoffs durch Röhren und das Hitzdrahtmanometer.
[8] GILLILAND, E. R.: *Ind. Eng. Chem.* **26** (1934) 681. Diffusion coefficients in gaseous systems.
[9] THIELE, E. W.: *Ind. Eng. Chem.* **31** (1939) 916. Relation between catalytic activity and size of particle.
[10] BOSANQUET, C. H.: *British TA Report* BR/507 (Sept. 27th 1944). The optimum pressure for a diffusion separation plant.
[11] DENBIGH, K. G.: *Trans. Faraday Soc.* **40** (1944) 352. Velocity and yield in continuous reaction systems.
[12] POLLARD, W. G. and PRESENT, R. D.: *Phys. Rev.* **73** (1948) 762. Gaseous self-diffusion in long capillary tubes.
[13] WHEELER, A.: *Adv. Catalysis* **3** (1951) 249. Reaction rates and selectivity in catalyst pores.
[14] BARRER, R. M. and GROVE, D. M.: *Trans. Faraday Soc.* **47** (1951) 826, 837. Flow of gases and vapours in a porous medium and its bearing on adsorption problems: I. Steady state of flow, II. Transient flow.
[15] ERGUN, S.: *Chem. Eng. Prog.* **48** No. 2 (Feb. 1952) 89. Fluid flow through packed columns.
[16] KRAMERS, H. and ALBERDA, G.: *Chem. Eng. Sci.* **2** (1953) 173. Frequency-response analysis of continuous-flow systems.
[17] VAN HEERDEN, C.: *Ind. Eng. Chem.* **45** (1953) 1242. Autothermic processes—properties and reactor design.
[18] WEISZ, P. B. and PRATER, C. D.: *Adv. Catalysis* **6** (1954) 143. Interpretation of measurements in experimental catalysis.
[19] WEHNER, J. F. and WILHELM, R. H.: *Chem. Eng. Sci.* **6** (1956) 89. Boundary conditions of flow reactor.
[20] BILOUS, O. and AMUNDSON, N. R.: *A.I.Ch.E.Jl.* **2** (1956) 117. Chemical reactor stability and sensitivity II. Effect of parameters on sensitivity of empty tubular reactors.
[21] ARIS, R. and AMUNDSON, N. R.: *A.I.Ch.E.Jl.* **3** (1957) 280. Longitudinal mixing or diffusion in fixed beds.
[22] McHENRY, K. W. and WILHELM, R. H.: *A.I.Ch.E.Jl.* **3** (1957) 83. Axial mixing of binary gas mixtures flowing in a random bed of spheres.

[23] EERKENS, J. W. and GROSSMAN, L. M.: *Tech. Report* HE/150/150 (Univ. Calif. Inst. Eng. Res., Dec. 5th 1957). Evaluation of the diffusion equation and tabulation of experimental diffusion coefficients.

[24] DENBIGH, K. G.: *Chem. Eng. Sci.* **8** (1958) 125. Optimum temperature sequences in reactors.

[25] EVERETT, D. H.: *Colston Res. Symp.* (Bristol, 1958). The structure and properties of porous materials: Some problems in the investigation of porosity by adsorption methods.

[26] MAY, W. G.: *Chem. Eng. Prog.* **55** No. 12 (Dec. 1959) 49. Fluidized-bed reactor studies.

[27] TAMARU, K.: *Trans. Faraday Soc.* **55** (1959) 824. Adsorption during the catalytic decomposition of formic acid on silver and nickel catalysts.

[28] THALLER, L. and THODOS, G.: *A.I.Ch.E.Jl.* **6** (1960) 369. The dual nature of a catalytic reaction: the dehydration of sec-butyl alcohol to methyl ethyl ketone at elevated pressures.

[29] DEANS, H. A. and LAPIDUS, L.: *A.I.Ch.E.Jl.* **6** (1960) 656, 663. A computational model for predicting and correlating the behavior of fixed-bed reactors: I. Derivation of model for nonreactive systems, II. Extension to chemically reactive systems.

[30] HORN, F.: *Z. Electrochem.* **65** (1961) 209. Optimum temperature regulation for continuous chemical processes.

[31] TINKLER, J. D. and METZNER, A. B.: *Ind. Eng. Chem.* **53** (1961) 663. Reaction rate in nonisothermal catalysts.

[32] CARBERRY, J. J.: *A.I.Ch.E.Jl.* **7** (1961) 350. The catalytic effectiveness factor under nonisothermal conditions.

[33] BEEK, J.: *Advances in Chemical Engineering* **3** (1962) 203. Design of packed catalytic reactors.

[34] WEISZ, P. B. and HICKS, J. S.: *Chem. Eng. Sci.* **17** (1962) 265. Behaviour of porous catalyst particles in view of internal mass and heat diffusion effects.

[35] SCOTT, D. S. and DULLIEN, F. A. L.: *A.I.Ch.E.Jl.* **8** (1962) 113. Diffusion of ideal gases in capillaries and porous solids.

[36] PETERSEN, E. E.: *Chem. Eng. Sci.* **17** (1962) 987. Nonisothermal chemical reaction in porous catalysts.

[37] ROWE, P. N.: *Chem. Eng. Prog.* **60** No. 3 (March 1964) 75. Gas-solid reaction in a fluidized bed.

[38] ØSTERGAARD, K.: *Proc. 3rd Int. Cong. Catalysis, Amsterdam* **2** (1964) 1348. The influence of intraparticle heat and mass diffusion on the selectivity of parallel heterogeneous catalytic reactions.

[39] SINFELT, J. H.: *Advances in Chemical Engineering* **5** (1964) 37. Bifunctional catalysis.

[40] GUNN, D. J. and THOMAS, W. J.: *Chem. Eng. Sci.* **20** (1965) 89. Mass transport and chemical reaction in multifunctional catalyst systems.

[41] PETERSEN, E. E.: *Chem. Eng. Sci.* **20** (1965) 587. A general criterion for diffusion influenced chemical reactions in porous solids.

[42] GUNN, D. J.: *Chem. Eng. Sci.* **21** (1966) 383. Nonisothermal reaction in catalyst particles.

[43] THOMAS, W. J. and JOHN, B.: *Trans. Inst. Chem. Eng.* **45** (1967) T119. Kinetics and catalysis of the reactions between sulphur and hydrocarbons.

[44] CHOU, A., RAY, W. H. and ARIS, R.: *Trans. Inst. Chem. Eng.* **45** (1967) T153. Simple control policies for reactors with catalyst decay.

[45] JACKSON, R.: *Trans. Inst. Chem. Eng.* **45** (1967) T160. An approach to the numerical solution of time-dependent optimisation problems in two-phase contacting devices.

[46] GUNN, D. J.: *Chem. Eng. Sci.* **22** (1967) 963. The optimisation of bifunctional catalyst systems.

[47] GUNN, D. J.: *Chem. Eng. Sci.* **22** (1967) 1439. Diffusion and chemical reaction in catalysis.

[48] THOMAS, W. J. and WOOD, R. M.: *Chem. Eng. Sci.* **22** (1967) 1607. Use of maximum principle to calculate optimum catalyst composition profile for bifunctional catalyst systems contained in tubular reactors.

[49] JACKSON, R.: *J. Optim. Theory Applic.* **2** (1968) 1. Optimal use of mixed catalysts.

[50] OGUNYE, A. F. and RAY, W. H.: *Trans. Inst. Chem. Eng.* **46** (1968) T225. Non-simple control policies for reactors with catalyst decay.

[51] GUNN, D. J. and PRYCE, C.: *Trans. Inst. Chem. Eng.* **47** (1969) T341. Dispersion in packed beds.

[52] JENKINS, B. and THOMAS, W. J.: *Can. J. Chem. Eng.* **48** (1970) 179. Optimum catalyst formulation for the aromatization of methylcyclopentane.

[53] McGREAVY, C. and THORNTON, J. M.: *Can. J. Chem. Eng.* **48** (1970) 187. Generalized criteria for the stability of catalytic reactors.

[54] MACDONALD, W. R. and HABGOOD, H. W.: *19th Canadian Chem. Eng. Conf., Edmonton* (1969) Paper 45. Measurement of diffusivities in zeolites by gas chromatographic methods.

LIST OF SYMBOLS USED IN CHAPTER 2

$\mathcal{A}$	Frequency factor in Arrhenius equation		$T^{-1}*$
A_c	Cross-sectional area		L^2
a	Interfacial area per unit volume		L^{-1}
a_B	Activity of component **B**		—
a'	Surface area for heat transfer per unit hydraulic radius		L
B	$\left(=\dfrac{-\Delta H C_\infty D_e \mathbf{R}}{2rhE}\right)$ Dimensionless parameter introduced by McGreavy and Thornton[53] (see Fig. 2.7)		
C	Total molar concentration		ML^{-3}
C_A	Molar concentration of component **A**		ML^{-3}
C_A	Surface concentration of **A**		L^{-2}
C_i	Molar concentration at interface		ML^{-3}
C_M	Molar concentration at centre of pellet		ML^{-3}
C_S	Concentration of available sites		L^{-2}
C_V	Concentration of vacant sites		L^{-2}
C_∞	Molar concentration at infinity		ML^{-3}
$\bar{c}_p$	Mean heat capacity of fluid per unit mass	$(HM^{-1}\theta^{-1})$	$L^2T^{-2}\theta^{-1}$
$\bar{c}_{pC}$	Mean heat capacity of coolant per unit mass	$(HM^{-1}\theta^{-1})$	$L^2T^{-2}\theta^{-1}$
D	Molecular bulk diffusion coefficient		L^2T^{-1}
D_e	Effective diffusivity		L^2T^{-1}
D_{eK}	Effective diffusivity in Knudsen regime		L^2T^{-1}
D_{eM}	Effective diffusivity in molecular regime		L^2T^{-1}
D_K	Knudsen diffusion coefficient		L^2T^{-1}
D_P	Diffusion coefficient for forced flow		L^2T^{-1}
D'_e	Effective diffusivity based on concentration expressed as Y		$ML^{-1}T^{-1}$
D'_L	Dispersion coefficient in longitudinal direction based on concentration expressed as Y		$ML^{-1}T^{-1}$
D'_r	Radial dispersion coefficient based on concentration expressed as Y		$ML^{-1}T^{-1}$
d_p	Particle diameter		L
d'_M	Volume of voids per total surface area		L
E	Activation energy per mol for chemical reaction		L^2T^{-2}
E_D	Apparent activation energy per mol in diffusion controlled region		L^2T^{-2}
e	Void fraction		—
F	Ratio of activity in poisoned catalyst pellet to activity in unpoisoned pellet		—
G	Mass flow rate		MT^{-1}
G'	Mass flow per unit area		$ML^{-2}T^{-1}$
G'_C	Mass flow rate of coolant per unit cross-section of reactor tube		$ML^{-2}T^{-1}$
ΔH	Difference of enthalpy between products and reactants (heat of reaction $= -\Delta H$)	(HM^{-1})	L^2T^{-2}
h	Heat transfer coefficient	$(HL^{-2}T^{-1}\theta^{-1})$	$MT^{-3}\theta^{-1}$
h_D	Mass transfer coefficient		LT^{-1}
j	Mass transfer factor		—
K	Thermodynamic equilibrium constant—for $\mathbf{A} + \mathbf{B} \rightleftharpoons \mathbf{P}$		$M^{-1}LT^2$
	—for $\mathbf{A} \rightleftharpoons \mathbf{P}$		—
K_A	Adsorption-desorption equilibrium constant for **A**		$M^{-1}LT^2$
K_B	Adsorption-desorption equilibrium constant for **B**		$M^{-1}LT^2$
K_P	Adsorption-desorption equilibrium constant for **P**		$M^{-1}LT^2$
K_S	Equilibrium constant for surface reaction		
k	Chemical rate constant for forward reaction, per unit volume of particle or reactor		$T^{-1}*$
k'	Chemical rate constant for reverse reaction, per unit volume of particle or reactor		$T^{-1}*$
$\bar{k}$	Overall rate or reactor constant per unit volume		$T^{-1}*$
$\mathbf{k}$	Chemical rate constant for forward surface reaction (molecules per unit area and unit time)		$L^{-2}T^{-1}$
$\mathbf{k}'$	Chemical rate constant for reverse surface reaction (molecules per unit area and unit time)		$L^{-2}T^{-1}$
$\mathbf{k}_a$	Chemical rate constant per unit area		$LT^{-1}*$
$\mathbf{k}_{Aa}$	Rate constant for adsorption of **A** (see equation 2.79)		$M^{-1}L^{-1}T$

---

* For first order reaction

k_{Ad}	Rate constant for desorption of **A** (see equation 2.79)		$L^{-2}T^{-1}$
k_e	Effective thermal conductivity	$(HL^{-1}T^{-1}\theta^{-1})$	$MLT^{-3}\theta^{-1}$
k_G	Gas film mass transfer coefficient		$L^{-1}T$
k_l	Thermal conductivity in longitudinal direction	$(HL^{-1}T^{-1}\theta^{-1})$	$MLT^{-3}\theta^{-1}$
k_{Pa}	Rate constant for adsorption of **P** (see equation 2.81)		$M^{-1}L^{-1}T$
k_{Pd}	Rate constant for desorption of **P** (see equation 2.81)		$L^{-2}T^{-1}$
k_r	Thermal conductivity in radial direction	$(HL^{-1}T^{-1}\theta^{-1})$	$MLT^{-3}\theta^{-1}$
k_S	Chemical rate constant for forward surface reaction (molecules per unit time)		T^{-1}
k'_S	Chemical rate constant for reverse surface reaction (molecules per unit time)		T^{-1}
k'_e	Rate constant for reverse surface reaction (molecules per unit time)		T^{-1}
L	Length of reactor		L
L	Half-thickness of slab *or* distance given by equation 2.70		L
M	Molecular weight		
N	Avogadro's number		M^{-1}
N	Molar flux of material		$ML^{-2}T^{-1}$
n	Order of chemical reaction		—
n_A	Number of adsorbed molecules of component **A**		—
n_D	Apparent order of chemical reaction in diffusion controlled regime		—
n_P	Number of adsorbed molecules of component **P**		—
n_S	Number of available active sites		—
P	Pressure		$ML^{-1}T^{-2}$
P_A	Partial pressure of component **A**		$ML^{-1}T^{-2}$
P_{Ai}	Partial pressure of component **A** at interface		$ML^{-1}T^{-2}$
Q	Heat transferred per unit volume and unit time to surroundings	$(HL^{-3}T^{-1})$	$ML^{-1}T^{-3}$
Q_1^M	Heat generated by chemical reaction per unit mass	(HM^{-1})	L^2T^{-2}
Q_2^M	Heat gain by feed to the reactor per unit mass	(HM^{-1})	L^2T^{-2}
R	Gas constant		$L^2T^{-2}\theta^{-1}$
R_1	Component of drag force per unit area in direction of flow		$ML^{-1}T^{-2}$
$\mathscr{R}^M$	Reaction rate (mols per unit mass of catalyst and unit time)		T^{-1}
$\mathscr{R}^M_{YT}$	Reaction rate (mols per unit mass of catalyst and unit time) when a function of Y and T		T^{-1}
$\mathscr{R}_V$	Reaction rate (mols per unit volume of reactor and unit time)		$ML^{-3}T^{-1}$
$\mathscr{R}'_V$	Reaction rate (mols per unit volume of particle and unit time)		$ML^{-3}T^{-1}$
$\mathscr{R}''$	Surface reaction rate (molecules per unit area and unit time)		$L^{-2}T^{-1}$
$\mathscr{R}''_{Aa}$	Rate of adsorption of **A** (molecules per unit area and unit time)		$L^{-2}T^{-1}$
$\mathscr{R}''_{AS}$	Rate of surface reaction of **A** (molecules per unit area and unit time)		$L^{-2}T^{-1}$
$\mathscr{R}''_{Pd}$	Rate of desorption of **P** (molecules per unit area and unit time)		$L^{-2}T^{-1}$
$\mathscr{R}''_S$	Rate of surface reaction (molecules per unit area and unit time)		$L^{-2}T^{-1}$
r	Radius		L
r_e	Radius of cylinder with same surface to volume ratio as pore		L
S	Catalyst selectivity		—
S_g	Specific surface area		$M^{-1}L^2$
S_x	External surface area		L^2
S'	Total catalyst surface area		L^2
T	Temperature		θ
T_C	Temperature of coolant stream		θ
T_{CL}	Temperature of coolant at reactor outlet		θ
T_{Co}	Temperature of coolant at reactor inlet		θ
T_i	Temperature at exterior surface of pellet		θ
T_L	Temperature at exit of reactor		θ
T_M	Temperature at particle centre		θ
T_{MX}	Maximum temperature		θ
T_W	Wall temperature		θ
ΔT	$(= T_0 - T_{CL})$ Temperature difference of the non-reacting fluid over length of reactor		θ
U	Overall heat transfer coefficient	$(HL^{-2}T^{-1}\theta^{-1})$	$MT^{-3}\theta^{-1}$
u	Gas velocity		LT^{-1}
u_1	Mean velocity in voids		LT^{-1}
V_g	Specific pore volume		$M^{-1}L^3$
V_p	Particle volume		L^3
v	Volumetric rate of flow		L^3T^{-1}
W	Mass of catalyst		M

x	Distance in x-direction	**L**
x_A	Mol fraction of **A**	—
Y	Mols of reactant per unit mass of fluid	—
Y_L	Mols per unit mass of reactant (or product) at reactor exit	—
Y_{MX}	Mols per unit mass of reactant (or product) for which temperature is maximum	—
Y_0	Mols per unit mass of reactant (or product) at reactor inlet	—
z	Length variable for reactor tube	**L**
$Pe.$	$(=u d_p/D)$ Peclet number	—
$Re.$	$(=u \rho d_p/\mu)$ Reynolds number	—
$Sc.$	$(=\mu/\rho D)$ Schmidt number	—
$Sh.$	$(=h_D d_p/D)$ Sherwood number	—
$Tr.$	$(=C_i/C_\infty)$ Transport factor	—
α	Factor characterising pore geometry	—
β	Parameter representing maximum temperature difference in particle relative to external surface (equation 2.32)	—
Γ	$(=G'\bar{c}_p/G'_C \bar{c}_{pC})$ Relative heat capacities of two fluid streams to exchange heat	—
Γ'	$(=U a L/G'_C \bar{c}_{pC})$ Heat transfer factor	—
δ	Film thickness	**L**
δ_A	Net difference in number of mols of product and reactant per mol of component **A**	—
ε	Dimensionless activation energy (equation 2.33)	—
ζ	Fraction of surface poisoned	—
η	Effectiveness factor	—
θ_A	Fraction of surface occupied by component **A**	—
λ	Defined as $\sqrt{\dfrac{k}{D_e}}$ for first order process (equation 2.20)	**L**$^{-1}$
μ	Viscosity	**ML**$^{-1}$**T**$^{-1}$
ρ	Fluid density	**ML**$^{-3}$
ρ_b	Bulk density	**ML**$^{-3}$
ρ_p	Particle density	**ML**$^{-3}$
ρ_s	True density of solid	**ML**$^{-3}$
τ	Tortuosity factor	—
ϕ	Thiele modulus (equation 2.23)	—
$\bar{\phi}$	Generalised Thiele modulus (equation 2.24)	—
ϕ'	Modified Thiele modulus (equation 2.39)	—
χ	Degree or percentage reaction	—
ψ	Particle porosity	—

Subscripts

A	Refers to component **A**
B	Refers to component **B**
P	Refers to product **P**
X	Refers to component **X**

CHAPTER 3

Process control

CONTROL in one form or another is an essential part of any chemical engineering operation. In all processes there arises the necessity of keeping flows, pressures, temperatures, compositions, etc., within certain limits for reasons of safety or specification. Such control is most often accomplished simply by measuring the variable it is required to control (the *controlled variable*), comparing this measurement with the value at which it is desired to maintain the controlled variable (the *desired value* or *set point*) and adjusting some further variable (the *manipulated variable*) which has a direct effect on the controlled variable until the desired value is obtained.

In order to design such a system to operate not only automatically but efficiently, it is necessary to obtain both the steady-state and dynamic (unsteady state) relationships between the particular variables involved. It can be seen that automatic operation is highly desirable, as manual control would necessitate continuous monitoring of the controlled variable by a human operator. The efficiency of observation of the operator would inevitably fall off with time. Furthermore, fluctuations in the controlled variable may be too rapid and frequent for manual adjustment to suffice.

THE CONTROL LOOP AND BLOCK DIAGRAM

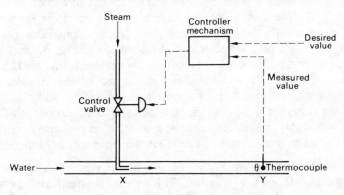

FIG. 3.1. Simple feedback control system

A simple control system, or loop, is illustrated in Fig. 3.1. The temperature (θ) of the water at Y is measured by means of a thermocouple, the output of

185

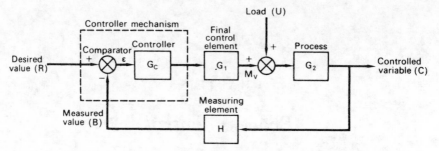

FIG. 3.2. Block diagram of the feedback control system in Fig. 3.1

which is fed to a controller mechanism. The latter can be divided into two sections (normally housed in the same unit). In the first (the *comparator*) the *measured value* (B) is compared with the desired value (R) to produce an *error* (ε), where:

$$\varepsilon = R - B \qquad \qquad ....(3.1)$$

The second section of the mechanism (the *controller*) produces an output which is a function of the magnitude of ε. This is fed to a control valve in the steam line, so that the valve closes when θ increases and vice versa. The system as shown may be used to counteract fluctuations in temperature due to extraneous causes such as variations in water flowrate or upstream temperature— termed *load* changes. It may also be employed to change the water temperature at Y to a new value by adjustment of the desired value.

A control system may be more simply represented in the form of a *block diagram* (Fig. 3.2). This shows how information flows around the control loop and the function of each constituent section. Each component is represented by a block which denotes the relationship between the variable entering the block and the variable leaving. The symbols used in Fig. 3.2 are widely employed by control engineers[1] although considerable variation does occur in the literature[2, 3, 4]. The control loop is generally made up of five essential parts, i.e. (a) the process, (b) the measuring element, (c) the comparator, (d) the controller, and (e) the final control element. Comparison of Figs. 3.1 and 3.2 shows that the final control element is the control valve on the steam line. The manipulated variable (M_v) is the steam flowrate or the flow of heat to the water. The load (U) enters the loop at this point as changes in load will affect the heat entering the system. The input to the process is the sum effect of both U and M_v. The process, in this case, is simply the movement of any change in temperature from X to Y. The controlled variable (C) is the temperature of the water at Y.

It can be seen that the term *control loop* is appropriate as information passes around a closed loop of components. This form of control is called *closed-loop* or *feedback* (referring to the feed-back of information from the controlled variable to the comparator). The control loop as shown in Figs. 3.1 and 3.2 could equally well consist of electronic or pneumatic components or a mixture of both. The choice of which to use is dictated by consideration of cost, accuracy and safety. Although pneumatic mechanisms were used almost universally for many years, electronic installations are now rapidly gaining in popularity. PALMER[70] has discussed the relative advantages of each.

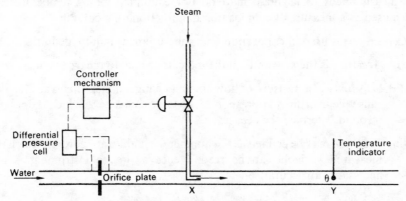

FIG. 3.3. Feed-forward control system

Another type of control is occasionally employed which does not require the feed-back of information concerning the controlled variable. This is termed *feed-forward*, *predictive* or *open-loop* control. A possible arrangement is shown in Fig. 3.3. It is assumed that the inlet water temperature remains constant. The heat input to the water is adjusted directly by measurement of the water flowrate. This method has the advantage of anticipating the effect on θ of variations in water flowrate and that θ will not have to change from its desired value before corrective action can be taken (as with the feedback arrangement). The difficulty is that in order to design such a predictive system it is necessary to determine first how the temperature at Y will respond to changes in both water and steam flowrates. This becomes a considerable problem with more complex systems[59, 64].

PROCESS CONTROL EQUIPMENT

Equipment used for monitoring and controlling process streams is very extensive. The physical construction and details of such equipment are generally found in books on instrumentation. Many excellent books have been published which give detailed descriptions of a great deal of the equipment in everyday use[6, 7, 8, 9, 11, 30]. A brief discussion of some of the more common control equipment appears below. For further details the reader is referred to the references quoted and to current manufacturers' literature.

The measuring element

This is probably the most important component of any control system. Without accurate measurement of the controlled variable it is impossible to obtain satisfactory control. Variables normally selected for measurement are temperature, pressure, flowrate, liquid level and composition. Frequently, a variable which is a function of the variable it is desired to control is selected for measurement to facilitate matters. For example, boiling temperature is often used as a measure of composition in a fractionating column.

Certain terms used in connection with measurement require definition, viz.:

(i) *Accuracy* is the proximity of the indicated value to the true value.

(ii) *Sensitivity* is a measure of how small a change in measured variable will cause the instrument to respond. The largest change to which it does not respond is termed the *dead-zone*.

(iii) *Precision* may be defined as the number of significant figures to which the measured variable can be read. Hence a precise instrument is not necessarily an accurate one.

Measurement of temperature

There are numerous devices used for measuring temperature and each has its own characteristics and limitations. JONES[6] has classified these instruments according to the nature of the change produced in the testing body by the change of temperature, i.e.:

(a) expansion thermometers,
(b) change of state thermometers,
(c) instruments exploiting electrical phenomena, and
(d) radiation and optical pyrometers.

Class (a) covers bi-metallic strip, liquid-in-glass, liquid-in-metal and gas thermometers.

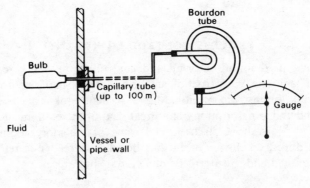

FIG. 3.4. Mercury-in-steel (pressure-spring) thermometer

A typical liquid-in-metal thermometer is shown in Fig. 3.4. As the temperature surrounding the bulb is raised, the mercury inside expands to a greater extent than the bulb, thus forcing the coiled Bourdon tube to open and move the pointer. The gas thermometer operates on a similar principle. Both are simple and reliable and are frequently used to record and control.

Class (b) principally covers vapour-pressure thermometers which are again similar in construction to Fig. 3.4. The liquid however only partly fills the bulb and the Bourdon tube is moved by variations in the vapour pressure above the liquid.

An important instrument in class (c) is the thermocouple. This is one of the most common temperature measuring devices used in the process industries. It makes use of the fact that an e.m.f. is generated when strips of two dissimilar metals are joined at their ends, one end being at a different temperature from the other. The e.m.f. generated is measured using a millivoltmeter or potentiometer (Fig. 3.5).

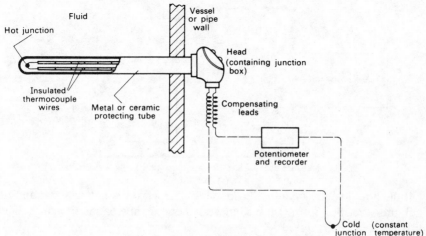

FIG. 3.5. Industrial thermocouple arrangement

One junction of the thermocouple is inserted into the medium whose temperature is to be measured whilst the temperature of the other junction is kept constant. Hence the e.m.f. generated varies with the temperature of the medium.

Another device in class (c) is the resistance thermometer which is the most accurate type of temperature measuring instrument normally used. It consists of a sensitive element whose resistance varies with temperature and a Wheatstone bridge network to measure the resistance. The industrial instrument is similar in appearance to the thermocouple but the detecting tube contains a coiled resistance winding. Whereas the thermocouple measures point temperatures, the resistance thermometer gives the average temperature of a larger sensitive element.

Instruments in class (d) are used for the measurement of high temperatures (> 400°C). The optical pyrometer usually compares the monochromatic radiation emitted by the hot object of unknown temperature with that from a standard source. The radiation pyrometer concentrates the radiant energy from a hot source on to a thermocouple junction.

Measurement of pressure and flowrate

Instruments commonly used for the measurement of both these variables are discussed in Volume 1, Chapter 5.

Measurement of liquid level

Liquid level is normally determined either by the use of a float, by measurement of the static head of liquid, or by a capacitance bridge method. The static pressure exerted by a head of liquid can be measured by any suitable pressure measuring device. A typical arrangement for a volatile liquid is shown in Fig. 3.6.

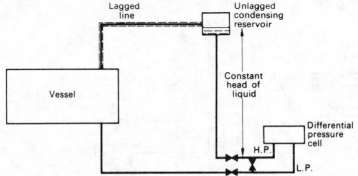

FIG. 3.6. Measurement of level of volatile liquid using a differential pressure cell

If the liquid is not volatile, the condensing reservoir is not required and the high pressure and low pressure connections on the differential pressure cell are disconnected and reversed.

Floats are independent of the static pressure and are either of the buoyant or the displacement type. The former is less dense than the liquid in which it is partially immersed and the latter is slightly denser than the liquid. Consequently the buoyant or *moving* float has a constant immersion and will rise and fall precisely the same distance as the liquid level. Its *position* at any instant is measured. The displacement or *static* type has a variable immersion and a variable upthrust upon it. The level is measured in this case in terms of the *net weight* of the float.

The capacitance method is generally employed with large cylindrical storage tanks. One electrode is inserted down the centre of the tank whilst the outer wall is used as the other. The tank then acts as a concentric cylindrical capacitor having a capacitance which alters with liquid level. The capacitance (and hence the liquid level) is measured using a stabilised capacitance bridge.

Measurement of composition

Although compositions of process streams are frequently monitored and controlled by means of temperature and pressure measurement, it is becoming increasingly common to measure compositions continuously by more direct methods. A variety of physical and chemical properties which are employed for continuous analysis have been listed by HOLZBOCK[7] and MAWSON[65]. Commercial instruments designed to measure such properties are described by SIGGIA[9]. The more frequently used methods are chromatography, ultraviolet analysis, infra-red analysis, refraction, specific gravity, pH, dielectric constant, and flash point. Several papers have been published on the use of composition analysers for control purposes[41, 43, 45].

The final control element

The only final control element of any significance is the automatic control valve.

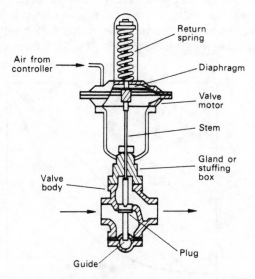

FIG. 3.7. Pneumatic control valve

A typical arrangement is shown in Fig. 3.7. It consists of two main sections, the valve motor and the valve body. The motor section may be electronic or pneumatic in operation. If it is pneumatic the air pressure may be applied above or below the diaphragm depending on whether, for reasons of safety, the valve should open or shut in the event that the air supply should fail.

Valves may be single- or double-seated (Fig. 3.8). With single-seated valves the valve plug is subjected to the total differential pressure force across the valve. Such valves are sensitive to pressure fluctuations and powerful motor elements are required for large pressure drops. Double-seated valves balance out the pressure differential but it is difficult to obtain complete shut-off.

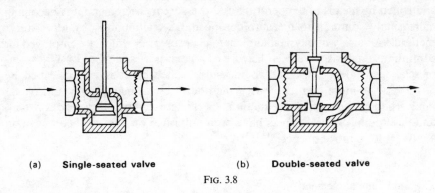

(a)   **Single-seated valve**   (b)   **Double-seated valve**

Fig. 3.8

The valve motor is generally constructed such that the valve-stem position is proportional to the force applied. The relation between the stem position and the flow through the valve at constant pressure drop is termed the *valve characteristic*. Two characteristics must be evaluated for valve selection, the *inherent* (Fig. 3.9) and the *installed* characteristics[10]. The former is the theoretical

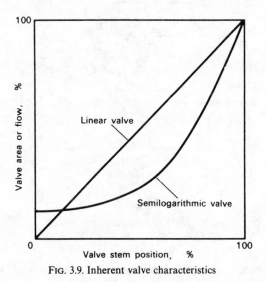

Fig. 3.9. Inherent valve characteristics

performance of the valve and is generally either *linear* or *semi-logarithmic* (frequently termed *equal percentage*). Hence, for a pneumatic motor element:

$$y_s - y_{s_0} = K_1(P_d - P_{d_0}) \qquad \qquad \ldots .(3.2)$$

where y_s is the distance travelled by the valve-stem,
   P_d is the pressure applied to the motor diaphragm,
   P_{d_0} is the applied pressure required to give a position y_{s_0} at $t = 0$, and
   K_1 is a constant.

With a linear valve:

$$y_s - y_{s_0} = K_2(Q - Q_0) \qquad \qquad \dots.(3.3)$$

where Q is the flowrate at constant pressure drop, and
Q_0 is the value of Q at $t = 0$.

For a semi-logarithmic valve, CONSIDINE[10] gives the relationship:

$$Q = Q_s \exp(jy_s) \qquad \qquad \dots.(3.4)$$

where Q_s is the flow at constant pressure drop at zero stroke, and
j is a constant which is characteristic of the valve.

The *turndown*, which is the ratio of the normal maximum flow through the valve to the minimum controllable flow, is usually about 70 per cent of the *rangeability*, which is the ratio of the maximum controllable flow through the valve to the minimum controllable flow. Thus standard practice is to size a control valve such that the maximum flow through it under normal operating conditions is approximately 70 per cent of the maximum possible flow.

The equal percentage valve is used where large rangeability is desired and where equal percentage characteristics are necessary to match process characteristics.

The installed characteristic depends upon the ratio of the pressure drop through the valve to the total pressure drop across the line and the valve. If a linear valve is handling all the system pressure drop, its installed characteristic will also be linear. As the percentage of the pressure drop handled falls, the characteristic changes rapidly to that of an on-off valve. The equal percentage valve also loses its characteristic in the same way, but to a far less extent. Hence, this type of valve is frequently chosen in preference to the linear type. The effects of choosing valves with incorrect characteristics have been illustrated by PETERS[35].

The controller

As indicated previously, the controller mechanism may be considered as consisting of two sections, viz. the comparator and the controller itself. The purpose of the first is to compare the measured and desired values of the controlled variable and to compute the difference between them as the error. The second section operates to alter the setting of the final control element in such a way as to minimise the error in the least possible time with the minimum disturbance to the system. The control action selected for the controller depends upon the dynamic behaviour of the other components in the control loop.

Types of control action

The simplest type of control which is commonly experienced is that having an *on-off* or *two-position* action. A typical example is the thermostatically controlled domestic immersion heater. Depending on the temperature of the water in the tank, the power supply is either connected to, or disconnected from, the heater. The relationship between controller input and output might appear as in Fig. 3.10.

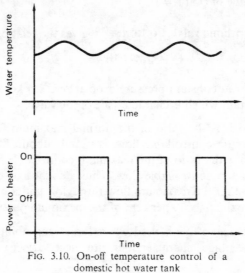

FIG. 3.10. On-off temperature control of a domestic hot water tank

Although such a system is inexpensive and extremely simple, the oscillatory nature of the control makes it suitable only for those applications where it can be used alone and close control is not essential. Consequently, its use in the control of industrial processes is severely limited and it will not be discussed further.

There are three principal types or *modes* of control action which are more generally employed, viz. *proportional* (**P**), *integral* (**I**) and *derivative* (**D**). In the first the controller produces an output signal which is proportional to the error. Thus, for a pneumatically operated mechanism:

$$P = P_0 + K_c \varepsilon \qquad \qquad \ldots (3.5)$$

where P is the output pressure signal,
K_c is the gain or sensitivity, and
P_0 is the controller output when there is no error.

(It is generally assumed when considering control system dynamics that the control system is at steady-state for $t < 0$, and that ε is zero for $t < 0$).

Hence, with proportional control, the greater the magnitude of the error the larger is the corrective action applied.

The integral and derivative modes are normally used in combination with the proportional mode. Integral action (frequently called *automatic reset*) gives an output which is proportional to the time integral of the error. Proportional plus integral (**P** + **I**) action may be represented thus:

$$P = P_0 + K_c\varepsilon + K_I \int_0^t \varepsilon \, dt \qquad \qquad \ldots(3.6)$$

where K_I is a constant.

Derivative action (often termed *rate control*) gives an output which is proportional to the derivative of the error. Hence, for **P + D** control:

$$P = P_0 + K_c\varepsilon + K_D \frac{d\varepsilon}{dt} \qquad \qquad \ldots(3.7)$$

where K_D is a constant.

Frequently all three modes are used together as **P + I + D** control, i.e.:

$$P = P_0 + K_c\varepsilon + K_I \int_0^t \varepsilon \, dt + K_D \frac{d\varepsilon}{dt} \qquad \qquad \ldots(3.8)$$

Reasons for use of particular control modes—offset

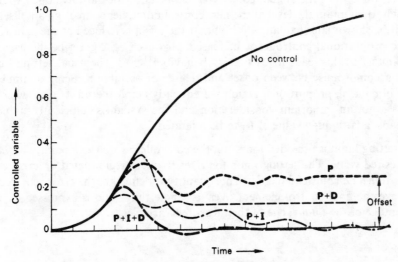

FIG. 3.11. Response of controlled variable to step disturbance in load using different control modes

Figure 3.11 shows typical responses of a controlled variable to a step disturbance in load for a simple control loop of the type shown in Fig. 3.2. The effects of different control actions can be summarised as follows.

(a) *Proportional control.* The response has a high maximum deviation and there is a significant time of oscillation. The period of this oscillation is moderate. For a sustained change in load, the controlled variable is not returned to its original value (the desired value) but attains a new equilibrium value termed the *control point*. This difference between desired value and control point is called the *offset* or *droop*. The reason for offset with proportional action can be seen if it is remembered that the control action is proportional to the error. Consider a simple level control system (Fig. 3.12). Suppose the flow of liquid

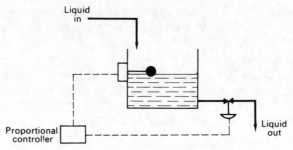

FIG. 3.12. Level control system

into the tank is increased. In order to maintain the level the valve on the outlet must be opened further. This will only occur if there is a continual output from the controller. This output itself can only exist if there is an error signal supplied to the controller. In order to maintain this error, the level will rise above the desired value until the system comes to equilibrium with the valve open wide enough to maintain the level at the new control point, hence creating an offset. It will be shown later (equation 3.109) that the offset is reduced as the gain K_c of the proportional controller is increased. However, K_c cannot be increased indefinitely as this leads to oscillatory behaviour[14, 36]. The usual setting for K_c is a compromise between offset and degree of oscillation. Being a simple form of control, proportional action is frequently employed on its own where offset is not an important consideration and the system is sufficiently stable to enable a fairly high value of K_c to be tolerated.

A setting knob is provided for K_c on the controller as well as a pointer to set the desired value. The setting knob for K_c is frequently graduated in terms of *proportional band* instead of gain. This quantity is defined as the error required to move the final control element over its whole range and is expressed as a percentage of the *total* range of the measured variable.

Example

In Fig. 3.12 the control system is used to control the liquid level within the range 1·85 m to 2·25 m. It is found that after adjustment the controller output pressure changes by 4000 N/m² for a 0·01 m variation in level with the desired value held constant. If a variation in output pressure of 80,000 N/m² moves the control valve from fully open to fully closed, determine the gain and proportional band.

Solution

From equation 3.5:

$$P_A = P_0 + K_c \varepsilon_A$$

$$P_B = P_0 + K_c \varepsilon_B$$

Subtracting gives:

$$(P_A - P_B) = K_c(\varepsilon_A - \varepsilon_B)$$

As the level changes by 0·01 m whilst the desired value is held constant, the error must change by 0·01 m with a change in output pressure of 4000 N/m².

$$\therefore\ K_c = \frac{4000}{0·01} = 4 \times 10^5 \text{ N/m}^2 \text{ per metre level}$$

If an error change of 0·01 m causes a change in output pressure of 4000 N/m^2 then the error required to move the valve from fully open to fully shut, assuming valve movement, output pressure and error to be linearly related, will be:

$$\frac{80,000}{4000} \times 0{\cdot}01 = 0{\cdot}2 \text{ m}$$

therefore, by definition:

$$\text{Proportional band} = \frac{0{\cdot}2}{2{\cdot}25 - 1{\cdot}85} \times 100$$

$$= \underline{50\%}$$

(b) P + I control

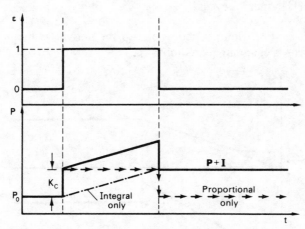

FIG. 3.13. Response of **P + I** controller to unit step change in error

It can be seen from equation 3.6 and Fig. 3.13 that the controller output will continue to increase as long as $\varepsilon > 0$. With proportional control an error (offset) had to be maintained in order to keep the level constant (Fig. 3.12) at a new control point after a step change in inlet flowrate. This error was necessary to produce an output pressure from the proportional controller to the control valve. However, with **P + I** control, the contribution from the integral action does not return to zero with the error but remains at the value it has reached at that time. This contribution provides the additional output pressure necessary to open the valve wide enough to keep the level at the desired value. No persisting error (i.e. no offset) is now necessary. Thus with integral action offset is eliminated. A theoretical discussion of this is given later (equation 3.111).

The disadvantages of **P + I** control are that it gives rise to a higher maximum deviation, a longer response time and a longer period of oscillation than with proportional action alone (see Fig. 3.11). This type of control action is therefore used where the above can be tolerated and offset is undesirable. It is a frequently used combination, especially where the responses of the other components in the control loop are rapid.

Equation 3.6 may be rewritten thus:

$$P = P_0 + K_c \left(\varepsilon + \frac{1}{\tau_I} \int_0^t \varepsilon \, dt \right) \qquad \qquad \ldots (3.9)$$

where τ_I is termed the *integral time*.

The maximum deviation etc. of the controlled variable is determined by the settings of both K_c and τ_I. Thus a second knob is provided on the controller and is calibrated either in terms of τ_I min or $1/\tau_I$ (min)$^{-1}$—the *reset rate*. PETERS[36] has illustrated how decreasing the integral time increases the tendency for the control system to cycle.

(c) **P + D** *control.* Control action due to the derivative mode occurs only when the error is changing (equation 3.7).

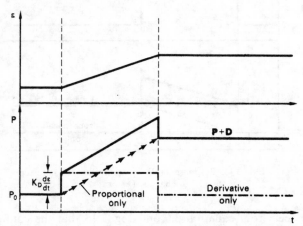

FIG. 3.14. Response of **P + D** controller to ramp change in error

The presence of the derivative mode contributes an additional output, $K_D(d\varepsilon/dt)$, to the final control element as soon as any change in error occurs. When the error ceases to change, derivative action no longer occurs (Fig. 3.14). The effect of this is similar to having a proportional controller with high gain when the measured variable is changing most rapidly, and low gain when the latter is varying slowly. Due to this action the controlled variable exhibits the least oscillation and lowest maximum deviation (Fig. 3.11). The same amount of offset occurs as with proportional control alone with the same proportional gain (page 234 *et seq.*). However, the addition of derivative action allows a higher gain to be used before the control system becomes unstable (page 248 *et seq.*). By this means a smaller offset can be obtained than with proportional action only. Derivative action is frequently termed *anticipatory* or *rate control* and is employed where excessive oscillations have to be eliminated.

Equation 3.7 may be rewritten:

$$P = P_0 + K_c\left(\varepsilon + \tau_D\frac{d\varepsilon}{dt}\right) \qquad \ldots.(3.10)$$

where τ_D is the *derivative time* measured normally in min. Means of setting both K_c and τ_D on the controller are provided.

(d) **P + I + D** *control*. This is essentially a compromise between the advantages and disadvantages of **P + I** and **P + D** control. Offset is eliminated by the presence of integral action and the derivative mode reduces the maximum deviation and time of oscillation although the latter are still greater than with **P + D** control alone (Fig. 3.11).

Here we have, from equation 3.8:

$$P = P_0 + K_c\left(\varepsilon + \frac{1}{\tau_I}\int_0^t \varepsilon\,dt + \tau_D\frac{d\varepsilon}{dt}\right) \qquad \ldots.(3.11)$$

Such controllers are frequently installed because of their versatility, and not because analysis of the system has indicated the need for the presence of all three modes of control.

Construction of pneumatic controllers

Brief descriptions of the pneumatic generation of the ideal control modes discussed above follows. It should be emphasised that the majority of industrial controllers give control actions which only approximate to these ideal cases. HOLZBOCK[7] and YOUNG[5] describe a number of industrial controllers—both pneumatic and electronic.

(a) *Proportional control. Narrow band action*. Figure 3.15 shows how a controller mechanism may be incorporated in a control loop. The pneumatic output from the differential pressure cell (ΔP_M) increases with flow through the orifice (see Volume 1, Chapter 5). This is transmitted to a recorder and controller usually located adjacently in the plant control room. The pneumatic signal is converted into a mechanical movement by means of a bellows and spring device A. The bellows movement is linked both to the recorder pen and to the flapper of the controller mechanism which pivots about X. If the fluid flowrate in the line increases, ΔP_M will rise, thus causing the flapper to move closer to the nozzle. This will restrict the flow of air through the nozzle and hence the pressure (P) upstream from the nozzle will rise. The latter is fed directly to the valve motor which acts in such a way that as P increases the valve shuts—thus reducing the flow in the line, and consequently ΔP_M. With this type of controller mechanism the movement of the flapper necessary to cause P to vary over its entire range is very small (approx. 2×10^{-5} m). GOULD and SMITH[40] have shown that

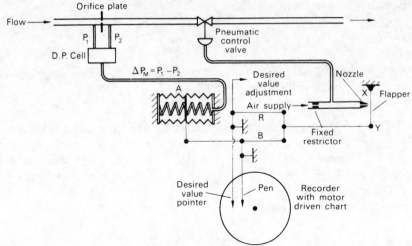

Fɪɢ. 3.15. Arrangement of simple pneumatic flow control loop with narrow band
proportional control

for such very small movements the change in output pressure (ΔP) is proportional
to the movement of the flapper relative to the nozzle. The controller linkage is
arranged so that the flapper movement is proportional to the difference between
the desired value (adjusted manually) and the measured value (ΔP_M). Thus:

$$\Delta P \propto R - B$$

i.e., as required to satisfy equation 3.5:

$$\Delta P \propto \varepsilon \qquad \qquad \dots(3.12)$$

This system has a very high gain due to the extreme sensitivity of P to very
small movements of the flapper. Because of this, it is termed *narrow-band
proportional action*—high gain corresponding to small or *narrow* proportional
band. The high gain leads to an undesirable control action which continually
cycles between maximum and minimum limits, i.e. on-off control.

(b) *Proportional control. Wide band action.* The extreme sensitivity of the

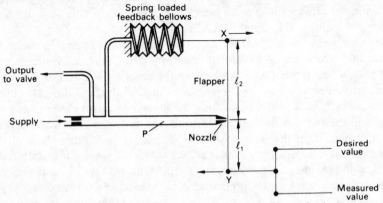

Fɪɢ. 3.16. Pneumatic wide band proportional controller mechanism

narrow band mechanism is reduced by introducing a feedback bellows as in Fig. 3.16. If P increases due to movement of the flapper towards the nozzle, the bellows will expand. Thus, as pivot Y is moved to the left by a change in measured or desired value, X will move to the right—so reducing the effective movement of the flapper relative to the nozzle. This results in a decrease in sensitivity which is dependent on the extension of the bellows per unit change in P and on the ratio of $l_1 : l_2$. Such an arrangement can be shown to give an action which approximates closely to equation 3.5 and proportional bands up to 600 per cent may be obtained[1].

(c) **P + I** *action*. Integral action is added by the insertion of a restrictor and a

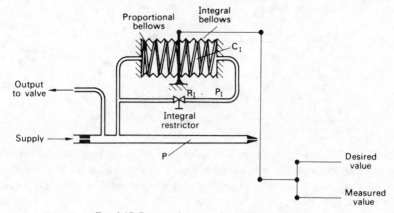

FIG. 3.17. Pneumatic generation of **P + I** action

further bellows as shown in Fig. 3.17. The rate of change of pressure in the integral bellows is proportional to the pressure driving force across the restrictor, i.e.:

$$\frac{dP_I}{dt} = K_1(P - P_I) \qquad \ldots.(3.13)$$

Suppose, for $t < 0$, $P_I = P = P_0$, and that, at $t = 0$, P_0 is suddenly increased to $P_0 + \Delta P$ by movement of the flapper, then initially:

$$\frac{dP_I}{dt} = K_1(P_0 + \Delta P - P_I) = K_1(P_0 + \Delta P - P_0) \qquad \ldots.(3.14)$$

therefore:

$$P_I = K_1 \int_0^t \Delta P \, dt$$

But $\Delta P \propto \varepsilon$ (equation 3.12); therefore, as required by the form of equation 3.9:

$$P_I = K_2 \int_0^t \varepsilon \, dt \qquad \ldots.(3.15)$$

Note that for $t > 0$, P_I is no longer equal to P_0 and therefore equation 3.15 only strictly applies at $t = 0$. The value of K_1 (and consequently τ_I) depends upon the capacity of the integral bellows (C_I) and the resistance to flow through the integral restrictor (R_I). C_I is assumed to change little and R_I is adjusted to obtain different values of τ_I.

(d) **P + D** *action*. By inserting a restrictor in the line to the proportional

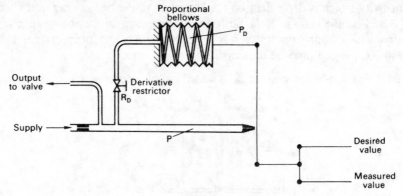

FIG. 3.18. Pneumatic generation of **P + D** action

bellows, any change in P will not be immediately transmitted to the feedback system. Thus, initially, the arrangement shown in Fig. 3.18 will act as a narrow-band proportional controller changing to a wide-band action as the pressure in the feedback bellows (P_D) approaches P. The rate at which $P_D \rightarrow P$ depends upon the resistance to flow (R_D) through the derivative restrictor. This mechanism thus simulates derivative action in that it is most sensitive when the error is changing the most rapidly. The derivative time (τ_D) is varied by adjustment of R_D.

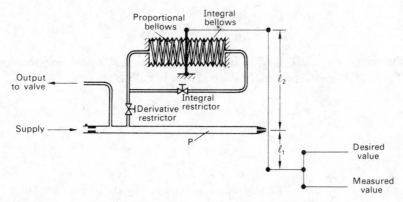

FIG. 3.19. Pneumatic generation of **P + I + D** action

(e) **P + I + D** *action*. This is obtained by simply coupling the mechanisms described in (c) and (d) above (see Fig. 3.19).

AIKMAN and RUTHERFORD[38] have shown that the output pressure from this type of controller follows the relation:

$$P = P_0 + \frac{l_2\beta}{l_1 b_s}\left\{\varepsilon + \frac{1}{\tau_I\beta}\int_0^t \varepsilon\,dt + \frac{\tau_D}{\beta}\frac{d\varepsilon}{dt}\right\} \qquad \dots \text{(3.16)}$$

$$= P_0 + K_3\beta\left\{\varepsilon + \frac{1}{T_I}\int_0^t \varepsilon\,dt + T_D\frac{d\varepsilon}{dt}\right\} \qquad \dots \text{(3.17)}$$

where b_s is the spring constant of the proportional bellows,

$\beta = 1 + \dfrac{2\tau_D}{\tau_I}$, the *interaction factor*,

$T_I = \tau_I\beta$,   the *effective* integral time, and

$T_D = \tau_D/\beta$,   the *effective* derivative time.

(cf. equation 3.11).

Thus the settings of the control modes are interdependent to some degree. The extent of this interaction is given by the value of β. The relation between β, τ_I and τ_D depends upon the type of controller mechanism used. The effect of β is to increase the controller gain and to limit the ratio of T_D/T_I obtainable. Thus, for this mechanism:

$$\frac{T_D}{T_I} = \frac{\tau_D/\tau_I}{(1 + 2\tau_D/\tau_I)^2}$$

therefore for $(T_D/T_I)_{max}$, $\tau_D/\tau_I = \frac{1}{2}$ and $\beta = 2$, i.e.:

$$(T_D/T_I)_{max} = \tfrac{1}{8}$$

The relation for **P + I** action can be obtained by imagining the derivative restrictor (Fig. 3.19) to be opened wide such that $\tau_D = 0$. Then from equation 3.16:

$$P = P_0 + \frac{l_2}{l_1 b_s}\left\{\varepsilon + \frac{1}{\tau_I}\int_0^t \varepsilon\,dt\right\} \qquad \dots \text{(3.18)}$$

(cf. equation 3.9).

For **P + D** action the integral restrictor is completely shut, i.e. $\tau_I = \infty$, then from equation 3.16:

$$P = P_0 + \frac{l_2}{l_1 b_s}\left\{\varepsilon + \tau_D\frac{d\varepsilon}{dt}\right\} \qquad \dots \text{(3.19)}$$

(cf. equation 3.10).

Note that there is no interaction in these cases.

These are examples of ideal controllers and in practice the actions of industrial instruments follow a variety of equations which only approximate to equations 3.17, 3.18 and 3.19. The degree of interaction between modes varies with the make of controller and frequently interaction will be present in the cases of **P + I** and **P + D** actions.

The transfer function

Before an efficient control system can be designed it is necessary to consider how all sections of the control loop will behave under the influence of variations in load and/or desired value. This requires experimental investigation or a mathematical analysis with respect to time, i.e. in the unsteady state. Although each section will necessitate a separate analysis, there are a number of basic simple physical systems which can be treated in much the same way. These quantitative procedures can be simplified by the use of the Laplace transform. This operational approach requires that the differential equations describing the behaviour of the sections be linear in form, which is not very often the case. Fortunately, non-linear relationships can frequently be represented by linear approximations with little error subject to certain limitations (e.g. see page 210). Once each section or system has been described in this way it is possible to form its appropriate *transfer function*, where:

$$\text{Transfer function} = \frac{\text{Laplace transform of output}}{\text{Laplace transform of input}} \qquad \dots (3.20)$$

(the Laplace transform of any variable x (say) with respect to the independent variable t being defined by the equation:

$$\bar{x} = \int_0^\infty x\, e^{-pt}\, dt$$

where p is a parameter).

Thus the transfer function is basically an input-output mathematical relationship. This is a most appropriate concept to use in conjunction with block diagrams which are themselves basically input-output schematic diagrams so that each block may be represented by the transfer function describing its behaviour.

The advantage of defining a transfer function in terms of Laplace transforms of input and output is that the differential equations developed to describe the unsteady-state behaviour of the system are reduced to simple algebraic relationships (e.g. cf. equations 3.25 and 3.27). Such relationships are much easier to deal with, and normal algebraic laws can be used to relate the various transfer functions of each component of the control loop (see page 233 *et seq.*). Furthermore, the output (or response) of the system to a variety of inputs may be obtained without classical integration (e.g. pages 209 and 210). Particularly simplified is the determination of the ultimate response to a sinusoidal input (page 227) which is of considerable use in control system design.

The transfer function approach will be used throughout the remainder of this chapter. An elementary knowledge of the Laplace transformation on the part of the reader is assumed and a list of the more useful transforms appears in the Appendix (page 601).

Linear systems and the principle of superposition

The majority of systems discussed in the remainder of this chapter are linear with respect to time, i.e. their time dependent properties can be described by linear differential equations (see Appendix 3.1, page 266). Such systems follow the *principle of superposition*[18]. This property is such that if the individual output of a system is known for each of several different inputs acting separately, then the total output for all the inputs acting together can be obtained by simply summing those individual outputs (see Appendix 3.1).

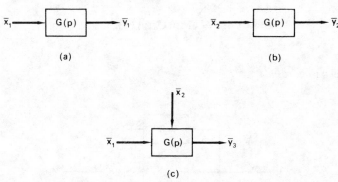

(a)   (b)

(c)

FIG. 3.20. Illustration of the principle of superposition

In Fig. 3.20 $G(p)$ is the transfer function of a particular linear system. For an input $\bar{x}_1$, an output $\bar{y}_1$ is obtained, both being expressed in terms of the Laplace transform:

$$G(p) = \frac{\bar{y}_1}{\bar{x}_1} \quad \text{(Fig. 3.20a)}$$

For an input $\bar{x}_2$:

$$G(p) = \frac{\bar{y}_2}{\bar{x}_2} \quad \text{(Fig. 3.20b)}$$

For an input $\bar{x}_1 + \bar{x}_2$, by the principle of superposition:

$$\bar{y}_3 = \bar{y}_1 + \bar{y}_2 = G(p)\bar{x}_1 + G(p)\bar{x}_2 = G(p)(\bar{x}_1 + \bar{x}_2) \qquad \text{(Fig. 3.20c)}$$

Block diagram algebra

(a) *Blocks in series.*

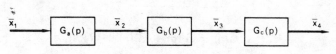

FIG. 3.21. Blocks in series

In Fig. 3.21 $G_a(p)$, $G_b(p)$ and $G_c(p)$ are transfer functions describing the input–output relationship for each block respectively:

$$G_a(p) = \frac{\bar{x}_2}{\bar{x}_1}, \quad G_b(p) = \frac{\bar{x}_3}{\bar{x}_2}, \quad G_c(p) = \frac{\bar{x}_4}{\bar{x}_3}.$$

The transfer function of the whole system is given by $G(p) = \bar{x}_4/\bar{x}_1$. This can also be obtained by multiplying together the three blocks in series, viz.:

$$\frac{\bar{x}_4}{\bar{x}_1} = \frac{\bar{x}_2}{\bar{x}_1}\frac{\bar{x}_3}{\bar{x}_2}\frac{\bar{x}_4}{\bar{x}_3}$$

i.e.

$$G(p) = G_a(p)\, G_b(p)\, G_c(p) \qquad\qquad \ldots(3.21)$$

(b) *Blocks in parallel.*

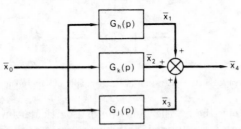

Fig. 3.22. Blocks in parallel

In Fig. 3.22:

$$G_h(p) = \frac{\bar{x}_1}{\bar{x}_0}, \quad G_k(p) = \frac{\bar{x}_2}{\bar{x}_0}, \quad G_j(p) = \frac{\bar{x}_3}{\bar{x}_0}$$

(signal $\bar{x}_0$ is applied equally as input to all three blocks).

The transfer function of the whole system is $G(p) = \bar{x}_4/\bar{x}_0$, and by the principle of superposition $\bar{x}_4$ is obtained from the additive effects of $\bar{x}_1$, $\bar{x}_2$ and $\bar{x}_3$. Therefore:

$$\frac{\bar{x}_4}{\bar{x}_0} = \frac{\bar{x}_1 + \bar{x}_2 + \bar{x}_3}{\bar{x}_0}$$

and

$$G(p) = G_h(p) + G_k(p) + G_j(p) \qquad\qquad \ldots(3.22)$$

(c) *Junctions of signals.* The summation of signals is represented on a block diagram as shown in Fig. 3.23.

Figure 3.23a represents the relationship:

$$\bar{x}_3 = \bar{x}_1 + \bar{x}_2$$

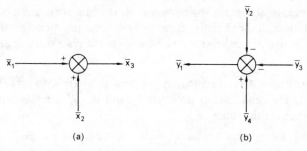

(a)   (b)

FIG. 3.23. Summation of signals

and Fig. 3.23*b*:

$$\bar{y}_1 = \bar{y}_4 - \bar{y}_3 - \bar{y}_2$$

Order of a control system component

If the unsteady-state behaviour of a component is described by a first order differential equation, it is termed a *first order system*. Other descriptions frequently used are *first order lag, single capacity system* and *single exponential stage*. Similarly, a component described by a second order differential equation is termed a *second order system*, and so on.

Transfer functions of capacity systems

First order systems

The behaviour of a large number of control loop components may be described by first order differential equations provided that a number of simplifying assumptions are made. Great care should be taken that the assumptions made are reasonable under the conditions to which the component is subjected. Two examples of a first order system are described—a measuring element and a process.

(a) *A measuring element—the thermocouple.* Consider a thermocouple junction immersed in a fluid whose temperature θ_0 varies with time (Fig. 3.24). Assume

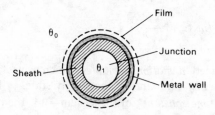

FIG. 3.24. Cross-section of thermo-couple junction

that all resistance to heat transfer resides in the film surrounding the thermocouple wall, that all the thermal capacity lies in the junction and that for $t < 0$ there is no change of temperature with time, i.e. the system is initially at steady-state.

The temperature of the thermocouple θ_1 will also vary with time. It is required to determine the relationship between θ_0 and θ_1. Writing an unsteady-state energy balance for the junction:

$$\left\{\begin{array}{l} \text{Energy input} \\ \text{per unit time} \end{array}\right\} - \left\{\begin{array}{l} \text{Energy output} \\ \text{per unit time} \end{array}\right\} = \left\{\begin{array}{l} \text{Rate of accumulation} \\ \text{of energy in junction} \end{array}\right\}$$

i.e.
$$h_1 A_1 (\theta_0 - \theta_1) - 0 = \frac{d}{dt}(m_1 C_1 \theta_1)$$

where h_1 is the film heat transfer coefficient,

A_1 is the mean area for heat transfer in the film,

C_1 is the average specific heat of the junction material, and

m_1 is its mass.

These are all assumed constant with respect to time, therefore:

$$h_1 A_1 (\theta_0 - \theta_1) = m_1 C_1 \frac{d\theta_1}{dt} \qquad \ldots.(3.23)$$

In the steady-state there is no accumulation; therefore from equation 3.23:

$$h_1 A_1 (\theta'_0 - \theta'_1) = 0 \qquad \ldots.(3.24)$$

or

$$\theta'_0 = \theta'_1 \qquad \ldots.(3.24a)$$

where θ'_0, θ'_1 are the steady-state values of θ_0 and θ_1 respectively. Subtracting equation 3.24 from equation 3.23 gives:

$$(\theta_0 - \theta'_0) - (\theta_1 - \theta'_1) = \frac{m_1 C_1}{h_1 A_1} \frac{d}{dt}(\theta_1 - \theta'_1)$$

But $\dfrac{d}{dt}(\theta_1 - \theta'_1) = \dfrac{d\theta_1}{dt}$ as θ'_1 is constant, so:

$$\vartheta_0 - \vartheta_1 = \tau_1 \frac{d\vartheta_1}{dt} \qquad \ldots.(3.25)$$

where $\vartheta_0 = \theta_0 - \theta'_0$,

$\vartheta_1 = \theta_1 - \theta'_1$, and

$$\tau_1 = \frac{m_1 C_1}{h_1 A_1} = (m_1 C_1)\left(\frac{1}{h_1 A_1}\right)$$

ϑ_0 and ϑ_1 are termed *deviation variables* which represent the difference between the values at any time and at the steady state. τ_1 has dimensions of time and is called the *time constant** of the system—it is the product of the heat

* It is the convention in Chemical Engineering process control to use minutes as the normal unit of time for a time constant. This convention is followed in this text.

capacity of the junction and the resistance to heat transfer of the surrounding film.

The Laplace transform of equation 3.25 is:

$$\bar{\vartheta}_0 - \bar{\vartheta}_1 = \tau_1 \left[p\bar{\vartheta}_1 - (\vartheta_1)_{t=0} \right]$$

since

$$\overline{\frac{d\vartheta_1}{dt}} = \int_0^\infty \frac{d\vartheta_1}{dt} e^{-pt} dt = \left[e^{-pt} \vartheta_1 \right]_0^\infty + p \int_0^\infty e^{-pt} \vartheta_1 \, dt$$

$$= -(\vartheta_1)_{t=0} + p\bar{\vartheta}_1$$

But

$$(\vartheta_1)_{t=0} = \theta'_1 - \theta'_1 = 0 \qquad \ldots(3.26)$$

therefore:

$$\frac{\bar{\vartheta}_1}{\bar{\vartheta}_0} = G(p) = \frac{1}{1 + \tau_1 p} \qquad \ldots(3.27)$$

$G(p)$ is the transfer function relating θ_0 and θ_1. It can be seen from equation 3.26 that the use of deviation variables is not only physically relevant but also eliminates the necessity of considering initial conditions. Equation 3.27 is typical of transfer functions of first order systems in that the numerator consists of a constant and the denominator a first order polynomial.

(b) *A process—liquid flowing through a tank*. Consider liquid flowing into a tank of uniform area of cross-section A with variable volumetric flowrate Q_1 (Fig. 3.25). Let the volumetric flowrate of liquid leaving and the volume of

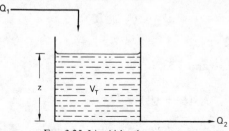

FIG. 3.25. Liquid level system

liquid in the tank at any time be Q_2 and V_T respectively. Assume the liquid density ρ to be constant. If the inflow remains constant, the liquid head z will adjust itself until:

$$Q'_1 = Q'_2 \qquad \ldots(3.28)$$

where Q'_1, Q'_2 are the steady-state values of Q_1 and Q_2 respectively. This is termed *self-regulation* and is a phenomenon exhibited by a number of process systems. The outflow is a function of the liquid head and the resistance to flow in the outlet line. If the head lost due to friction is neglected, then from an energy balance for turbulent flow (equation 2.67, Volume 1) in the outlet pipe:

$$Q_2 = K_1 z^{\frac{1}{2}} \qquad \ldots(3.29)$$

H

where K_1 is a constant for a particular pipe.

A mass balance over the tank in the unsteady state yields:

$$Q_1\rho - Q_2\rho = \frac{d}{dt}(\rho V_T) = \rho A \frac{dz}{dt} \qquad \ldots\ldots(3.30)$$

Substituting from equation 3.29:

$$Q_1 - K_1 z^{\frac{1}{2}} = A \frac{dz}{dt} \qquad \ldots\ldots(3.31)$$

Equation 3.31 cannot be transformed in the usual way as it contains the non-linear term $z^{\frac{1}{2}}$ for which there is no simple transform. This term can be approximated, however, by a linear expression (i.e. "linearised"). Variations (*perturbations*) in level are considered to occur around the steady-state level z'. Q_2 can then be expanded as a Taylor's series in terms of these variations[20], thus:

$$Q_2 = Q_2' + \left[\frac{dQ_2}{dz}\right]_{z'} (z - z') + \left[\frac{d^2Q_2}{dz^2}\right]_{z'} \frac{(z-z')^2}{2} + \text{higher order terms}$$

If the variation in level is small, it is possible to neglect powers greater than unity with little error, i.e.:

$$Q_2 = Q_2' + \left[\frac{dQ_2}{dz}\right]_{z'} (z - z') \qquad \ldots\ldots(3.32)$$

Substituting from equation 3.29, we have:

$$Q_2 = Q_2' + K_2(z - z') \qquad \ldots\ldots(3.33)$$

where $K_2 = K_1/(2\sqrt{z'})$.

From equations 3.30 and 3.33, we obtain:

$$(Q_1 - Q_2') - K_2(z - z') = A \frac{dz}{dt}$$

Substituting from equation 3.28 and writing deviation variables:

$$\mathcal{Q}_1 - K_2 \varkappa = A \frac{d\varkappa}{dt}$$

where $\mathcal{Q}_1 = Q_1 - Q_1'$ and $\varkappa = z - z'$.
Transforming yields the transfer function between inflow and liquid level:

$$G_1(p) = \frac{\bar{\varkappa}}{\bar{\mathcal{Q}}_1} = \frac{1/K_2}{1 + \tau p} \qquad \ldots\ldots(3.34)$$

where $\tau = A/K_2$ (cf. equation 3.27).

The constant $1/K_2$ in the numerator of this transfer function represents the steady-state relationship between flowrate and level. It is termed the *steady-state gain* of the system for reasons which are discussed later (page 224).

Differentiation of equation 3.33 gives:

$$\frac{dz}{dt} = \frac{1}{K_2} \frac{dQ_2}{dt}$$

Substituting for dz/dt in equation 3.30 leads to the transfer function between inflow and outflow, i.e.:

$$G_2(p) = \frac{\bar{\mathcal{Q}}_2}{\bar{\mathcal{Q}}_1} = \frac{1}{1 + \tau p} \qquad \dots (3.35)$$

In this case the steady-state gain is unity. Similarly the steady-state gain of the thermal system described by equation 3.27 is unity.

The degree of variation around the steady state to which systems can be subjected in order to be approximated by a linear relationship (e.g. equation 3.32) differs from system to system. The dynamics of a highly non-linear reactor might be described satisfactorily by a linear analysis for perturbations of up to ± 3 per cent[69]. On the other hand the dynamics of some distillation columns have been shown to remain reasonably linear in the face of variations of ± 25 per cent in some forcing functions[84].

First order systems in series

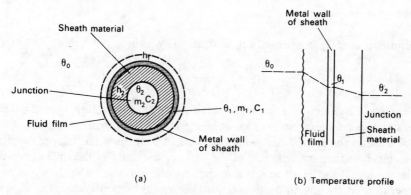

(a)

(b) Temperature profile

FIG. 3.26. Thermocouple junction including resistance of sheath

Suppose the resistance to heat transfer of the sheath surrounding the thermocouple described above is not negligible. The unsteady-state heat transfer mechanism must then be considered in two stages.

An energy balance over the fluid film surrounding the metal wall (Fig. 3.26), neglecting the heat capacity of the sheath material and the fluid film, gives:

$$h_1 A_1 (\theta_0 - \theta_1) - h_2 A_2 (\theta_1 - \theta_2) = \frac{d}{dt}(m_1 C_1 \theta_1) \qquad \dots (3.36)$$

and over the sheath material:

$$h_2 A_2(\theta_1 - \theta_2) - 0 = \frac{\mathrm{d}}{\mathrm{d}t}(m_2 C_2 \theta_2) \qquad \ldots(3.37)$$

where θ_1, m_1 and C_1 are the temperature, mass, and specific heat of the metal wall,
h_1 is the heat transfer coefficient for the film,
θ_2, m_2 and C_2 are the temperature, mass, and specific heat of the junction,
h_2 is the heat transfer coefficient for the sheath, and
A_1 and A_2 are the mean areas for heat transfer in the film and sheath respectively.

Introducing deviation variables and transforming, we obtain from equation 3.36:

$$\overline{\vartheta}_0 - \overline{\vartheta}_1 - \frac{h_2 A_2}{h_1 A_1}[\overline{\vartheta}_1 - \overline{\vartheta}_2] = \tau_1 p \overline{\vartheta}_1 \qquad \ldots(3.38)$$

where $\tau_1 = \dfrac{m_1 C_1}{h_1 A_1}$ and C_1, C_2, h_1 and h_2 are assumed constant with respect to time.

From equation 3.37:

$$\overline{\vartheta}_1 - \overline{\vartheta}_2 = \tau_2 p \overline{\vartheta}_2 \qquad \ldots(3.39)$$

where $\tau_2 = \dfrac{m_2 C_2}{h_2 A_2}$.

Elimination of $\overline{\vartheta}_1$ from equations 3.38 and 3.39 gives:

$$G(p) = \frac{\overline{\vartheta}_2}{\overline{\vartheta}_0} = \frac{1}{\tau_1 \tau_2 p^2 + [\tau_1 + \tau_2(1 + h_2 A_2/h_1 A_1)]\, p + 1} \qquad \ldots(3.40)$$

This transfer function represents two first order interacting systems. The interaction occurs because the rate of heat transfer from the first capacity (the metal wall) to the second (the junction) is dependent upon the temperature of the latter. τ_1 is the time constant of the first capacity and τ_2 the time constant of the second. If the heat transfer rate through the film was independent of the temperature of the junction the two systems would be non-interacting. Equation 3.36 would then be:

$$h_1 A_1(\theta_0 - \theta_1) - 0 = \frac{\mathrm{d}}{\mathrm{d}t}(m_1 C_1 \theta_1)$$

which leads to the transfer function:

$$G_1(p) = \frac{\overline{\vartheta}_1}{\overline{\vartheta}_0} = \frac{1}{1 + \tau_1 p} \qquad \ldots(3.41)$$

From equation 3.39:

$$G_2(p) = \frac{\overline{\vartheta}_2}{\overline{\vartheta}_1} = \frac{1}{1 + \tau_2 p} \qquad \ldots(3.42)$$

Therefore:

$$G(p) = \frac{\overline{\vartheta}_2}{\overline{\vartheta}_0} = G_1(p)\, G_2(p)$$

$$= \frac{1}{\tau_1 \tau_2 p^2 + (\tau_1 + \tau_2)\, p + 1} \qquad \ldots\,(3.43)$$

(cf. equation 3.40).

It is physically impossible for the above thermal system to exist without interaction. However, examples of non-interacting systems do occur in practice. For instance, the behaviour of vapour and liquid streams in any stagewise process can usually be approximated by a number of non-interacting first order systems in series.

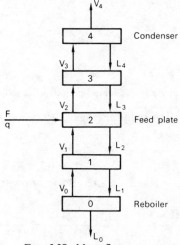

FIG. 3.27. Mass flows in continuous plate distillation column

ROSE and WILLIAMS[44] used a first order transfer function to represent the dynamics of liquid and vapour flow in a 5-stage continuous distillation column. Thus for stage n in Fig. 3.27:

$$\frac{\overline{\mathcal{V}}_n}{\overline{\mathcal{V}}_{n-1}} = \frac{1}{1 + \tau_1 p} \quad \text{and} \quad \frac{\overline{\mathcal{L}}_n}{\overline{\mathcal{L}}_{n+1}} = \frac{1}{1 + \tau_2 p} \qquad \ldots\,(3.44)$$

where τ_1 is the time required for vapour to pass through the liquid on tray n and τ_2 the time for liquid to pass across the plate. These time constants are frequently termed the *hold-up times* of liquid and vapour on the plate. Alternatively reference is made often to the liquid and vapour stage *hold-ups* in terms of the volumes or masses of liquid and vapour existing at any instant in a stage.

For the feed plate:

$$\left. \begin{aligned} \overline{\mathcal{V}}_F &= \frac{\overline{\mathcal{V}}_{F-1}}{1 + \tau_1 p} + (1 - q)\,\frac{\overline{\mathscr{F}}}{1 + \tau_3 p} \\[2mm] \overline{\mathcal{L}}_F &= \frac{\overline{\mathcal{L}}_{F+1}}{1 + \tau_2 p} + q\,\frac{\overline{\mathscr{F}}}{1 + \tau_4 p} \end{aligned} \right\} \qquad \ldots\,(3.45)$$

Equation 3.45 takes into account the effect of variations in feed, vapour and liquid rates on the column vapour and liquid flows. The feed quality q is assumed to remain constant. (q is the fraction of the feed which flows down the column as liquid).

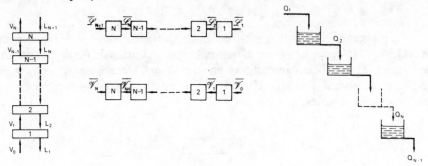

(a) Stripping column   (b) Block diagram for stripping column   (c) Tanks in series

FIG. 3.28. Non-interacting stages in series

For an N-stage stripping column (Figs. 3.28 a and b), from equation 3.21:

$$\frac{\bar{\mathscr{L}}_1}{\bar{\mathscr{L}}_{N+1}} = \frac{\bar{\mathscr{L}}_N}{\bar{\mathscr{L}}_{N+1}} \frac{\bar{\mathscr{L}}_{N-1}}{\bar{\mathscr{L}}_N} \cdots \frac{\bar{\mathscr{L}}_2}{\bar{\mathscr{L}}_3} \frac{\bar{\mathscr{L}}_1}{\bar{\mathscr{L}}_2}$$

If the liquid phase time constants are all the same, from equation 3.44:

$$\frac{\bar{\mathscr{L}}_1}{\bar{\mathscr{L}}_{N+1}} = \frac{1}{(1+\tau_2 p)}\frac{1}{(1+\tau_2 p)} \cdots \frac{1}{(1+\tau_2 p)}\frac{1}{(1+\tau_2 p)} = \frac{1}{(1+\tau_2 p)^N} \qquad \ldots\text{(3.46)}$$

Similarly:

$$\frac{\bar{\mathscr{V}}_N}{\bar{\mathscr{V}}_0} = \frac{1}{(1+\tau_1 p)^N} \qquad \ldots\text{(3.47)}$$

Similar transfer functions to equations 3.46 and 3.47 can be obtained for any number of non-interacting first order systems in series. For N tanks in series (Fig. 3.28c) having the same time constant τ, from equation 3.35:

$$\frac{\bar{\mathscr{Q}}_{N+1}}{\bar{\mathscr{Q}}_1} = \frac{1}{(1+\tau p)^N}$$

Second order systems

Consider the U-tube manometer shown in Fig. 3.29. Suppose the liquid level to be initially at BB. If the pressure differential $(P_1 - P_2)$ increases the liquid in the manometer will move as indicated. During this movement the difference in the force exerted on each limb will provide momentum to the column of liquid which will be opposed by inertia, frictional drag and the force due to gravity.

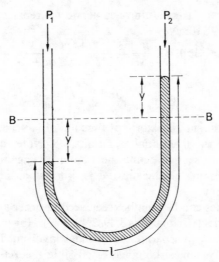

Fig. 3.29. U-tube manometer

If y is the distance moved in time t,

$$\text{Inertia force} = \text{Mass of liquid column} \times \text{Acceleration}$$

$$= \frac{\pi d^2}{4} l\rho \frac{d^2 y}{dt^2}$$

where d is the diameter of the manometer tube and ρ is the density of the liquid.

From Volume 1, Chapter 3:

$$\text{Frictional drag per unit cross-sectional area} = -\Delta P_f = 4\left(\frac{R}{\rho u^2}\right)\frac{l}{d}\rho u^2$$

If laminar flow is assumed, then (from Volume 1, Chapter 3):

$$\frac{R}{\rho u^2} = 8Re.^{-1} = \frac{8\mu}{\rho u d}$$

Therefore:

$$-\Delta P_f = \frac{32\mu l}{d^2} u = \frac{32\mu l}{d^2}\frac{dy}{dt}$$

$$\text{Force due to gravity} = 2\frac{\pi d^2}{4}\rho y g$$

Hence force balance gives:

$$(P_1 - P_2)\frac{\pi d^2}{4} = \frac{\pi d^2}{4} l\rho \frac{d^2 y}{dt^2} + \frac{32\mu l}{d^2}\frac{\pi d^2}{4}\frac{dy}{dt} + \frac{\pi d^2}{2}\rho y g$$

Therefore:

$$\frac{d^2 y}{dt^2} + \frac{32\mu}{d^2\rho}\frac{dy}{dt} + \frac{2g}{l}y = \frac{-\Delta P}{l\rho} \qquad \dots(3.48)$$

where $\Delta P = P_2 - P_1$.

Although y is already a deviation variable, we will replace it by $\mathscr{y}$, before transforming equation 3.48 to give:

$$G(p) = \frac{\bar{\mathscr{y}}}{-\overline{\Delta P}} = \frac{1/(2\rho g)}{\tau^2 p^2 + 2\zeta\tau p + 1} \qquad \ldots (3.49)$$

where
$$\tau = \sqrt{\frac{l}{2g}} \quad \text{and} \quad \zeta = \frac{8\mu}{d^2\rho}\sqrt{\frac{2l}{g}}.$$

Equation 3.49 is the standard form of a second order transfer function arising from the second order differential equation 3.48. Note that two parameters, τ and ζ, are now necessary to define the system. τ is again the time constant; ζ is termed the damping coefficient and is dimensionless; $1/(2\rho g)$ represents the steady-state gain.

Distinct similarities can be seen between second order systems and two first order systems in series (cf. equations 3.49, 3.40, 3.43). However, in the latter case it is possible physically to separate the two lags involved. This is not so with a true second order system, although it is possible theoretically to separate it into two first order lags having the same time constant by factorising the denominator of the transfer function; e.g. from equation 3.49, for a system with unit steady-state gain:

$$G(p) = \frac{1}{\tau^2 p^2 + 2\zeta\tau p + 1} = \frac{1}{\tau^2(p - \beta_1)(p - \beta_2)} \qquad \ldots (3.50)$$

where
$$\beta_1 = -\frac{\zeta}{\tau} + \frac{\sqrt{(\zeta^2 - 1)}}{\tau}$$

and
$$\beta_2 = -\frac{\zeta}{\tau} - \frac{\sqrt{(\zeta^2 - 1)}}{\tau}$$

β_1 and β_2 will be real or complex depending on the value of ζ. The same factors will be obtained by equating the denominator of the transfer function (equation 3.49) to zero, i.e.:

$$\left.\begin{array}{r} \tau^2 p^2 + 2\zeta\tau p + 1 = 0 \\[2mm] (p - \beta_1)(p - \beta_2) = 0 \end{array}\right\} \ldots (3.51)$$

or

Equation 3.51 is termed the *characteristic equation* of the system and is shown later to be of considerable importance in determining the system stability.

Distance–velocity lag

A typical example of this type of lag is experienced in the continuous sampling of a process stream (Fig. 3.30). It can be seen that a change in the composition x of the main stream will be detected by the analyser only after a time τ, where τ is the time required for the change to travel through the sample line (Fig. 3.30b). Hence, for plug flow:

$$\tau = \frac{\text{Volume of sample line}}{\text{Volumetric flowrate of sample}}$$

and
$$y_{t=t} = x_{t=t-\tau} \qquad \ldots (3.52)$$

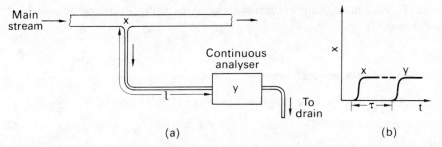

FIG. 3.30. Distance–velocity lag in sample line

where y is the response of the analyser. At steady state:

$$y_{t=0} = x_{t=-\tau} \qquad \dots (3.53)$$

Subtracting equation 3.53 from equation 3.52 and introducing deviation variables:

$$\mathcal{y}_{t=t} = x_{t=t-\tau} \qquad \dots (3.54)$$

From the translation theorem*:

$$\bar{x}_{t=t-\tau} = \exp(-p\tau)\,\bar{x}_{t=t}$$

therefore transforming equation 3.54 gives:

$$G(p) = \frac{\bar{\mathcal{y}}_{t=t}}{\bar{x}_{t=t}} = \exp(-p\tau) \qquad \dots (3.55)$$

Of course, x could be any type of function, then y will be precisely the same function but occurring τ units of time later.

This type of lag is frequently encountered in flow systems and can be termed *distance–velocity* (D/V) *lag*, *dead-time* or *transportation lag*. The presence of

---

* The theorem of translation may be established as follows:

$$\bar{x}_{t=t-\tau} = \int_0^\infty x_{t=t-\tau} \exp(-pt)\,dt$$

Putting $t - \tau = t_0$:

$$\bar{x}_{t=t-\tau} = \int_{-\tau}^\infty x_{t=t_0} \exp\{-p(t_0 + \tau)\}\,dt_0$$

But for $t_0 \leqslant 0$, $x_{t=t_0} = 0$ (Fig. 3.30b), therefore:

$$\bar{x}_{t=t-\tau} = \int_0^\infty x_{t=t_0} \exp\{-p(t_0 + \tau)\}\,dt_0$$

$$= \exp(-p\tau)\int_0^\infty x_{t=t_0} \exp(-pt_0)\,dt_0$$

$$= \exp(-p\tau)\,\bar{x}_{t=t}$$

much D/V lag in any control loop can lead to instability in control action[19,67] (see Fig. 3.51).

Transfer functions of controllers

From equation 3.5, for proportional control:

$$P - P_0 = K_c \varepsilon \qquad \qquad \ldots (3.56)$$

$(P - P_0)$ represents the deviation of the controller output pressure from its steady-state value and may therefore be replaced by a deviation variable $\mathscr{P}$. At time $t = 0$ we assume that there is no error, thus ε is itself a deviation variable. Hence equation 3.56 becomes:

$$\mathscr{P} = K_c \varepsilon$$

Transforming gives:

$$G_c(p) = \overline{\mathscr{P}}/\overline{\varepsilon} = K_c \qquad \qquad \ldots (3.57)$$

which is the transform of an ideal proportional controller.

By similar procedures we obtain from equations 3.9, 3.10 and 3.11, for **P + I** control:

$$G_c(p) = K_c \left(1 + \frac{1}{\tau_I p} \right) \qquad \qquad \ldots (3.58)$$

for **P + D** control:

$$G_c(p) = K_c(1 + \tau_D p) \qquad \qquad \ldots (3.59)$$

and for **P + I + D** control:

$$G_c(p) = K_c \left(1 + \frac{1}{\tau_I p} + \tau_D p \right) \qquad \qquad \ldots (3.60)$$

Response of control loop components to forcing functions

Forcing function is a term given to any disturbance which is externally applied to a system. A number of simple functions are of considerable use in both the theoretical and experimental analysis of control systems and their components. Note that the response to a forcing function of a system or component without feedback is termed the *open-loop* response. (This should not be confused with the term *open-loop* control used to describe feed-forward control.) The response of a system incorporating feedback is referred to as the *closed-loop* response. Only three of the most useful forcing functions will be discussed at this point.

(a) *The step function.* This is defined (Fig. 3.31) by:

$$F(t) = \begin{cases} 0, & t < 0 \\ M, & t \geqslant 0 \end{cases} \qquad \qquad \ldots (3.61)$$

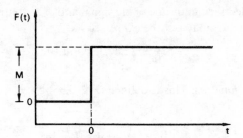

FIG. 3.31. Step function of magnitude M

or

$$F(t) = M\,u(t) \qquad\qquad \ldots.(3.62)$$

where $u(t)$ represents a step function of unit magnitude. The transform of $F(t)$ is given by

$$F(p) = \frac{M}{p} \qquad\qquad \ldots.(3.63)$$

(b) *The sinusoidal function.* For this function (Fig. 3.32):

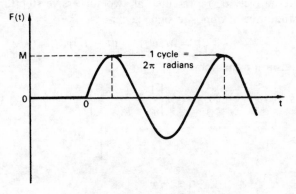

FIG. 3.32. Sinusoidal function of amplitude M

$$F(t) = \begin{cases} 0, & t < 0 \\ M \sin \omega t, & t \geqslant 0 \end{cases} \qquad\qquad \ldots.(3.64)$$

or

$$F(t) = M u(t) \sin \omega t \qquad\qquad \ldots.(3.65)$$

Therefore:

$$F(p) = \frac{M\omega}{p^2 + \omega^2} \qquad\qquad \ldots.(3.66)$$

Here M is termed the amplitude of the signal and ω its frequency in radians/unit time. The frequency may also be expressed as:

$$f = \omega/2\pi \text{ cycles/unit time} \qquad \ldots\text{(3.67)}$$

(c) *The pulse function.* This is defined (Fig. 3.33) by:

$$F(t) = \begin{cases} 0, & t < 0 \\ M, & 0 \leqslant t \leqslant t_0 \\ 0, & t > t_0 \end{cases} \qquad \ldots\text{(3.68)}$$

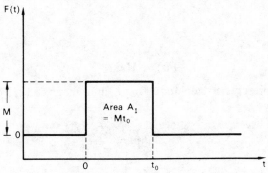

FIG. 3.33. Pulse function of magnitude M

This may alternatively be represented as two successive step functions of the same magnitude but of opposite sign; i.e.:

$$F(t) = M[u(t) - u(t - t_0)] \qquad \ldots\text{(3.69)}$$

Using the translation theorem:

$$F(p) = M\left[\frac{1}{p} - \frac{1}{p}e^{-pt_0}\right]$$

$$= \frac{M}{p}\left[1 - e^{-pt_0}\right] \qquad \ldots\text{(3.70)}$$

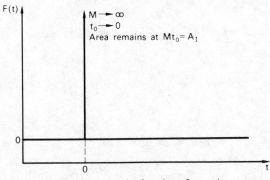

FIG. 3.34. Impulse function of area A_I

The *impulse* is a special case of the pulse function in which $t_0 \to 0$ but the area A_I under the impulse curve remains constant and finite (Fig. 3.34). Thus, for the impulse function:

$$F(p) = \lim_{t_0 \to 0} \left\{ \frac{M}{p} (1 - e^{-pt_0}) \right\}$$

$$= \lim_{t_0 \to 0} \left\{ \frac{A_I}{t_0 p} (1 - e^{-pt_0}) \right\} = \frac{0}{0}$$

By l'Hôpital's rule[20] the limit of an indeterminate quantity is the same as that of the function obtained by independently differentiating numerator and denominator with respect to the variable which is approaching the limit. Therefore:

$$F(p) = \lim_{t_0 \to 0} \left\{ \frac{d/dt_0 [A_I(1 - e^{-pt_0})]}{d/dt_0 [t_0 p]} \right\}$$

$$= \lim_{t_0 \to 0} \left\{ \frac{A_I p e^{-pt_0}}{p} \right\} = A_I \qquad \qquad \dots (3.71)$$

When $A_I = 1$ we have the special case of the unit impulse.

Response to step function

(i) *First order system.* From equation 3.27:

$$G(p) = \frac{\vartheta_1}{\vartheta_0} = \frac{1}{1 + \tau_1 p} \qquad \qquad \text{(equation 3.27)}$$

For step input (equation 3.63):

$$\vartheta_0 = \frac{M}{p}$$

therefore:

$$\vartheta_1 = \frac{1}{1 + \tau_1 p} \frac{M}{p} \qquad \qquad \dots (3.72)$$

$$= -\frac{M}{p + 1/\tau_1} + \frac{M}{p}$$

Inversion of equation 3.72 gives the response of the thermocouple to a step change in temperature of the surrounding fluid, i.e.:

$$\vartheta_1 = \begin{cases} M\left[1 - \exp\left(-\frac{t}{\tau_1}\right)\right], & t \geqslant 0 \\ \\ 0, & t < 0 \end{cases} \qquad \dots (3.73)$$

Plotting ϑ_1/M vs t/τ_1 shows the distinctive characteristics of the step response of a first order system (Fig. 3.35). The response is immediate and the slope is a

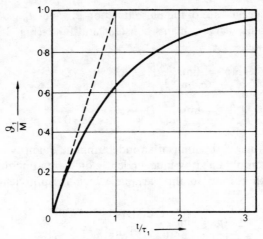

FIG. 3.35. Response of first order system to step
forcing function

maximum at $t = 0$ and decreases with the increasing time. If the response is measured experimentally, the time constant may be evaluated thus:

(a) Put $t/\tau_1 = 1$ in equation 3.73. Then:

$$\vartheta_1/M = 1 - \exp(-1) = 0.632$$

Hence the response will have reached 63.2 per cent of its ultimate value when $t = \tau_1$.

(b) From equation 3.73:

$$\frac{d}{dt}(\vartheta_1/M) = \frac{1}{\tau_1}\exp\left(-\frac{t}{\tau_1}\right)$$

Therefore at $t = 0$, slope of graph $= 1/\tau_1$. Thus, if the initial rate of change were maintained, the response would be completed in a time equivalent to one time constant.

If the value of the time constant obtained by both methods is not the same within the limits of experimental error, the response is not truly first order.

(ii) *Second order system.* Consider a step change in pressure differential applied to a manometer (Fig. 3.29). From equation 3.49 the movement of the column of liquid is given by:

$$\bar{y} = \frac{1/(2\rho g)}{\tau^2 p^2 + 2\zeta\tau p + 1}\frac{M}{p}$$

$$= \frac{MK}{\tau^2 p(p - \beta_1)(p - \beta_2)} \qquad \ldots(3.74)$$

where $\beta_{1,2} = -\frac{\zeta}{\tau} \pm \frac{\sqrt{(\zeta^2 - 1)}}{\tau}, \quad K = 1/(2\rho g)$

Three particular cases of equation 3.74 must be considered for inversion (see Appendix 3.2)*:

(a) $\zeta < 1$. The roots of the characteristic equation (equation 3.51) are complex conjugates. Inversion gives[2]:

$$y = MK\left\{1 - \frac{1}{\phi}\sin\left(\frac{\phi t}{\tau} + \tan^{-1}\frac{\phi}{\zeta}\right)\exp\left(-\frac{\zeta t}{\tau}\right)\right\} \qquad \ldots(3.75)$$

where $\phi = \sqrt{(1 - \zeta^2)}$.

(b) $\zeta > 1$. The characteristic equation has real roots, and:

$$y = MK\left\{1 - \left(\cosh\frac{vt}{\tau} + \frac{\zeta}{v}\sinh\frac{vt}{\tau}\right)\exp\left(-\frac{\zeta t}{\tau}\right)\right\} \qquad \ldots(3.76)$$

where $v = \sqrt{(\zeta^2 - 1)} = i\phi$.

(c) $\zeta = 1$. The roots are repeated, and:

$$y = MK\left\{1 - \left(1 + \frac{t}{\tau}\right)\exp\left(-\frac{t}{\tau}\right)\right\} \qquad \ldots(3.77)$$

Note that the term MK appears in each case. This is simply the product of the steady-state gain of the system K and the magnitude M of the forcing function. If these quantities were both unity the term would vanish, i.e. be replaced by unity. No steady-state term appears in equation 3.73 as the steady-state gain in this case is unity (cf. equation 3.27).

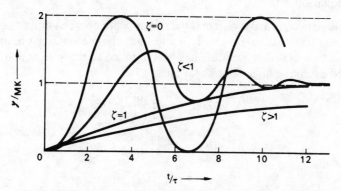

FIG. 3.36. Response of second order system to step forcing function

The effect of the value of the damping coefficient ζ on the response is shown in Fig. 3.36. For $\zeta < 1$ the response is seen to be oscillatory or *underdamped*; when $\zeta > 1$ it is sluggish or *overdamped*; and when $\zeta = 1$ it is said to be *critically damped*, i.e. the ultimate value is approached with the greatest speed without oscillation. When $\zeta = 0$ there is no damping and the system output oscillates continuously.

* In the following, the condition for $t < 0$ is understood to be zero response (steady-state) unless otherwise stated. Response equations are for $t \geqslant 0$. See Appendix 3.2 for details.

Equation 3.76 may be expressed in the form:

$$y = MK\{1 - A_3 \exp(-B_1 t/\tau) - A_4 \exp(-B_2 t/\tau)\}$$

A similar expression is obtained for the response of two first order systems in series, e.g. from equation 3.43.

Initial and final value theorems. Steady-state gain

Frequently it is required to determine the initial or final value of the system response to some forcing function. It is possible to evaluate this information without inverting the appropriate transform into the time domain.

The value of the response at $t = 0$ is given by the *initial value* theorem[a] which states that:

$$\lim_{t \to 0} \{F(t)\} = \lim_{p \to \infty} \{pF(p)\} \qquad \ldots(3.78)$$

where $F(p)$ is the transform of $F(t)$.

The *final value* theorem[b] gives the value of the response at $t = \infty$:

$$\lim_{t \to \infty} \{F(t)\} = \lim_{p \to 0} \{pF(p)\} \qquad \ldots(3.79)$$

provided that $pF(p)$ does not become infinite for any value of p satisfying Re $(p) \geqslant 0$, where Re denotes the *real part* of a function. If this condition does not hold, $F(t)$ does not approach a limit as $t \to \infty$. This condition does not apply to equation 3.78.

The steady-state gain of a system is the relationship between input and output in the steady-state, i.e. as $t \to \infty$. It is the final change obtained after a step forcing function of unit magnitude has been applied to the system and may be determined using the final value theorem.

(a)   Now
$$\frac{\overline{dF(t)}}{dt} = pF(p) - F(0) = \int_0^\infty \frac{dF(t)}{dt} e^{-pt} dt$$

∴
$$\lim_{p \to \infty} \{pF(p) - F(0)\} = \lim_{p \to \infty} \left\{ \int_0^\infty \frac{dF(t)}{dt} e^{-pt} dt \right\}$$

∴
$$\lim_{p \to \infty} \{pF(p)\} - F(0) = 0$$

∴
$$F(0) = \lim_{t \to 0} \{F(t)\} = \lim_{p \to \infty} \{pF(p)\}$$

(b)
$$\lim_{p \to 0} \{pF(p) - F(0)\} = \lim_{p \to 0} \left\{ \int_0^\infty \frac{dF(t)}{dt} e^{-pt} dt \right\}$$

∴
$$\lim_{p \to 0} \{pF(p)\} - F(0) = \int_0^\infty dF(t) = \left[F(t)\right]_0^\infty = F(\infty) - F(0)$$

∴
$$F(\infty) = \lim_{t \to \infty} \{F(t)\} = \lim_{p \to 0} \{pF(p)\}$$

provided $pF(p)$ does not become infinite for any value of p for which the real part of p is greater or equal to zero.

Example

Determine the steady-state gain of the U-tube manometer described in Fig. 3.29. Also calculate the initial value of the response of the liquid level when a step change of magnitude M in pressure differential is applied across it.

Solution

From equation 3.74, transformed form of response in level to unit step change in pressure differential is:

$$\bar{y} = \frac{1/(2\rho g)}{\tau^2 p^2 + 2\zeta\tau p + 1}\frac{1}{p}$$

Using final value theorem (equation 3.79) ($pF(p)$ does not become infinite for $\mathrm{Re}(p) > 0$):

$$\text{Steady-state gain} = \lim_{t \to \infty}\{F(t)\} = \lim_{p \to 0}\{pF(p)\}$$

$$= \underline{1/(2\rho g)}$$

Again from equation 3.74, transform of response to step change of magnitude M is given by:

$$\bar{y} = \frac{1/(2\rho g)}{\tau^2 p^2 + 2\zeta\tau p + 1}\frac{M}{p}$$

Using initial value theorem (equation 3.78):

$$\text{Initial value of liquid level response} = \lim_{t \to 0}\{F(t)\} = \lim_{p \to \infty}\{pF(p)\}$$

$$= \underline{\underline{0}}.$$

Response to sinusoidal function

(i) *First order system.* From equations 3.27 and 3.66

$$\bar{\vartheta}_1 = \frac{1}{(1+\tau_1)}\frac{M\omega}{(p^2+\omega^2)}$$

Expanding by means of partial fractions gives:

$$\bar{\vartheta}_1 = \frac{M\omega\tau_1{}^2}{(1+\omega^2\tau_1^2)}\frac{1}{(1+\tau_1 p)} - \frac{M\omega\tau_1}{(1+\omega^2\tau_1^2)}\frac{p}{(p^2+\omega^2)} + \frac{M\omega}{(1+\omega^2\tau_1^2)}\frac{1}{(p^2+\omega^2)}$$

The inverse transform follows:

$$\vartheta_1 = \frac{M}{1+\omega^2\tau_1^2}\{\omega\tau_1\exp(-t/\tau_1) - \omega\tau_1\cos\omega t + \sin\omega t\}$$

A more useful form is obtained by using the identity:

$$x\sin\alpha + y\cos\alpha = z\sin(\alpha + \psi) \qquad \dots(3.80)$$

where $z^2 = x^2 + y^2$, $\tan\psi = y/x$.

Hence:

$$\vartheta_1 = \frac{M\omega\tau_1}{1+\omega^2\tau_1^2}\exp(-t/\tau_1) + \frac{M}{\sqrt{(1+\omega^2\tau_1^2)}}\sin(\omega t + \psi) \quad \dots(3.81)$$

where $\psi = \tan^{-1}(-\omega\tau_1)$.

When $t \to \infty$ the *ultimate* periodic response is obtained, i.e.:

$$\lim_{t \to \infty} \vartheta_1 = \frac{M}{\sqrt{(1 + \omega^2 \tau_1^2)}} \sin(\omega t + \psi) \qquad \ldots(3.82)$$

This type of function is said to be *stationary*, i.e. its value varies with time but in a regularly repeating pattern. Stationary conditions should be not confused with the steady-state which is constant in time and can only be obtained in response to a constant forcing function.

Comparison of equations 3.64 and 3.82 shows that when $t \to \infty$:
(a) both input and output are sinusoidal functions of frequency ω,
(b) the amplitude of the output is always less than that of the input (it is frequently said that the signal is attenuated), and
(c) the output differs in phase from the input by an angle ψ.

If $\psi < 0$ the output is said to *lag* the input, and if $\psi > 0$ it is said to *lead* the input. The relationship between ultimate input and output for a sinusoidal forcing function constitutes an important tool in the analysis and design of control systems termed *frequency response analysis*. Of particular importance are the *amplitude ratio* (A.R.) and the *phase shift* ψ. The A.R. represents the relationship between the output and input amplitudes as $t \to \infty$, i.e.:

$$\text{A.R.} = \left(\frac{\text{amplitude of output signal}}{\text{amplitude of input signal}} \right)_{t \to \infty} \qquad \ldots(3.83)$$

The phase shift is the difference in angle measured in radians between input and output as $t \to \infty$ (Fig. 3.37).

For the above first order system, from equations 3.64 and 3.82:

$$\text{A.R.} = \frac{1}{\sqrt{(1 + \omega^2 \tau_1^2)}} \qquad \ldots(3.84)$$

and

$$\psi = \tan^{-1}(-\omega \tau_1) \qquad \ldots(3.85)$$

Both A.R. and ψ are functions of ω and for a first order system A.R. < 1 and the output always lags the input ($\psi < 0$). As $\omega \to \infty$, A.R. $\to 0$ and $\psi \to -90°$, thus the maximum lag obtainable with a first order system is $90°$.*

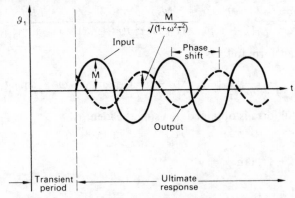

FIG. 3.37. Ultimate response of first order system to sinusoidal forcing function

* It is the convention in Chemical Engineering process control to measure phase shift in degrees. This convention is followed in this text.

The substitution rule. The frequency response of a system may be more easily determined by effecting the substitution $p = i\omega$ in the appropriate transfer function (where $i = \sqrt{-1}$). Hence substituting into equation 3.27:

$$G(i\omega) = \frac{1}{1 + \tau_1 i\omega}$$

$$= \frac{1}{1 + \tau_1^2 \omega^2} + i\frac{(-\tau_1\omega)}{1 + \tau_1^2 \omega^2} \qquad \ldots\ldots(3.86)$$

The A.R. and phase shift are obtained as the magnitude and argument of equation 3.86. Hence:

$$\text{A.R.} = \frac{1}{\sqrt{(1 + \omega^2 \tau_1^2)}} \qquad \text{(equation 3.84)}$$

and

$$\psi = \tan^{-1}(-\omega\tau_1) \qquad \text{(equation 3.85)}$$

It can be shown[1] that this method may be applied to any system described by a linear differential equation or to a distance–velocity lag in order to obtain the frequency response characteristics.

(ii) *Second order system.* The frequency response is of principal interest. Hence, substituting $p = i\omega$ in equation 3.49:

$$G(i\omega) = \frac{1/(2\rho g)}{-\tau^2\omega^2 + 2\zeta\tau i\omega + 1}$$

Hence:

$$\text{A.R.} = \text{magnitude of } G(i\omega)$$

$$= \frac{1/(2\rho g)}{\sqrt{[(1 - \omega^2\tau^2)^2 + (2\zeta\omega\tau)^2]}} \qquad \ldots\ldots(3.87a)$$

For a second order system with unit steady-state gain, from equation 3.50:

$$\text{A.R.} = \frac{1}{\sqrt{[(1 - \omega^2\tau^2)^2 + (2\zeta\omega\tau)^2]}} \qquad \ldots\ldots(3.87b)$$

$$\psi = \text{argument of } G(i\omega)$$

$$= \tan^{-1}\left(\frac{-2\zeta\omega\tau}{1 - \omega^2\tau^2}\right) \qquad \ldots\ldots(3.88)$$

In this case, for certain values of $\omega\tau$ with $\zeta < 1/\sqrt{2}$, the amplitude ratio is greater than unity (equation 3.87b and Fig. 3.48). Also $\psi \to -90°$ as $\omega\tau \to 1$ (equation 3.88). As $\omega\tau$ increases above unity $\tan\psi$ approaches zero but is always positive. Thus, as $\omega\tau \to \infty$, $\psi \to -180°$. The maximum phase lag is therefore 180°, whereas it is 90° for a first order system. For an nth order system the maximum lag obtainable is $90n$ degrees ($n\pi/2$ radians).

(iii) *Distance–velocity lag.* In order to determine the frequency response characteristics substitute $p = i\omega$ in equation 3.55:

$$G(i\omega) = \exp(-i\omega\tau)$$

$$= \cos \omega\tau - i \sin \omega\tau$$

therefore:

$$\text{A.R.} = \sqrt{(\cos^2 \omega\tau + \sin^2 \omega\tau)} = 1 \qquad \ldots(3.89)$$

and

$$\psi = \tan^{-1}\left\{-\frac{\sin \omega\tau}{\cos \omega\tau}\right\} = -\omega\tau \qquad \ldots(3.90)$$

Response to pulse function

Only the special case of the impulse will be considered (but see also Fig. 3.46). This is a very useful function for testing the dynamic behaviour of equipment as it does not introduce any further p terms into the analysis (equation 3.71). The determination of the response of any system in the time domain to an impulse forcing function is facilitated by noting that:

$$\{F(p)\}_{\text{impulse}} = p\,\{F(p)\}_{\text{step}} \qquad \ldots(3.91)$$

i.e. the transform of the impulse response is simply the transform of the step response multiplied by the operator p (cf. equations 3.72 and 3.93a).

Inverting equation 3.91 gives:

$$\{F(t)\}_{\text{impulse}} = \frac{\mathrm{d}}{\mathrm{d}t}\{F(t)\}_{\text{step}} \qquad \ldots(3.92)$$

Thus if the step response in the time domain is known, the impulse response can be determined by differentiating the former.

(i) *First order system.* From equations 3.27 and 3.71:

$$\overline{\vartheta}_1 = \frac{1}{1 + \tau_1 p}\,A_I = \frac{A_I/\tau_1}{p + 1/\tau_1} \qquad \ldots(3.93a)$$

$$\therefore \qquad \vartheta_1 = A_I/\tau_1 \exp(-t/\tau_1) \qquad \ldots(3.93b)$$

For unit impulse $A_I = 1$, then:

$$\vartheta_1 = 1/\tau_1 \exp(-t/\tau_1) \qquad \ldots(3.94)$$

Alternatively, for unit step change:

$$\{\vartheta_1\}_{\text{step}} = 1 - \exp(-t/\tau_1) \qquad \text{(equation 3.73)}$$

Therefore, using equation 3.92:

$$\{\vartheta_1\}_{\text{impulse}} = \frac{\mathrm{d}}{\mathrm{d}t}\{1 - \exp(-t/\tau_1)\}$$

$$= 1/\tau_1 \exp(-t/\tau_1) \qquad \text{(equation 3.94)}$$

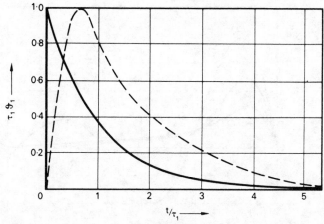

FIG. 3.38. Response of first order system to unit impulse forcing
function

Figure 3.38 shows that the theoretical response to an impulse immediately
rises to its maximum value and then decays exponentially. The broken line
shows the probable response obtained in practice to an experimental impulse
having a finite time scale.

(ii) *Second order system.* There are three cases to consider dependent on the
form of the roots of the characteristic equation (see Fig. 3.39).

(a) $\zeta < 1$. From equation 3.75 with $M = A_I$:

$$\{y\}_{\text{impulse}} = \frac{d}{dt}\{y\}_{\text{step}} = A_I K \left\{\frac{1}{\tau\phi}\exp\left(-\frac{\zeta t}{\tau}\right)\sin\frac{\phi t}{\tau}\right\} \quad \ldots .(3.95)$$

(b) $\zeta > 1$. From equation 3.76:

$$\{y\}_{\text{impulse}} = A_I K \left\{\frac{1}{\tau v}\exp\left(-\frac{\zeta t}{\tau}\right)\sinh\frac{v t}{\tau}\right\} \quad \ldots .(3.96)$$

(c) $\zeta = 1$. From equation 3.77:

$$\{y\}_{\text{impulse}} = A_I K \left\{\frac{t}{\tau^2}\exp\left(-\frac{t}{\tau}\right)\right\} \quad \ldots .(3.97)$$

Again for unit impulse put $A_I = 1$.

Response of more complex systems to forcing functions

Transfer functions involving polynomials of higher degree than two and
decaying exponentials (distance–velocity lags) may be dealt with in the same
manner as above, i.e. by the use of partial fractions and inverse transforms if

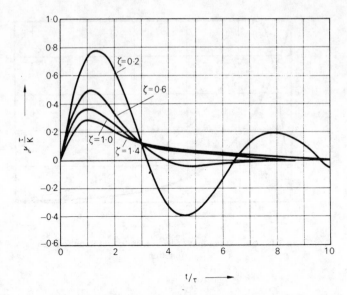

FIG. 3.39. Response of second order system to unit impulse
forcing system

the step response or the transient part of the sinusoidal response is required,
or by the substitution method if the frequency response is desired.

Inverting the transfer function becomes increasingly difficult with increase
in order if it cannot easily be divided into lower order functions in series. It is
frequently necessary to use numerical methods to determine the roots of the
high order polynomials that result requiring the use of a digital computer[50],
although FLEMMING[49] has reported an iterative method, which is simple if a
computer is available, of obtaining the time domain response directly from the
transformed function. Determination of the frequency response is considerably
facilitated by division into lower degree transfer functions. For instance, it is
possible to write the third order transfer function:

$$G(p) = \frac{8}{18p^3 + 27p^2 + 10p + 1}$$

as

$$G(p) = 8 \left\{ \frac{1}{(1 + 3p)(1 + 6p)(1 + p)} \right\}$$

$$= K \{G_1(p) \, G_2(p) \, G_3(p)\}$$

where $K = 8$ is the system steady-state gain.

If the amplitude ratio and phase shift due to $G_1(p)$ are $(\text{A.R.})_1$ and ψ_1 respectively, etc. then the amplitude ratio and phase shift of $G(p)$ are given by:

$$\text{A.R.} = K \,(\text{A.R.})_1 \,(\text{A.R.})_2 \,(\text{A.R.})_3 \qquad \ldots .(3.98)$$

and

$$\psi = \psi_1 + \psi_2 + \psi_3 \qquad \ldots .(3.99)$$

These relationships may be extended to include any number of terms on the right-hand side.

Hence from equations 3.84 and 3.85:

$$\text{A.R.} = 8 \,\frac{1}{\sqrt{(1 + 9\omega^2)}} \frac{1}{\sqrt{(1 + 36\omega^2)}} \frac{1}{\sqrt{(1 + \omega^2)}}$$

and

$$\psi = \tan^{-1}(-3\omega) + \tan^{-1}(-6\omega) + \tan^{-1}(-\omega).$$

Equation 3.99 may be established as follows:

Consider $G(p) = K G_1(p) \, G_2(p)$. Putting $p = i\omega$ we obtain a complex number relationship of the form:

$$\frac{a + ib}{K} = (a_1 + ib_1)(a_2 + ib_2)$$

$$= a_1 a_2 - b_1 b_2 + i\{b_1 a_2 + b_2 a_1\}$$

Equating real and imaginary parts:

$$\frac{a}{K} = a_1 a_2 - b_1 b_2 \quad \text{and} \quad \frac{b}{K} = b_1 a_2 + b_2 a_1$$

$$\text{A.R.} = \text{magnitude of } (a + ib) = \sqrt{(a^2 + b^2)}$$

$$= K\sqrt{(a_1^2 a_2^2 + b_1^2 b_2^2 + b_1^2 a_2^2 + a_1^2 b_2^2)}$$

$$= K\sqrt{(a_1^2 + b_1^2)} \sqrt{(a_2^2 + b_2^2)} = K \,(\text{A.R.})_1 \,(\text{A.R.})_2.$$

Also

$$\psi = \tan^{-1}\frac{b}{a} = \tan^{-1}\left\{\frac{K(b_1 a_2 + b_2 a_1)}{K(a_1 a_2 - b_1 b_2)}\right\}$$

and

$$\psi_1 + \psi_2 = \tan^{-1}\frac{b_1}{a_1} + \tan^{-1}\frac{b_2}{a_2}$$

$\therefore$

$$\tan(\psi_1 + \psi_2) = \frac{b_1/a_1 + b_2/a_2}{1 - (b_1/a_1)(b_2/a_2)} = \frac{b_1 a_2 + b_2 a_1}{a_1 a_2 - b_1 b_2}$$

$\therefore$

$$\tan\psi = \tan(\psi_1 + \psi_2)$$

Hence

$$\psi = \psi_1 + \psi_2.$$

This may be extended to any number of functions.

Example

In the heat exchanger arrangement shown in Fig. 3.40 the following are known:

(a) The response of the temperature θ_2 of stream 2 leaving exchanger A to a change in that of stream 1 entering (θ_1) is first order with a time constant of 0·67 min and a steady-state gain of 0·33.

(b) The response of θ_4 to a change in θ_3 is underdamped second order with a time constant of 3·2 min and a damping coefficient of 0·48. The steady-state gain is unity.

(c) It takes 0·5 min for any change in θ_2 to affect θ_3.

Determine (i) the response of θ_4 to a step change of unit magnitude in θ_1, and (ii) the frequency response of θ_4 to θ_1. Assume all flows and the inlet temperature of stream 2 to A and of stream 3 to B to remain constant.

Solution

The system may be drawn as a block diagram as shown in Fig. 3.41.

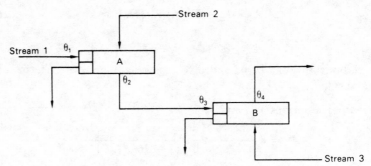

FIG. 3.40. Heat exchanger arrangement

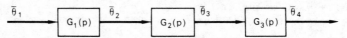

FIG. 3.41. Block diagram of heat exchanger system

(a) $\qquad G_1(p) = \dfrac{\bar{\vartheta}_2}{\bar{\vartheta}_1} = \dfrac{0.33}{1 + 0.67p}$ $\qquad$ (equation 3.27)

(b) $\qquad G_2(p) = \dfrac{\bar{\vartheta}_3}{\bar{\vartheta}_2} = \exp(-0.5p)$ $\qquad$ (distance–velocity lag—equation 3.55)

(c) $\qquad G_3(p) = \dfrac{\bar{\vartheta}_4}{\bar{\vartheta}_3} = \dfrac{1}{10.2p^2 + 3.1p + 1}$ $\qquad$ (equation 3.50)

$\therefore \qquad G(p) = \dfrac{\bar{\vartheta}_4}{\bar{\vartheta}_1} = G_1(p)\,G_2(p)\,G_3(p)$ $\qquad$ (equation 3.21)

$$= \frac{0.33 \exp(-0.5p)}{(1 + 0.67p)(10.2p^2 + 3.1p + 1)}$$

(i) For step change of unit magnitude in θ_1:

$$\bar{\vartheta}_4 = \frac{0.33 \exp(-0.5p)}{p(1 + 0.67p)(10.2p^2 + 3.1p + 1)}$$

For step response the distance–velocity lag simply gives rise to a dead-time of 0·5 min before the response commences (equation 3.52) and therefore need only be taken into account in the final solution. Thus, leaving out the latter:

$$\bar{\vartheta}_4' = \frac{0.33}{p(1 + 0.67p)(10.2p^2 + 3.1p + 1)}$$

$$= \frac{0.33}{p} - \frac{0.0087}{1 + 0.67p} - \frac{3.24p + 1.2}{10.2p^2 + 3.1p + 1}$$

$$= \frac{0.33}{p} - \frac{0.013}{p + 1.5} - 0.317\left\{\frac{p + 0.153}{(p + 0.153)^2 + 0.075} + \frac{0.222}{(p + 0.153)^2 + 0.075}\right\}$$

$\therefore \quad \vartheta_4' = 0.33 - 0.013\exp(-1.5t) - 0.317\exp(-0.153t)\{\cos(0.274t) + 0.811\sin(0.274t)\}$

(ii) The frequency response may be determined by applying the substitution rule to $G(p)$. It is easier, however, to make use of equations 3·98 and 3·99. If $(A.R.)_1$, $(A.R.)_2$ and $(A.R.)_3$ are the amplitude ratios of $G_1(p)$, $G_2(p)$ and $G_3(p)$ respectively, then:

$$(A.R.)_1 = \frac{0.33}{\sqrt{\{1 + (0.67\omega)^2\}}} = \frac{0.33}{\sqrt{(0.45\omega^2 + 1)}}$$ (equation 3.84)

$$(A.R.)_2 = 1$$ (equation 3.89)

$$(A.R.)_3 = \frac{1}{\sqrt{\{[1 - (3.2\omega)^2]^2 + [3.1\omega]^2\}}} = \frac{1}{\sqrt{(104\omega^4 - 10.8\omega^2 + 1)}}$$ (equation 3.87)

$$\therefore \quad A.R. = \frac{0.33}{\sqrt{\{(0.45\omega^2 + 1)(104\omega^4 - 10.8\omega^2 + 1)\}}}$$

Similarly $\quad \psi = -\tan^{-1}(0.67\omega) - 0.5\omega - \tan^{-1}\left(\dfrac{3.1\omega}{1 - 10.2\omega^2}\right)$

Transfer functions of feedback control systems

Once each element in the control loop has been described in terms of its transfer function, the behaviour of the closed loop can be determined by the formulation of *overall transfer functions*. Two such are of importance, i.e. those relating the controlled variable (C) to the desired value (R) and to the load (U).

(i) *Overall transfer function between C and R*. This is obtained by assuming that no changes occur in U, i.e. if $\mathscr{U}$ represents a deviation variable, then $\mathscr{U} = 0$. As $\mathscr{U}$ will have no effect on the control loop it may be left out of the block diagram. Thus, Fig. 3.2 becomes as shown in Fig. 3.40.

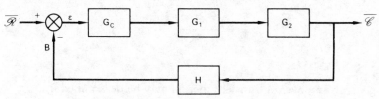

FIG. 3.42. Block diagram for changes in desired value only

From Fig. 3.42 and equation 3.21:

$$\bar{\varepsilon} = \overline{\mathscr{R}} - \overline{\mathscr{B}}$$ ....(3.100)

$$\overline{\mathscr{C}} = G_c G_1 G_2 \bar{\varepsilon}$$ ....(3.101)

$$\overline{\mathscr{B}} = H\overline{\mathscr{C}}$$ ....(3.102)

Eliminating ε and B from equations 3.100, 3.101 and 3.102 the overall transfer function is obtained, i.e.:

$$\frac{\overline{\mathscr{C}}}{\overline{\mathscr{R}}} = \frac{G}{1 + GH}$$ ....(3.103)

where $G = G_c G_1 G_2$.

Thus Fig. 3.42 may be replaced by an equivalent single block, as shown in Fig. 3.43.

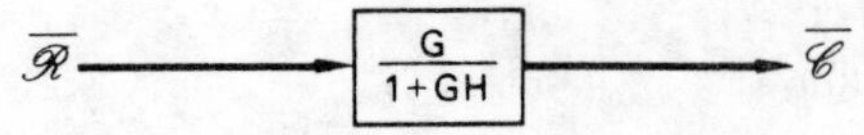

$$\mathscr{R} \longrightarrow \boxed{\dfrac{G}{1+GH}} \longrightarrow \mathscr{C}$$

FIG. 3.43. Equivalent single block

(ii) *Overall transfer function between C and U*. In this case $\mathscr{R} = 0$, i.e. the desired value is held constant.

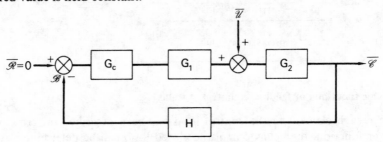

FIG. 3.44. Block diagram for changes in load only

From Fig. 3.44:

$$\bar{\varepsilon} = -\overline{\mathscr{B}}$$

$$\overline{\mathscr{C}} = G_2(\overline{\mathscr{U}} + G_cG_1\bar{\varepsilon})$$

$$\overline{\mathscr{B}} = H\overline{\mathscr{C}}$$

$$\therefore \qquad \frac{\overline{\mathscr{C}}}{\overline{\mathscr{U}}} = \frac{G_2}{1 + GH} \qquad \qquad \dots.(3.104)$$

where $G = G_cG_1G_2$.

In general, any overall transfer function may be determined from:

$$\frac{\overline{\mathscr{X}}_o}{\overline{\mathscr{X}}_i} = \frac{G_a}{1 + G_b} \qquad \qquad \dots.(3.105)$$

where G_a is the multiple of the transfer functions of all blocks between $\overline{\mathscr{X}}_i$ (input) and $\overline{\mathscr{X}}_o$ (output), and G_b is the multiple of all blocks in the whole loop— termed the *open-loop transfer function* of the control system.

The effect of simultaneous changes in load and desired value can be determined by the principle of superposition, i.e. by summing the separate variations due to each type of disturbance alone.

Calculation of offset from the overall transfer function

(i) *Load change with proportional control*. Referring to Fig. 3.42—for a proportional controller:

$$G_c = K_c \qquad \qquad \dots.(3.106)$$

Assume for simplicity that the time constants of G_1 (final control element) and H (measuring element) are negligible compared to that of G_2 (the process), i.e. that G_1 and H are constants. Suppose also that G_2 is first order, then:

$$G_2 = \frac{1}{1 + \tau_2 p}$$

Thus, from equation 3.105—for a change in load only:

$$\frac{\overline{\mathscr{C}}}{\overline{\mathscr{U}}} = \frac{1/(1 + \tau_2 p)}{1 + K_c G_1 H/(1 + \tau_2 p)} = \frac{1}{1 + \tau_2 p + K_c G_1 H}$$

For unit step change in load:

$$\overline{\mathscr{C}} = \frac{1}{p}\left(\frac{1}{1 + \tau_2 p + K_c G_1 H}\right) \qquad \ldots.(3.107)$$

Offset is defined as the difference between the desired and actual response of the output as $t \to \infty$, i.e. in this case,

$$\text{Offset} = \lim_{t \to \infty} \mathscr{R}(t) - \lim_{t \to \infty} \mathscr{C}(t) = \mathscr{R}(\infty) - \mathscr{C}(\infty) \qquad \ldots.(3.108)$$

the desired change being the same as the set-point change. But, for a load change the desired requirement of the control loop is to keep the output steady, i.e. the set-point is kept fixed and $\mathscr{R}(\infty) = 0$. $\mathscr{C}(\infty)$ is given by the final value theorem, i.e. from equation 3.107,

$$\lim_{t \to \infty} \mathscr{C}(t) = \lim_{p \to 0}\left\{p\frac{1}{p}\left(\frac{1}{1 + \tau_2 p + K_c G_1 H}\right)\right\}$$

$$= \frac{1}{1 + K_c G_1 H}$$

Hence, from equation 3.108:

$$\text{Offset} = -\frac{1}{1 + K_c G_1 H} \qquad \ldots.(3.109)$$

Thus the offset is reduced as the controller gain K_c is increased.

(ii) *Set-point change with proportional control.* For change in set-point only:

$$\frac{\overline{\mathscr{C}}}{\overline{\mathscr{R}}} = \frac{K_c G_1/(1 + \tau_2 p)}{1 + K_c G_1 H/(1 + \tau_2 p)}$$

$$= \frac{K_c G_1}{1 + \tau_2 p + K_c G_1 H}$$

For unit step change in set-point:

$$\text{Offset} = \mathcal{R}(\infty) - \mathcal{C}(\infty)$$

$$= \frac{1 + K_c G_1(H - 1)}{1 + K_c G_1 H} \qquad \ldots.(3.110)$$

Again the offset is reduced as K_c increases.

(iii) *Load change with* **P** + **I** *control.* From equation 3.58:

$$G_c = K_c \left(1 + \frac{1}{\tau_1 p}\right)$$

Keeping the remaining transfer functions as before:

$$\frac{\overline{\mathcal{C}}}{\overline{\mathcal{U}}} = \frac{1/(1 + \tau_2 p)}{1 + K_c \left(1 + \dfrac{1}{\tau_1 p}\right) G_1 \left(\dfrac{1}{1 + \tau_2 p}\right) H}$$

$$= \frac{\tau_1 p}{\tau_1 p(1 + \tau_2 p) + K_c G_1 H(\tau_1 p + 1)}$$

For unit step change in load:

$$\text{Offset} = \mathcal{R}(\infty) - \mathcal{C}(\infty) = 0 \qquad \ldots.(3.111)$$

Thus the presence of integral action has removed the offset indicated by equation 3.109. Similar procedures may be employed with other control actions.

It is possible to determine the step response of simple control systems by inverting the transfer functions obtained as in equation 3.107. However, this becomes increasingly difficult (as it does with open-loop transfer functions) when higher order systems are present.

The experimental application of forcing functions to process systems

In practice many processes are very complex and it is extremely difficult to describe their operation with any degree of accuracy by a suitable mathematical model. Thus the nature of the response of the process to various forcing functions must be determined experimentally. The classical methods are to obtain the response to step changes—*step testing*, or to sinusoidal changes—*frequency testing*. Both methods have particular advantages and disadvantages which have been comprehensively listed by WILLIAMS and LAUHER[14].

(i) Step testing

This is easy to apply experimentally, e.g. by rapidly opening a valve or quickly changing the set-point of a controller. However, it is a severe type of disturbance and damage to the system under test may result if too large an upset is injected. Frequently it is necessary to determine an approximate system transfer function from such a test in order to examine the stability of the process (which is discussed later). In general the transfer function is obtained by fitting responses given by assumed transfer functions to the response using a least squares procedure. Often it is difficult to fit the curves with accuracy and smaller time constants may not be detected[49]. This will cause the fitted transfer functions to be in considerable error at high frequencies (i.e. where changes are rapid) when such time constants become effective.

(ii) *Frequency testing*[25]

This is much more tedious to carry out as it must be performed at a number of different frequencies. Usually the time of testing can be equated to the time the unit is off production. Consequently this is also a more expensive procedure than step experimentation. However, it is not such a severe disturbance and time constants of all sizes can be determined with reasonable accuracy if a sufficiently large range of frequencies is used. Frequency testing of feedback control systems is carried out with the controller disconnected and on open loop (Fig. 3.45). Systematic methods have been developed which lead to the desired settings of the controller to be inserted. These methods are discussed later.

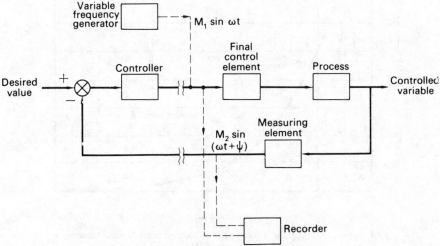

FIG. 3.45. Experimental frequency testing

Frequency response procedures are only applicable to linear systems and any non-linearities which may occur when the controller is connected into the system will lead to poor control with the predicted controller settings. The experimental step response, being applied to a closed loop system, automatically takes non-linear effects into account.

The system transfer function is usually determined by plotting amplitude ratio data as a function of frequency using logarithmic scales (*Bode* plot). The plot is approximated by a series of straight lines with integer values of slope. Working from low to high frequencies the various changes in slope indicate the terms constituting the transfer function[22]. An alternative procedure[60] is to fit the transfer function in the form of a polynomial ratio to the frequency response data expressed in complex number form. Although the latter method necessitates the use of a digital computer it automatically takes the phase shift data into account which the other procedure does not.

(iii) *Pulse testing*

This is a popular and practical means of obtaining frequency response data. The pulse applied can be of many forms, e.g. rectangular, triangular, displaced cosine. HOUGEN[66] has discussed the applications of pulse testing at length.

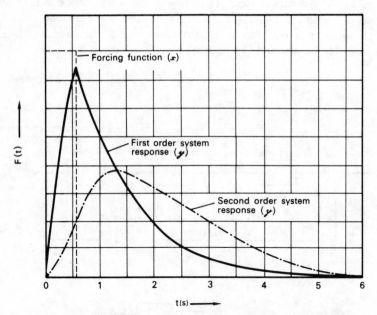

FIG. 3.46. Rectangular pulse forcing function and typical system response[66]

The main requirements are that the system dynamics be excited but the system should not be driven so hard that it is forced to the material limit of its output (i.e. *saturated*). The single pulse is designed to excite the system with all frequencies at once and hence by suitable computation the entire frequency response characteristics may be extracted from the results of a single pulse test.

If the forcing function is a closed pulse x the system output will also be a closed pulse of some form y (Fig. 3.46) unless pure integration terms are present. Input and output are related in the form of a transfer function:

$$G(p) = \frac{\overline{y}}{\overline{x}} = \frac{\displaystyle\int_0^\infty y\,e^{-pt}\,dt}{\displaystyle\int_0^\infty x\,e^{-pt}\,dt} \qquad \ldots\ldots(3.112)$$

Provided the input and output pulses close, the integrals in equation 3.112 are both finite and the frequency response may be computed by substituting $i\omega$ for p. Numerical integration is necessary and a digital computer is required for the calculations. LEES and HOUGEN[47] have compared frequency and pulse testing methods, concluding that the experimental advantages of the latter outweigh the computational disadvantages.

(iv) Stochastic testing

It is theoretically possible to obtain the dynamics of a system from normal operating records[52]. A process is always subject to random variations in certain forcing functions. Such variations will excite the system dynamics giving an output dependent upon the random input. Frequency response data can be extracted from such random input/output relationships by statistical methods[21, 78] but the practical problems are considerable. Often the magnitude of the disturbance is insufficient to obtain a measurable response. (The value of an input signal to which a system will not respond is termed the *dead-band* or *dead-zone*.) Also a number of external variables may be exciting the system at the same time. These disadvantages are usually overcome by applying a random sequence of pulses as the desired forcing function, i.e. a stochastic process. The frequency response characteristics of the system can be extracted from a relatively short experimental test run[78]. WOOD and ROBBINS[79] have compared step, frequency and stochastic testing of an industrial fractionating column. The statistical method proved to be the most unreliable due to the effect of *noise* in the measuring instruments and perturbations in heat losses from the column.

CONTROL SYSTEM DESIGN

The normal function of any control system is to ensure that the controlled variable attains its desired value as rapidly as possible after a disturbance has occurred and with the minimum of oscillation. Determination of the response of a system to a given forcing function will show what final value the controlled variable will attain and the manner in which it will arrive at that value. This latter is a function of the stability of the response. For example, in considering the response of a second order system to a step change, it can be seen that oscillation increases as ζ decreases (Fig. 3.36). The stability is therefore considered to decrease with ζ. The limiting case occurs when $\zeta = 0$ and the response oscillates with constant amplitude—it is then said to be *conditionally stable*[13]. With more complex systems it is possible to obtain oscillations of increasing amplitude such that the output never attains its desired value but increases to its physical limits. This type of response is termed *unstable*. The majority of fluid flow and heat transfer processes give overdamped or critically damped responses when no controller is attached to the process (i.e. on open loop). Whenever a closed loop system is formed incorporating a controller, however, there is always the possibility of an unstable response occurring.

Tests for unstable systems

The characteristic equation

Consider the transfer function given by equation 3.103. For a unit step change in the desired value $\mathscr{R}$, the Laplace transform of the controlled variable $\mathscr{C}$ is given by:

$$\bar{\mathscr{C}} = \frac{1}{p} \frac{G}{1 + GH}$$

$$= \frac{GF(p)}{p(p - \alpha_1)(p - \alpha_2)\ldots(p - \alpha_n)} \qquad \ldots.(3.113)$$

where $\alpha_1, \alpha_2, \ldots, \alpha_n$ are the n roots of the equation:

$$1 + GH = 0 \qquad \ldots.(3.114)$$

and $F(p)$ is some function of p.

Equation 3.114 is the *characteristic equation* of the control loop of Fig. 3.42 and is dependent only on the open-loop transfer function GH. The determination of the nature of the roots of the characteristic equation forms the basis of all techniques used to establish system stability. Now, in order to calculate the step response, equation 3.113 must be split into partial fractions for inversion, thus:

$$\bar{\mathscr{C}} = \frac{B_1}{p} + \frac{B_2}{p - \alpha_1} + \frac{B_3}{p - \alpha_2} + \ldots + \frac{B_{n+1}}{p - \alpha_n}$$

Inversion gives:

$$\mathscr{C} = B_1 + B_2 e^{\alpha_1 t} + B_3 e^{\alpha_2 t} + \ldots + B_{n+1} e^{\alpha_n t} \qquad \ldots.(3.115)$$

The roots of the characteristic equation may be real and/or complex, depending upon the form of the open-loop transfer function. Suppose α_1 to be complex, such that:

$$\alpha_1 = \beta + i\gamma$$

where β and γ are real. Then:

$$e^{\alpha_1 t} = e^{\beta t} e^{i\gamma t} = e^{\beta t} (\cos \gamma t + i \sin \gamma t) \qquad \ldots.(3.116)$$

It can be seen from equations 3.113, 3.115 and 3.116 that should any root of the characteristic equation have a positive real part, the resulting step response (equation 3.115) will contain an exponentially increasing term, i.e. it will be unstable. Thus for stability, the roots of the characteristic equation must have negative real parts. Some idea can be obtained of the degree of stability of the system from the magnitude of the real parts of the roots. If the latter are negative and large, the transients will decay rapidly and the response will be more stable than for a system having roots with small negative real parts.

Note that the characteristic equation is the same for both desired value and load changes (equations 3.103 and 3.104).

The Routh–Hurwitz criterion

It is often difficult to determine the roots of the characteristic equation. HURWITZ[32] and ROUTH[15] developed an algebraic procedure for finding the number of roots with positive real parts and consequently whether the system is unstable or not.

(a) The characteristic equation is first written in the form:

$$c_n p^n + c_{n-1} p^{n-1} + \ldots\ldots\ldots + c_1 p + c_0 = 0 \qquad \ldots.(3.117)$$

where c_n is positive. If any of the remaining coefficients are negative, the system is unstable. If, however, all are positive, the Hurwitz criterion is applied.

(b) The Hurwitz criterion that the roots shall not have positive real parts requires that the following determinant of order n, and the $n-1$ other determinants obtained by omitting successively the last row and column of the previous one, shall all be positive (m is the row number):

$$\begin{vmatrix} c_{n-1} & c_n & 0 & 0 & 0\ldots & 0 \\ c_{n-3} & c_{n-2} & c_{n-1} & c_n & 0\ldots & 0 \\ c_{n-5} & c_{n-4} & c_{n-3} & c_{n-2} & c_{n-1}\ldots & 0 \\ \ldots & & & & & \\ c_{n-2m+1} & c_{n-2m+2} & c_{n-2m+3} & \ldots & & \\ \ldots & & & & & \\ 0 & 0 & 0 \ldots & & & c_0 \end{vmatrix}$$

Example

A system has the characteristic equation:

$$p^3 + 6p^2 + 11p + 6 = 0$$

Determine whether or not the system is stable.

I

Solution

The characteristic equation is already in the form of equation 3.117. All the coefficients are positive, thus the Hurwitz criterion must be applied to determine the stability. The characteristic equation is third order, thus $n = 3$ and the first determinant to be tested is of order 3. Therefore:

$$\Delta_1 = \begin{vmatrix} 6 & 1 & 0 \\ 6 & 11 & 6 \\ 0 & 0 & 6 \end{vmatrix} = 6 \begin{vmatrix} 11 & 6 \\ 0 & 6 \end{vmatrix} - \begin{vmatrix} 6 & 6 \\ 0 & 6 \end{vmatrix}$$

$$= 6 \times 66 - 36 = +\text{ve}$$

The second is of order 2:

$$\Delta_2 = \begin{vmatrix} 6 & 1 \\ 6 & 11 \end{vmatrix} = 66 - 6 = +\text{ve}$$

And the third of order 1:

$$\Delta_3 = 6 = +\text{ve}$$

Hence the roots of the characteristic equation do not have any positive real parts and the system is stable.

The Bode stability criterion

The disadvantages of the Routh–Hurwitz test are that it is necessary to know the system transfer function and that it gives no information concerning the *degree* of stability of the system. For cases where it is desired to know the latter and only experimental response data are available, the *Nyquist* and *Bode* stability criteria are of considerable use. The Nyquist criterion[33] can be universally applied to all systems and involves plotting the system frequency response in the complex plane. This procedure is discussed later. A rather simpler approach is supplied by the criterion of BODE[34], which can be shown to be an extension of the Nyquist procedure[16] but applies only to systems for which the amplitude ratio and phase shift decrease continuously with frequency. Fortunately most control systems are of this type and consequently the Bode method will be treated in some detail.

The Bode diagram

In order to apply the Bode criterion the system frequency response must be represented graphically in the form of a Bode diagram or plot. This consists of *two* graphs which normally are drawn with the axes

(a) $\log_{10}$ (amplitude ratio) versus $\log_{10}$ (frequency)
(b) phase angle versus $\log_{10}$ (frequency)

with frequency plotted as the abscissa in both cases.

(i) *First order system.* From equations 3.84 and 3.85 and putting $\tau = \tau_1$:

$$\text{A.R.} = \frac{1}{\sqrt{(1 + \omega^2\tau^2)}} \qquad \text{(equation 3.84)}$$

$$\psi = \tan^{-1}(-\omega\tau) \qquad \text{(equation 3.85)}$$

Thus:

$$\log (\text{A.R.}) = -\tfrac{1}{2} \log \left[1 + (\omega\tau)^2\right]$$

therefore:

$$\lim_{\omega\tau \to \infty} \left[\log(\text{A.R.})\right] = -\log(\omega\tau) \qquad \ldots (3.118)$$

since $(\omega\tau)^2 \gg 1$ as $\omega\tau \to \infty$.

Equation 3.118 is termed the *high frequency asymptote* (H.F.A.) and is a straight line of slope -1 passing through the point $(1, 1)$ on a plot of $\log (\text{A.R.})$ versus $\log (\omega\tau)$ ($\omega\tau$ is employed in order that the diagram may be used for a first order system with any time constant). The *low frequency asymptote* (L.F.A.) is given by:

$$\lim_{\omega\tau \to 0} \left[\log (\text{A.R.})\right] = 0$$

i.e.

$$\lim_{\omega\tau \to 0} (\text{A.R.}) = 1 \qquad \ldots (3.119)$$

Hence the two asymptotes, as shown in Fig. 3.47, intersect at the point $(1,1)$. The frequency at this point is given by:

$$\omega_c = \frac{1}{\tau} \qquad \ldots (3.120)$$

where ω_c is termed the *corner* or *break* frequency. From equation 3.84, when $\omega = \omega_c$:

$$\text{A.R.} = \frac{1}{\sqrt{(1 + \omega_c^2 \tau^2)}} = \frac{1}{\sqrt{2}} = 0 \cdot 707$$

An A.R. plot may be sketched from its asymptotes and the A.R. at ω_c. This is sufficiently accurate for most purposes.

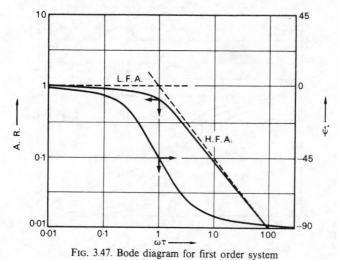

FIG. 3.47. Bode diagram for first order system

The phase curve may also be easily sketched, for from equation 3.85:

$$\lim_{\omega\tau\to 0} (\psi) = 0° \quad \text{and} \quad \lim_{\omega\tau\to\infty} (\psi) = -90°$$

and when $\omega_c = \dfrac{1}{\tau}$, $\psi = -45°$.

Intermediate points must be calculated. It may be noted that the curve is symmetrical about ψ at ω_c.

(ii) *Second order system.* From equations 3.87b and 3.88:

$$\text{A.R.} = \frac{1}{\sqrt{[(1 - \omega^2\tau^2)^2 + (2\zeta\omega\tau)^2]}} \qquad \text{(equation 3.87b)}$$

$$\psi = \tan^{-1}\left(\frac{-2\zeta\omega\tau}{1 - \omega^2\tau^2}\right) \qquad \text{(equation 3.88)}$$

The Bode plot in this case (Fig. 3.48) is distinguished by the fact that ζ is a parameter which affects both the A.R. and ψ curves. However, the asymptotes may be determined in precisely the same manner as for the first order system. It is found that, for all ζ, the A.R. high frequency asymptote is a straight line of slope -2 passing through the point (1, 1) and the L.F.A. is given by the line A.R. = 1. The ψ curves all tend to zero degrees as $\omega\tau \to 0$ and to $-180°$ as $\omega\tau \to \infty$. When $\omega_c = 1/\tau$, $\psi = -90°$ independently of ζ.

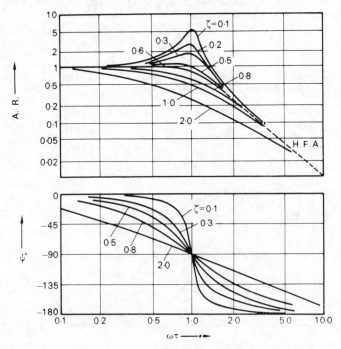

FIG. 3.48. Bode diagram for second order system

For $\zeta < 1/\sqrt{2}$ the A.R. curves exhibit maxima in the region of $\omega\tau = 1$ and also give values of A.R. > 1. The frequency at which the A.R. is a maximum for any given value of ζ ($<1/\sqrt{2}$) is termed the resonant frequency.

Other functions may be plotted in a similar manner (Fig. 3.49).

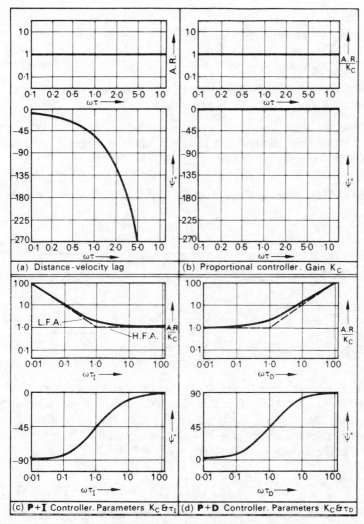

FIG. 3.49. Bode plots of some common functions

(iii) *Systems in series.* The usefulness of the logarithmic plot becomes apparent when it is desired to determine the frequency response of systems in series. The resultant amplitude ratio and phase shift may be obtained using equations 3.98 and 3.99, i.e.:

$$A.R. = K(A.R.)_1 \, (A.R.)_2 \, (A.R.)_3 : \ldots \qquad \text{(equation 3.98)}$$

$\therefore \quad \log(\text{A.R.}) = \log K + \log(\text{A.R.})_1 + \log(\text{A.R.})_2 + \log(\text{A.R.})_3 + \ldots$

$$\ldots(3.121)$$

and

$$\psi = \psi_1 + \psi_2 + \psi_3 + \ldots\ldots \qquad \text{(equation 3.99)}$$

The overall Bode plot of a number of systems in series may be obtained therefore as follows:

(a) To determine the resultant A.R. add the individual A.R.s on the log/log diagram, treating values above unit A.R. as positive and below unit A.R. as negative. It is normally sufficiently accurate simply to add the asymptotes.

(b) Add the individual phase shifts on the linear/log plot to obtain the resultant phase shift.

(c) If the resultant transfer function is multiplied by some constant, e.g. the steady-state gain, the entire A.R. curve is moved vertically by a constant amount. The phase shift is unaffected.

System stability from the Bode diagram. Consider the control loop shown in

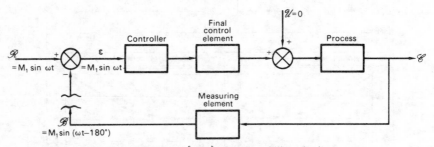

Fig. 3.50. Establishment of Bode stability criterion

Fig. 3.50. Suppose the loop to be broken after the measuring element, and that a sinusoidal forcing function $M_1 \sin \omega t$ is applied to the desired value setting $\mathcal{R}$. Suppose also that the open-loop gain (or amplitude ratio) of the system is unity and that the phase shift is $-180°$. Then the output $\mathcal{B}$ from the measuring element (i.e. the system open-loop response) will have the form:

$$\mathcal{B} = M_1 \sin(\omega t - 180°)$$
$$= -M_1 \sin \omega t$$

Now, if at some instant of time $\mathcal{R}$ is reset to zero and the loop instantaneously closed, we will have:

$$\varepsilon = \mathcal{R} - \mathcal{B}$$
$$= 0 - (-M_1 \sin \omega t)$$
$$= M_1 \sin \omega t.$$

This indicates that the oscillation, once set in motion, will be maintained with constant amplitude around the closed loop for $\mathcal{U} = \mathcal{R} = 0$. If, however, the open-loop gain or A.R. of the system is greater than unity, the amplitude of the sinusoidal signal will increase around the control loop, whilst the phase shift will be unaffected. Thus the amplitude of the signal will grow indefinitely, i.e. the system will be unstable.

This heuristic argument forms the basis of the *Bode stability criterion*[24, 34] which states that *a control system is unstable if its open-loop frequency response exhibits an A.R. greater than unity at the frequency for which the phase shift is* $-180°$. This frequency is termed the cross-over frequency (ω_{co}). The degree of stability of the system can be estimated from its Bode diagram. If the open-loop A.R. is unity when $\psi = -180°$ the closed loop control system will oscillate with constant amplitude, i.e. it will be on the verge of instability. The greater the difference between the open-loop A.R. (<1) at ω_{co} and A.R. $= 1$, the more stable the closed loop system will be. This difference is normally measured in terms of *gain margin* where, if $(A.R.)_{co}$ is the open-loop A.R. at ω_{co} measured on the Bode plot:

$$\text{Gain margin} = \frac{1}{(A.R.)_{co}} \qquad \dots (3.122)$$

Hence a gain margin < 1 signifies an unstable system.

A *phase margin* has also been defined for the purpose of design, i.e. on the open-loop Bode plot:

$$\text{Phase margin} = 180° - \text{Phase lag in degrees at frequency for which the A.R. is unity} \qquad \dots (3.123)$$

A negative phase margin indicates an unstable system. Normally, design specifications for a control system require a gain margin > 1.7 and a phase margin $> 30°$.

The effect of the various control modes on the stability of a control loop can be examined using the concept of gain and phase margin.

Example

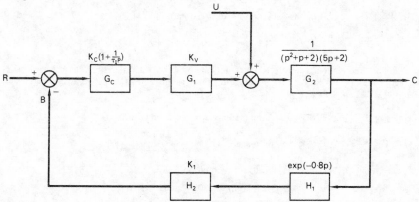

Fig. 3.51. Block diagram of control loop

A control system using **P + I** control is represented by the block diagram shown in Fig. 3.51. The transfer functions describing each block are as shown with $K_c = 10$, $\tau_I = 1$ min, $K_1 = 0.8$, $K_v = 0.5$. By determination of gain and phase margins, show the effect on the stability of the control system of introducing derivative action with $\tau_D = 1$ min.

Solution

The open-loop transfer function of the control system without derivative action is given by:

$$G(p) = \frac{10\left(1 + \dfrac{1}{p}\right) \times 0.5 \times \exp(-0.8p) \times 0.8}{(p^2 + p + 2)(5p + 2)}$$

$$= \frac{\left(1 + \dfrac{1}{p}\right)\exp(-0.8p)}{(0.5p^2 + 0.5p + 1)(2.5p + 1)}$$

$$= G_1(p)\,G_2(p)\,G_3(p)\,G_4(p)$$

The Bode diagram of $G(p)$ is obtained by breaking down the transfer function into its constituent parts, plotting each separately and performing a graphical summation.

(a) $G_1(p) = \dfrac{1}{2.5p + 1}$ is the transfer function of a first order system with a time constant of 2.5 min. Hence from equations 3.84, 3.85, 3.120:

$$\text{A.R.} = \frac{1}{\sqrt{(1 + 6.25\omega^2)}}, \quad \psi = \tan^{-1}(-2.5\omega)$$

and

$$\omega_c = 0.4 \text{ radians/min.}$$

(b) $G_2(p) = \dfrac{1}{0.5p^2 + 0.5p + 1}$ is the transfer function of a second order system with a time constant of $1/\sqrt{2}$ min and a damping coefficient of $1/(2\sqrt{2})$. From equations 3.87b and 3.88:

$$\text{A.R.} = \frac{1}{\sqrt{\{(1 - \tfrac{1}{2}\omega^2)^2 + (\tfrac{1}{2}\omega)^2\}}}, \quad \psi = \tan^{-1}\left(-\frac{\tfrac{1}{2}\omega}{1 - \tfrac{1}{2}\omega^2}\right)$$

The asymptotes intersect at the point A.R. $= 1$, $\omega = \sqrt{2}$.

(c) $G_3(p) = \exp(-0.8p)$ represents a distance-velocity lag. From equations 3.89 and 3.90:

$$\text{A.R.} = 1 \text{ for all } \omega, \quad \psi = -0.8\omega.$$

(d) $G_4(p) = (1 + 1/p)$. Using the substitution rule we obtain:

$$\text{A.R.} = \sqrt{\left(1 + \frac{1}{\omega^2}\right)}, \quad \psi = \tan^{-1}\left(-\frac{1}{\omega}\right)$$

The A.R. low frequency asymptote is a straight line of slope -1. The H.F.A. is A.R. $= 1$. These intersect at the point A.R. $= 1$. The phase shift plot is the inverse of that for a first order system, i.e. $\psi \to -90°$ as $\omega \to 0$ and $\psi \to 0°$ as $\omega \to \infty$ (see Fig. 3.49c).

From Fig. 3.52, $\omega_{co} = 0.77$ radians/min. Hence from equation 3.122:

$$\text{Gain margin} = \frac{1}{A} = \frac{1}{0.96} = \underline{\underline{1.04}}$$

At A.R. $= 1$, $\omega = 0.74$ radians/min, thus from equation 3.123:

$$\text{Phase margin} = 180° - 177° = \underline{\underline{3°}}$$

Although by definition the system is stable, both gain and phase margins are so small that a slight variation in any of the control system parameters could cause instability.

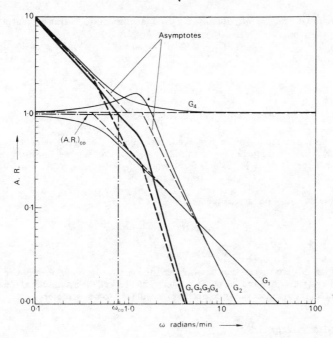

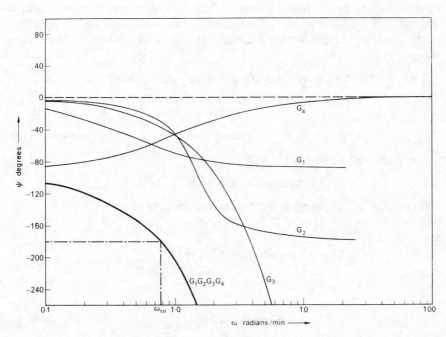

FIG. 3.52. Open-loop Bode diagram for control system shown in
Fig. 3.51. (No derivative action)

(e) When derivative action is introduced with $\tau_D = 1$ min the controller transfer function becomes:

$$G_5(p) = \left(1 + \frac{1}{p} + p\right)$$

giving

$$\text{A.R.} = \frac{1}{\omega}\sqrt{(\omega^4 - \omega^2 + 1)}$$

and

$$\psi = \tan^{-1}\left(\omega - \frac{1}{\omega}\right)$$

The low frequency asymptote on the amplitude ratio plot is a straight line of slope -1 passing through $(1,1)$. The high frequency asymptote is a straight line of slope $+1$ passing through $(1,1)$. The phase shift approaches $-90°$ as $\omega \to 0$ and tends to $+90°$ as $\omega \to \infty$. At $\omega = 1.0$ radian/min, $\psi = 0°$.

From Fig. 3.53, $\omega_{co} = 1.3$ radians/min.

$$\text{Gain margin} = \frac{1}{0.48} = \underline{\underline{2.08}}$$

and

$$\text{Phase margin} = \underline{\underline{29°}}.$$

The addition of derivative action has stabilised the control loop considerably, increasing the gain margin by a factor of 2 and the phase margin by 26°. In fact, the magnitude of the gain margin is now such that the response of the control system will tend to be overdamped.

The Nyquist criterion[33]

The polar plot (Nyquist diagram). This is a frequently used alternative to the Bode diagram for representing frequency response data. It is the locus of all points occupied by the tip of a vector in the complex plane whose magnitude and direction are determined by the amplitude ratio and phase shift respectively, as the frequency of the forcing function applied to the system is varied from zero to infinity.

Thus, for a first order system with unit steady-state gain:

$$\text{A.R.} = \frac{1}{\sqrt{(1 + \omega^2\tau^2)}} \qquad \text{(equation 3.84)}$$

$$\psi = \tan^{-1}(-\omega\tau) \qquad \text{(equation 3.85)}$$

If V_c is the vector, then for $\omega = 0$:

$$\text{magnitude of } V_c = \text{A.R.}_{\omega=0} = 1$$

and direction of V_c relative to the real axis is given by:

$$\psi_{\omega=0} = 0°$$

For $\omega = \infty$:

$$\text{magnitude of } V_c = 0$$

and

$$\psi = -90°$$

and when $\omega = 1/\tau$:

$$\text{magnitude of } V_c = \frac{1}{\sqrt{2}}$$

and

$$\psi = -45°.$$

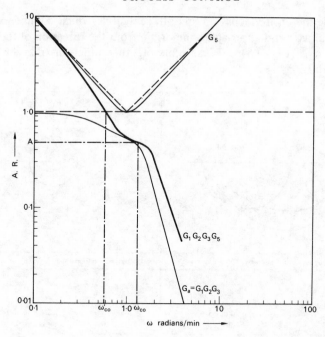

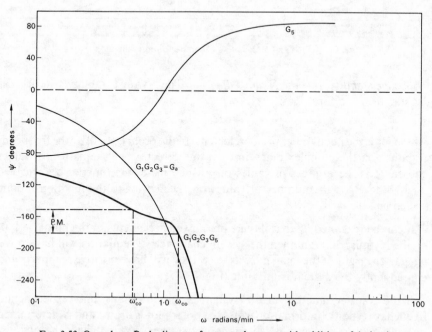

FIG. 3.53. Open-loop Bode diagram for control system with addition of derivative action

In this case the entire locus is a smooth semicircle which commences on the positive real axis and approaches the origin the direction of the negative imaginary axis (Fig. 3.54a). Polar plots of other functions may be sketched similarly (Fig. 3.54).

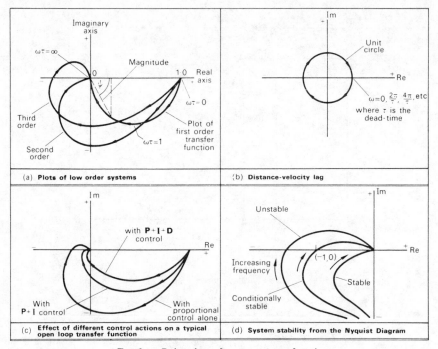

FIG. 3.54. Polar plots of some common functions

System stability from the Nyquist diagram. The Nyquist stability criterion can be stated in the form[18, 26]:

$$N_E = Z - P_E \qquad \qquad ....(3.124)$$

where N_E is the net number of encirclements of the point $(-1, 0)$ by the frequency response on the complex plane in the same direction as the path of p values chosen, Z is the number of values of p which make the function zero (called *zeros*), and P_E is the number of values of p which make the function infinite (called *poles*).

It can be reasoned[14] that if there are any encirclements of the point $(-1, 0)$ on the Nyquist diagram then the system characteristic equation will have roots lying to the right of the imaginary axis, i.e. they will have positive real parts and consequently the system is unstable (Fig. 3.54d).

If the transfer function of the system is unknown and only experimental frequency response information exists the polar plot is drawn and N determined by examination. For a fuller discussion of system analysis using the Nyquist diagram the reader is referred to work by WILLIAMS and LAUHER[14].

Methods of adjusting feedback controller settings

Many procedures exist for estimating optimum settings for controllers and the most common are described here. One of the usual bases employed is that the system response should have a *decay ratio* of $\frac{1}{4}$, i.e. the ratio of the overshoot of the first peak to the overshoot of the second peak is 4:1 (Fig. 3.55). There is no direct mathematical justification for this[17] but it is a compromise between a rapid initial response and a short *response time*. [The response time, *settling time* or *line-out time* is the time required for the absolute value of the system response to come within a small specified amount of the final value of the response (Fig. 3.55)].

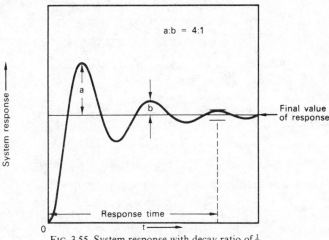

FIG. 3.55. System response with decay ratio of $\frac{1}{4}$

(a) *Determination of settings from frequency response data*

If the frequency response characteristics of the control system are known then it is possible to use the concept of gain and phase margin to estimate controller parameters required to give particular sized margins. However this necessitates trial and error procedures. The empirical method of ZIEGLER and NICHOLS[37] is more easily applied as follows.

The open-loop Bode diagram for all components in the control loop, excepting the controller, is plotted and the cross-over frequency determined. If the overall amplitude ratio at ω_{co} is α, then, by the Bode criterion, the gain of a proportional controller which would cause the system to be on the verge of instability will be:

$$K_u = \frac{1}{\alpha} \qquad \qquad \ldots . (3.125)$$

where K_u is termed the *ultimate gain*. The *ultimate period* is defined as the period of sustained cycling which would occur if a proportional controller of gain K_u were used, thus:

$$T_u = \frac{2\pi}{\omega_{co}} \qquad \qquad \ldots . (3.126)$$

The Ziegler–Nichols controller settings are derived from K_u and T_u on the basis of gain and phase margins of 2 and 30° respectively for proportional control alone. The addition of integral action introduces more phase lag at all frequencies and hence a lower value of proportional gain (K_c) is required to maintain the same phase margin. Adding derivative action introduces phase lead and thus a greater value of K_c can be tolerated. The controller settings recommended are shown in Table 3.1.

TABLE 3.1. *Ziegler–Nichols controller settings*

Control action	Controller settings		
	K_c	τ_I	τ_D
P	0·5 K_u	—	—
P + I	0·45 K_u	$T_u/1\cdot2$	—
P + I + D	0·6 K_u	$T_u/2$	$T_u/8$

Example

Determine the settings of the **P + I** and **P + I + D** controllers used in Fig. 3.51 by the method of Ziegler and Nichols.

Solution

The open-loop Bode diagram without the controller is given by G_a in Fig. 3.53.

Hence
$$\omega_{co} = 1\cdot12 \text{ radians/min.}$$

$$\therefore \quad K_u = \frac{1}{\alpha} = \frac{1}{0\cdot48} = 2\cdot08$$

and
$$T_u = \frac{2\pi}{1\cdot12} = 5\cdot61 \text{ min}$$

For a **P + I** controller, from Table 3.1, desired settings are:
$$K_c = \underline{0\cdot94}, \quad \tau_I = \underline{4\cdot7 \text{ min.}}$$

For a **P + I + D** controller, desired settings are:
$$K_c = \underline{1\cdot25}, \quad \tau_I = \underline{2\cdot8 \text{ min.}}, \quad \tau_D = \underline{0\cdot7 \text{ min.}}$$

Note that no settings are given for the **P + D** controller. Other methods must be used to obtain these.

(b) *Loop tuning*[37]

This procedure may be used when the frequency response of the system is unknown. The system is tested as a closed loop with any integral or derivative action of the controller removed. The controller proportional gain is carefully increased until continuous cycling occurs in the control loop. It may be necessary to inject small pulses in order to cause the onset of cycling because of process inertia. The gain at which this occurs corresponds to the K_u of the Ziegler–Nichols method. The period of oscillation is taken to be T_u. The controller settings are determined then using Table 3.1.

(c) *Cohen–Coon method*[39]

This is an experimental procedure based on measuring the open-loop step response.

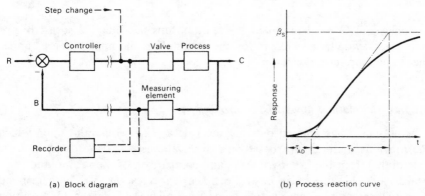

(a) Block diagram   (b) Process reaction curve

FIG. 3.56. Cohen–Coon method

The controller is placed on manual control (i.e. effectively removing it from the control loop) and the response of the measured variable to a small step change in manipulated variable is recorded as shown in Fig. 3.56a. This response is termed the *process reaction curve*. A tangent is drawn to the curve at the point of inflexion (Fig. 3.56b). The intercept of this tangent on the abscissa is termed the *apparent dead-time* (τ_{AD}) of the system. The slope of the tangent is given by:

$$m = \beta_s/\tau_a$$

where τ_a and β_s are the *apparent time constant* and final steady-state value of the response respectively.

If u_s is the steady-state gain of the system, then the controller settings are given by Table 3.2.

TABLE 3.2. *Cohen–Coon controller settings*

Control action	Controller settings		
	K_c	τ_I	τ_D
P	$\dfrac{1}{u_s}\dfrac{\tau_a}{\tau_{AD}}\left(1+\dfrac{\tau_{AD}}{3\tau_a}\right)$	—	—
P + I	$\dfrac{1}{u_s}\dfrac{\tau_a}{\tau_{AD}}\left(\dfrac{9}{10}+\dfrac{\tau_{AD}}{12\tau_a}\right)$	$\tau_{AD}\left\{\dfrac{30+3\tau_{AD}/\tau_a}{9+20\tau_{AD}/\tau_a}\right\}$	—
P + D	$\dfrac{1}{u_s}\dfrac{\tau_a}{\tau_{AD}}\left(\dfrac{5}{4}+\dfrac{\tau_{AD}}{6\tau_a}\right)$	—	$\tau_{AD}\left\{\dfrac{6-2\tau_{AD}/\tau_a}{22+3\tau_{AD}/\tau_a}\right\}$
P + I + D	$\dfrac{1}{u_s}\dfrac{\tau_a}{\tau_{AD}}\left(\dfrac{4}{3}+\dfrac{\tau_{AD}}{4\tau_a}\right)$	$\tau_{AD}\left\{\dfrac{32+6\tau_{AD}/\tau_a}{13+8\tau_{AD}/\tau_a}\right\}$	$\tau_{AD}\left\{\dfrac{4}{11+2\tau_{AD}/\tau_a}\right\}$

The Cohen–Coon settings are based on the assumption that the open-loop system behaves in the same way as the transfer function:

$$G(p) = \frac{u_s \exp(-\tau_{AD}p)}{1 + \tau_a p}$$

Cohen and Coon determined the relationships in Table 3.2 so as to give responses having large decay ratios, minimum offset and minimum area under the closed loop response curve.

(d) *Integral relations for setting controllers*

MURRILL[17] has pointed out that the $\frac{1}{4}$ decay ratio constraint upon which the previous methods are based has disadvantages, e.g. the decay ratio is determined from the first two peaks and comparison of the second and third peaks may yield a different decay ratio. Since perfect response would correspond to there being no error at any time, some criterion based upon the minimisation of the total error under the response curve would be logical. Workers at Louisiana State University[83] have presented three such criteria, the most successful of which appears to be the integral of time multiplied by the absolute value of the error (ITAE)

$$\text{ITAE} = \int_0^\infty t |C(t) - R(t)| \, dt \qquad \dots\text{(3.127)}$$

Errors that exist for long periods of time are penalised particularly. Controller settings based on this criterion are given in Table 3.3 and are based on a tuning relation of the form:

$$Y = A \left(\frac{\tau_{AD}}{\tau_a}\right)^B \qquad \dots\text{(3.128)}$$

where τ_{AD} and τ_a are obtained from the process reaction curve. The controller settings are obtained from Y by the relationships:

$Y = u_s K_c$ for the proportional mode,

$Y = \tau_a/\tau_I$ for the reset mode, and

$Y = \tau_D/\tau_a$ for the rate mode.

TABLE 3.3. *Constants for equation 3.128 based upon the ITAE criterion*[83]

Control action	Controller mode					
	Proportional		Reset		Rate	
	A	B	A	B	A	B
P	0·490	−1·084	—	—	—	—
P + I	0·859	−0·977	0·674	−0·680	—	—
P + I + D	1·357	−0·947	0·842	−0·738	0·381	0·995

There appears to be no general agreement as to which of the above methods will lead to the "best" control, and the "best" settings will vary with the application. Certainly the values obtained by the first three methods should be treated as first estimates only. Finer tuning of the controllers will be accomplished on site with the process in operation.

ADDITIONAL AND MORE ADVANCED CONTROL TECHNIQUES

In this section further frequently used techniques will be described briefly. Space does not allow a comprehensive discussion, but the intention is to make the reader aware of their existence. Some are recent developments which involve mathematical procedures with which the reader is probably unfamiliar. It is not the function of this section to describe such procedures, and the interested student of control is referred to the extensive literature that exists.

Cascade control [27, 42, 51, 53]

Up to this point only single loop systems have been considered with one input and one output variable. In practice both multiloop and multivariable systems frequently occur. Cascade control is a simple example of a multiloop system in which the set point of one controller (the *secondary* or *slave* controller) is adjusted by the output of another (the *primary* or *master* controller).

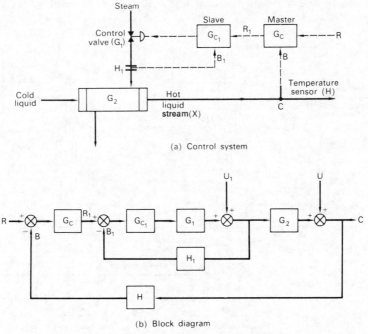

(a) Control system

(b) Block diagram

FIG. 3.57. Heat exchanger with cascade control

Consider the heat exchanger arrangement shown in Fig. 3.57. In the normal way the temperature of the outlet stream X would be controlled directly by measuring the temperature of X and adjusting the steam flow to the exchanger accordingly, i.e. without the slave controller. Suppose, however, that the steam supply pressure fluctuates. The variations in steam flow which will result from these fluctuations would have to produce a change in the temperature of X before corrective action could be taken. With the cascade arrangement shown the variations in flow will immediately be damped out by the slave flow control loop and there will be no detectable change in the temperature of X. The block diagram for this multiloop system appears in Fig. 3.57b where U represents changes in flow of cold liquid and U_1 fluctuations in steam flow

Feed-forward control

A simple case of this has been described already (Fig. 3.3). We are now in a position to extend the technique to a more practical situation. Consider a process control loop in which a large time-lag exists between the measurement of the controlled variable and the effect of the control action, e.g. product composition control in distillation. The feedback system cannot operate until the controlled variable moves away from its desired value. By the time the control action takes effect it may augment the disturbance rather than reduce it. At the least the controlled variable will be off specification for a considerable time. Feed-forward control can be applied in such instances causing control action to be taken before the controlled variable deviates from its specified value.

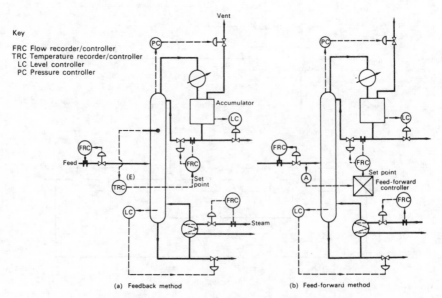

FIG. 3.58. Control of overhead product composition

Figure 3.58a shows a typical method of controlling overhead product composition using a feedback system when there are fluctuations in feed composition. The boiling temperature is measured at some point in the rectifying section of the column and the output of the temperature controller is connected in cascade with the reflux flow control loop. The position of the temperature measuring element is determined by the boiling temperature profile in the column and the proximity of the positions of entry of the disturbance and the controlling variable[29, 46]. Whatever the point of measurement there is bound to be some delay before control action can be taken. To circumvent this we may apply the predictive system of Fig. 3.58b where variations in feed composition are measured by some type of continuous analyser A. The signal from A is fed directly to the feed-forward controller, the output of which is cascaded on to the reflux flow control loop. If the transfer functions relating feed composition (x_F), reflux (l_{N+1}) and overhead product composition (x_D) are known, then we can write[57]:

$$G_1(p) = \bar{x}_D/\bar{x}_F \quad \text{and} \quad G_2(p) = \bar{x}_D/\bar{l}_{N+1}$$

i.e.
$$\bar{x}_D = G_1(p)\,\bar{x}_F \quad \text{and} \quad \bar{x}_D = G_2(p)\,\bar{l}_{N+1}$$

If variations in feed composition and reflux flow occur simultaneously, then by the principle of superposition the resultant variation in overhead product composition will be:

$$\bar{x}_D = G_1(p)\,\bar{x}_F + G_2(p)\,\bar{l}_{N+1} \qquad \ldots\ldots(3.129)$$

The desired control criterion is that variation in overhead composition should be zero, i.e. that:

$$x_D = 0$$

Substitution of this condition in equation 3.129 gives:

$$0 = G_1(p)\,\bar{x}_F + G_2(p)\,\bar{l}_{N+1}$$

$$\therefore \qquad G(p) = \bar{l}_{N+1}/\bar{x}_F = -\frac{G_1(p)}{G_2(p)} \qquad \ldots\ldots(3.130)$$

$G(p)$ is the transfer function of the feed-forward controller which is obtained from the known transfer functions $G_1(p)$ and $G_2(p)$ using equation 3.130. Other feed-forward control systems may be designed similarly. All should give perfect control in theory. However, in practice, due to errors in measurement, extraneous effects and errors in determining the basic transfer functions (e.g. $G_1(p)$ and $G_2(p)$), control is not perfect and considerable drifting of the controlled variable can occur. These difficulties have been emphasised by WARDLE and WOOD[81] in work carried out on an industrial fractionating plant. The addition of feedback control is necessary to correct such drifting. For example, in the system shown in Fig. 3.58 the outputs of the feedback temperature controller E and the feed-forward controller would be added and the combined signal applied to the set point of the reflux flow controller.

Multivariable systems and interaction

The distillation process is also a good example of a multivariable system, i.e. one in which there is more than one forcing variable and more than one output. For instance, variations may occur in feed composition, feed flow, feed quality, condenser cooling-water flow, steam to reboiler flow, etc., as input signals. Output variables will be overhead and bottom product compositions and flows, etc. The control problem can be simplified considerably by individually controlling as many of the input variables as possible, e.g. feed flow by a simple flow control loop, feed quality by a preheater, etc. Furthermore the regulation of one output variable (e.g. x_D) might be of much greater importance than another (x_w). In the latter case x_D would be controlled (e.g. as in Fig. 3.58a) and x_w would be allowed to vary at will.

One particular difficulty experienced with multivariable systems is the effect of *interaction* between particular control loops. Suppose that, in addition to the control of x_D shown in Fig. 3.58a, x_w were also controlled using a temperature sensor installed in the stripping section coupled (via a controller) to the set point of the steam-to-reboiler flow controller. Variations in rectifying section compositions due to some disturbance would lead to corrective action through adjustment of the reflux flow. This would affect compositions in the stripping section necessitating further corrective action through adjustment of the boil-up rate which, in turn, would interact with the rectifying section control system, etc. Such interaction can lead to oscillation in both controlled variables.

The general analytical treatment of linear interacting multivariable systems is a complex one requiring the use of matrix algebra[28]. This complexity increases rapidly with the number of variables considered. Both RIJNSDORP[76] and ROSENBROCK[80] have recently reported some success in the design of non-interacting control schemes for such systems.

Adaptive control[54, 55, 56, 59]

This represents one of the most promising developments in the control field. It is a form of self-optimising control in which the control system continuously monitors its own performance and compensates itself for disturbances within itself or in its surroundings by modifying its own parameters[17].

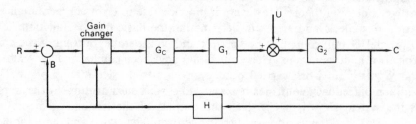

FIG. 3.59. Block diagram of adaptive control system[14]

A typical adaptive system is shown in Fig. 3.59[14]. The simple feedback loop (Fig. 3.2) may contain some component which is non-linear in character. In order to maintain a constant overall system gain under such circumstances, it will be necessary to compensate the controller gain K_c in an inverse manner with respect to the remaining gains in the loop. The basis upon which an adaptive system determines its performance and the necessary compensation is called the *performance index*. In Fig. 3.59 the performance index is a constant overall system gain.

Sampled data systems

Often a given measured variable is sampled at discrete time intervals when continuous measurement is either unnecessary or undesirable. The sharing of one composition analyser by a number of control systems is a frequent case. Also the digital computer is a sampled data device and when used as a controller operates as a sampled data controller. In such systems the sampling interval is of considerable importance. It can be shown[23] that if the sampling frequency is higher than twice the open-loop *critical* frequency (i.e. the frequency for which the phase shift is $-180°$) the design methods for continuous linear systems previously discussed are entirely adequate with the sampler treated as a D/V lag equal to the sampling interval. If the sampling frequency is less, then it is necessary to use the z transform instead of the Laplace transform, where the variable z is defined by:

$$z = \exp(pT) \qquad \qquad ....(3.131)$$

and T is the sampling interval. Thus:

$$p = \frac{1}{T} \ln z \qquad \qquad ....(3.132)$$

The z transformation is effected by substituting equation 3.132 in a Laplace-transformed expression.

The z transform method is of considerable use, as normally each successive signal from the sampler is held until the next signal replaces it (Fig. 3.60b).

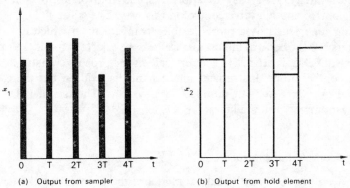

(a) Output from sampler   (b) Output from hold element

FIG. 3.60. Sampled data signals

This operation is performed by a *hold element*. The output from the hold element will be a succession of positive steps followed by negative steps of equal magnitude at time T later. Hence the transformed form of the output will be:

$$\bar{x}_2 = A_I \left[\frac{1}{p} - \frac{1}{p} \exp(-pT) \right] \qquad \ldots (3.133)$$

for a step of magnitude A_1.

The input is an impulse of magnitude A_I. Therefore:

$$\bar{x}_1 = A_I$$

Thus the hold element transfer function is:

$$G(p) = \bar{x}_2 / \bar{x}_1 = \frac{1}{p} [1 - \exp(-pT)] \qquad \ldots (3.134)$$

The corresponding z transform is obtained by substituting equation 3.132 to give:

$$G(z) = \left(\frac{z-1}{z} \right) \frac{T}{\ln z} \qquad \ldots (3.135)$$

thus removing the D/V lag term. The use of the z transform generally in the analysis of sampled data systems will produce equations of a more convenient form for solution. Excellent brief summaries of this class of system have been given by JOHNSON[4] and by BREWSTER and BJERRING[82].

Computer control

(a) *Direct digital control (D.D.C.)*

In this type of control, conventional pneumatic or electronic controllers are replaced by signals generated by a digital computer. The set points are entered into the computer by the operator or are created internally from some programmed strategy. It is easy to apply cascade control, ratio control, feedforward control and adaptive control in this way. The static D.D.C. system is now being superseded by a more flexible arrangement in which not only the parameters but the control linkages as well can be selected to suit the operating conditions. D.D.C. is attractive economically if the process contains a hundred or more control loops. Instrument maintenance is reduced but the control is sequential in nature as with sampled data systems. The use of D.D.C. is becoming widespread and the literature is extensive[31,61,68,72,74].

(b) *Supervisory digital control*[31,61,62,71,75]

(i) *Off-line control.* A mathematical model of the process may be developed by theoretical reasoning, empirical experiments or by a combination of both.

The word "process" could mean one plant, a number of plants, a whole works or business enterprise. The independent variables present in the model are chosen so as to maximise a *coefficient of performance* (C.O.P.) for the process. This coefficient of performance is related generally to the overall profit obtained from the process and is optimised using various standard computational methods, e.g. linear programming, hill climbing, dynamic programming, etc. Thus a computer away from (or off-line from) the process is used to determine the optimum means of running the process. In the oil industry, because of changes in market requirements and prices, a linear programme determining how to run a company's tankers and refineries for given crude supplies and markets might be run on a computer every month[58].

(ii) *On-line control*. This may be open or closed loop in form. In the first case variables which enable the coefficient of performance to be measured are converted to a form which a digital computer can accept. From these signals the computer calculates the C.O.P. and also may print out the values of the variables required for the improvement of the process model. Furthermore, given the current fixed operating conditions, the computer will optimise the C.O.P. using one of the methods described above, and will inform the process operator if any changes in process operating variables should be made. The computer may also use the process measurements to "up-date" the model of the process, e.g. by appropriately determining catalyst activity.

In such a scheme the operator will be informed at regular intervals how to run the process—the frequency with which he is presented with this information will, of course, depend on the process, frequency of disturbances, etc., but could quite possibly be every half-hour.

For continuous complex processes or processes subject to frequent disturbances or, in some cases, batch processes, it will not be possible for the operator to "keep up" with the computer. Thus the loop, instead of being closed by the operator, may be closed by the computer, i.e. the computer will generate signals for the required desired values which will be applied to the appropriate controllers. This will be achieved most conveniently if the controllers are electronic. Furthermore, it might be attractive to use the computer even further to dispense with the conventional electronic or pneumatic controllers and apply D.D.C. in such a control scheme (Fig. 3.61).

For on-line control, one can either use a digital computer for each process, or alternatively a larger computer may be time shared between several processes. If there is sufficient scope for such a computer, the latter scheme will usually be more attractive because the cost of computing decreases with increasing size (and capital cost) of computers. The first scheme is more flexible, although if the model of the process improves and becomes more sophisticated, then having access to a larger computer having large store and greater speed means that one can cope with this situation. (N.B. The possibility of breakdown of the computer must be taken into account.)

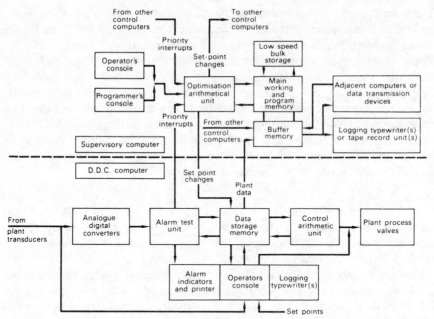

FIG. 3.61. Supervisory digital control combined with D.D.C.[61]

(c) *Economic justification for supervisory digital control*

The great majority of processes will repay a study which is undertaken with a view to computer control. This is because, in order to apply computer control, it is necessary to develop a mathematical model of the plant and an improved understanding of the process will be obtained. This will enable the process to be run in a more profitable manner.

The purchase of a digital computer for open or closed loop on-line control will be profitable if even a small saving—say a reduction in operating costs of 1 per cent—can be achieved, provided that the throughput of the process is large. If the plant is operating on a small scale, then even a saving of 10 per cent may not justify the cost of installing a computer which is of the order of £30,000 upwards.

The expected percentage reduction in costs with computer control on-line will depend on the frequency with which it is necessary to change the process variables. Obviously, if this is required very infrequently, say every day or week— then there will be very little change in the coefficient of performance and very little scope for the application of on-line control even for a plant with large throughput. The situation will be much more attractive if changes have to be made frequently because of, say, changes in imposed operating variables or process characteristics. A type of process that has been widely considered is a reactor with a decaying catalyst[62,63]. As well as such continuous processes, batch operations show considerable potential in that it is possible to calculate the optimum way to process the whole of a batch and this results in a substantial improvement over instantaneous optimisation of a batch process. However, the scale of batch operations is often too small to enable the cost of a computer to be justified.

APPENDIX 3.1

LINEARITY AND THE SUPERPOSITION PRINCIPLE

If $\mathbf{F}$ is an operator which satisfies the conditions:

$$\mathbf{F}(u + v) = \mathbf{F}(u) + \mathbf{F}(v) \qquad \dots\text{(i)}$$

$$\mathbf{F}(K_1 u) = K_1 \mathbf{F}(u) \qquad \dots\text{(ii)}$$

for any functions u and v and for any constant K_1, then the operator $\mathbf{F}$ is said to be linear.

One particular example of a linear operation as defined is that of taking a derivative. Differential equations of the type:

$$a_0 \frac{d^n x}{dt^n} + a_1 \frac{d^{n-1} x}{dt^{n-1}} + a_2 \frac{d^{n-2} x}{dt^{n-2}} + \dots + a_n x = \mathbf{F}(t)$$

satisfy conditions (i) and (ii) and consequently are termed "linear". Systems whose time dependent properties are described by such equations are said to be linear with respect to time.

Problems involving linear operators are simplified by the use of the *principle of superposition* which may be stated thus:

If $w = u$ and $w = v$ are both solutions of $\mathbf{F}(w) = 0$ (where $\mathbf{F}$ is any linear operator), then the same is true of $w = K_1 u + K_2 v$ for any constants K_1 and K_2.

The proof of this follows immediately from (i) and (ii), because then for a linear operator:

$$\mathbf{F}(K_1 u + K_2 v) = \mathbf{F}(K_1 u) + \mathbf{F}(K_2 v) \quad \text{(from (i))}$$

$$= K_1 \mathbf{F}(u) + K_2 \mathbf{F}(v) \quad \text{from (ii))}$$

But $\mathbf{F}(u) = 0$ and $\mathbf{F}(v) = 0$, as $w = u$ and $w = v$ are both solutions of $\mathbf{F}(w) = 0$.

$$\therefore \quad \underline{\mathbf{F}(K_1 u + K_2 v) = 0}$$

APPENDIX 3.2

DETERMINATION OF THE STEP RESPONSE FUNCTION OF A SECOND ORDER SYSTEM FROM ITS TRANSFER FUNCTION

From equation 3.74:

$$\overline{y} = \frac{MK}{\tau^2 p(p - \beta_1)(p - \beta_2)}$$

where:

$$\beta_{1,2} = -\frac{\zeta}{\tau} \pm \frac{\sqrt{(\zeta^2 - 1)}}{\tau}$$

Therefore:

$$\frac{\tau^2}{MK} \overline{y} = \frac{1}{p(p - \beta_1)(p - \beta_2)}$$

$$= \frac{A}{p} + \frac{B}{p - \beta_1} + \frac{C}{p - \beta_2} \qquad \dots\text{(i)}$$

Inversion of equation (i) gives:

$$\frac{\tau^2}{MK} y = A + B \exp(\beta_1 t) + C \exp(\beta_2 t) \qquad \dots\text{(ii)}$$

(a) $\zeta < 1$

Now
$$\beta_1 = -\frac{\zeta}{\tau} + \frac{\surd(\zeta^2 - 1)}{\tau}$$

Thus, for $\zeta < 1$:

$$\beta_1 = -\frac{\zeta}{\tau} + i\frac{\surd(1 - \zeta^2)}{\tau}$$

$$= -\frac{\zeta}{\tau} + i\frac{\phi}{\tau} \qquad \qquad \dots\text{(iii)}$$

where $\phi = \surd(1 - \zeta^2)$ and is real. Therefore:

$$\exp(\beta_1 t) = \exp\left(-\frac{\zeta}{\tau}t + i\frac{\phi}{\tau}t\right)$$

$$= \exp\left(-\frac{\zeta}{\tau}t\right)\exp\left(i\frac{\phi}{\tau}t\right)$$

$$= \exp\left(-\frac{\zeta}{\tau}t\right)\left(\cos\frac{\phi}{\tau}t + i\sin\frac{\phi}{\tau}t\right) \qquad \dots\text{(iv)}$$

Similarly:

$$\exp(\beta_2 t) = \exp\left(-\frac{\zeta}{\tau}t\right)\left(\cos\frac{\phi}{\tau}t - i\sin\frac{\phi}{\tau}t\right) \qquad \dots\text{(v)}$$

From equation (i) by partial fractions:

$$A = \frac{1}{\beta_1 \beta_2}$$

i.e.
$$\frac{1}{A} = \frac{\zeta^2}{\tau^2} - \frac{\zeta^2 - 1}{\tau^2} = \frac{1}{\tau^2}$$

$\therefore$
$$A = \tau^2$$

$$B = \frac{1}{\beta_1(\beta_1 - \beta_2)}$$

i.e.

$$\frac{1}{B} = \beta_1\left(\frac{2\surd(\zeta^2 - 1)}{\tau}\right) = \beta_1\left(\frac{2i\phi}{\tau}\right)$$

Substituting from equation (iii):

$$\frac{1}{B} = \left(-\frac{\zeta}{\tau} + i\frac{\phi}{\tau}\right)\frac{2i\phi}{\tau} = -\frac{2\phi}{\tau}\left(\frac{\phi}{\tau} + i\frac{\zeta}{\tau}\right)$$

$\therefore$
$$B = -\frac{\tau^2}{2\phi(\phi + i\zeta)}$$

Similarly:
$$C = \frac{1}{\beta_2(\beta_2 - \beta_1)}$$

$$= -\frac{\tau^2}{2\phi(\phi - i\zeta)}$$

Substituting for A. B and C in equation (iii) and using equations (iv) and (v):

$$\frac{\tau^2}{MK}y = \tau^2 - \frac{\tau^2}{2\phi(\phi + i\zeta)}\exp\left(-\frac{\zeta}{\tau}t\right)\left(\cos\frac{\phi}{\tau}t + i\sin\frac{\phi}{\tau}t\right)$$

$$- \frac{\tau^2}{2\phi(\phi - i\zeta)}\exp\left(-\frac{\zeta}{\tau}t\right)\left(\cos\frac{\phi}{\tau}t - i\sin\frac{\phi}{\tau}t\right)$$

$$\therefore \qquad \frac{1}{MK} y = 1 - \exp\left(-\frac{\zeta}{\tau} t\right)\left(\frac{1}{\phi^2 + \zeta^2}\cos\frac{\phi}{\tau} t + \frac{\zeta/\phi}{\phi^2 + \zeta^2}\sin\frac{\phi}{\tau} t\right)$$

But

$$\frac{1}{\phi^2 + \zeta^2} = \frac{1}{1 - \zeta^2 + \zeta^2} = 1$$

and

$$\frac{\zeta/\phi}{\phi^2 + \zeta^2} = \frac{\zeta}{\sqrt{(1 - \zeta^2)}}$$

$$\therefore \qquad \frac{1}{MK} y = 1 - \exp\left(-\frac{\zeta}{\tau} t\right)\left(\cos\frac{\phi}{\tau} t + \frac{\zeta}{\sqrt{(1 - \zeta^2)}}\sin\frac{\phi}{\tau} t\right) \qquad \ldots\text{(vi)}$$

Equation (vi) can be put into a more useful form by using the trigonometric identity:

$$p\cos\alpha + q\sin\alpha = r\sin(\alpha + \theta)$$

where

$$r = \sqrt{(p^2 + q^2)}$$

and

$$\tan\theta = \frac{p}{q}$$

Thus

$$r = \sqrt{\left\{1 + \frac{\zeta^2}{1 - \zeta^2}\right\}} = \frac{1}{\sqrt{(1 - \zeta^2)}} = \frac{1}{\phi}$$

and

$$\theta = \tan^{-1}\left(\frac{\sqrt{(1 - \zeta^2)}}{\zeta}\right) = \tan^{-1}\frac{\phi}{\zeta}$$

Thus equation (vi) becomes:

$$y = MK\left\{1 - \frac{1}{\phi}\sin\left(\frac{\phi t}{\tau} + \tan^{-1}\frac{\phi}{\zeta}\right)\exp\left(-\frac{\zeta t}{\tau}\right)\right\} \qquad \text{(equation 3.75)}$$

(b) $\zeta > 1$

$$\beta_{1,2} = -\frac{\zeta}{\tau} \pm \frac{\sqrt{(\zeta^2 - 1)}}{\tau}$$

$$= -\frac{\zeta}{\tau} \pm \frac{v}{\tau}$$

where $v = \sqrt{(\zeta^2 - 1)}$ and is real.

As for $\zeta < 1$, from equation (i), by partial fractions:

$$A = \tau^2$$

$$B = \frac{1}{\beta_1(\beta_1 - \beta_2)}$$

$$\therefore \qquad \frac{1}{B} = \left(-\frac{\zeta}{\tau} + \frac{v}{\tau}\right)\frac{2v}{\tau}$$

$$\therefore \qquad B = \frac{\tau^2}{2v(v - \zeta)}$$

Similarly

$$C = \frac{\tau^2}{2v(v + \zeta)}$$

Substituting for A, B and C in equation (ii):

$$\frac{\tau^2}{MK} y = \tau^2 + \frac{\tau^2}{2v(v - \zeta)}\exp\left(-\frac{\zeta t}{\tau} + \frac{vt}{\tau}\right) + \frac{\tau^2}{2v(v + \zeta)}\exp\left(-\frac{\zeta t}{\tau} - \frac{vt}{\tau}\right)$$

$$\therefore \qquad \frac{1}{MK} y = 1 - \exp\left(-\frac{\zeta t}{\tau}\right)\left[-\frac{1}{2v(v - \zeta)}\exp\left(\frac{vt}{\tau}\right) - \frac{1}{2v(v + \zeta)}\exp\left(-\frac{vt}{\tau}\right)\right] \qquad \ldots\text{(vii)}$$

Putting
$$\cosh\frac{vt}{\tau} = \frac{1}{2}\left[\exp\left(\frac{vt}{\tau}\right) + \exp\left(-\frac{vt}{\tau}\right)\right]$$

and
$$\sinh\frac{vt}{\tau} = \frac{1}{2}\left[\exp\left(\frac{vt}{\tau}\right) - \exp\left(-\frac{vt}{\tau}\right)\right]$$

in equation (vii), leads to:

$$y = MK\left\{1 - \left(\cosh\frac{vt}{\tau} + \frac{\zeta}{v}\sinh\frac{vt}{\tau}\right)\exp\left(-\frac{\zeta t}{\tau}\right)\right\} \qquad \text{(equation 3.76)}$$

(c) $\zeta = 1$

In this case

$$\beta_1 = \beta_2 = -\frac{\zeta}{\tau} = -\frac{1}{\tau}$$

Thus equation (i) becomes:

$$\frac{\tau^2}{MK}\bar{y} = \frac{1}{p(p + 1/\tau)^2}$$

$$= \frac{A}{p} + \frac{B}{(p + 1/\tau)^2} + \frac{C}{p + 1/\tau}$$

Equating coefficients etc. gives:

$$A = \tau^2, \quad B = -\tau \quad \text{and} \quad C = -\tau^2.$$

$\therefore$
$$\frac{1}{MK}\bar{y} = \frac{1}{p} - \frac{1/\tau}{(p + 1/\tau)^2} - \frac{1}{p + 1/\tau} \qquad \dots\text{(viii)}$$

Inverting equation (viii):

$$\frac{1}{MK}y = 1 - \frac{t}{\tau}\exp\left(-\frac{t}{\tau}\right) - \exp\left(-\frac{t}{\tau}\right)$$

$\therefore$
$$y = MK\left\{1 - \left(1 + \frac{t}{\tau}\right)\exp\left(-\frac{t}{\tau}\right)\right\} \qquad \text{(equation 3.77)}$$

FURTHER READING

COUGHANOWR, D. R. and KOPPEL, L. B.: *Process Systems Analysis and Control* (McGraw-Hill 1965)

KOPPEL, L. B.: *Introduction to Control Theory* (Prentice-Hall 1968)

KUO, B. C.: *Discrete Data Control Systems* (Prentice-Hall 1970)

MURRILL, P. W.: *Automatic Control of Processes* (International Textbook Co. 1967)

RAVEN, F. H.: *Automatic Control Engineering* (McGraw-Hill 1961)

SMITH, C. L.: *Digital Computer Process Control* (Intext Educational Publishers 1972)

REFERENCES TO CHAPTER 3

[1] COUGHANOWR, D. R. and KOPPEL, L. B.: *Process Systems Analysis and Control* (McGraw-Hill 1965)
[2] CEAGLSKE, N. H.: *Automatic Process Control for Chemical Engineers* (Wiley 1956)
[3] HADLEY, W. A. and LONGOBARDO, G.: *Automatic Process Control* (Pitman 1963)
[4] JOHNSON, E. F.: *Automatic Process Control* (McGraw-Hill 1967)
[5] YOUNG, A. J.: *Introduction to Process Control System Design* (Longmans 1955)
[6] JONES, E. B.: *Instrument Technology*, 2nd ed. (Butterworths 1965)
[7] HOLZBOCK, W. G.: *Instruments for Measurement and Control*, 2nd ed. (Reinhold 1962)
[8] COXON, W. F.: *Flow Measurement and Control* (Heywood 1959)

(9) SIGGIA, S.: *Continuous Analysis of Chemical Process Systems* (Wiley 1959)

(10) CONSIDINE, D. M.: *Process Instruments and Control Handbook* (McGraw-Hill 1957)

(11) RHODES, T. J.: *Industrial Instruments for Measurement and Control* (McGraw-Hill 1941)

(12) REDDICK, H. W. and MILLER, F. H.: *Advanced Mathematics for Engineers* (Wiley 1955)

(13) THALER, G. J. and BROWN, R. G.: *Servomechanism Analysis* (McGraw-Hill 1953)

(14) WILLIAMS, T. J. and LAUHER, V. A.: *Automatic Control of Chemical and Petroleum Processes* (Gulf 1961)

(15) ROUTH, E. J.: *Advanced Rigid Dynamics* (Macmillan 1884)

(16) TAKAHASHI, T.: *Mathematics of Automatic Control* (Holt, Rinehart & Winston 1966)

(17) MURRILL, P. W.: *Automatic Control of Processes* (International Textbook Co. 1967)

(18) TORO, V. DEL and PARKER, S. R.: *Principles of Control Systems Engineering* (McGraw-Hill 1960)

(19) SHILLING, G. D.: *Process Dynamics and Control* (Holt, Rinehart & Winston 1963)

(20) MICKLEY, H. S., SHERWOOD, T. K. and REED, C. E.: *Applied Mathematics in Chemical Engineering* (McGraw-Hill 1957)

(21) BLACKMAN, R. B. and TUKEY, J. W.: *The Measurement of Power Spectra* (Dover 1958)

(22) TRUXAL, J. G.: *Automatic Feedback Control System Synthesis* (McGraw-Hill 1955)

(23) RAGAZZINI, J. R. and FRANKLIN, G. F.: *Sampled Data Control Systems* (McGraw-Hill 1958)

(24) BODE, H. W.: *Network Analysis and Feedback Amplifier Design* (Van Nostrand 1945)

(25) CALDWELL, W. I., COON, G. A. and ZOSS, L. M.: *Frequency Response for Process Control* (McGraw-Hill 1959)

(26) WYLIE, C. R.: *Advanced Engineering Mathematics* (McGraw-Hill 1960)

(27) HARRIOT, P.: *Process Control* (McGraw-Hill 1964)

(28) KOPPEL, L. B.: *Introduction to Control Theory* (Prentice-Hall 1968)

(29) ANISIMOV, I. V.: *Automatic Control of Rectification Processes* (Consultants Bureau 1961)

(30) CARROLL, G. C.: *Industrial Process Measuring Instruments* (McGraw-Hill 1962)

(31) SAVAS, E. S.: *Computer Control of Industrial Processes* (McGraw-Hill 1965)

(32) HURWITZ, A.: *Math. Annln.* **46** (1895) 273. Über die Bedingungen, unter welchen eine Gleichung nur Wurzeln mit negativen realen Theilen besitzt.

(33) NYQUIST, H.: *Bell System Tech. J.* **11** (1932) 126. Regeneration theory.

(34) BODE, H. W.: *Bell System Tech. J.* **19** (1940) 421. Relations between attenuation and phase in feedback amplifier design.

(35) PETERS, J. C.: *Ind. Eng. Chem.* **33** (1941) 1095. Getting the most from automatic control.

(36) PETERS, J. C.: *Trans. A.S.M.E.* **64** (1942) 247. Experimental studies of automatic control.

(37) ZIEGLER, J. G. and NICHOLS, N. B.: *Trans. A.S.M.E.* **64** (1942) 759. Optimum settings for automatic controllers.

(38) AIKMAN, A. R. and RUTHERFORD, C. I.: *D.S.I.R. Conference on Automatic and Manual Control* (July 1951) 175. Characteristics of air-operated controllers.

(39) COHEN, G. H. and COON, G. A.: *Trans. A.S.M.E.* **75** (1953) 827. Theoretical consideration of retarded control.

(40) GOULD, L. A. and SMITH, P. E.: *Instr.* **26** (1953) 1026. Dynamic behaviour of pneumatically operated control equipment (Part II).

(41) CAMPBELL, G. G. and GODIN, J. B.: *Ind. Eng. Chem.* **46** (1954) 1413. Ultraviolet spectrophotometer for automatic control.

(42) ZIEGLER, J. G.: *Tech. Data Sheet* TDS 10A 103. *Taylor Instr. Co.* (1954). Cascade control systems.

(43) BERGER, D. E. and CAMPBELL, G. G.: *Chem. Eng. Prog.* **51** (1955) 348. Experience in controlling a large separations column with the continuous infrared analyzer.

(44) ROSE, A. and WILLIAMS, T. J.: *Ind. Eng. Chem.* **47** (1955) 2284. Automatic control in continuous distillation.

(45) BERGER, D. E. and SHORT, G. R.: *Ind. Eng. Chem.* **48** (1956) 1027. Sampling and control characteristics of an analysis-controlled pentane fractionator.

(46) WILLIAMS, T. J., HARNETT, R. T. and ROSE, A.: *Ind. Eng. Chem.* **48** (1956) 1008. Automatic control in continuous distillation.

(47) LEES, S. and HOUGEN, J. O.: *Ind. Eng. Chem.* **48** (1956) 1064. Pulse testing a model heat exchange process.

(48) LASPE, C. G.: *Instr. Soc. Amer. Jl.* **3** (1956) 134. A practical application of transient response techniques to process control systems analysis.

(49) FLEMMING, D. P.: *Mathl Tabl. Natn. Res. Coun. Wash.* **10** (1956) 73. Iterative procedure for evaluating a transient response through its power series.

[50] MULLER, D. E.: *Mathl. Tabl. Natn. Res. Coun. Wash.* **10** (1956) 208. A method for solving algebraic equations using an automatic computer.

[51] GOLLIN, N. W.: *Tech. Data Sheet TDS* 10A 109, *Taylor Instr. Co.* (1958). Cascade control systems.

[52] WOODROW, R. A.: *Trans. Soc. Instr. Tech.* **10** (1958) 101. Closed loop dynamics from normal operating records.

[53] WILLS, D. M.: *Tech. Bull.* TX 119–1, *Min.-Honeywell Regulator Co.* (1960). Cascade control applications and hardware.

[54] GIBSON, J. E.: *Control Eng.* **7** (Aug. 1960) 113. Making sense out of the adaptive principle.

[55] GIBSON, J. E.: *Control Eng.* **7** (Oct. 1960) 109. Mechanizing the adaptive principle.

[56] GIBSON, J. E.: *Control Eng.* **7** (Dec. 1960) 93. Generalizing the adaptive principle.

[57] RIPPIN, D. W. T. and LAMB, D. E.: *A.I.Ch.E. Meeting Washington, D.C.* (Dec. 1960). A theoretical study of the dynamics and control of binary distillation.

[58] NEWBY, W. J. and DEAM, R. J.: *Trans. Inst. Chem. Eng.* **40** (1962) 350. Optimisation and operational research.

[59] LUPFER, D. E. and PARSONS, J. R.: *Chem. Eng. Prog.* **58** (Sept. 1962) 37. A predictive control system for distillation columns.

[60] SANATHANAN, C. K. and KOERNER, J.: *I.E.E.E. Trans. (Automatic Control)* **AC-8** (1963) 56. Transfer function synthesis as a ratio of two complex polynomials.

[61] WILLIAMS, T. J.: *Chem. Eng., Albany* **71** (March 2nd 1964) 97. What to expect from direct digital control.

[62] LEE, W. T.: *Control* **48** (April 1964) 174. Computer control of a catalytic chemical process (Part I).

[63] LEE, W. T.: *Control* **48** (May 1964) 237. Computer control of a catalytic chemical process (Part II).

[64] MACMULLAN, E. C. and SHINSKEY, F. G.: *Control Eng.* **11** (March 1964) 69. Feedforward analog computer control of a superfractionator.

[65] MAWSON, J.: *Soc. Chem. Industry Conference on Automatic Control in the Chemical Process and Allied Industries* (April 1964) 15. On-line analysis of process streams.

[66] HOUGEN, J. O.: *Chem. Eng. Prog.* Monograph Ser. No. 4 **60** (1964). Experiences and experiments with process dynamics.

[67] HARRIOT, P.: *Chem. Engt Prog.* **60** (Aug. 1964) 81. Theoretical analysis of components.

[68] CUNDALL, C. M.: *Control Eng.* **12** (April 1965) 98. Back-up methods of D.D.C.

[69] CRANDALL, E. D. and STEVENS, W. F.: *A.I.Ch.E.Jl.* **11** (1965) 930. An application of adaptive control to a continuous stirred tank reactor.

[70] PALMER, O. J.: *Hydrocarb. Process Petrol. Refin.* **44** (May 1965) 241. Which: air or electric instruments?

[71] HODGE, B.: *Chem. Eng., Albany* **72** (June 7th 1965) 177. Company control via computer.

[72] WARE, W. E.: *Instr. and Control Syst.* (June 1965) 79. Direct digital control.

[73] LUYBEN, W. L.: *Chem. Eng. Prog.* **61** (Aug. 1965) 74. Feedforward control of distillation columns.

[74] BERNARD, J. W. and CASHEN, J. F.: *Instr. and Control Syst.* (Sept. 1965) 51. Direct digital control.

[75] WILLIAMS, T. J.: *Automatica* **3** (1965–66) 1. Economics and the future of process control.

[76] RIJNSDORP, J. E.: *Automatica* **3** (1965–6) 15. Interaction in two-variable control systems for distillation columns.

[77] WOOLVERTON, P. F. and MURRILL, P. W.: *Instr. Tech.* **14** (Jan. 1967) 35. An evaluation of four ideas in feed-forward control.

[78] HUTCHINSON, A. W. and SHELTON, R. J.: *Trans. Inst. Chem. Eng.* **45** (1967) T334. Measurement of dynamic characteristics of full-scale plant using random perturbing signals: an application to a refinery distillation column.

[79] WOOD, R. M. and ROBBINS, T.: *Final Report ABCM/BCPMA Distillation Panel* (1968) 257. Dynamics and control of large scale distillation columns.

[80] ROSENBROCK, H. H.: *Proc. Instn elect. Engrs* **116** (1969) 1929. Design of multivariable control systems using the inverse Nyquist array.

[81] WARDLE, A. P. and WOOD, R. M.: *Inst. Chem. Eng./European Federation of Chemical Engineering International Symposium on Distillation* (1969), *Session VI*, 66. Problems of application of theoretical feed-forward control models to industrial scale fractionating plant.

[82] BREWSTER, D. B. and BJERRING, A. K.: *British Paper and Board Manufacturers' Association Symposium on Papermaking Systems and their Control* (1969) 561. Measurements for sampled data control.

[83] LOPEZ, A. M., SMITH, C. L. and MURRILL, P. W.: *Brit. Chem. Eng.* **14** (1969) 1533. An advanced tuning method.
[84] EHRENFELD, J. R.: M.I.T., ScD. Thesis (1957). Dynamics and control of chemical processes.
[85] WARDLE, A. P.: University of Wales, Ph.D. Thesis (1968). The dynamics and control of distillation columns.

LIST OF SYMBOLS USED IN CHAPTER 3

A	Cross-sectional area		L^2
A_I	Area under impulse function		L^2
A, A_3, A_4	Constants		—
A_1, A_2	Mean areas for heat transfer		L^2
A.R.	Amplitude ratio		—
a, a_1, a_2	Real parts of complex numbers		—
B	Measured value		—
B, B_1, B_2 etc.	Constants		—
b_s	Spring constant of proportional bellows		—
b, b_1, b_2	Imaginary parts of complex numbers		—
C	Controlled variable		—
C	Capacity of integral bellows		L^3
C_1, C_2	Average specific heats	$(HM^{-1}\theta^{-1})$	$L^2T^{-2}\theta^{-1}$
c_n, c_{n-1}, c_0	Coefficients in characteristic equation		—
D	Derivative control action		—
d	Diameter of manometer tube		L
F	Feed flowrate		MT^{-1}
$F(p)$	Laplace transform of $F(t)$		—
$F(t)$	Function of t		—
G	$G_c\, G_1\, G_2$		—
G_1	Transfer function of final control element		—
G_2	Transfer function of process		—
G_a, G_b	Open loop transfer functions		—
G_c	Transfer function of controller		—
$G(p)$	Transfer function of a particular linear system		—
$G_1(p), G_a(p)$	Transfer function of block 1, a, etc.		—
g	Acceleration due to gravity		LT^{-2}
H	Transfer function of measuring element		—
h_1, h_2	Heat transfer coefficients	$(HL^{-2}T^{-1}\theta^{-1})$	$MT^{-3}\theta^{-1}$
I	Integral control action		—
i	$\sqrt{-1}$		—
K	Steady state gain		—
K_1, K_2, K_3	Constants		—
K_c	Proportional gain		—
K_D, K_I	Constants		—
K_u	Ultimate gain		—
K_v	Gain of final control element		—
L_1, L_2, L_{N-1}	Liquid flowrates		MT^{-1}
l	Length of liquid in manometer tube		L
l_1, l_2	Distances from ends of flapper to nozzle		L
l_{N+1}	Reflux flowrate		MT^{-1}
M, M_1, M_2	Magnitude or amplitude of signal		—
M_v	Manipulated variable		—
m	Slope of tangent		—
m_1	Mass		M
N	Number of stages		—
N_E	Net number of encirclements of point $(-1, 0)$		—
P	Proportional control action		—
P	Output pressure signal		$ML^{-1}T^{-2}$
P_0	Controller output pressure for zero error		$ML^{-1}T^{-2}$
P_1, P_2	Pressures applied to D.P. cell		$ML^{-1}T^{-2}$
P_A, P_B	Output pressure signal at A, B		$ML^{-1}T^{-2}$

P_D	Pressure in feedback bellows	$ML^{-1}T^{-2}$
P_d	Pressure applied to motor diaphragm	$ML^{-1}T^{-2}$
P_{d_0}	Applied pressure to give motor diaphragm a position y_0 at $t = 0$	$ML^{-1}T^{-2}$
P_E	Number of values of p which make function infinite	—
P_f	Frictional drag per unit area	$ML^{-1}T^{-2}$
p	Laplace transform parameter	T^{-1}
Q	Flowrate at constant pressure drop	MT^{-1}
Q_0	Value of Q at $t = 0$	MT^{-1}
Q_s	Flow at constant pressure drop at zero stroke (per cent)	MT^{-1}
q	Heat to vaporise 1 mol of feed divided by molar latent heat	—
R	Desired value	—
R_b	Rangeability	—
T	Sampling interval	T
T_u	Ultimate period	T
T_D	Effective derivative time	T
T_I	Effective integral time	T
t	Time	T
t_0	Time at end of pulse	T
U	Load	—
u	Velocity	LT^{-1}
u_s	Steady state gain	—
V	Vapour flowrate	MT^{-1}
V_c	Vector	—
V_T	Volume of liquid in tank	L^3
X	Outlet stream	—
x	Input to or output from control loop component	—
x_D	Overhead product composition	—
x_F	Feed composition	—
x_T	Turndown	—
Y	Coefficient in tuning relationship	—
y	Response of liquid in manometer	L
y_s	Distance travelled by valve stem	L
y_{s_0}	Position of valve stem at $t = 0$	L
y_1, y_2, y_3, y_4	Input to or output from control loop component	—
Z	Number of values of p which make function infinite	—
z	z transformation	—
z	Liquid head	L
α	Angle	—
$\alpha_1, \alpha_2, \alpha_n$	Roots of characteristic equation	—
β	Real part of complex root	—
β	Interaction factor	—
β_s	Final steady state value of response	—
γ	Imaginary part of complex root	—
Δ	Finite change in a quantity	—
$\Delta_1, \Delta_2, \Delta_3$	Determinants	—
ε	Error	—
$\varepsilon_A, \varepsilon_B$	Error at A and B	—
ζ	Damping coefficient	—
θ	Temperature	θ
μ	Viscosity	$ML^{-1}T^{-1}$
ρ	Liquid density	ML^{-3}
τ	Time constant	T
τ_a	Apparent time constant	T
τ_{AD}	Apparent dead time	T
τ_D	Derivative time	T
τ_I	Integral time	T
ν	$\sqrt{(\zeta^2 - 1)}$	—
ϕ	$\sqrt{(1 - \zeta^2)}$	—
ψ	Phase shift	—
ω	Frequency	T^{-1}
ω_c	Corner or break frequency	T^{-1}

K

ω_{co}	Cross-over frequency	$\mathbf{T}^{-1}$
$Re.$	Reynolds number	—

Primed symbols represent the steady-state value of the variable (e.g. θ')

Symbols in script denote the deviation from the steady-state value of the variable (e.g. $\vartheta = \theta - \theta'$)

Symbols with bar represent the Laplace transform of the variable (e.g. $\bar{\theta}$ is the Laplace transform of θ)

CHAPTER 4

Computers and methods for computation

THE advent of the modern digital computer has profoundly influenced the chemical engineer by greatly extending his capacity for solving problems. Not only can analytical equations be worked out with facility, but equations for which an analytical solution cannot be found may be solved by numerical methods. Also, the scale of problem that can be solved has greatly increased. Three decades ago, for example, to find the solution to, say, ten linear simultaneous equations was an extremely tedious and time-consuming task. Today, sets of a hundred or more such equations may be solved without extraordinary difficulty.

The mode of operation of the automatic digital computer has influenced the choice of the method of solution to be adopted. The computer represents an alphabetical or numerical character by the physical state of a small group of components and a code that identifies different physical states with different characters. The group of components is said to provide *storage* for a character, and the groups (of which there may be many thousands) are referred to as the store of the computer. The computer carries out operations upon characters in the store by means of control circuits; a sequence of instructions that specifies the operations to be performed is known as a *computer program,** and the set of rules that describes the way in which instructions must be formulated before presentation to the computer specifies what is known as a *computer language*. These characteristics of the digital computer have imposed requirements upon methods of computation, and changed values and judgements that were based upon the experience of the human computor who was aided only by a mechanical calculating machine.

The methods of computation given in this chapter are thus suited to the automatic digital machine, and in some cases not suitable when the problem is to be solved by hand. Only a small number of methods for any one type of problem has been given. In the experience of the author the methods described are the most appropriate for the particular class of problem for which the method is intended to find a solution. There are always pitfalls for the unwary, and where possible, these have been pointed out, sometimes by example.

---

* Editors' note: Since about 1960 the North American spelling has become widely used even in the U.K., and it is now predominant here. We shall use "program" in the specific sense of a computer program(me).

The first part of the chapter is devoted to a computer language known as Fortran, and a small number of computer programs has been written to illustrate some of the capabilities of the language. The latter part of the chapter includes a discussion of the following important topics: root-finding in non-linear equations, the solution to systems of linear simultaneous equations, the calculus of finite differences, interpolation, numerical differentiation and integration, ordinary differential equations, parabolic partial differential equations, elliptic partial differential equations, linear programming and optimisation of non-linear functions.

The physical problems discussed in other parts of the book and in earlier volumes have to be put into mathematical form before the methods of solution can be applied. It should be emphasised that the physical problem has to be well posed before the mathematical problem is properly defined. If a mathematical solution is found to be meaningless, it is possible that the definition of the physical problem is deficient!

THE DIGITAL COMPUTER

The early history of computers

Developments that have given us modern designs for digital and analogue computers were first started more than one hundred years ago, but the remarkable advance of both types of machine in the recent past was stimulated by the war of 1939. This showed a clear need for machines of large computing power to organise the calculations of inventory control, to solve problems of tracking and intercepting aircraft and missiles, and to explore the consequences of military strategies. At the same time the designers of computing systems were able to draw upon a rapidly growing range of electronic devices and techniques.

In the first half of the nineteenth century toothed wheels, gears and mechanical linkages were the only possible components that could form the basis of a design for an automatic computer, and yet Charles Babbage in England, using the idea of punched card coding developed earlier by Jacquard in France, designed an analytical engine that anticipated many of the features of modern digital computers. Jacquard had developed an automatic loom that could weave many varied patterns of cloth because the sequence of operations was controlled by means of punched cards. Babbage proposed that the sequence of operations of his engine should be controlled by punched cards with storage for numbers provided by toothed wheels. The analytical engine was intended to carry out tabulation of mathematical functions and numerical integration, and to solve equations of many types, for by that time a number of appropriate mathematical methods had been developed. But the state of technology was not sufficiently advanced, and the analytical engine was not completed.

In the period 1940–1950 many electronic devices were well established, and the construction of a digital computer was practicable. Many of the early ideas of Babbage found expression both in the computers of that period and in the digital computer of today.

FIG. 4.1. Central processor unit and random access memory
for a Digico M16E computer

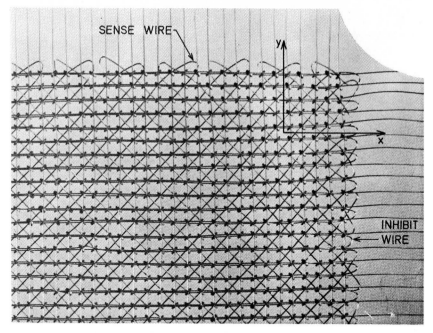

Fig. 4.2a. Close-up of a small Mullard matrix plane containing 32 × 32 ferrite cores
(FX 2551, 0·75 mm)

Fig. 4.2b. Visual display unit connected to a Digico M16E computer

The electronic computer

The digital computers of the period 1940–1960 included large numbers of thermionic valves. This meant that faults were frequent and the machines were not very reliable. The use of transistors and other methods of storage has greatly increased reliability and today digital computers are in active use in such critical applications as the control of chemical plant. The reliability of modern electronic devices can be supplemented by operating computers in pairs, and arranging that each computer checks the functioning of the other, or arranging that both are checked by a third, standby computer.

Schemes of this type have been proposed for particularly critical applications, and can now be seriously considered because the cost of modern computers is perhaps an order of magnitude less than the cost of comparable predecessors of fifteen years ago. Figure 4.1 is a photograph of the circuit boards that comprise the central processor, store, converters and interface devices of a Digico M16E minicomputer. The computer includes 64,000 words of store in a space of dimensions $0 \cdot 45$ m $\times$ $0 \cdot 35$ m $\times$ $0 \cdot 40$ m; the reduction in physical size has been even more remarkable than the reduction in cost

The heart of the digital computer is the *central processor*, comprising the *arithmetic unit* and the *control circuits*. Information necessary for the types and sequence of operations, and the numbers to be operated on, are held in storage. The central processor organises the transfer of numbers between the store and the arithmetic unit, and controls the type of operation by referring to the program held in store.

Early storage devices consisted of electronic valves and ultrasonic delay lines, later superseded in part by magnetic drum storage. The storage locations in the latter device corresponded to positions on a magnetic drum that rotated at high speed. The individual storage location could be recognised to be in one of two states of magnetisation and information could be represented by one or by a number of storage locations by means of a binary code. Information could then be read from the drum, or written on to it, by means of *read* and *write* heads fixed in close proximity to the drum, linking it to the control circuits of the computer.

The identification of information held in a storage location depends upon the electrical recognition of the state of the storage location. It is possible to hold an electrical device in one of ten different states, say grid voltage levels on a thermionic valve, and to identify each state by a decimal digit. However, the circuits that are required to do this are complex and not very reliable, so that in all modern computers simple storage devices are used that can each be recognised in one of two possible states. Straightforward read and write circuits can be used in conjunction with the storage elements and therefore the accuracy of reading or writing is not critically dependent upon the magnitude of the electrical pulses employed in reading and writing.

Because a storage element is recognised as existing in one of only two states, it is usual to carry out all of the internal computations in binary arithmetic, and the numbers are therefore expressed in powers of 2 instead of as decimal in powers of 10. For example, the number $5 = 5 \times 10^0$ in decimal is expressed as

$1 \times 2^2 + 0 \times 2^1 + 1 \times 2^0 = 101$ in binary mode, while the number $25 = 2 \times 10^1 + 5 \times 10^0$ in decimal is expressed as $1 \times 2^4 + 1 \times 2^3 + 0 \times 2^2 + 0 \times 2^1 + 1 \times 2^0 = 11001$ in binary. The fractional number $0.75 = 7 \times 10^{-1} + 5 \times 10^{-2}$ is given by the binary expression $1 \times 2^{-1} + 1 \times 2^{-2} = 0.11$. The position of the point in the latter number shows the partition between negative and positive powers of 2, just as the decimal point shows the partition between negative and positive powers of 10.

A storage element thus corresponds to one binary digit, or *bit* in the common abbreviation. A given number or instruction requires several digits for definition, and therefore several digits are associated to form a *word* of storage. A typical word may consist of 24 binary digits, and because the digits are always associated, in that the electronic control circuits carry out operations on whole words, the word is usually regarded as the basic element of computer storage.

There is now a variety of storage media, including the ferrite core matrix and magnetic drums, tapes and discs. The ferrite core matrix is an important example of a *random access* store in that information can be read from store in a time that is independent of the location of storage. The reading or writing of information on magnetic drum, disc or tape storage is not truly random in that reading from or writing to a given location cannot take place until the required location passes under the read or write head. The delay in access depends upon the relative position of the storage location and read/write heads at the time of the computer command. For a magnetic drum, a typical delay is 5 milliseconds, for a magnetic disc 0.1 to 0.3 seconds, while for a magnetic tape the delay may be as long as one or two minutes. For many purposes magnetic drums and magnetic discs may be effectively regarded as random access devices because read/write heads are arranged to scan perhaps 100 recording tracks instead of the single scan of magnetic tape, so that the access time is acceptably short.

The ferrite core matrix is arranged so that any core can be located by two wires rather in the manner of specifying a point on a two-dimensional diagram by x and y coordinates.

Figure 4.2a is a photograph of a small matrix plane containing ferrite cores each 0.03 in diameter. Each core is threaded by x and y wires, a looped diagonal wire known as the *sense* wire, and a looped wire parallel to the x direction known as the *inhibit* wire.

The ferrite core has a *square loop* magnetisation characteristic in that the passage of a current smaller than a critical value will not disturb the magnetisation of the core. Information can be written into a given core by passing currents each slightly greater than half the critical value through the appropriate x and y wires. The magnetisation of the selected core is driven into one state by passage of the currents in one direction, and into the second state by passage of currents in the reverse direction.

Reading of information from the store depends upon the sense wire. If currents slightly greater than half the critical value are passed through a given core in one direction a voltage will be induced in the sense wire if there is a change in magnetisation, otherwise there will be no induced voltage. The presence or absence of an induced voltage will indicate the state of magnetisation of the core.

As the passage of current may have changed the state of magnetisation the reading has to be followed by the passage of currents in the *opposite* direction to restore the lost information. If no signal is received from the sense wire during reading, the state of magnetisation of a core has not been changed; a pulse of current co-incident with the restore pulse is therefore passed through the inhibit wire to nullify the effect of the restore pulse. If a signal is received from the sense wire no inhibit pulse is delivered. In this way the state of the matrix is restored to the original.

A particular core in a matrix may be selected by *decoding matrices* for the x and y wires that depend upon the characteristics of the *diode;* a diode is a semi-conductor device that shows a small resistance to current flow in one direction but a large resistance to current flow in the reverse direction. Voltages that correspond to the binary code of the chosen x wire, for example, are applied to the input nodes of a diode network, and the manner of connecting the diodes in the network allows the pulse to be applied to the selected wire. In this way information is read from, or written on to, particular memory locations under the control of the central processor.

A typical ferrite core plane will contain $64 \times 64 = 4096$ cores each representing a bit, and the planes will be associated so that there is one plane for each bit of the computer word. A 4096 24 bit word computer will contain a stack of 24 planes each containing 4096 cores. A typical time of access to a store location lies within the range of 0·2–2 microseconds (µs), and the time is substantially the same for each location within the matrix. The ferrite core matrix is now being replaced by semiconductor devices that are cheaper and more compact. However, the ferrite core matrix will retain information even when power is switched off, so that ferrite cores are used for some applications in spite of greater expense.

Although this time is a great improvement over that offered by the magnetic drum, for a few key units in the central processor it is desirable to have even faster devices. For example, the location of the program instruction under execution is known as the *address* of the instruction. The address is held in a register of the control unit, and it is desirable that the time required to change to the address of the next instruction should be very small. Fast transistor circuits may be used for this and other registers of the arithmetic and control unit and the time required to change the contents of the register may be 0·1 µs or less.

The organisation of computer storage

The modern computer has several types of storage[5]. The fastest and most expensive is the transistor register, which may cost £10 per word of storage capacity, but this type of storage is reserved for key locations of which there may be a dozen or less. Working storage for the computer may be provided by the ferrite core matrix at a cost of about £0·50 per word, or by semiconductor memory at a cost in the region of £0·25 per word. The total storage capacity ranges from 8000 words for small computers to one million words for large computers. The program under execution and the data currently being processed are held in core, and the program is interpreted and the data are manipulated by means of the central processor and the associated control circuits and fast registers.

In addition to immediate storage, computing systems require space for the storage of programs and data blocks awaiting execution, space for the storage of information that has been processed but not yet printed by the output device, and space for the storage of files of information that will be referred to by the computer. Magnetic discs, both of the permanent and interchangeable variety, can provide several million words of storage for this type of application at a cost in the region of £0·0006 per word; these units are cheaper and more flexible than magnetic drums and are now the most favoured way of providing backing storage for operating systems. For program and data storage, however, the cheapness and flexibility of magnetic tape often outweighs the disadvantage of long access time. Although access times may be of the order of 1 minute or even longer, a magnetic tape can provide one million words of storage at a cost as low as £5 $\times$ 10^{-7} per word, and therefore many computer systems have both magnetic tape and magnetic disc facilities.

A computer of intermediate size is organised so that the program, or that part which is under execution, together with its numerical data, is held in core storage; other sections not currently required, areas reserved for the input and output of information and other programs may be held on magnetic discs, while further programs and data files may be held on magnetic tapes or magnetic discs. The organisation of the flow of work in the computer is controlled by an *executive* or *supervisory* program usually permanently held in core storage.

Figure 4.3 is a block diagram illustrating a typical method of computer organisation.

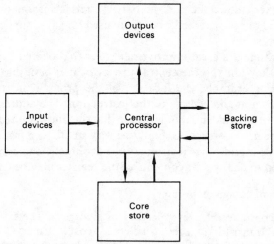

FIG. 4.3. A method of computer organisation

Communication with the computer

Information to be processed by the computer may be read in by means of terminals, paper tape, punched cards, or magnetic tape. A typical punched card is shown in Fig. 4.4.

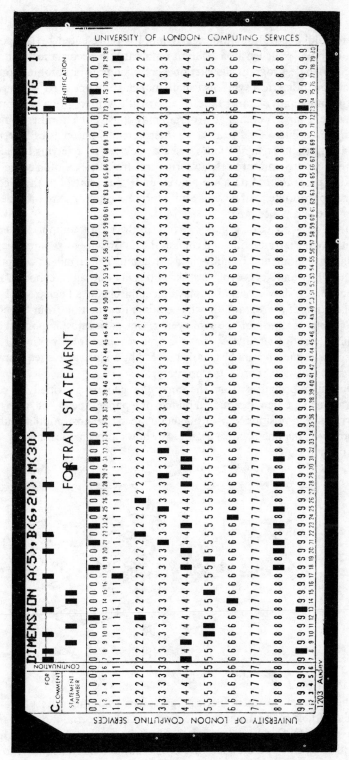

FIG. 4.4. A punched card (Fortran version)

281

Each alphabetical or numerical symbol is represented by the position of a hole or holes punched in a column on a card, in accordance with a specified code. A single program instruction corresponds to a combination of alphabetical and numerical symbols usually punched on one card, or punched into a continuous length of paper tape, or electrically written into a continuous length of magnetic tape. A complete program may consist of a number of cards, which when placed into a card reader, are interpreted by automatically feeding cards, one at a time, into a bank of photoelectric devices. By referring to the card code, control circuits interpret the information from the photoelectric reader, so that the program and data are read in under the control of the executive program. Appropriate read devices and interpretive circuits are provided for magnetic tape and paper tape.

In the past few years a major change in the use of computers has developed because of the introduction of terminals that allow a user to communicate directly during the editing and running of his program. The terminals comprise either a small printer and keyboard, or a visual display unit and keyboard. Figure 4.2b (facing p. 277) shows a visual display unit connected to a Digico M16E computer; in operation of a program, or while editing, information from the computer to the operator is shown on the display unit. One of the most important advantages of the interactive terminal is that the time required from writing of a program to successful running can be greatly reduced by an interactive sequence of correction and short test runs of the program.

Output information from the computer can be provided by causing the operation of a typewriter, a card punch, or a paper tape punch, or by activating a writing head positioned over magnetic tape. The most popular output device is, however, the line printer. Figure 4.5 shows a mechanical arrangement for a line printer.

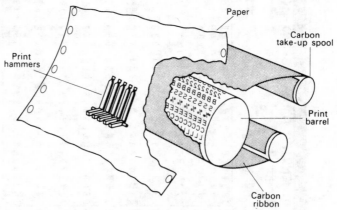

FIG. 4.5. The line printer drum

The alphabetical and numerical figures used in output are set as raised figures on the surface of the drum. As the drum rotates paper and carbon ribbon pass over the surface and hammers under computer control strike at the drum so that the imprint of the underlying drum character is transmitted to the paper.

The rate of output of information is very great. A typical fast line printer is capable of printing 1,200 lines per minute $(20\,s^{-1})$, each line containing 120 alphabetical or numerical characters.

Programs and languages

A computer program comprises input instructions, instructions to control the processing of the input data, and output instructions that specify the output information and the output device. Instructions that may be directly interpreted by the computer are directly related to the function of the control circuits of the computer. For example, a control circuit that will carry out the addition of a binary digit in a location of store to a binary digit in a register is known as an *add circuit*, and this circuit is activated by an *add instruction*. Control circuits in a computer are organised in groups of word size, so that an add instruction in a 24 binary digit computer will cause the contents of a 24 binary digit location in store to be added to a particular register. Instructions of this type, in which there is a direct correspondence between instruction and control circuits are known as *machine language* instructions. A machine language instruction will include the address of the location in store holding the number that is to be operated upon.

As the control circuits of a computer accomplish relatively simple operations such as single additions, subtractions, transfers and tests, programs written in machine language often contain a large number of simple instructions, each instruction specifying the function to be performed and the address of the operand. All programs for the early computers were written in machine language, and because of the large number of instructions, the necessity of keeping track of individual addresses, and the necessity of maintaining arithmetic accuracy and guarding against overflow in arithmetic operations, the development of a computer program often took a long time and required a great many trials.

This difficulty led to the development of higher level languages for computer programming. This type of language allows the computer programmer to specify a sequence of operations in terms that are similar to those of everyday writing or mathematical notation. For example, *Fortran* and *Algol* are two high level languages that allow mathematical equations to be written down in a manner that is closely related to the original form of the equation. The existence of these two languages, and of others suitable for a variety of applications including business and accounting, has had an enormous influence upon the growth of both the numbers of users and applications.

A computer can process programs written in high level languages by means of *compilers*. A compiler is a program, itself written in machine language, that translates statements written in that high level language into machine language commands, and organises the allocation of storage for the translated program instructions, and for data used in the program. The validity of statements in the program is checked by the computer under the control of the compiler, and if an error is found, a signal of the error is given.

The use of high level languages relieves the programmer of the task of organising storage for his program and data in store, and frequently reduces the number of computer instructions by a factor of four or five. The machine language programs produced by the compiler are not as efficient as those that could be written by an experienced programmer, but it is generally accepted that the saving in time during the development of the program when written in a high level language more than offsets the disadvantages of possible inefficiency in the final program.

The use of a high level language relieves the programmer of the necessity of following the detailed operation of his program by the computer. Fortran is one language that is now well tried, and has been successfully applied to many engineering and mathematical problems. Unfortunately there are many variants of the language, not only by different computer manufacturers, but also for different computers of the same manufacturer. However, the form of many of the Fortran instructions that we will use in our discussion of applications will be acceptable to many compilers now in current use. Reference should always be made to the manual for the particular compiler being used.

FORTRAN

A Fortran program is composed of a series of *statements* or instructions that define operations upon the *variables* of the program. A statement is usually represented by one punched card according to a code that identifies alphabetical and numerical symbols with the position of holes punched in the card. Figure 4.4 shows a typical 80 column card prepared for Fortran statements.

The statement is punched into columns 7–72 of the punched card; spaces are left where desired, because the compiler usually disregards blank columns. Columns 1–5 are used for the statement number; this is necessary if the computation is to be directed to that statement at any time. For long statements 66 columns are not sufficient, and many compilers allow the statement to be continued on other cards known as *continuation cards*. A continuation card is identified by a non-zero number punched into column 6, and this is the only occasion column 6 is used in a Fortran statement. Columns 73–80 are not processed by the compiler, but may be used by the programmer to identify and sequence the cards of a program.

The punching of a C into column 1 identifies a *comment* card. Comment cards are not processed by the compiler, but appear in the output record as part of the program list. Their principal purpose is to identify a portion of the program by verbal description.

Fortran variables and constants

A *variable* that occurs in an algebraic equation may be represented in a computer program by a sequence of alphabetical or numerical characters of which the first must be alphabetical. The sequence of characters is known as the *name* of the Fortran variable. Some examples of names are* A, DIET, I, KOUNT, B(I, J) and D(4, 5, 6). The computer identifies a location in main storage

---

* The type face which follows will be used for words, etc., which are or may be part of an actual Fortran program.

with the name of a Fortran variable, and the value of the variable is held in that location. An operation upon the variable in a computer program will correspond to the operation carried out by the computer upon the number held in the corresponding location in store.

Fortran variables may be subscripted and be either *fixed point* (*integer*) or *floating point* (*real*). Fixed point variables always take integer values and, unless declared otherwise, the name of the variable starts with one of the letters IJKLMN. The value of a floating point variable includes a fractional part, and unless declared otherwise, the name must start with one of the other letters of the alphabet ABCDEFGH OPQRSTUVWXYZ. For example, the value of the fixed point variable M may be $-1, 1, 500, -399$ or 781, while the value of the floating point variable B may be $-5 \cdot 0, 677 \cdot 29, 378 \cdot 6$ or $7 \cdot 32 \times 10^{-6}$. If a floating point or a fixed point number appears in a computer program it is known as a floating point or fixed point *constant*.

The value of a floating point variable or a floating point constant normally includes a decimal point. The value of a fixed point variable or a fixed point constant is always an integer, without a decimal point or exponent. Most arithmetic computing is performed with floating point variables or constants, but subscripts of variables, counts of iterations and other quantities of an integer nature are usually processed in fixed point arithmetic because the round-off error that occurs only with floating point arithmetic is avoided.

The subscripts may appear as fixed point variables or expressions in a computer program, but the subscripted variables are not identified until the subscript has been identified as a fixed point constant. For example $A(I, J)$ is a general expression for a variable that is part of an array A, but $A(1, 2)$ is a particular variable whose value is kept in a specially reserved area of main store. Repetitive operations on subscripted variables may be defined by a repetitive incrementation of a subscript, and ·this is a valuable feature of high level languages.

The appearance of a subscripted variable in a program must always be preceded by a DIMENSION statement in which the maximum number of each subscripted variable is defined. The statement appears at the head of the program and takes the form:

$$\text{DIMENSION } A(5), B(6, 20), M(30)$$

so that computer space is reserved for 5 A variables, 120 B variables and 30 M variables.

The DIMENSION statement is an example of a *non-executable* statement, in that the effect is to reserve storage for the dimensional variables, and therefore the statement gives directions to the compiler. First, however, we will consider *executable* statements, of which the two main types are *arithmetic* statements, which cause the execution of mathematical operations, and *control* statements that are used to steer the course of a computation.

Arithmetic statements

This type of statement defines variables in terms of arithmetic or functional operations. Five common operations are available. They are addition (symbolised by $+$), subtraction ($-$), multiplication ($*$), division ($/$) and exponentiation ($**$). Some simple examples of arithmetic statements are:

addition	Y = A + B	
multiplication	Y = A $*$ B	
exponentiation	Y = A $**$ B	i.e. A^B

The first example instructs the computer to add the floating point variable stored in location A to the variable stored in location B, and to place the sum in location Y. The previous contents of location Y are destroyed. The last statement raises the number in location A to the power of the number in location B, and places the result in location Y.

Arithmetic statements can include a large number of arithmetic and functional operations. For example the expression:

$$A + BX + CX^2$$

when written in Fortran may appear as:

$$Y = A + B * X + C * X **2 \qquad\qquad ....(4.1)$$

or as:

$$Y = A + X * (B + C * X) \qquad\qquad ....(4.2)$$

When there is more than one operation in a statement, each is executed by the computer in a fixed order of priority. The first priority is exponentiation followed by multiplication and division, with the lowest priority given to addition and subtraction. Within this hierarchy, operations are executed from left to right starting with the innermost brackets of an expression, and working outwards. Brackets are therefore used to guide the course of computation in an arithmetic statement. If no brackets are present, as in statement 4.1, the straightforward hierarchy is applied. The exponentiation X^2 is calculated first, then B is multiplied by X and C is multiplied by X^2; finally BX is added to A, CX^2 is added to the sum $A + BX$ and the result is placed in Y. In statement 4.2 the inner bracket is first processed so that CX is determined and then added to B, followed by the other operations in hierarchy. The manner of expression of 4.2 is more economical; fewer arithmetic operations are performed because the exponentiation is eliminated.

The raising of a floating point variable to a fixed point exponent is always

TABLE 4.1. *Fortran designation of functions*

Function	Fortran designation
Exponential	EXP(X)
Natural logarithm	ALOG(X)
Common logarithm	ALOG10(X)
Sine	SIN(X)
Cosine	COS(X)
Arctangent	ATAN(X)
Hyperbolic tangent	TANH(X)
Square root	SQRT(X)
Absolute value or modulus	ABS(X)

allowed, but otherwise there is a diversity among Fortran compilers concerning the permitted combinations of fixed and floating point quantities in the same arithmetic expression; the combinations are defined in the manual for a particular compiler.

Some functions occur in arithmetic expressions so often that routines are provided for calculation on demand. Table 4.1 gives a list of routines commonly available. Here X is an arithmetic expression that may consist of any number of arithmetic operations including functions, for example SIN(A), SIN(A + B), SIN(A + B * ALOG(X)).

Control statements

Executable statements in a Fortran program are executed in the order of appearance in the program, unless the order of execution is changed by a *control statement*. The order of execution can only be directed to a numbered or labelled statement, or to the start of a DO loop from the last statement in the loop. Some common examples of control statements follow.

The unconditional GO TO *statement.* This statement has the form GO TO n where n is a statement number. Control of the program is unconditionally transferred to n.

The computed GO TO *statement.* This statement has the form GO TO (j, k, l, m), I where $j, k, \ldots$ are statement numbers, and I is a non-subscripted fixed point variable. This instruction transfers program control to the Ith statement number in the bracket. For example, if I is set to 3, program control is transferred to the third statement in the list.

The arithmetic IF *statement.* This statement has the form IF(X) j, k, l where X is a fixed or floating point variable, or a fixed or floating point expression, and $j, k,$ and l are statement numbers. Control is transferred to statement j if X is negative, to statement k if X is zero and to statement l if X is positive.

The DO *statement.* This statement has the form DO n I $= k, l, m$ or DO n I $= k, l$ where n is a statement number and k, l and m are unsubscripted fixed point variables or fixed point constants. The meaning of the statement is: execute each statement after this one, up to and including statement number n, until the index I, which starts at k and is incremented in steps of m each time the loop of instructions to n is completed, equals or exceeds l. If the increment m is 1 the second form of the statement may be used. The facility of repeating a cycle of instructions given by the DO statement is very important.

DO loops may be placed within the range of other DOs; this arrangement is known as a *nesting* of DO loops. Transfer type instructions that allow transfer out of a loop may be included, but there are several rules concerning the transfer of control and nesting of DO loops that must be followed. The rules are subject to variations between different compilers. However, many features are common although there are important differences concerning the conditions under which control may be transferred back into a DO loop.

The index I and the parameters k, l and m, may always be referred to within the range of a DO loop but can never be redefined. The terminal statement of a DO loop must be executable, but cannot be a GO TO, RETURN, STOP, PAUSE, DO or arithmetic IF statement.

DO loops may be nested to any depth, but the range of an inner must be completely contained within the range of an outer DO. Nested DO loops may share the same terminal statement but transfer of control to that statement is usually not allowed, except from the innermost loop sharing the terminal statement. Many Fortran compilers do not allow transfer into the range of a DO loop, but at least one modern series of compilers (the ICL 1900) will allow a transfer of control into the range of a DO loop if there has previously been a transfer out of the range, and if the values of the loop index and parameters have remained unchanged. Except for this feature, the other rules given for DO loops apply to most Fortran compilers in current use.

The CONTINUE *statement.* This takes the form CONTINUE, and is a dummy statement that causes no action. It is often used as the terminal statement in a DO loop that would otherwise end in an impermissible statement, or would share its terminal statement with an inner loop.

The STOP *statement.* This statement, taking the simple form STOP, stops a program. The statement may also take the form STOP n, where n is an integer constant. When the program reaches this instruction STOP n is printed and the position of the STOP in the program is then identified.

The PAUSE *statement.* This takes the form PAUSE or PAUSE n. The program stops on reaching this point and PAUSE, or in the second form PAUSE n, will be printed out. On restarting, the computer will execute the statements following PAUSE.

The END *statement.* This statement has the form END and marks the end of a program or sub-program. It acts as an instruction to the compiler.

Input and output instructions

The input devices most frequently used are the paper tape reader, the card reader and the magnetic tape unit. Output units include the paper tape punch, the card punch, the magnetic tape unit and the most popular, the line printer. Output on paper tape, cards and magnetic tape has to be transcribed from code using further devices if a line printer is not available.

Input and output instructions are often referred to a record of input or output. A record contains a unit of information to be transferred. The physical significance of a record depends upon the device; for card input and output a record is one card, while a line of print corresponds to a record on the line printer. An input or output instruction designates the device of transfer, the names of the variables to be transmitted and the format in which the variables are to be read or written. The form of instructions available differs amongst Fortran compilers, but those described here are common to most, if not all, versions of Fortran IV. Input and output instructions take the form:

$$\left.\begin{array}{l} \text{READ } (i, n) \text{ list} \\ \text{WRITE } (j, n) \text{ list} \end{array}\right\} \dots (4.3)$$

The integer numbers i and j identify devices used for input and output, n is the number of the statement that specifies the format in which the values of the variables are to be read or written, while the list identifies the variables to be read from the input device or written on to the output device. Device i might be a card reader while device j may be a line printer. The allocation of the numbers i and j is done in a small job description segment that is separate from the program. For example, a job description segment for an ICL 1900 computer might read:

```
MASTER (CODE)
INPUT 1 = CR1
OUTPUT 2 = LP1
END
```

This segment identifies input device $i = 1$ as a card reader, while the output device $j = 2$ is identified as a line printer, both to be used in a program of which the principal segment is known as CODE. This is one method of allocating input and output devices. Other methods function in a similar way, but as the job descriptions are often specific to a particular installation they will not be discussed further.

The list of variables given in input or output instructions is read in conjunction with a FORMAT statement in which the external form of the alphabetical or numerical data is specified.

The FORMAT specification: Numerical fields

The characters forming a number or alphabetical expression are known as a *field*. A specification defines the type of field, the number of characters within the field, including blank characters, and the position of a decimal point where this is included.

There are three principal types of numerical field.

I fields contain integers only. A single field specification takes the form In where n is the number of characters in the field or the width of the field. The sign of the field is included in the width; for example, -724 is of format I4. A multiple field specification when fields are the same width may be written kIn where k is an integer that specifies the number of fields of format In.

F fields contain floating point numbers without an exponent. A field specification takes the form $l\text{P}\text{F}m.n$, where the number m defines the width of the field including sign and decimal point, n specifies the number of digits to the right of the decimal point and $l\text{P}$ is a scale factor that may include a negative sign. The scale factor may be omitted to give the form $\text{F}m.n$, so that the number $+751.28$ for example is of format F7.2. The function of the scale factor when used on input differs from the function on output. If the scale factor l is in operation on input the external number is divided by 10^l, if in operation on output the internal number is multiplied by 10^l. As for I fields multiple field specifications may be written $k(l\text{PF}m.n)$.

E fields describe floating point numbers, but an exponent field is included. The specification takes the form $l\,\text{P}\text{E}m.n$, where the number m defines the width of the field including sign and exponent, n defines the number of digits to the right of the decimal point and l is a scale factor that may be omitted. The exponent field takes the form $\text{E} \pm i$, where i is a two digit integer that defines the magnitude of the exponent. Two examples of this type of field, the numbers $+123.45\text{E}+05$ and $-3761.1\text{E}-02$, are of format E11.2 and E11.1 respectively.

Many Fortran compilers produce an output in E format in which the number is scaled within the range 0·1 to 1·0 together with an exponent, so that the number of digits to the right of the decimal point is also the number of significant digits. If a scale factor is in operation on output the decimal part is multiplied by 10^l on output and l is subtracted from the exponent, so that a format specification 1 PE11.4 might give rise to $-7.8963\text{E} - 12$ instead of $-0.7896\text{E} - 11$ given by the format specification E11.4. A scale factor has no effect on input if the number includes an exponent field, otherwise the external number is divided by 10^l. A scale factor placed in a format specification affects all subsequent E or F descriptors in that specification until the scale factor is changed.

Multiple field specifications of the form $k(l\text{P}\text{E}m.n)$ may also be used where appropriate.

The E format is generally used when there is some uncertainty in the magnitude of the computed result. If the F format is used on output there is a danger that some significant figures may be lost by truncation, and this cannot happen if an E specification is used.

Most Fortran compilers will allow some abbreviation of data on input. Thus, if a number is positive no sign need be given, also, blanks anywhere are taken to be zeros. For E type fields the exponents $\text{E} + 09$, E9, E09 or $+9$ are acceptable and equivalent alternatives. Exponents may be included in F fields on input and exponents may be omitted from E fields on input. The position of the decimal point punched into a number on input will override the position of the decimal point given by the format specification. If no decimal point is punched in a number read in under E or F format, the position of the decimal point will be interpreted according to the format specification.

One particularly useful feature of a current compiler is the *free format* facility of ICL 1900 Fortran. If E, F or I specifications take the form E0.0, F0.0 or I0 then the numbers can be entered in free format, in which the field is taken to be terminated by the first space within it, or by the end of a record. Obviously spaces are not then allowed within a number on input except after E of an exponent, and all characters must be punched.

Some Fortran compilers may not accept some of the abbreviations, and the reader should become familiar with the particular compiler he is using, but in many cases the compiler will follow the pattern of abbreviations and conventions just described.

The FORMAT specification: Alphanumerical fields

For most types of work there is a need for the output of alphabetical characters to serve as table headings and other literal information in the output record. The H or Hollerith specification allows the writing of any character within the Fortran set, including numerical, blank and alphabetical (alphanumerical) characters. The specification has the form wH followed by the w characters that are to be written into the output record. For example, the instructions:

$$\left.\begin{array}{l} \text{WRITE (2,100)} \\ 100 \quad \text{FORMAT (16H STREAM FUNCTION)} \end{array}\right\} \dots (4.4)$$

will cause the writing of the 16 characters STREAM FUNCTION following H into the output record. Blank characters, as before the S and F in this example, are counted as part of the field width.

Hollerith specifications may be mixed with numerical specifications on output. Thus the instructions:

$$\left.\begin{array}{l} \text{WRITE (2,60) W,X,Y,Z} \\ 60 \quad \text{FORMAT (3H W=,F7.1,3H X=,E11.4,} \\ \qquad\qquad\qquad \text{3H Y=,1 PE11.4,3H Z=,2 PE12.4)} \end{array}\right\} \dots (4.5)$$

could produce the output record:
W=   123.2 X= 0.5231E–06 Y=–2.6158E 01 Z= 23.1612E–01 in which the characters W = X = Y = Z = and four blank characters have been written by means of the H specification. The reader will also notice that the last two numbers have been scaled into the range 1–10 and 10–100 by means of 1 P and 2 P scale factors.

The Hollerith specification can be used on input. For example, the instruction:

$$\left.\begin{array}{l} \text{READ (1,100)} \\ 100 \quad \text{FORMAT (16H STREAM FUNCTION)} \end{array}\right\} \dots (4.6)$$

will cause 16 characters to be read in to overwrite the 16 characters STREAM FUNCTION. A subsequent WRITE statement using this FORMAT will cause the writing, not of the original characters, but of the 16 characters read in by this instruction.

Special features of format control in line printer output

When the output device is a line printer, control of the line spacing and page position is taken from the first character of the line. The first character is a paper control character which is not normally printed in the output record. The functions of the various control characters are given in Table 4.2.

TABLE 4.2. *Effect of standard carriage control characters*

Control character	Effect
Blank	Single spaced lines
0	Double spaced lines
1	Line is written on top of new page
+	Line printer not advanced and overprinting occurs

The H or X specification may be used to designate the control character. In the examples of 4.4 and 4.5, the first characters are spaces, so that the printer will operate with single line feed.

The FORMAT specification: Blank fields and new records

The conversion code X is used to write blank characters into the output record, or to skip characters in the input record. The specification takes the form wX. When included in a format specification on output w blank characters are placed into the output record, and when included in a format specification on input, w characters in the input record are skipped.

The solidus / is used to change the record on input or output. When the solidus is used in multiples of n, $n-1$ complete records are skipped on input, or $n-1$ blank records are produced on output.

The instructions:

$$\left. \begin{array}{l} \text{WRITE (2,50)X,Y,Z} \\ \text{50   FORMAT (5X,F7.1/5X,F7.1//5X,F7.1)} \end{array} \right\} \quad \ldots (4.7)$$

would produce 4 blank characters and a number in format F7.1 on one line, 4 blank characters and a number in F7.1 on the next line, a blank line, and a final line of 4 blank characters and a number in format F7.1.

The FORMAT specification: Relationship to input and output lists

The lists of variables that appear in input and output instructions are interpreted in conjunction with a format specification. The instructions 4.5 provide a simple example of an output list and the associated format statement. A new record is started at the left outermost bracket of the format specification, and the quantities inside the outermost bracket are scanned from left to right until the right outermost bracket is reached where the record is ended. If there are still items in the list that have not been reached, and the format specification contains no inner brackets, a new record is started and the format specification is again scanned. This is repeated until no more items remain in the list.

If the format specification contains inner brackets and items in the list have not been completely processed after a first scan of the specification from left to right, scanning is repeated, usually by starting from the left bracket that corresponds to the right internal bracket furthermost to the right in the specification.

In the examples 4.5 and 4.7 the number of variables in the list exactly equals the number of field descriptors in the format specification, so that there would be no items unprocessed in the list when the first scanning of the specification had been completed. Suppose, however, that we have the instructions:

$$\left.\begin{array}{l} \text{WRITE (2,60)X,Y,Z,W, P,Q} \\ \text{60   FORMAT (2(5X,F7.1), (5X,E11.4))} \end{array}\right\} \dots (4.8)$$

There are three points worthy of note in 4.8. First, there are more items in the output list than appear in the format specification, and therefore the specification has to be scanned more than once. Second, the field descriptor (5X,F7.1) is prefaced by the integer 2 so that this field occurs twice. Statement 60 is thus equivalent to:

60   FORMAT (5X, F7.1, 5X, F7.1, (5X, E11.4))

Third, internal brackets appear in the format specification, and as scanning is repeated from the left bracket corresponding to the rightmost internal bracket, repetition will take place from the bracket shown by the arrow.

The first scan of statement 60 relates X and Y to F7.1 specifications and Z to an E11.4 specification, so that these three numbers are written out as one record. The second and subsequent scans will start from the arrowed left bracket and cause the writing of the values of W, P and Q in format E11.4 on separate records. The action of statement 60 in 4.8 is equivalent to that given by the format specification:

60 FORMAT (5X,F7.1,5X,F7.1,5X,E11.4/5X,E11.4/5X,E11.4/5X,E11.4)

$$\dots (4.9)$$

The rules of association between lists and format statements are the same for both input and output. If the WRITE statement of 4.8 is replaced by a READ statement, the numbers X, Y and Z are read from the first record, while W, P and Q are each read from subsequent separate records. The X fields denote the characters skipped on input.

Input and output of arrays

A comparison between the format statements of 4.8 and 4.9 gives a suggestion of the considerable economies in format specification that arise from the rules of association. Sometimes there is a need for the input or output of complete arrays of dimensioned variables. This is most effectively done either by an *implicit* DO instruction that can be used for input or output, or by the association of an array name in an input or output list with a FORMAT statement. In the latter case it is necessary to relate the data on input or output to the precise way in which the elements of an array are held in computer storage when the array is stored in double or higher subscript form.

The elements of an array $A(i,j,k)$ are stored in consecutive storage locations with the first subscript varying the most rapidly, the second following, and so on. For example, a matrix dimensioned as $A(2,2,2)$ is stored in the succession $A(1,1,1)$, $A(2,1,1)$, $A(1,2,1)$, $A(2,2,1)$, $A(1,1,2)$, $A(2,1,2)$, $A(1,2,2)$, $A(2,2,2)$.

The implicit DO loop

The list of variables in READ or WRITE statements may be replaced by a DO-implied list. A simple example of such a list and the associated format statement is:

$$\left.\begin{array}{l} \text{READ}\ (1,30)\ (A(I), I = 1,20) \\ 30\quad \text{FORMAT}\ (10F7.1) \end{array}\right\} \ \ldots (4.10)$$

This causes the reading of 20 numbers in format F7.1, from two cards as the specification is scanned twice. The numbers are written into the storage locations reserved for the variables $A(1), \ldots A(20)$.

Implicit DOs may be nested as in the following example. The innermost loop is processed first:

$$\left.\begin{array}{l} \text{READ}\ (1,30)\ ((A(I,J), J = 1,10), B(I), C(I), I = 1,10) \\ 30\quad \text{FORMAT}\ (10F7.1) \end{array}\right\} \ \ldots (4.11)$$

The instructions have the same effect as the following:

$$\left.\begin{array}{l} \text{READ}\ (1,30)\ (A(1,1), A(1,2), A(1,3),\ etc.\ to \\ \quad A(1,10), B(1), C(1), A(2,1),\ etc.\ to \\ \quad A(2,10), B(2), C(2),\ etc.\ to \\ \quad A(10,1),\ etc.\ to\ A(10,10), B(10), C(10)) \end{array}\right\} \ \ldots (4.12)$$

30   FORMAT (10F7.1)

In this example the format specification is scanned 12 times.

Array lists

The second method of input or output of arrays is illustrated by the instructions:

$$\left.\begin{array}{l} \text{DIMENSION}\ A(10,10) \\ \text{READ}\ (1,\ 11)\ A \\ 11\quad \text{FORMAT}\ (5E11.4) \end{array}\right\} \ \ldots (4.13)$$

In many examples of the usage of DIMENSION statements, the statement reserves space for the *maximum* number of dimensional variables in the program, and operation upon arrays of dimensions *smaller* than the stated maximum creates no problems. For instructions of the type 4.13, however, there must be exact correspondence between the dimensions of the array given by the DIMENSION statement, and the dimensions of the array to be input or output. It is also necessary to relate the sequence of numbers in the input record to the manner of storage in the computer. The fields in the input record should correspond to the sequence A(1,1), A(2,1), A(3,1),...A(10,1), A(1,2), A(2,2),...A(1,10),...A(10,10), otherwise the variables will not be referred to correctly in the program.

Implicit DOs and the array facilities of 4.13 are also available on output.

Program organisation: Functions and subroutines

The exposition of the Fortran language up to this point is not complete, but it is sufficiently comprehensive to cover the writing of a wide range of programs. In practice, for problems of some complexity, it is usually desirable to organise the program into a *main* program segment and other *subprograms* or subroutines that can be called in as desired.

The breaking-up of a program into smaller constituent programs is sometimes necessary because the core storage of a computer is not large enough to hold the program as a single entity. But as a general rule, even for smaller jobs, it is good practice to distinguish logically distinct parts of a program as subroutines or functions. The complete computer program will then consist of a *master* or *steering* subprogram that controls the course of the computation and calls in subroutines or functions as required. The term subprogram is used here to describe portions of the program as a whole, including master subprograms, subroutines or functions.

A function can be any sequence of statements or arithmetic operations. If the function is sufficiently short to be written as one Fortran statement it may take the form of a *statement function* that can be included in the master or subroutine subprogram.

The statement function

A statement function consists of a single statement that must appear in the program before its first point of use. It has the form:

$$f(a_1, a_2 \ldots a_n) = e$$

where f is the name of the function chosen in accordance with the rules for the naming of variables, $a_1 \ldots a_n$ are the names of *dummy variables* chosen according to the same rules and said to be the *formal arguments* of the function, and e is the arithmetic expression defining the function in terms of the dummy variables. The names of the dummy variables have no connection with other variables of the same name appearing in the program.

When the value of the function is to be computed, the dummy or formal arguments of the function are replaced by *actual* arguments specified by the program.

As an example consider the calculation of an arithmetic expression by means of the following statement function:

$$AREA(X,Y,Z) = SQRT(X*X+Y*Y+Z*Z)*3.1417*Y$$

A subsequent statement in the program may take the form:

$$ALPHA = AREA(3., B+C/D, D)*CHI/OMEGA$$

This statement will cause the statement function to be calculated with the arguments 3., $B+C/D$, D substituted for the dummy arguments X, Y and Z in the statement function. In effect, the dummy arguments specify the number and type of the arguments to be used ·in the function.

Storage allocation and subroutines

Subroutines or functions, other than statement functions, are allocated storage in the computer for variables and data blocks quite independently of one another and of the allocation of storage for variables and blocks of the master segment. At least some of the data variables must be common to the master segment and subroutines, so that it is necessary to arrange for separate parts of the program to have access to common areas of data storage. Access of the program to common areas of data storage can be arranged either by specifying arguments for the function or subroutine, or by use of a COMMON specification in both the master and ancillary programs.

Subroutines

A subroutine is entered from the master program by means of a CALL statement. This statement has the form:

$$CALL \ s(a_1, a_2 \ldots a_n)$$

where s is the name chosen for the subroutine, and $a_1, \ldots a_n$ are dummy variables that specify the number and type of the dummy arguments in the same way as for the statement function. The name of the subroutine is made up of alphabetical and numerical symbols of which the first is alphabetical. The first statement of a subroutine has the form:

$$SUBROUTINE \ s(a_1, a_2 \ldots a_n)$$

or simply:

$$SUBROUTINE \ s$$

if the subroutine has no arguments.

On entering the subroutine statements following the first will be executed sequentially until the following statement is reached:

$$RETURN$$

when control is transferred back to the master program at the point immediately following the CALL statement. Values of the arguments $a_1, \ldots a_n$ may be used during the processing of instructions in the subroutine and, in addition, the values of arguments may be changed in the subroutine, so that if a result is to be obtained it can be returned to the main (calling) program by this method. The method of returning a result is shown in the following example of the partial listing of instructions in a subroutine named TIME:

SUBROUTINE TIME (X,Y,TASK)

TASK = *expression*

RETURN

END

A subroutine should not contain other subroutine or function statements, and should not call itself directly or indirectly.

Function subprograms

A function subprogram is entered from the master program when the function is specified in a statement of the master program. Thus if a function SEARCH (A,B,C,D) appears in a statement of the master program:

$$Y = SEARCH (X(I+1),X(I+2),X(I+3),X(I+4))/(U*U*D)$$

values of the variables $X(I+1)$, $X(I+2)$, $X(I+3)$ and $X(I+4)$ are substituted for the dummy arguments and the function SEARCH is entered.

The first statement of the function subprogram has the form:

FUNCTION SEARCH (A,B,C,D)

The names of the function and dummy variables are chosen in accordance with the rules for the naming of variables. Thus the type of function may be specified by the first letter of the name to be integer if IJKLM or N, and real if ABCDEFGH OPQRSTUVWXY or Z. This is an important distinction between the function and subroutine subprograms because the type of a subroutine is not specified by the first letter of its name.

Following the first statement of a function, statements are executed in sequence or by direction until a RETURN instruction is encountered when control is transferred back to the master program.

As with subroutines, results may be returned to the main (calling) program by changing the values of arguments. However, particular care is required, for if a statement in the main program has the form:

$$EN = A/SEARCH(RO,RE,A,B) * A/B$$

in which A and B are variables defined in the main program (note that they are *not* related to the dummy variables A and B of the function subprogram), and the value of either A or B is changed in the function, the result of this expression may be unpredictable.

A function subprogram should not contain subroutine or other function subprograms, and should not call itself directly or indirectly.

The COMMON *statement*

The common block is an area of data storage in the computer that is available to the master program and subprograms. If access to the common block is required the master program or subprogram must contain a COMMON

statement that is usually placed immediately after the DIMENSION statements. The COMMON statement has the form:

$$COMMON/n_1/a_1/n_2/a_2/ \ etc.$$

where n_1, n_2 ... are the names of common blocks, and a_1, a_2 ... are the lists of
variables stored in these common blocks. The name of the block obeys the
rules for the names of variables, but in this particular instance, the name may
be shared with a variable, array or statement function and the initial-letter
rule does not apply. However, the name of the block must be distinct from those
of subprograms or standard functions. If a name is not given to a block,
variables are placed in the unlabelled common block. Thus the statement:

$$COMMON/ZERO/P,Q,R//A,B,C$$

or the statement:

$$COMMON \ A,B,C/ZERO/P,Q,R$$

causes the computer to allocate storage for the variables A, B and C in the
unlabelled common block, while space for the variables P, Q and R is provided
in the common block known as ZERO.

The computer allocates storage space to variables in the order of appearance
in a given common block. Thus if one subprogram contains the statement:

$$COMMON \ ALT,AAD,ARC,W(120)/ALPHA/A,B,C$$

and another subprogram contains the statement:

$$COMMON \ ALT,EAR,ELI,F(6,20)/ALPHA/A,B,CASH$$

then the first three variables in the common block ALPHA are known as
A, B and C in the first program, and as A, B and CASH in the second, while
the first three variables in the unlabelled common block are known as ALT,
AAD and ARC in the first subprogram and as ALT, EAR and ELI in the second.
The matrix W of the first subprogram occupies 120 positions in the unlabelled
common block, but these locations are identified with the 120 positions of the
matrix F(6,20) of the second subprogram.

It should be noted that the dimensions of any subscripted variables may be
given in a common statement, but if this is done, the same variable should not
also appear in a dimension statement.

The COMMON statement provides efficient intercommunication between
all subprograms by allowing each subprogram to address a particular location
in a common block.

Order of statements in a program or subprogram

The recommended order is:

DIMENSION statements
COMMON statements
Statement function definitions
Executable statements.

FORMAT statements may appear anywhere in the program.

A computer program for sorting

For our first program we will consider the sorting of 50 positive numbers into a sequence in which the first number is the largest, and succeeding numbers decrease in magnitude. The procedure may be divided into the following four steps, according to the block diagram of the program logic shown on Fig. 4.6:

1. Read the numbers into the computer.
2. Start the comparison with the first number that has not been sorted.
3. Compare this number with succeeding numbers that have not been sorted. If a larger number is found, compare succeeding numbers with the larger number. At the end of one chain of inspections the largest number is found. Interchange this number with the first unsorted number.
   Repeat 2 and 3 until no numbers remain unsorted.
4. Print the numbers out. The numbers will appear in the desired sequence.

Each step may be represented as statements in a Fortran program. Thus step 1 may be accomplished by:

$$\left. \begin{array}{ll} & \text{DIMENSION A(50)} \\ & \text{READ (1,1) A} \\ 1 & \text{FORMAT (10F7.1)} \end{array} \right\} \dots (4.14)$$

The 50 numbers, each in the format F7.1, may be punched on 5 cards, ten numbers at a time. At the end of the READ statement the numbers occupy the 50 locations $A(1), \dots A(50)$ in unsorted form.

The sorting process is started by comparing $A(1)$ with $A(2)$ by means of an arithmetic IF statement. The larger is then compared with $A(3)$ and so on. For each comparison the value of the largest number found is kept in location B, and the position that this number occupies in the array A is identified by the fixed point variable L. At the end of the first sort the largest number found is interchanged with the number in location $A(1)$ and the next chain of sorting is started by a comparison between $A(2)$ and $A(3)$. The process is repeated until the complete array has been sorted. The Fortran instructions are:

$$\left. \begin{array}{ll} & \text{DO2 I} = 1,49 \\ & \text{B} = \text{A(I)} \\ & \text{L} = \text{I} \\ & \text{DO3 K} = \text{I,49} \\ & \text{IF } (\text{B} - \text{A(K}+1)) \ 4,3,3 \\ 4 & \text{L} = \text{K}+1 \\ & \text{B} = \text{A(L)} \\ 3 & \text{CONTINUE} \\ & \text{C} = \text{A(I)} \\ & \text{A(I)} = \text{A(L)} \\ 2 & \text{A(L)} = \text{C} \end{array} \right\} \dots (4.15)$$

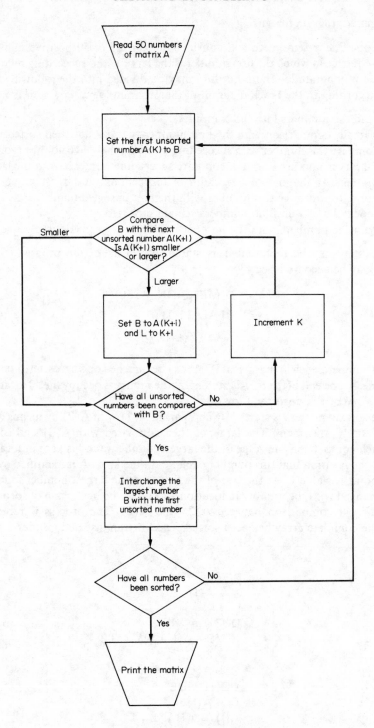

FIG. 4.6. A block diagram for the sorting program

The comparison between two numbers is effected by the arithmetic IF statement. If $A(K+1)$ is greater than B, the comparison is continued with $A(K + 1)$. As $(K + 1)$ identifies the location of the new number this is stored in L, and the old value of B is replaced by $A(L)$. The DO 3 loop ends in a CONTINUE statement, because an IF statement is not permissible as a terminal statement.

The interchange of numbers at the end of each chain is given by the last three instructions. The largest number is found in $A(L)$ and this is placed in $A(I)$. But as the contents of $A(I)$ are thus destroyed, the previous value of $A(I)$ is first placed in location C until the contents of $A(L)$ have been transferred.

Finally, the Fortran instructions for step 4 are:

$$\left.\begin{array}{l} \text{WRITE (2,1) A} \\ \text{STOP} \\ \text{END} \end{array}\right\} \quad \ldots . (4.16)$$

The format specification 1 may be used because format statements can appear anywhere in the program. If the output is written by a line printer the numbers will appear on 5 lines each containing 70 characters, but the paper control character is uncertain and it is better to use the instructions:

$$\left.\begin{array}{ll} & \text{WRITE (2, 5) A} \\ 5 & \text{FORMAT (10(5X,F7.1))} \\ & \text{STOP} \\ & \text{END} \end{array}\right\} \quad \ldots . (4.17)$$

which will have the effect of placing 4 blank characters initially and 5 blank characters between each number, with 10 numbers to each line in single spacing (assuming that there are at least 119 print positions in output).

When an appropriate job description is added, the complete program is given by the instructions 4.14, 4.15 and 4.17.

METHODS FOR COMPUTATION

The roots of transcendental and polynomial equations

The roots of equations containing polynomials or transcendental functions cannot usually be found in terms of analytical expressions and it is necessary to find them by means of numerical calculations. In physical applications it is sometimes of interest to find the successive roots of certain transcendental equations, and on other occasions, to find a particular root of a polynomial or transcendental equation. In the latter cases it is necessary to distinguish between the desired root and others, and in all cases the nature of the roots should be investigated before computing.

Two direct ways of finding the roots of equations are the method of linear interpolation, and the method of Newton. Both are suitable for automatic computing and are often preferred to more elaborate methods when the computation is to be carried out by an automatic digital machine.

The method of linear interpolation

In the method of linear interpolation successive values of the function $y = f(x)$ are calculated by systematically incrementing x until the value y is found to change sign between two successive estimates, say at x_0 and at $x_0 + H$. A typical relationship between x_0, H, $f(x_0)$ and $f(x_0 + H)$ is shown in Fig. 4.7.

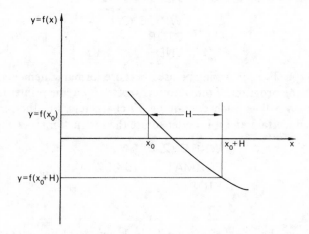

Fɪɢ. 4.7. A root of a polynomial or transcendental equation

A better approximation to the root is given by the intersection between the x axis and the line joining the point $(x_0, f(x_0))$ to the point $(x_0 + H, f(x_0 + H))$. From Fig. 4.7 it can be seen that the new estimate is:

$$x_0 + \frac{H\,f(x_0)}{|f(x_0 + H) - f(x_0)|}$$

If necessary, further trial calculations may be made from this point with a reduced increment H_1 until the function again changes sign and a new interpolated estimate of the root is made. This procedure is repeated until the root has been estimated to the required degree of accuracy.

A variation of the method simply repeats the calculation of successive values of the function $y = f(x)$. When y changes sign between two successive estimates, at $x_0 + (n - 1)H$ and at $x_0 + nH$, the value of the increment is reduced and the interval $x_0 + (n - 1)H$ to $x_0 + nH$ is scanned until the function again changes sign. The procedure is repeated until sufficient accuracy has been obtained. This variation is illustrated in the second example of a Fortran program.

A computer program to find the root of a polynomial

Suppose that a particular root of the equation:

$$y = ax^4 + bx^3 + cx^2 + dx + e \qquad \dots (4.18)$$

lies within the range $X_1 < x < X_2$. In the example it is known that only one root lies within this range. The steps in the scheme of computation are:

1. The range $X_1 - X_2$ is divided into 10 equal increments called H.
2. The function y is calculated at $x = X_1$ and the sign of the function is tested. If negative, K is set to 1, or if positive, K is set to 2.
3. The function y is calculated at $x = X_1 + H$, and the sign of the function is again tested. If the sign of the function has not changed, x is again incremented and y is calculated and tested, until y is found to change sign.
4. When the sign of y is found to change, say between $x = X_1 + (n - 1)H$ and $x = X_1 + nH$, the root is assumed to lie between these values. The size of the increment H is then compared with ERROR, a variable that is set to the desired accuracy.
5. If H is smaller than, or equal to ERROR, the current value of x is printed as the estimate of the root. ERROR is the maximum inaccuracy. If H is greater than ERROR, a new value of increment equal to $H/10$ is set and the above steps are repeated over the range $X_1 + (n - 1)H$ to $X_1 + nH$ until the increment chosen becomes equal to or smaller than ERROR.
6. If the sign of y does not change over the range $X_1 - X_2$, the line printer is caused to print NO ROOT FOUND.
7. At the close of step 5 or step 6 the computation stops.

A listing of Fortran statements for the program is shown, and a block diagram is given as Fig. 4.8.

```
      DIMENSION Y(10)
      READ (1,1) A,B,C,D,E,ERROR,X1,X2
   1  FORMAT (4E11.4)
```

The parameters a, b, c, d, e of the polynomial, the maximum allowed error of the solution (ERROR), and the range defined by X_1 and X_2 are read in. The data are punched into two cards, and the format statement is scanned twice.

The initial values of x and of H are set by the next two instructions:

```
      X = X1
      H = 0.1 * (X2 − X1)
```

Next follow the trial computations. The value of y is calculated, and if positive K is set to 2, or if negative K is set to 1. The value of x is incremented and further

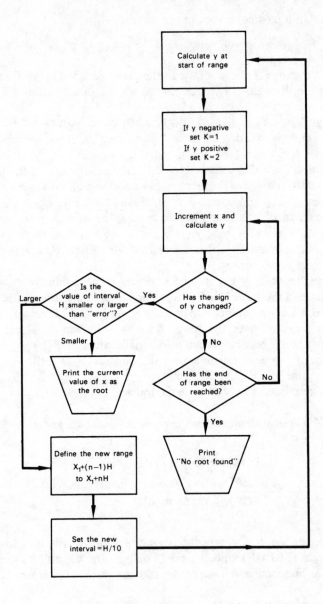

FIG. 4.8. Block diagram for the program to find the root of a polynomial equation

values of y are calculated until the function is found to change sign.

```
11    Y(1) = E+X*(D+X*(C+X*(B+A*X)))
      X = X+H
      IF (Y(1)) 2,3,4
2     K = 1
      GO TO 5
4     K = 2
5     DO 6 I = 2,10
      Y(I) = E+X*(D+X*(C+X*(B+A*X)))
      WRITE (2, 15) X,ERROR,H,Y(I),I
15    FORMAT (4(5X,E11.4),I3)
      GO TO (7,8), K
7     IF (Y(I)) 6,3,9
9     X = X−H
      IF (H−ERROR) 3,3,10
10    H = 0.1 * H
      GO TO 11
8     IF (Y(I)) 9,3,6
6     X = X + H
```

Once a change of sign has been found H is compared with ERROR. If the accuracy is not sufficient the interval H is reduced and the procedure is repeated. If the function y does not change sign in the first 10 calculations, the next instructions cause the line printer to show that no root has been found:

```
      WRITE (2,13)
13    FORMAT (14H NO ROOT FOUND)
      STOP
```

If the value of the root is found to the desired accuracy, the next instructions will be reached:

```
3     WRITE (2,14) X
14    FORMAT (5H X = ,E 11.4)
      STOP
      END
```

In one application the program was used to find the root of the polynomial:

$$y = x^4 − 7x^3 + 17x^2 − 17x + 6 \qquad\qquad ....(4.19)$$

that lies within the range $1\cdot5 < x < 2\cdot5$. The output obtained during execution of this program is shown below:

0.1600E 01	0.1000E–02	0.1000E 00	0.2016E 00 2
0.1700E 01	0.1000E–02	0.1000E 00	0.1911E 00 3
0.1800E 01	0.1000E–02	0.1000E 00	0.1536E 00 4
0.1900E 01	0.1000E–02	0.1000E 00	0.8910E–01 5
0.2000E 01	0.1000E–02	0.1000E 00	0.4075E–09 6
0.2100E 01	0.1000E–02	0.1000E 00	−0.1089E 00 7
0.2010E 01	0.1000E–02	0.1000E–01	−0.1010E–01 2
0.2001E 01	0.1000E–02	0.1000E–02	−0.1009E–02 2
0.2000E 01	0.1000E–02	0.1000E–03	−0.1000E–03 2

X = 0.2000E 01

The interested reader may like to follow the way in which the output from the line printer is controlled by the program.

The result is correct because the polynomial 4.19 may be arranged in the form of 4.20:

$$y = (x - 1)^2 (x - 2)(x - 3) \qquad \dots (4.21)$$

Newton's method

Newton's method produces a better approximation to a root from an estimate $x = a$, by use of the linear terms in a Taylor series expansion. An expansion about the point $x = a$ has the form:

$$f(a + h) = f(a) + hf'(a) + (h^2/2!)f''(a)\dots \qquad \dots (4.21)$$

As the condition for the root is given by $f(a + h) = 0$, the increment h to be added to the previous estimate of the root is $-f(a)/f'(a)$. Figure 4.9 shows the

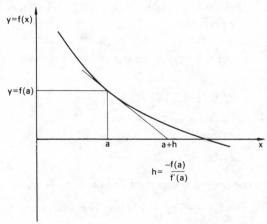

FIG. 4.9. Newton's method of root finding

way in which the value of a derivative at a point on the curve is used to predict the value of the root. The increment $-f(a)/f'(a)$ is shown on the diagram.

Although examples may be found in which either Newton's method or linear interpolation converges to a root even though the initial estimate is wildly astray, it is usually necessary to place the initial estimate with some care. When the method of root estimation is completely automatic in a digital computer program, a preliminary study of the properties of the roots may ensure that many of the dangers of root-finding are avoided. For example, the method of root-finding described as an application of linear interpolation is appropriate when the roots are real and distinct, but not when the roots are real and equal, or imaginary.

There are also dangers in the application of Newton's method that may be illustrated by finding the second root of the equation:

$$(\tan x) - x = 0 \qquad \qquad \ldots\ldots(4.22)$$

The functions $\tan x$ and x are sketched in Fig. 4.10. From this diagram it is evident that the roots of the equation correspond to the intersections of the graph of $\tan x$ with the graph of x. The consecutive roots lie within the ranges

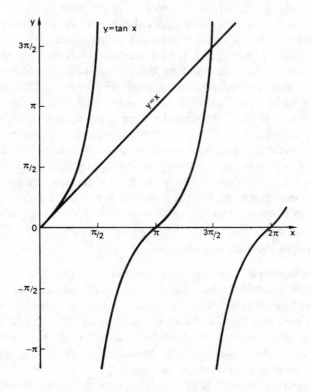

FIG. 4.10. The functions $\tan x$ and x

$0-\pi/2$, $\pi-3\pi/2$, $2\pi-5\pi/2$, ... $n\pi-(2n+1)\pi/2$. From Newton's formula an approximation ϕ_i to a given root is related to the previous approximation by the equation:

$$\phi_i = \phi_{i-1} - \frac{\tan(\phi_{i-1}) - \phi_{i-1}}{\sec^2(\phi_{i-1}) - 1} \qquad \ldots\ldots(4.23)$$

Suppose that a first approximation to the second root is π, then it is found by substitution into equation 4.23 that the value predicted for the root is ∞. This is due to the fact that the derivative of $\tan x - x$ in the vicinity of $x = n\pi$ approaches zero, and approximations to the root that start in a wide neighbourhood of $x = n\pi$ will not converge to the root within the range $n\pi-(2n+1)\pi/2$.

If the first approximation to the second root is $1\cdot4\pi$, the Newton formula converges and the first few estimates are shown below:

$$\phi_0 = 4\cdot39822 \quad (= 1\cdot4\pi)$$
$$\phi_1 = 4\cdot52433$$
$$\phi_2 = 4\cdot49880$$
$$\phi_3 = 4\cdot493546519$$
$$\phi_4 = 4\cdot493416002$$
$$\phi_5 = 4\cdot493409767$$

$$\Delta\phi_0 = +0\cdot126100939$$
$$\Delta\phi_1 = -0\cdot025525893$$
$$\Delta\phi_2 = -0\cdot005258242$$
$$\Delta\phi_3 = -0\cdot000130517$$
$$\Delta\phi_4 = -0\cdot000006235$$

The last estimate is probably correct to 7 or 8 significant figures.

A study of the quantities $\Delta\phi$ suggests that the accuracy of root estimation increases more rapidly in the later than the earlier estimates because the gain in significant figures is larger in the later estimates. If the gain in significant figures in a number of steps is independent of the earlier number of steps, the method of approximation is said to show *first order* convergence. Linear interpolation often gives first order convergence while the method of Newton gives a better rate of convergence that is approximately *second order*.

Both methods may be used to find the multiple roots of polynomials when these are real. Newton's method may also be used when the roots are imaginary. If a root at $x = a$ is found, a polynomial may be reduced by dividing by the factor $(x - a)$. If this is done it is necessary to determine the position of the root with a high degree of accuracy because errors introduced by a lack of precision may severely affect the coefficients of the reduced polynomial.

Systems of linear simultaneous equations

Systems of linear equations occur so frequently in numerical work that their solution is fundamental to many branches of numerical analysis. The solution does not usually involve a difficulty in principle, because there is a fair selection of well-tried methods, but difficulties can arise because of the propagation of errors in the computations. Sometimes it is necessary to use a particular method of solution, or a greater precision of computation, in order to minimise errors found to be too large for the desired accuracy.

Operations upon systems of linear equations are most conveniently described by matrix notation. The relationship between simultaneous equations and their matrix representation may be shown by considering the system of equations 4.24:

$$
\left.
\begin{aligned}
a_{11}x_1 + a_{12}x_2 + a_{13}x_3 \ldots + a_{1n}x_n &= b_1 \\
a_{21}x_1 + a_{22}x_2 + a_{23}x_3 \ldots + a_{2n}x_n &= b_2 \\
\cdot \quad \cdot \quad\quad\quad\quad\quad \\
\cdot \quad \cdot \quad\quad\quad\quad\quad \\
a_{n1}x_1 + a_{n2}x_2 + a_{n3}x_3 \ldots + a_{nn}x_n &= b_n
\end{aligned}
\right\} \quad \ldots (4.24)
$$

This system of equations may be represented in the explicit matrix form of equation 4.25:

$$
\begin{bmatrix}
a_{11} & a_{12} & a_{13} & \cdots & a_{1n} \\
a_{21} & a_{22} & a_{23} & \cdots & a_{2n} \\
& \cdot & \cdot & \cdot & \\
& \cdot & \cdot & \cdot & \\
a_{n1} & a_{n2} & a_{n3} & \cdots & a_{nn}
\end{bmatrix}
\begin{bmatrix}
x_1 \\ x_2 \\ \cdot \\ \cdot \\ x_n
\end{bmatrix}
=
\begin{bmatrix}
b_1 \\ b_2 \\ \cdot \\ \cdot \\ b_n
\end{bmatrix}
\quad \ldots (4.25)
$$

or more conveniently still by the matrix equation:

$$\mathbf{Ax = b} \qquad\qquad \ldots (4.26)$$

The matrix $\mathbf{A}$ is composed of the coefficients $a_{11}, \ldots$ known as the elements of the matrix. $\mathbf{A}$ is an $n \times n$ matrix having n rows and n columns and in this particular example is the matrix of coefficients. The matrices $\mathbf{x}$ and $\mathbf{b}$ are $n \times 1$ matrices, sometimes known as *column vectors*.

The position of an element in the matrix is shown by the subscripts. The first subscript indicates the row in which the element is found, and the second indicates the column; by means of this convention any element can be located in the matrix.

The choice of the method of solution depends upon the nature of the matrix of coefficients. This matrix is often a square matrix that can be classified by the values of its elements.

Types of matrix

1. A *symmetric matrix* has elements that satisfy the condition:

$$a_{ij} = a_{ji}$$

2. If all of the elements are zero the matrix is a *null matrix*.
3. A *diagonal matrix* has zero elements everywhere except on the principal diagonal, composed of the elements $a_{11}, a_{22}, a_{33}, \ldots a_{nn}$.
4. A *unit matrix* has zero elements everywhere except on the principal diagonal, where each element has the value 1.

5. A *lower triangular matrix* has zero elements everywhere above the principal diagonal.

6. An *upper triangular matrix* has zero elements everywhere below the principal diagonal.

7. A matrix that contains zero elements everywhere except on the principal and adjacent diagonals is known as a *banded matrix*. If the principal and two flanking diagonals are occupied, the matrix is *tridiagonal*; if the principal and four flanking diagonals are occupied, the matrix is *pentadiagonal*.

8. *Column* and *row matrices* are sometimes known as column and row *vectors*.

Matrix operations

Operations upon a matrix may be defined by a number of rules in a manner similar to the definition of operations upon ordinary real numbers. The principal operations include *addition, subtraction, transposition, multiplication* and *inversion*.

Addition and subtraction. Matrix addition and subtraction are defined by the equation:

$$\mathbf{A} \pm \mathbf{B} = \begin{bmatrix} a_{11} \pm b_{11}, & a_{12} \pm b_{12} \cdots \\ a_{21} \pm b_{21}, & a_{22} \pm b_{22} \cdots \\ \cdot & \cdot & \cdot \\ \cdot & \cdot & \cdot \\ a_{n1} \pm b_{n1}, & \cdots & a_{nn} \pm b_{nn} \end{bmatrix} \qquad \dots (4.27)$$

From the definition it is clear that matrix addition is defined only for matrices of the same number of rows and the same number of columns. It is also clear that:

$$\mathbf{A} + \mathbf{B} = \mathbf{B} + \mathbf{A} \qquad \dots (4.28)$$

and matrix addition is therefore *commutative*.

The general element of the matrix $\mathbf{C}$ that is the sum of $\mathbf{A}$ and $\mathbf{B}$ is:

$$c_{ij} = a_{ij} + b_{ij} \qquad \dots (4.29)$$

Matrix operations are particularly convenient in Fortran because the subscripts of an element of a matrix can be identified with the subscripts of the Fortran array that represents the matrix. A Fortran program for matrix addition can make use of nested DO loops. The computation kernel of such a program has the simple form:

$$\left. \begin{array}{ll} & \text{DO 1 I} = 1, m \\ & \text{DO 1 J} = 1, n \\ 1 & \text{C(I,J)} = \text{A(I,J)} + \text{B(I,J)} \end{array} \right\} \quad \dots (4.30)$$

The program will include appropriate DIMENSION, input, output and FORMAT statements.

Transposition. The matrix is rotated about the leading diagonal so that the columns of the matrix are the rows of the transpose. The elements a_{ij} and a_{ji} are interchanged. The kernel of a suitable Fortran program is:

$$\left.\begin{array}{l} \text{DO 1 I} = 1, m \\ \text{DO 1 J} = 1, n \\ \text{1} \qquad \text{B(I,J)} \;\; = \text{A(J,I)} \end{array}\right\} \quad \dots.(4.31)$$

—instructions that set matrix **B** as the transpose of **A**.

Multiplication. The multiplication of two matrices **A** and **B** is defined:

$$\begin{bmatrix} a_{11}a_{12}a_{13}\dots a_{1m} \\ a_{21}a_{22} \qquad a_{2m} \\ \cdot \quad \cdot \quad \cdot \\ \cdot \quad \cdot \quad \cdot \\ a_{n1}\,a_{n2} \qquad a_{nm} \end{bmatrix} \times \begin{bmatrix} b_{11}b_{12}b_{13}\dots b_{1l} \\ b_{21}b_{22}b_{23}\dots b_{2l} \\ \cdot \quad \cdot \quad \cdot \\ \cdot \quad \cdot \quad \cdot \\ \cdot \quad \cdot \quad \cdot \\ b_{m1} \qquad b_{ml} \end{bmatrix} =$$

$$\qquad n \times m \qquad\qquad\qquad m \times l$$

$$\begin{bmatrix} a_{11}b_{11} + a_{12}b_{21}\dots + a_{1m}b_{m1}, a_{11}b_{12}\dots + a_{1m}b_{m2},\dots, & a_{11}b_{1l}\dots + a_{1m}b_{ml} \\ a_{21}b_{11} +\dots \qquad\qquad + a_{2m}b_{m1},\dots \\ \cdot \quad \cdot \quad \cdot \\ \cdot \quad \cdot \quad \cdot \\ a_{n1}b_{11} +\dots \qquad\qquad + a_{nm}b_{m1},\dots \quad\dots, & a_{n1}b_{1l} +\dots a_{nm}b_{ml} \end{bmatrix}$$

$$\qquad\qquad n \times l \qquad\qquad\qquad\qquad \dots.(4.32)$$

If **C** is the product of two matrices **A** and **B**, the general element of **C** is:

$$c_{jk} = \sum_{i=1}^{m} a_{ji}b_{ik} = a_{j1}b_{1k} + a_{j2}b_{2k}\dots + a_{jm}b_{mk} \qquad \dots.(4.33)$$

Matrix multiplication is defined only when the matrix on the left has a number of columns equal to the number of rows of the matrix on the right. The product of the two has rows equal to the number of rows of the matrix on the left and columns equal to the number of columns of the matrix on the right. In general matrix multiplication is not commutative:

$$\mathbf{AB} \neq \mathbf{BA} \qquad\qquad \dots.(4.34)$$

indeed only one product is defined, except when the number of rows and the number of columns of one matrix are respectively equal to the number of columns and the number of rows of the second matrix.

Matrix multiplication is also easily executed by using the facility of arrays in Fortran. The computation kernel of a matrix multiplication is:

$$\left.\begin{array}{l} \text{DO 1 J} = 1, n \\ \text{DO 1 K} = 1, l \\ \text{C(J,K)} \ = 0 \\ \text{DO 1 I} = 1, m \\ 1 \quad \text{C(J,K)} = \text{C(J,K)} + \text{A(J,I)} * \text{B(I,K)} \end{array}\right\} \quad \dots (4.35)$$

The location $C(J,K)$ acts as an accumulator that is first set equal to zero, and then successively accumulates the products of the elements from the jth row of matrix $\mathbf{A}$ and the kth row of matrix $\mathbf{B}$. This is a good illustration of the convenience of Fortran arrays in matrix operations.

Elementary row and column operations. These operations are extremely important in the successive matrix operations that are used to solve a system of linear equations. The elementary operations comprise an interchange of rows or columns, a multiplication of a row or column by a constant, and the replacement of a row or column, say the ith, by the sum of the ith and k times the jth row or column. The operations are just those that are used in the solution of linear simultaneous equations by the method of elimination. For example, the two linear simultaneous equations 4.36 are reduced to the upper triangular form 4.37 by subtracting the first row from the second:

$$\begin{bmatrix} 1 & 1 \\ 1 & -1 \end{bmatrix} \begin{bmatrix} x_1 \\ x_2 \end{bmatrix} = \begin{bmatrix} 2 \\ 0 \end{bmatrix} \qquad \dots (4.36)$$

$$\begin{bmatrix} 1 & 1 \\ 0 & -2 \end{bmatrix} \begin{bmatrix} x_1 \\ x_2 \end{bmatrix} = \begin{bmatrix} 2 \\ -2 \end{bmatrix} \qquad \dots (4.37)$$

The elementary operations may be defined in terms of matrix multiplication. Thus an elementary row operation upon an $n \times s$ matrix $\mathbf{A}$ may be performed by calculating the product $\mathbf{EA}$ where $\mathbf{E}$ is the matrix obtained by performing the row operation on the unit matrix $\mathbf{I}_n$. For example, if the first row of $\mathbf{I}_2$ is subtracted from the second, the elementary matrix $\mathbf{E}$ has the form:

$$\mathbf{I}_2 = \begin{bmatrix} 1 & 0 \\ 0 & 1 \end{bmatrix} \rightarrow \mathbf{E} = \begin{bmatrix} 1 & 0 \\ -1 & 1 \end{bmatrix} \qquad \dots (4.38)$$

If the matrix of coefficients of the equations 4.36 is multiplied on the left by $\mathbf{E}$:

$$\begin{bmatrix} 1 & 0 \\ -1 & 1 \end{bmatrix} \begin{bmatrix} 1 & 1 \\ 1 & -1 \end{bmatrix} = \begin{bmatrix} 1 & 1 \\ 0 & -2 \end{bmatrix} \qquad \dots (4.39)$$

the matrix of coefficients of equation 4.37 is obtained.

Column operations may also be related to matrix multiplication. Each elementary column operation upon an $n \times s$ matrix $\mathbf{A}$ can be achieved by multiplying $\mathbf{A}$ by the matrix $\mathbf{E}$ on the right that is obtained by performing the column operation on the unit matrix $\mathbf{I}_s$.

Matrix inversion

This is an operation that is closely linked to the solution of a system of linear equations. A matrix can have a unique inverse only if the matrix is square, and if its *rank* is equal to the number of rows. A matrix with these properties is said to be *non-singular*, and the value of the determinant of the matrix is not zero. The condition for a unique inverse is very important, because if the rank of the matrix of coefficients of a system of linear equations is not equal to the number of rows, a unique solution cannot be obtained.

If all of the equations used in forming the matrix of coefficients are independent, the rank of the matrix will be equal to the number of rows, but if one or more of the equations have been obtained from others by a combination of elementary row operations, the rank of the matrix will be less than the number of rows. In other words, the rank of a matrix is the number of *independent* rows or *independent* equations used in forming the matrix. The equations

$$
\left.
\begin{aligned}
x_1 + 2x_2 + \;\;x_3 &= \;\;5 \\
x_1 + \;\;x_2 + 2x_3 &= \;\;7 \\
3x_1 + 4x_2 + 5x_3 &= 19
\end{aligned}
\right\} \qquad \dots (4.40)
$$

are not independent because the third equation is obtained from the first two. The matrix of coefficients, although 3×3 in size, is only of rank 2.

The question of independence of the equations in a set is not usually resolved in such a simple manner. However, dependent equations may be completely eliminated by row operations and therefore a systematic reduction of a matrix will leave rows equal to the number of independent equations, while the dependent rows have been replaced by null rows. For example, the matrix of coefficients of equations 4.40 may be reduced by subtracting the first row and twice the second row from the third:

$$
\begin{bmatrix} 1 & 2 & 1 \\ 1 & 1 & 2 \\ 3 & 4 & 5 \end{bmatrix} \rightarrow \begin{bmatrix} 1 & 2 & 1 \\ 1 & 1 & 2 \\ 0 & 0 & 0 \end{bmatrix} \qquad \dots (4.41)
$$

A square matrix of row rank less than the number of rows is said to be *singular*. The determinant of a singular matrix is zero. A condition of compatibility between the matrix of coefficients and the *augmented* matrix of a system of linear equations is important when the matrix of coefficients is singular. The augmented matrix comprises the matrix of coefficients augmented by the right-hand side of the equations as an additional column to give a dimension of $n \times (n + 1)$. The augmented matrix of the system 4.25 is:

$$\begin{bmatrix} a_{11} & a_{12} & a_{13} & \cdots & a_{1n} & b_1 \\ a_{21} & a_{22} & a_{23} & \cdots & a_{2n} & b_2 \\ \cdot & \cdot & \cdot \\ \cdot & \cdot & \cdot \\ a_{n1} & a_{n2} & a_{n3} & \cdots & a_{nn} & b_n \end{bmatrix}$$

The condition of compatibility is that the rank of the matrix of coefficients must equal the rank of the augmented matrix; otherwise no solution can be obtained. If the equations 4.40 are slightly changed to 4.42:

$$\left. \begin{array}{r} x_1 + 2x_2 + x_3 = 5 \\ x_1 + x_2 + 2x_3 = 7 \\ 3x_1 + 4x_2 + 5x_3 = 25 \end{array} \right\} \qquad \ldots (4.42)$$

it can easily be verified that the rank of the matrix of coefficients is 2, but the rank of the augmented matrix is three and therefore equations 4.42 do not have a solution.

Conditions for a unique inverse. A unique inverse can only be obtained from a square matrix having a rank equal to the number of rows. The inverse A^{-1} is related to A by the matrix equation:

$$AA^{-1} = I \qquad \ldots (4.43)$$

where I is the unit matrix. The order of multiplication of the matrices A and A^{-1} is not important.

The relationship of the inverse to the solution of a set of simultaneous equations may be shown by considering again:

$$Ax = b \qquad \text{(equation 4.26)}$$

when both sides of the equation are multiplied on the left by the inverse of the matrix A:

$$A^{-1}Ax = A^{-1}b \qquad \ldots (4.44)$$

and on substituting equation 4.43 into 4.44:

$$Ix = A^{-1}b \qquad \ldots (4.45)$$

The multiplication of a matrix by the unit matrix gives the matrix itself, and therefore the column vector x is equated to a right-hand side that requires only a matrix multiplication to put the solution into an explicit form.

The computation of the inverse of the matrix of coefficients allows the solution, to the system of equations 4.26, to be written down in terms of a matrix multiplication for any right-hand side. If the equations are to be solved for a number of right-hand sides but an unchanged matrix of coefficients, it is only necessary to invert the matrix of coefficients once. The inverse A^{-1} may then be stored for the subsequent multiplication of any right-hand side when desired.

The inverse of a matrix may be computed in a manner shown by the following theorem:

"If a matrix A can be reduced to the unit matrix by a series of elementary operations, the same series of elementary operations acting upon the unit matrix will generate the inverse A^{-1}."

Proof. Consider equation 4.43, and let the sequence of elementary operations $E_1 E_2 \ldots E_i$ reduce a matrix A to the unit matrix. If the sequence is applied to both sides of equation 4.43, equation 4.46 is obtained:

$$E_1 E_2 \ldots E_i AA^{-1} = E_1 E_2 \ldots E_i I \qquad \ldots\ldots(4.46)$$

By hypothesis the result of $E_1 E_2 \ldots E_i A$ is the unit matrix I and:

$$IA^{-1} = E_1 E_2 \ldots E_i I \qquad \ldots\ldots(4.47)$$

Since the multiplication of a matrix by a unit matrix gives the matrix itself:

$$A^{-1} = E_1 E_2 \ldots E_i I \qquad \ldots\ldots(4.48)$$

establishing the theorem. The inverse may be computed by systematically reducing a matrix to the unit matrix, and performing the same operations upon a unit matrix held elsewhere in the store of the computer.

It is clear from this that the computation of an inverse requires more work than solving a set of linear simultaneous equations. The inverse is not usually computed unless a solution is required for a large number of systems, each with the same matrix of coefficients, but in which the right-hand sides are different.

Methods of solution for systems of linear simultaneous equations

It is now established that some methods of solution are better adapted to automatic digital computers than others. The method of Gaussian elimination with selection of pivots, or pivotal condensation, is particularly suitable and this method will be discussed in detail.

Gaussian elimination: Computer program for pivotal condensation. The operations may be described by referring to equation 4.24:

$$\left.\begin{array}{l} a_{11}x_1 + a_{12}x_2 + a_{13}x_3 + a_{14}x_4 \ldots + a_{1n}x_n = b_1 \\ a_{21}x_1 + a_{22}x_2 + a_{23}x_3 + \qquad \ldots + a_{2n}x_n = b_2 \\ \qquad \cdot \quad \cdot \quad \cdot \\ \qquad \cdot \quad \cdot \quad \cdot \\ a_{n1}x_1 + a_{n2}x_2 + a_{n3}x_3 \ldots \qquad \ldots + a_{nn}x_n = b_n \end{array}\right\} \text{(equation 4.24)}$$

The largest coefficient of x_1 in the set of equations is first chosen, and used to eliminate the coefficients of x_1 from all of the other equations. If the largest coefficient is a_{k1}, equation k is multiplied by a_{11}/a_{k1} and subtracted from equation 1 so that the coefficient of x_1 is eliminated. In the same manner, the coefficient of x_1 in equation 2 is eliminated and the procedure is repeated for all equations except the kth, the *pivotal* equation. The element a_{k1} is known as the *pivotal element*. The pivotal equation takes no further part in the elimination.

The coefficients of x_2 are next scanned; the largest is selected as the new pivotal element, and by means of the new pivotal equation the coefficients of

x_2 are eliminated from the remaining $n - 2$ equations. After the coefficients of x_1 to x_{n-1} have been eliminated in turn, the equations may be placed in order of elimination to give the upper triangular form:

$$
\left.
\begin{aligned}
c_{11}x_1 + c_{12}x_2 + \ldots \quad\quad\quad\quad + c_{1n}x_n &= d_1 \\
c_{22}x_2 + \ldots \quad\quad\quad\quad + c_{2n}x_n &= d_2 \\
\cdot \quad \cdot \quad \cdot \quad\quad\quad \\
\cdot \quad \cdot \quad \cdot \quad\quad\quad \\
c_{nn}x_n &= d_n
\end{aligned}
\right\} \quad \ldots\text{(4.49)}
$$

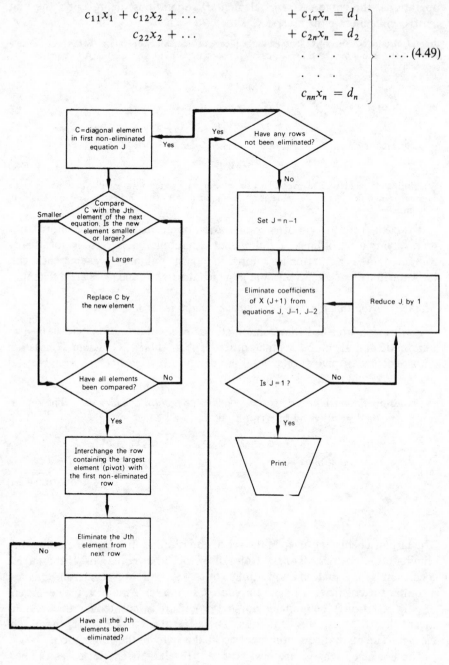

FIG. 4.11. A block diagram for the program on pivotal condensation

The last equation is solved for x_n. The value of x_n can then be substituted into the second-last equation to give x_{n-1}, and the process may be continued to give each x_i, ending with the value of x_1. This is the process of *back-substitution*. The values of x_i are a complete solution to equation 4.24.

A computer program for the solution of a system of equations by the method of pivotal condensation will include elimination and back-substitution as integral but distinct parts of the program. The program will also find the largest element of a column by a search procedure and record the sequence of elimination.

Suppose that the augmented matrix of coefficients has been read into the computer. The sorting program 4.15 may be adapted to the task of finding the pivotal element. The instructions 4.50 to 4.53 arrange that the pivotal element is first found and the whole of the pivotal row is interchanged with an earlier row such that the pivotal element now lies on the principal diagonal of the matrix of coefficients. The coefficients in the pivotal column in the rows following the pivotal row are now eliminated, and the search and elimination procedure is repeated until the matrix of coefficients is placed in the upper triangular form of 4.49.

The Fortran instructions for this program are shown below. The instructions for the reading and writing of information have been omitted, but reading, writing, FORMAT and DIMENSION statements may be added as appropriate. A block diagram of the program is shown as Fig. 4.11.

The statements of 4.50 define the search procedure for the pivotal element. At the end of the search the address of the pivotal row is stored as K1. The number of equations is N.

```
        NA = N + 1
        NB = N - 1
        DO 1  J = 1,NB
        K2 = J + 1
        K1 = J
        C = ABS(A(J,J))                    ....(4.50)
        DO 2 I = K2,N
        IF(C - ABS(A(I,J))) 3,2,2
   3    C = ABS(A(I,J))
        K1 = I
   2    CONTINUE
```

In the next stage the pivotal row is interchanged with row J by means of the next four instructions.

```
        DO 4 K = J,NA
        B = A(J,K)                         ....(4.51)
        A(J,K) = A(K1,K)
   4    A(K1,K) = B
```

Note that the contents of $A(J,K)$ are stored in B before interchange. The next four instructions placed in a nested DO loop eliminate the coefficients in column J by means of the pivotal row now held in row J:

$$
\left.\begin{array}{ll}
& \text{DO 1 I = K2,N} \\
& \text{B = A(I,J)/A(J,J)} \\
& \text{DO 1 K = J,NA} \\
\text{1} & \text{A(I,K) = A(I,K) − A(J,K)∗B}
\end{array}\right\} \quad \dots (4.52)
$$

The matrix $\mathbf{A}$ is now in the upper triangular form of equation 4.49. The final instructions control the back-substitution so that the matrix is reduced to the unit matrix, leaving the values of x_1 to x_n in column $N+1$.

$$
\left.\begin{array}{ll}
& \text{DO 5 I = 1,NB} \\
& \text{NI = NA − I} \\
& \text{B = A(NI,NA)/A(NI,NI)} \\
& \text{L = 1 + I} \\
& \text{DO 6 K = L,N} \\
& \text{J = NA − K} \\
\text{6} & \text{A(J,NA) = A(J,NA) − A(J,NI)∗B} \\
& \text{A(NI,NI) = 1.} \\
\text{5} & \text{A(NI,NA) = B} \\
& \text{A(1,NA) = A(1,NA)/A(1,1)} \\
& \text{A(1,1) = 1.}
\end{array}\right\} \quad \dots (4.53)
$$

The reader should follow the operation of the DO loops and compare with the schemes of search, elimination and back substitution.

As an example four simultaneous equations were solved by means of this computer program. The four equations are shown as the matrix equation 4.54:

$$
\begin{bmatrix}
1\cdot00 & 1\cdot25 & 0\cdot15 & 0\cdot38 \\
1\cdot39 & 0\cdot16 & 0\cdot90 & 0\cdot76 \\
0\cdot12 & 3\cdot69 & 0\cdot50 & 1\cdot98 \\
3\cdot10 & 2\cdot90 & 1\cdot00 & 1\cdot00
\end{bmatrix}
\begin{bmatrix}
x_1 \\ x_2 \\ x_3 \\ x_4
\end{bmatrix}
=
\begin{bmatrix}
5\cdot00 \\ 9\cdot00 \\ 14\cdot00 \\ 22\cdot00
\end{bmatrix}
\quad \dots (4.54)
$$

In the first stage of the program the equation is placed into the upper triangular form of 4.55. The positions of some rows have been transformed during this stage:

$$
\begin{bmatrix}
3\cdot1000 & 2\cdot9000 & 1\cdot0000 & 1\cdot0000 \\
0\cdot0000 & 3\cdot5777 & 0\cdot4613 & 1\cdot9413 \\
0\cdot0000 & 0\cdot0000 & 0\cdot5986 & 0\cdot9304 \\
0\cdot0000 & 0\cdot0000 & 0\cdot0000 & 0\cdot2180
\end{bmatrix}
\begin{bmatrix}
x_1 \\ x_2 \\ x_3 \\ x_4
\end{bmatrix}
=
\begin{bmatrix}
22\cdot0000 \\ 13\cdot1484 \\ 3\cdot3262 \\ -2\cdot0684
\end{bmatrix}
\quad \dots (4.55)
$$

In the final stage of back-substitution the matrix of coefficients has formally become the unit matrix, and the values of the variables are given by the vector on the right.

$$\begin{bmatrix} 1 & 0 & 0 & 0 \\ 0 & 1 & 0 & 0 \\ 0 & 0 & 1 & 0 \\ 0 & 0 & 0 & 1 \end{bmatrix} \begin{bmatrix} x_1 \\ x_2 \\ x_3 \\ x_4 \end{bmatrix} = \begin{bmatrix} -2{\cdot}1970 \\ 6{\cdot}2058 \\ 20{\cdot}3023 \\ -9{\cdot}4883 \end{bmatrix} \qquad \ldots . (4.56)$$

The use of the largest element in a column as the pivot in elimination is important because the growth of errors in computation is considerably restricted. If elimination is carried out in sequence it will be found that when the pivotal element is small the equations obtained after elimination will be dominated by the eliminating equation and considerable inaccuracies can develop[2, 4, 6].

Since numbers in the computer are carried to a limited fixed precision the calculations of elimination and back-substitution can cause the growth of errors due to rounding off of numbers in the intermediate stages. If the elements of A and b in the original equations are correct to a precision of k digits, it has been estimated[2] that it is necessary to retain $k + 1$ digits for 10 equations, $k + 2$ for 100 equations and $k + \log_{10}n$ digits for n equations if the accuracy of the solution is to be comparable with the precision of the data. This is true for *well-conditioned* equations.

Ill-conditioned equations are sometimes found in which the solution is abnormally sensitive to small changes in the coefficients of A, and it may be difficult or impossible to obtain an accurate solution. If this happens it is advisable to ensure that the physical problem defining the equations is well-conditioned, because an error or a fault in the specification of the problem can lead to ill-conditioning in the derived equations.

Gaussian elimination is known as a direct method of solution because all of the operations are completely specified. The method of *triangular decomposition* is a second direct method of some popularity that is less susceptible to round-off error. The matrix of coefficients is expressed as the product of an upper and a lower triangular matrix. When decomposed the triangular matrices can be used with any right-hand side, and this is sometimes more convenient than the calculation of an inverse. The method is described elsewhere[2,4].

Iterative methods: *The Gauss–Seidel method*. In addition to direct methods of solution, there are several indirect methods that usually take the form of a cyclic iteration, in which estimates of the solution are systematically modified until the complete solution is obtained. An important example is the Gauss–Seidel method. Consider the equations 4.24 rearranged in the following way, in which the superscripts indicate the number of estimates of the variable:

$$x_1^{(i+1)} = \frac{1}{a_{11}}(b_1 - a_{12}x_2^{(i)} - a_{13}x_3^{(i)} \quad \ldots - a_{1n}x_n^{(i)})$$

$$x_2^{(i+1)} = \frac{1}{a_{22}}(b_2 - a_{21}x_1^{(i+1)} - a_{23}x_3^{(i)} \ldots - a_{2n}x_n^{(i)})$$

$$\ldots$$

$$x_n^{(i+1)} = \frac{1}{a_{nn}}(b_n - a_{n1}x_1^{(i+1)} - a_{n2}x_2^{(i+1)} \ldots - a_{n,n-1}x_{n-1}^{(i+1)})$$

$$\qquad\qquad\qquad\qquad\qquad\qquad\qquad\qquad\qquad\qquad\qquad\qquad\ldots (4.57)$$

The method is started by a first estimate $x_1^{(1)}, x_2^{(1)} \ldots x_n^{(1)}$. The values are inserted in the first equation of 4.57 and the group of equations is solved in cycle to give a further set of estimates $x_1^{(2)}, x_2^{(2)} \ldots x_n^{(2)}$. As the cycle proceeds the new estimates are placed on the right-hand side of 4.57 as soon as they have been computed, and the cycle is repeated until the estimates converge, or until it is evident that convergence cannot be obtained.

The Gauss–Seidel method has the merit of being self-correcting because errors in estimation are removed as the computation progresses. Convergence of the method cannot be guaranteed and, in fact, failure to converge is quite common. However, when the coefficients on the principal diagonal of the matrix of coefficients are large compared with others in the matrix, the method is often found to converge in a satisfactory manner. This characteristic is important, because the linear equations obtained by some numerical representations of partial differential equations often possess dominating diagonals with many null elements elsewhere. Direct methods of solution require storage for elements whether these are zero or not, but the Gauss–Seidel method requires storage for non-zero elements only and therefore the solution of large systems of equations may be attempted by this method.

Acceleration of convergence. If the calculation converges, it is sometimes possible to accelerate convergence. One technique that is often useful is variously known as successive over-relaxation, extrapolated Liebmann and extrapolated Gauss–Seidel. The method uses an over-relaxation factor ω and its use may be illustrated by introducing ω into the equations 4.57:

$$x_1^{(i+1)} = x_1^{(i)} + \frac{\omega}{a_{11}}(b_1 - a_{11}x_1^{(i)} - a_{12}x_2^{(i)} - a_{13}x_3^{(i)} \ldots - a_{1n}x_n^{(i)})$$

$$x_2^{(i+1)} = x_2^{(i)} + \frac{\omega}{a_{22}}(b_2 - a_{21}x_1^{(i+1)} - a_{22}x_2^{(i)} - a_{23}x_3^{(i)} \ldots - a_{2n}x_n^{(i)})$$

$$\ldots$$

$$x_n^{(i+1)} = x_n^{(i)} + \frac{\omega}{a_{nn}}(b_n - a_{n1}x_1^{(i+1)} - a_{n2}x_2^{(i+1)} - a_{n3}x_3^{(i+1)} \ldots - a_{nn}x_n^{(i)})$$

$$\qquad\qquad\qquad\qquad\qquad\qquad\qquad\qquad\qquad\qquad\qquad\qquad\ldots (4.58)$$

The best value of ω to be used should be found by trial. For some matrices that arise in numerical representations of partial differential equations, the best value of the over-relaxation factor has been found to lie between 1 and $2^{(2)}$. The use of the over-relaxation factor should be seriously considered for large systems when convergence is slow, but it is not necessary for smaller problems when trials to find the best value of the factor may not be a sensible way of using computation time.

The Gauss–Seidel method gives good convergence when the elements on the principal diagonal are large. For example, the method may be used to solve equations that constitute a tridiagonal matrix of coefficients, thus:

$$\left.\begin{array}{rcl} 4x_1 + x_2 & = & 5 \\ x_1 + 4x_2 + x_3 & = & 10 \\ x_2 + 4x_3 + x_4 & = & 6 \\ x_3 + 4x_4 & = & 1 \end{array}\right\} \quad \dots\dots(4.59)$$

Starting from an initial guess, the method gave a solution that was substantially correct after six cycles of iterations:

	Initial	1st cycle	2nd cycle	3rd cycle	4th cycle	5th cycle	6th cycle
x_1	1·0	1·00	0·75	0·71875	0·72949	0·73163	0·731987
x_2	1·0	2·00	2·1250	2·08203	2·07349	2·07205	2·071816
x_3	1·0	0·75	0·9531	0·97656	0·98016	0·98075	0·980842
x_4	1·0	0·0625	0·01171	0·005859	0·004959	0·004813	0·0047894

The solution found by elimination is:

$$x_1 = 0·7320574$$
$$x_2 = 2·0717703$$
$$x_3 = 0·9808612$$
$$x_4 = 0·00478469$$

This example shows the method in its most favourable light. An example that carries a warning is given by the set of equations 4.60:

$$\left.\begin{array}{rcl} x_1 + 1·25x_2 + 0·15x_3 + 0·38x_4 & = & 5 \\ 1·39x_1 + 0·16x_2 + 0·90x_3 + 0·76x_4 & = & 9 \\ 0·12x_1 + 3·69x_2 + 0·50x_3 + 1·98x_4 & = & 14 \\ 3·1x_1 + 2·9x_2 + 1·0x_3 + 1·0x_4 & = & 22 \end{array}\right\} \quad \dots\dots(4.60)$$

The iterations starting from the initial guess $x_1 = x_2 = x_3 = x_4 = 1·0$ are shown below.

	Initial	1st cycle	2nd cycle	3rd cycle
x_1	1·0	3·22	−46·396	−1,914
x_2	1·0	28·28	953·97	35,554
x_3	1·0	−185·4	−7,458	−281,140
x_4	1·0	115·4	4,857	183,989

The attempt is seen to diverge rapidly because the matrix of coefficients of equations 4.60 does not contain large elements on the principal diagonal that dominate other elements. Indeed some of the elements are rather small. However, in some applications of practical importance equations having dominating diagonals and many zero elements occur, and the use of the Gauss–Seidel method here allows the solution of large systems of equations to be found where direct methods are not possible because of the amount of computer storage and time required by them.

Finite differences

In many applications of numerical analysis equally spaced values of a function are to be interpolated, integrated, differentiated or extrapolated. The properties of functions presented in this way may conveniently be studied by means of a difference table drawn up in one of three forms:

y_{-2}

$\qquad\Delta_{-2}$

$y_{-1}\qquad\Delta_{-2}^2$

$\qquad\Delta_{-1}\qquad\Delta_{-2}^3$

$y_0\qquad\Delta_{-1}^2\qquad\Delta_{-2}^4$

$\qquad\Delta_0\qquad\Delta_{-1}^3$

$y_1\qquad\Delta_0^2$

$\qquad\Delta_1$

y_2

Table of forward
differences

y_{-2}

$\qquad\nabla_{-1}$

$y_{-1}\qquad\nabla_0^2$

$\qquad\nabla_0\qquad\nabla_1^3$

$y_0\qquad\nabla_1^2\qquad\nabla_2^4$

$\qquad\nabla_1\qquad\nabla_2^3$

$y_1\qquad\nabla_2^2$

$\qquad\nabla_2$

y_2

Table of backward
differences

y_{-2}

$\qquad\delta_{-\frac{3}{2}}$

$y_{-1}\qquad\delta_{-1}^2$

$\qquad\delta_{-\frac{1}{2}}\qquad\delta_{-\frac{1}{2}}^3$

$y_0\qquad\delta_0^2\qquad\delta_0^4$

$\qquad\delta_{\frac{1}{2}}\qquad\delta_{\frac{1}{2}}^3$

$y_1\qquad\delta_1^2$

$\qquad\delta_{\frac{3}{2}}$

y_2

Table of central
differences

A first order difference is the difference between neighbouring values of the function in which the upper value is subtracted from the lower, a second order difference is the difference between neighbouring values of the first order differences of the function, a third order difference is the difference between neighbouring values of second order differences and so on. From the diagram above it can be seen that forward, backward and central differences are related,

$$\Delta y_n = \nabla y_{n+1} = \delta y_{n+\frac{1}{2}} = y_{n+1} - y_n \qquad \ldots (4.61)$$

For well-behaved functions the higher order differences approach zero. The $(n + 1)$th differences of a polynomial of order n are zero and the nth order differences are constant. Minor extensions of a function may be made by forming a difference table and determining the differences that are zero. By assuming that the differences remain zero the function can be extended, and this procedure is equivalent to fitting a polynomial to the function over an interval. The use of the finite difference procedures discussed here effectively assumes that the functions can be approximated by polynomials over an interval.

Mistakes in tabulation are quickly shown up by a sharp divergence in high order differences, and this is one reason why difference tables are useful when computation is carried out by hand. The increasing effect of an error with high order differences is shown by the tabulation below, in which it is assumed that an error ε has been made in the tabulated value of y_0.

$$
\begin{array}{llll}
y_{-2} \\
& \Delta_{-2} \\
y_{-1} & & \Delta^2_{-2} + \varepsilon \\
& \Delta_{-1} + \varepsilon & & \Delta^3_{-2} - 3\varepsilon \\
y_0 + \varepsilon & & \Delta^2_{-1} - 2\varepsilon & & \Delta^4_{-2} + 6\varepsilon \\
& \Delta_0 - \varepsilon & & \Delta^3_{-1} + 3\varepsilon \\
y_1 & & \Delta^2_0 + \varepsilon \\
& \Delta_1 \\
y_2
\end{array}
$$

Inspection of differences is obviously useful in locating errors of tabulation, but it is also clear that errors due to normal round-off are magnified in high order differences.

The formation of a difference table

An example of a difference table is given in Table 4.3, in which the differences

TABLE 4.3. *The error function (at interval* 0·1)

x	erf(x)	Δ^1	Δ^2	Δ^3	Δ^4	Δ^5	Δ^6	Δ^7
0	0·00000							
		11246						
0·1	0·11246		−222					
		11024		−209				
0·2	0·22270		−431		23			
		10593		−186		15		
0·3	0·32863		−617		38		−15	
		9976		−148		0		20
0·4	0·42839		−765		38		5	
		9211		−110		5		−8
0·5	0·52050		−875		43		−3	
		8336		−67		2		−7
0·6	0·60386		−942		45		−10	
		7394		−22		−8		13
0·7	0·67780		−964		37		3	
		6430		15		−5		−3
0·8	0·74210		−949		32		0	
		5481		47		−5		−9
0·9	0·79691		−902		27		−9	
		4579		74		−14		24
1·0	0·84270		−828		13		15	
		3751		87		1		−33
1·1	0·88021		−741		14		−18	
		3010		101		−17		36
1·2	0·91031		−640		−3		18	
		2370		98		1		
1·3	0·93401		−542		−2			
		1828		96				
1·4	0·95229		−446					
		1382						
1·5	0·96611							

of tabulated values of the error function, erf(x), have been calculated:

$$\text{erf}(x) = \frac{2}{\sqrt{\pi}} \int_0^x e^{-u^2} \, du \qquad \qquad \dots (4.62)$$

In accordance with the usual practice the significant figures of the differences are shown as integers. The 5th, 6th and 7th differences are irregular and range about zero. The greater effect of round-off errors upon high order differences is shown by the increasing amplitude of the fluctuations in the 6th and 7th differences. The 4th differences are constant over small ranges of the tabulation, but there is a significant decline over the whole of the tabulated range, and therefore a fourth order polynomial would give a satisfactory approximation to the function over a small range of x, but not over a wide range.

Although difference tables are necessary for hand computation, their formation is not a sound application of an automatic computer, because errors are much less frequent, and because the time spent in examining differences could more profitably be devoted to further computing. It is better either to compute with high order differences without inspection, or to operate with a smaller interval so that only the lower order differences need be considered.

Further finite difference operators

In addition to the finite difference operators Δ, ∇ and δ, two operators of importance are the displacement operator E, and the averaging operator μ. The displacement operator E is defined by the equation:

$$E y_n = y_{n+1} \qquad \qquad \dots (4.63)$$

The difference operators are related to E by the equations:

$$\left. \begin{aligned} \Delta &= E - 1 \\ \nabla &= 1 - E^{-1} \\ \delta &= E^{\frac{1}{2}} - E^{-\frac{1}{2}} \end{aligned} \right\} \qquad \dots (4.64)$$

where the inverse of the operator E has the significance:

$$E^{-1} y_{n+1} = y_n \qquad \qquad \dots (4.65)$$

The averaging operator μ is defined by the equation:

$$\mu y_n = \tfrac{1}{2}(y_{n+\frac{1}{2}} + y_{n-\frac{1}{2}}) \qquad \qquad \dots (4.66)$$

The differential operator D may be related to the displacement operator E by Taylor's theorem. A suitable form of the equation is:

$$y(x + h) = y(x) + h y'(x) + \frac{h^2}{2!} y''(x) \dots \qquad \dots (4.67)$$

where $y(x + h)$ is the value of the function at $x + h$ and the primes indicate the order of the derivative. As the definition of E gives:

$$E y(x) = y(x + h) \qquad \qquad \dots (4.68)$$

and the differential operator D is defined:

$$Dy = \frac{dy}{dx} \qquad \ldots (4.69)$$

when equation 4.67 is expressed in terms of D and E:

$$Ey(x) = \left(1 + hD + \frac{h^2}{2!}D^2 \ldots\right)y(x) \qquad \ldots (4.70)$$

The bracketed terms on the right-hand side form the exponential series of hD so that equation 4.70 establishes the relation between the operators E and D:

$$E = \exp(hD) \qquad \ldots (4.71)$$

Of course, equation 4.71 is simply a restatement of equation 4.67, but the development illustrates the remarkable analogy between the algebra of arithmetic variables and the algebra of finite difference operators.

Interpolation

Interpolation is a common operation when applied to tabulations of functions, but otherwise it is not often used. The position of the point of interpolation in the table determines the most satisfactory interpolation formula. By preference, central difference formulae are used because of the improved rapidity of convergence, but their use is not convenient near the ends of the tabulated interval, where the appropriate forward or backward difference formulae are required.

Newton's formula for interpolation using forward differences may be obtained for the point α located between the points 0 and 1 by the use of equations 4.63 and 4.64:

$$y_\alpha = E^\alpha y_0 = (1 + \Delta)^\alpha y_0 \qquad \ldots (4.72)$$

On expanding the right-hand side by means of the binomial theorem, equation 4.72 is expressed in terms of forward differences:

$$y_\alpha = y_0 + \alpha\Delta y_0 + \frac{\alpha(\alpha - 1)}{2!}\Delta^2 y_0 \ldots \qquad \ldots (4.73)$$

This formula is suitable for interpolation near the beginning of an interval.

A corresponding formula suitable for interpolation near the end of an interval is obtained by writing equation 4.67 in terms of backward differences, by means of equation 4.64:

$$y_\alpha = E^\alpha y_0 = (1 - \nabla)^{-\alpha} y_0 \qquad \ldots (4.74)$$

which on expansion by the binomial theorem gives:

$$y_\alpha = y_0 + \alpha\nabla y_0 + \frac{\alpha(\alpha + 1)}{2!}\nabla^2 y_0 \ldots \qquad \ldots (4.75)$$

The advantage of central difference formulae is that the interpolated point is estimated in terms of function values that are evenly distributed about the point of interpolation, and therefore there is an improved rapidity of convergence. Everett's formula is an important example of a central difference formula that uses only even order differences for interpolation. The formula is:

$$y_\alpha = \beta y_0 + \frac{(\beta + 1)\,\beta(\beta - 1)}{3!}\,\delta^2 y_0 + \frac{(\beta + 2)\,(\beta + 1)\,\beta(\beta - 1)\,(\beta - 2)}{5!}\,\delta^4 y_0 \ldots$$

$$+ \alpha y_1 + \frac{(\alpha + 1)\,\alpha(\alpha - 1)}{3!}\,\delta^2 y_1 + \frac{(\alpha + 2)\,(\alpha + 1)\,\alpha(\alpha - 1)\,(\alpha - 2)}{5!}\,\delta^4 y_1 \ldots \text{(4.76)}$$

where β is $(1 - \alpha)$. The coefficients of $\delta^2 y_0$, $\delta^4 y_0$, $\delta^2 y_1$, $\delta^4 y_1$ are known as Everett coefficients.

The accuracy of the Newton and Everett formulae may be compared by forming a difference table (Table 4.4) for the error function at intervals of 0·2, instead of the interval of 0·1 shown in Table 4.3.

TABLE 4.4. *The error function (at interval 0·2)*

x	erf (x)	Δ^1	Δ^2	Δ^3	Δ^4
0	0·00000				
		22270			
0·2	0·22270		−1701		
		20569		−1321	
0·4	0·42839		−3022		620
		17547		−701	
0·6	0·60386		−3723		660
		13824		−41	
0·8	0·74210		−3764		506
		10060		465	
1·0	0·84270		−3299		
		6761			
1·2	0·91031				

The interpolated values of $x = 0·5$, computed by Everett's formula and Newton's forward difference equation, are shown in Table 4.5.

TABLE 4.5. *Comparison of formulae for interpolation*

Order of differences	0	1	2	3	4
Everett's formula	0·516125	—	0·52034	—	0·52049
Newton's forward difference equation	0·42839	0·516125	0·52077	0·52075	0·52056

Tabulated value = 0·52050

The superior convergence of the central difference formula is clearly shown. The interval of tabulation is too wide to obtain the accuracy of the tabulated figures by interpolation with second order differences, but the Everett formula gives five-figure accuracy with fourth differences, while Newton's formula, although not inaccurate, is obviously inferior.

Numerical differentiation

Differentiation is one of the least accurate of numerical procedures, because the expressions for a first derivative in either backward or forward differences converge very slowly. If this is allowed for by taking a small interval of tabulation there is a tendency for a loss of significant figures, because functional values of similar magnitude are then used to form differences. There are related difficulties in the application of central difference formulae to find first derivatives.

An expression for the value of a derivative at a tabulated point is given by combination of equations 4.64 and 4.71:

$$hDy_0 = [\ln(1 + \Delta)] y_0 \qquad \qquad \dots (4.77)$$

When the logarithm is expanded in series:

$$Dy_0 = \frac{1}{h}(\Delta - \tfrac{1}{2}\Delta^2 + \tfrac{1}{3}\Delta^3 \dots) y_0 \qquad \dots (4.78)$$

This formula converges very slowly unless h is small, when however there is a serious loss of significant figures in forming the differences. The corresponding formula in backward differences is:

$$hDy_0 = -[\ln (1 - \nabla)] y_0 \qquad \qquad \dots (4.79)$$

a formula that gives on expansion in series:

$$Dy_0 = \frac{1}{h}(\nabla + \tfrac{1}{2}\nabla^2 + \tfrac{1}{3}\nabla^3 \dots) y_0 \qquad \dots (4.80)$$

Equations 4.78 and 4.80 must be used when the value of the derivative at the end of a tabulated interval is wanted, for example as a boundary condition to a differential equation. As with interpolation, a better convergence is found when central difference formulae are used, but these formulae are restricted to the central regions of the tabulated interval.

A central difference formula for the first derivative is given by equations 4.64 and 4.71. Since equation 4.71 shows that:

$$E^{\frac{1}{2}} = \exp\left(\frac{hD}{2}\right) \qquad \qquad \dots (4.81)$$

the last equation of 4.64 becomes:

$$\delta = \exp\left(\frac{hD}{2}\right) - \exp\left(-\frac{hD}{2}\right) \qquad \dots (4.82)$$

so that:

$$\delta = 2\sinh\left(\frac{hD}{2}\right) \qquad \qquad \dots (4.83)$$

and

$$hD = 2\sinh^{-1}\left(\frac{\delta}{2}\right) \qquad \qquad \dots (4.84)$$

On expansion of the right-hand side:

$$hD = \left(\delta - \frac{1}{24}\delta^3 + \frac{3}{640}\delta^5 \ldots\right) \qquad \ldots (4.85)$$

The differences $\delta y_0, \delta^3 y_0, \ldots$ do not appear in the difference table, and therefore when equation 4.85 is applied to the tabulated differences the result is:

$$Dy_{\frac{1}{2}} = \frac{1}{h}\left(\delta - \frac{1}{24}\delta^3 + \frac{3}{640}\delta^5 \ldots\right)y_{\frac{1}{2}} \qquad \ldots (4.86)$$

and this equation gives the value of the derivative at the middle of a tabulated interval. The first approximation to the derivative is δ/h, a good approximation if h is small, when however there may be a difficulty in the loss of significant figures by forming δ. On the other hand too large a value of h gives poor convergence for equation 4.86. The best choice of h lies between the extremes.

A central difference formula that gives first derivatives at a tabular point may be found by applying the properties of the averaging operator μ of equation 4.66. On introducing the displacement operator E into equation 4.66, the relationship between μ and E is found to be:

$$\mu = (E^{\frac{1}{2}} + E^{-\frac{1}{2}})/2 \qquad \ldots (4.87)$$

On squaring this equation, and also the last equation of 4.64, μ is found to be related to δ by:

$$\mu^2 = 1 + \tfrac{1}{4}\delta^2 \qquad \ldots (4.88)$$

The averaging operator is now introduced to give an expression for the derivative in the following way:

$$Dy_0 = \frac{D}{\sqrt{\left(1 + \dfrac{\delta^2}{4}\right)}}\mu y_0 = \frac{2}{h}\frac{\sinh^{-1}\left(\dfrac{\delta}{2}\right)}{\sqrt{\left(1 + \left(\dfrac{\delta}{2}\right)^2\right)}}\mu y_0 \qquad \ldots (4.89)$$

Finally on expanding the right-hand side in series, the derivative is expressed in the form of averaged odd order central differences:

$$Dy_0 = \frac{1}{h}\left(\mu\delta - \frac{1}{6}\mu\delta^3 + \frac{1}{30}\mu\delta^5 \ldots\right)y_0 \qquad \ldots (4.90)$$

The alternative formulae for the first derivative have been compared by calculating the derivative of the error function erf(x), for $x = 0.5$, by means of the differences given in Table 4.3. The three equations 4.78, 4.80 and 4.90 have been

TABLE 4.6. *Comparison of finite difference formulae for first derivatives*

Order of differences	1	2	3	4
Forward difference, equation 4.78	0·8336	0·8807	0·8800	0·8791
Backward difference, equation 4.80	0·9211	0·8828	0·8779	0·8789
Mean central difference, equation 4.90	0·8774	—	0·8788	—

Tabulated value = 0·87878

compared at each order of difference, and the comparison is shown in Table 4.6. The interval of tabulation in Table 4.3 is 0·1, an interval that is sufficiently small to cause the loss of one significant figure in all three formulae. In fact the loss of one significant figure at least cannot usually be avoided. Of the three formulae the central difference formula is clearly the best, since the prediction agrees with the tabulated value to four figures. The backward and forward difference formulae are less reliable, and the reliability cannot be improved by taking higher order differences into account, because wider fluctuations are introduced.

In making finite difference approximations to differential equations, a finite difference equation for the second derivative is often required. A useful central difference formula may be derived by squaring equation 4.85:

$$D^2 y_0 = \frac{1}{h^2}\left(\delta^2 - \frac{1}{12}\delta^4 + \frac{1}{90}\delta^6 \ldots\right)y_0 \qquad \ldots (4.91)$$

The second derivatives are computed at tabular intervals by means of this formula.

Numerical integration—Simpson's and the trapezoidal rule

There are many formulae, but most of the requirements for numerical integration procedures by means of an automatic digital computer can be met by only a few of them. One of the most useful of the simple formulae is Simpson's rule. The rule may be derived in a number of ways including one based upon the algebraic properties of finite difference operators.

Consider the integral over two intervals of a table, and its relationship to the integrating operator D^{-1}:

$$I = \int_{x_{-1}}^{x_1} y(x)\,dx = D^{-1}y_1 - D^{-1}y_{-1} = D^{-1}(y_1 - y_{-1})$$

$$= D^{-1}(E^1 - E^{-1})y_0 \qquad \ldots (4.92)$$

On substituting the relationship between E and D given by equation 4.71 into equation 4.92:

$$I = 2\sinh(hD)\,D^{-1}y_0$$

The series expansion for $\sinh(hD)$ can now be introduced:

$$I = \frac{2h}{hD}\left(hD + \frac{(hD)^3}{3!} + \frac{(hD)^5}{5!}\ldots\right)y_0$$

and the operator D itself may be expressed in terms of central differences as in equation 4.85 to give:

$$I = 2h\left[1 + \frac{\delta^2}{3!}\left(1 - \frac{1}{24}\delta^2 + \frac{3}{640}\delta^4\ldots\right)^2 + \frac{\delta^4}{5!}\left(1 - \frac{1}{24}\delta^2 + \frac{3}{640}\delta^4\ldots\right)^4\ldots\right]y_0$$

$$= 2h\left[1 + \frac{\delta^2}{6} - \frac{\delta^4}{180} + \frac{\delta^6}{1512}\ldots\right]y_0 \qquad \ldots (4.93)$$

The estimate of the integral may be regarded as a first, crude estimate of $2hy_0$, followed by difference corrections. The expression truncated after the second difference does not include points tabulated outside the interval of integration. When written down in terms of tabular values the equation truncated after second differences is:

$$\int_{x_{-1}}^{x_1} y(x)\, dx = \frac{h}{3}(y_{-1} + 4y_0 + y_1) \qquad \ldots (4.94)$$

This is Simpson's rule for integration, a very important and useful formula.

Simpson's rule and a number of others may be found more directly by integrating Newton's formula for interpolation, equation 4.73. The definite integral is taken over the range x_0 to $x_0 + nh$:

$$\int_{x_0}^{x_0+nh} y(x)\, dx = h \int_0^n \left\{ y_0 + \alpha\, \Delta y_0 + \frac{\alpha(\alpha-1)}{2!} \Delta^2 y_0 \right.$$
$$\left. + \frac{\alpha(\alpha-1)(\alpha-2)}{3!} \Delta^3 y_0 \ldots \right\} d\alpha$$
$$= h\left[ny_0 + \frac{n^2}{2}\Delta y_0 + \left(\frac{n^3}{3} - \frac{n^2}{2}\right)\frac{\Delta^2}{2!} y_0 + \left(\frac{n^4}{4} - n^3 + n^2\right)\frac{\Delta^3}{3!} y_0 \ldots \right] \qquad \ldots (4.95)$$

When n is set equal to 1 and differences above the first are neglected, another rule of integration is derived:

$$\int_{x_0}^{x_0+h} y(x)\, dx = \frac{h}{2}(y_0 + y_1) \qquad \ldots (4.96)$$

This is the trapezoidal rule, in which the graph of the function is replaced by a straight line joining the ordinates. This procedure is generally accurate only when h is very small.

When n is set equal to 2 and all differences above the second are neglected, equation 4.95 is transformed into Simpson's rule:

$$\int_{x_0}^{x_0+2h} y(x)\, dx = \frac{h}{3}(y_0 + 4y_1 + y_2) \qquad \ldots (4.97)$$

By putting $n = 6$ and neglecting differences above the sixth, another formula known as Weddle's rule is obtained.

When integration is to be performed by an automatic digital computer, Simpson's rule is generally favoured. If the range of integration is to be interpolated by a fixed number of points, Simpson's rule is certainly not the most accurate, since this distinction is usually awarded to Gauss's formula for integration. But Simpson's rule is exact for polynomials of degree 3 and smaller because third differences do not appear in equation 4.95; it is easy to program for automatic computation, and the accuracy of the integration can be easily checked by changing h, the width of subdivided intervals. The check of an automatic integration is very important since, by comparing successive estimates of the integral when h is changed, an excellent assessment of the accuracy of the integration is obtained.

Equation 4.97 gives the rule for integration over two intervals only. When the integration is carried out over $2m$ intervals, Simpson's rule becomes:

$$\int_{x_0}^{x_0 + 2mh} y(x)\, dx = \frac{h}{3}(y_0 + 4y_1 + 2y_2 + 4y_3 + 2y_4 + 4y_5 \ldots + y_{2m})$$

$$= \frac{h}{3}[y_0 + y_{2m} + 2(y_2 + y_4 \ldots + y_{2m-2})$$

$$+ 4(y_1 + y_3 + y_5 \ldots y_{2m-1})] \quad \ldots . (4.98)$$

To illustrate the relative accuracy of Simpson's and the trapezoidal rules some calculations of the definite integral:

$$I = \int_0^{1 \cdot 2} \operatorname{erf}(x)\, dx$$

have been made with both rules at different values of h. The results are shown in Table 4.7:

TABLE 4.7. *A comparison of numerical integration by Simpson's and the trapezoidal rule.*
Values of $\int_0^{1 \cdot 2} \operatorname{erf}(x) dx$ *from five figure tables of the error function*

h	0·2	0·1	0·05
Trapezoidal rule	0·65898	0·66141	0·66168
Simpson's rule	0·661889	0·6618617	0·6618595

Values of the integral may be obtained by integration by parts:

$$\int_0^X \operatorname{erf}(x)\, dx = X \operatorname{erf}(X) + \frac{1}{\sqrt{\pi}}(e^{-X^2} - 1)$$

and ten-figure error function and exponential tables give the value

$$\int_0^{1 \cdot 2} \operatorname{erf}(x)\, dx = 0 \cdot 661859364$$

Table 4.7 shows clearly that Simpson's rule is the better since the estimate of the integral at $h = 0 \cdot 2$ agrees more closely with the true value than the estimate obtained by the trapezoidal rule for $h = 0 \cdot 05$. The values of the integral are shown to six or seven figures even though the table used for the calculation gave five figures only, and it is clear that the numerical values of the integral for $h = 0 \cdot 05$ by Simpson's rule are *more* accurate than the basic tabulation. This is due to the averaging out of rounding errors in the summation, and is a very important feature of numerical integration.

Double integration

In addition to single integrals, double and higher integrals are sometimes required. Simpson's rule may be applied to the estimation of a double integral over the 9 ordinates shown in Fig. 4.12:

$$I = \int\limits_{x_0}^{x_0+2h} \int\limits_{z_0}^{z_0+2k} y \, dz \, dx \qquad \qquad \ldots \ (4.99)$$

First apply Simpson's rule to the inner integral:

$$I = \int\limits_{0}^{x_0+2h} \frac{k}{3}(y_{0\alpha} + 4y_{1\alpha} + y_{2\alpha}) \, dx$$

an integral that may be considered as three distinct parts each of which can be determined by further applications of Simpson's rule:

$$I = \frac{hk}{9}[y_{00} + 4y_{01} + y_{02} + 4y_{10} + 16y_{11} + 4y_{12} + y_{20} + 4y_{21} + y_{22}]$$

$$\ldots \ (4.100)$$

Equation 4.100 gives the value of the double integral 4.99 taken over the rectangular element of Fig. 4.12. The complete integral over a given area is found by

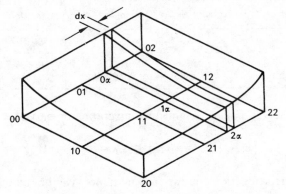

FIG. 4.12. Double integration of a function over an area

dividing the area into rectangular elements, integrating over each rectangular element, and finally summing the integrals of every element. In many cases the area of integration is irregular, and it is not worthwhile developing formulae here based upon equation 4.100 for rectangular, square or other regular areas of integration.

Ordinary differential equations

The dimensionless form of ordinary differential equations

When a differential equation is given as the description of a physical situation the dependent and independent variables have dimensions. For example, a temperature distribution may be expressed in degrees Centigrade, a diffusion coefficient in cm^2/sec, a length in cm, and so on. The carrying of dimensions into a differential equation and its solution will introduce a number of extraneous numerical factors and so complicate the calculations. The differential equation and its solution are made more concise by transforming all of the variables into dimensionless form. Other important advantages may also be introduced particularly when the equations are linear since the solution in dimensionless form will be independent of the numerical values of the boundary conditions when these can be incorporated into the dimensionless variables.

As an example of a differential equation and its dimensionless form, consider the diffusion of a substance into a flat film of liquid where it decomposes at a rate that is proportional to the concentration C. The differential equation is:

$$D\frac{d^2C}{dy^2} - KC = 0 \qquad \ldots(4.101)$$

where D is a diffusion coefficient and K is known as a reaction velocity constant. The boundary conditions for the film of thickness L are:

$$C = C_0 \text{ at } y = L$$

$$\frac{dC}{dy} = 0 \text{ at } y = 0$$

The variables y and C are now related to the following dimensionless variables u and x:

$$u = \frac{C}{C_0}, \quad x = \frac{y}{L}$$

so that the equation now becomes:

$$\frac{D}{KL^2}\frac{d^2u}{dx^2} - u = 0 \qquad \ldots(4.102)$$

with the boundary conditions:

$$u = 1 \text{ at } x = 1$$

$$\frac{du}{dx} = 0 \text{ at } x = 0$$

The differential equation and the boundary conditions are now entirely defined in terms of the dimensionless variables u, x and KL^2/D. The convenience of computation in terms of dimensionless variables, and the greater generality of the solutions, are the main reasons for the writing of differential equations in dimensionless form.

The solution of ordinary differential equations

The methods of Euler. Taylor's theorem, equation 4.67, is of fundamental importance in establishing and examining the accuracy of various numerical methods for the solution of ordinary and partial differential equations. One of

the earliest methods of solution was suggested by Euler. The method is inaccurate, although simple, but it provides a good illustration of the use of
Taylor's theorem in analysis.

Consider the ordinary, first order differential equation:

$$\frac{dy}{dx} = y' = f(x, y) \qquad \qquad \dots (4.103)$$

By Taylor's theorem the value of y in the neighbourhood of x, $y(x + h)$ is
given by:

$$y(x + h) = y(x) + hy'(x) + O(h^2)$$

where $O(h^2)$ is an error due to truncation of the Taylor series, and $y'(x)$ is the
derivative at x. If the starting value of y at $x = x_0$ is y_0, the value y at $x_0 + h$ is:

$$y_1 = y_0 + hf(x_0, y_0)$$

where the truncation error $O(h^2)$ has been neglected.

The value of y is advanced by means of the general formula:

$$y_{i+1} = y_i + hf(x_i, y_i) \qquad \qquad \dots (4.104)$$

It is usual to refer the accuracy of integration formulae to the neglected terms
of the Taylor series. The error in each step of Euler's formula is of order h^2,
and it is therefore known as a first order method.

A modification to Euler's method introduces an iterative correction to
equation 4.103. Once 4.103 has been applied, a better estimate of y_{i+1} is given
by:

$$y_{i+1}^{(2)} = y_i + \frac{h}{2}\left[f(x_i, y_i) + f(x_{i+1}, y_{i+1}^{(1)})\right] \qquad \qquad \dots (4.105)$$

where $y_{i+1}^{(1)}$ is the first estimate of y_{i+1} calculated from equation 4.104. The
second term on the right-hand side of 4.105 has replaced the value of the derivative at (x_i, y_i) by an estimate of the arithmetic mean of the derivative at either
end of the interval. The estimate of the mean derivative may be improved by
replacing $f(x_{i+1}, y_{i+1}^{(1)})$ by $f(x_{i+1}, y_{i+1}^{(2)})$ and applying 4.105 again to find $y_{i+1}^{(3)}$.
The procedure can be iterated until successive values of $y_{i+1}^{(k)}$ agree to the accuracy
desired. Equations 4.104 and 4.105 are then applied again to find y_{i+2}, and
the computation is progressed until the range of integration has been completed.

The operation of the modified Euler method is typical of *predictor–corrector*
methods of solving ordinary differential equations, in which one formula
of limited accuracy known as the *predictor* is used to estimate the succeeding
value of y, and this estimate is refined by the iteration of a second formula of
higher accuracy or improved convergence known as the *corrector*. Here
equation 4.104 is the predictor and 4.105 is the corrector.

The accuracy of the modified Euler formula may be found from a study of the Taylor series expansion for both y and y':

$$y_{i+1} = y_i + hy_i' + \frac{h^2}{2!} y_i'' + \frac{h^3}{3!} y_i''' \dots$$

$$y_{i+1}' = y_i' + hy_i'' + \frac{h^2}{2!} y_i''' + \frac{h^3}{3!} y_i'''' \dots$$

The last formula is multiplied by $h/2$, and on subtracting from the first, the expression for y_{i+1} becomes:

$$y_{i+1} = y_i + \frac{h}{2}(y_i' + y_{i+1}') + O(h^3)$$

The error in the modified Euler formula is of order h^3 so that the method is one of second order.

Once two consecutive values of y_i have been found the computation may be progressed more rapidly by use of a formula derived from a comparison of the Taylor series for $y(x + h)$ and $y(x - h)$:

$$y(x + h) = y(x) + hy'(x) + \frac{h^2}{2!} y''(x) + \frac{h^3}{3!} y'''(x) \dots$$

$$y(x - h) = y(x) - hy'(x) + \frac{h^2}{2!} y''(x) - \frac{h^3}{3!} y'''(x) \dots$$

On subtraction:

$$y(x + h) - y(x - h) = 2hy'(x) + O(h^3)$$

or, on expressing this equation in the manner of equation 4.105:

$$y_{i+1} = y_{i-1} + 2h\,f(x_i, y_i) + O(h^3) \qquad \dots \text{(4.106)}$$

Equation 4.106 is also a second order formula, but without the iterated corrections of the modified Euler. The computation is started from y_0, and y_1 is found by the modified Euler; $y_2, y_3 \dots$ may then be calculated by use of equation 4.106. When applied in this way, the Euler equations are known as starter formulae.

Of course, the Taylor series itself may be used to solve differential equations by advancing the integration from y_0 to y_1, but it is usually necessary that the form of equation 4.103 be suitably simple, otherwise the expressions for the derivatives become intractable.

The Runge–Kutta method. The relatively large error of second order formulae is a disadvantage in many applications. The fourth-order formula of Runge and Kutta has a step error of the order of h^5; this may be shown[7, 8] by use of Taylor series as for the modified Euler, but the proof is too long to show here. The Runge–Kutta method is easy to program and makes light demands upon the computer store. Backward values of the function are not required, so that when the step length is to be changed, interpolation of backward values is not necessary, and for the same reason the method can be used to start a computation.

The Runge–Kutta equations to advance the integration of equation 4.103 from y_0 to y_1 are:

$$k_0 = hf(x_0, y_0)$$

$$k_1 = hf\left(x_0 + \frac{h}{2}, y_0 + \frac{k_0}{2}\right)$$

$$k_2 = hf\left(x_0 + \frac{h}{2}, y_0 + \frac{k_1}{2}\right) \qquad \left. \right\} \quad \dots (4.107)$$

$$k_3 = hf(x_0 + h, y_0 + k_2)$$

$$y_1 = y_0 + \tfrac{1}{6}(k_0 + 2k_1 + 2k_2 + k_3)$$

The integration is completed by successive applications of these equations. It is interesting to note that, when the derivative in equation 4.103 is a function of x only, the Runge–Kutta method reduces to Simpson's rule for integration.

Round-off and truncation errors. The integration of a differential equation by a numerical method incorporates two distinct types of error as the solution is advanced. The first type is round-off error that arises because intermediate calculations are made to a fixed number of digits, and therefore further digits that arise in the computation are lost when rounded off to this fixed number. For example, the multiplication of a single length number by another single length number gives rise to a double length product, while a division can give rise to any number of digits in the result. Calculations should be programmed to give the minimum number of round-offs since the total round-off error increases with the number of round-off operations.

The second type of error is measured by the accuracy with which the function is approximated over an interval of integration, and is known as truncation error. For a given nth order formula the truncation error is of order h^{n+1}, but it should be clearly understood that this is simply an estimate of the order of magnitude, and it is not inconsistent for two formulae of different order to have similar truncation errors for a particular value of the interval h. However, the truncation error of an nth order formula is reduced by a factor of 2^{n+1} when the interval is halved. Thus the truncation error of a first order formula is reduced by a factor of 4, but the error of a fourth order formula is reduced by a factor of 32 when the interval is halved.

The total error of a numerical solution is the sum of the errors due to truncation, and to round-off. A reduction of the interval reduces the truncation error, but as the number of intervals and therefore the number of round-off operations in the range of integration is thereby increased, the round-off error is increased. As the interval is reduced from a value at which truncation error dominates, the total error may first be found to decrease to a minimum value, and then to increase as round-off error begins to dominate. If the minimum error is not acceptable, it is necessary either to use a more accurate formula, or to increase the precision of the computation, say by using double precision arithmetic.

Instabilities in numerical solutions. A number of types of instabilities may arise in the numerical solution of differential equations. For example, suppose that the general solution to a second order differential equation contains an increasing and a decreasing part and it is desired that the solution should contain only the decreasing part. The increasing part of the solution is set equal to zero, but as the integration proceeds through a number of intervals, rounding errors grow and introduce a portion of the unwanted solution that may even dominate the small, wanted solution. Thus the equation $y'' - y = 0$ has a general solution composed of two parts e^x and e^{-x}; if e^{-x} is desired, the part containing e^x is first set to zero, but rounding errors will introduce an element of e^x after integrating in the direction of increasing x. The remedy is either to integrate in a stable direction, or to work to a sufficiently high precision so that the unwanted solution does not grow to a significant level. This type of instability is a property of the differential equation.

A second type of instability may arise from the replacement of the differential equation by a finite difference expression. The instability arises either because some parts of the solution to the differential equation are poorly represented by the finite difference expression, or because solutions to the finite difference equation are introduced that are not compatible with the differential equation. If instabilities of this type are found, the remedy is either to use a more stable method, or to use a value of the interval that is small enough to contain the growth of the instability.

The Runge–Kutta method can show this type of instability, particularly when the differential equation has some solutions that decrease very rapidly compared with the others. As an alternative to Runge–Kutta a fourth-order, predictor-corrector method due to Hamming has good stability characteristics.

Hamming's method. In the common presentation of Hamming's method[11], the integration of equation 4.103 is advanced from y_i to y_{i+1} by the sequential application of four equations, a predictor, a modifier, a corrector and a final corrector:

predictor
$$y^{(1)}_{i+1} = y_{i-3} + \frac{4h}{3}\left[2f(x_i, y_i) - f(x_{i-1}, y_{i-1})\right.$$

$$\left. + 2f(x_{i-2}, y_{i-2})\right]$$

modifier
$$y^{(2)}_{i+1} = y^{(1)}_{i+1} - \tfrac{112}{121}(y^{(1)}_i - y^{(3)}_i)$$

$$\left.\begin{array}{c}\end{array}\right\} \dots (4.108)$$

corrector
$$y^{(3)}_{i+1} = \tfrac{1}{8}\{9y_i - y_{i-2} + 3h[f(x_{i+1}, y^{(2)}_{i+1})$$

$$+ 2f(x_i, y_i) - f(x_{i-1}, y_{i-1})]\}$$

final corrector
$$y_{i+1} = y^{(3)}_{i+1} + \tfrac{9}{121}(y^{(1)}_{i+1} - y^{(3)}_{i+1})$$

In some cases the corrector is iterated until the sequential estimates of y_{i+1}

M

agree to a desired accuracy before the final corrector is applied. Equation 4.108 requires a starter formula, and the fourth order Runge–Kutta or the modified Euler methods should be used. In calculating the first few values it is necessary to work at a precision that is at least as great as that employed in the body of the integration. When the modified Euler formulae are used this requirement means a smaller value of h than that used in the Hamming method.

Estimation of errors

In the process of completing a solution to a differential equation it is desirable to estimate the accuracy of the result in some way. One method that is always available is simply to repeat the procedure using a smaller value of h and compare the results. If they are comparable over the range, the solution is accepted. If not the value of h is changed until agreement is found. This method is effective, but not economical of computer time.

A more economical method is to estimate the local error at intervals throughout the computation, and to change the interval h so that the local error is kept between fixed limits. Hamming's method gives a ready estimate of the local error from the final corrector as $\frac{9}{121}(y_{i+1}^{(1)} - y_{i+1}^{(3)})$. This estimate is available every interval so that h may be kept near the optimum value throughout the calculation.

A ready estimate of local error is not available for the Runge–Kutta method, but a local error can be estimated at the expense of a little extra computation. At points throughout the calculation, say for every ten steps, a portion of the integration is repeated by doubling h and estimating y_i a second time. The local error may be calculated from a comparison of the value calculated with a single interval $y_i^{(1)}$, with the value calculated with a double interval $y_i^{(2)}$. By applying the method of extrapolation due to L. F. Richardson (see BUCKINGHAM[3]) it may be shown that the error in $y_i^{(1)}$ is $(y_i^{(1)} - y_i^{(2)})/15$ to an approximation of $O(h^6)$. The interval h may be changed according to this estimate.

Sets of simultaneous first-order equations. Both the Runge–Kutta method and Hamming's method are easily extended to solve sets of equations such as:

$$\left. \begin{aligned} \frac{dy}{dx} &= f(x, y, t) \\[2mm] \frac{dt}{dx} &= g(x, y, t) \end{aligned} \right\} \quad \ldots \text{(4.109)}$$

where t and y are dependent variables, and x is independent. The scheme of the Runge–Kutta method when used to solve equations 4.109 is changed from

4.107 to 4.110 in obvious ways:

$$k_0 = hf(x_0, y_0, t_0)$$

$$m_0 = hg(x_0, y_0, t_0)$$

$$k_1 = hf\left(x_0 + \frac{h}{2}, y_0 + \frac{k_0}{2}, t_0 + \frac{m_0}{2}\right)$$

$$m_1 = hg\left(x_0 + \frac{h}{2}, y_0 + \frac{k_0}{2}, t_0 + \frac{m_0}{2}\right)$$

$$k_2 = hf\left(x_0 + \frac{h}{2}, y_0 + \frac{k_1}{2}, t_0 + \frac{m_1}{2}\right)$$

$$m_2 = hg\left(x_0 + \frac{h}{2}, y_0 + \frac{k_1}{2}, t_0 + \frac{m_1}{2}\right) \qquad \bigg\} \quad \dots (4.110)$$

$$k_3 = hf(x_0 + h, y_0 + k_2, t_0 + m_2)$$

$$m_3 = hg(x_0 + h, y_0 + k_2, t_0 + m_2)$$

$$y_1 = y_0 + \tfrac{1}{6}(k_0 + 2k_1 + 2k_2 + k_3)$$

$$t_1 = t_0 + \tfrac{1}{6}(m_0 + 2m_1 + 2m_2 + m_3)$$

The stepwise development of the calculation shown in equations 4.110 may also be applied to Hamming's method (equations 4.108).

Initial-value and boundary-value problems. The solution to the equations 4.109 may be defined by values of y and t given at the same point. Integration starts from this point and proceeds until the range is completed. The differential equations are said to be defined by *initial conditions.* Sometimes, however, the value of y is given at one end of the range of integration and the value of t is given at the other end. This is referred to as a *boundary-value* problem defined by *two-point boundary conditions.*

The finite difference methods of Euler, Runge–Kutta and Hamming may still be applied, but a solution by trial and iteration is usually necessary. The integration is started at one end of the range of integration where, say, y is known. A value of t at this point is assumed and both equations are integrated. When completed, values of y and t at the other end have been found, and the calculated value of t is compared with the boundary value. If the values agree, the assumption of the value of t at the start of the range of integration was correct, and the calculation is completed. However, if the values do not agree, the first assumption was wrong; a new estimate of t is made and the calculations are repeated until agreement is found. A good knowledge of the properties of the solution is necessary to guide the computation to convergence.

Solutions by trial and iteration are necessary when the differential equations are non-linear. When the differential equations are linear the finite difference approximations take the form of a system of banded linear simultaneous equations that can be solved by direct methods.

Second-order differential equations

Differential equations of the second order and higher may be replaced by an equivalent system of first-order simultaneous equations that can be solved by the methods of Runge and Kutta or Hamming. For example, the second-order differential equation:

$$y'' = f(x, y, y') \qquad \dots (4.111)$$

may be expressed as the equivalent pair of equations:

$$\left. \begin{array}{l} y' = t \\ t' = f(x, y, t) \end{array} \right\} \qquad \dots (4.112)$$

Hamming's formulae (equations 4.108) may be applied to a pair of simultaneous first-order equations such as equations 4.112. Each stage of the estimation is applied first to one variable and then to the second, as shown in equations 4.113:

predictor

$$y_{i+1}^{(1)} = y_{i-3} + \frac{4h}{3}(2t_i - t_{i-1} + 2t_{i-2})$$

$$t_{i+1}^{(1)} = t_{i-3} + \frac{4h}{3}[2f(x_i, y_i, t_i) - f(x_{i-1}, y_{i-1}, t_{i-1}) + 2f(x_{i-2}, y_{i-2}, t_{i-2})]$$

modifier

$$y_{i+1}^{(2)} = y_{i+1}^{(1)} - \tfrac{112}{121}(y_i^{(1)} - y_i^{(3)})$$

$$t_{i+1}^{(2)} = t_{i+1}^{(1)} - \tfrac{112}{121}(t_i^{(1)} - t_i^{(3)})$$

corrector

$$\left. \dots (4.113) \right.$$

$$y_{i+1}^{(3)} = [9y_i - y_{i-2} + 3h(t_{i+1}^{(2)} + 2t_i - t_{i-1})]/8$$

$$t_{i+1}^{(3)} = [9t_i - t_{i-2} + 3h\{f(x_{i+1}, y_{i+1}^{(2)}, t_{i+1}^{(2)}) \\ + 2f(x_i, y_i, t_i) - f(x_{i-1}, y_{i-1}, t_{i-1})\}]/8$$

final corrector

$$y_{i+1} = y_{i+1}^{(3)} + \tfrac{9}{121}(y_{i+1}^{(1)} - y_{i+1}^{(3)})$$

$$t_{i+1} = t_{i+1}^{(3)} + \tfrac{9}{121}(t_{i+1}^{(1)} - t_{i+1}^{(3)})$$

This procedure may be applied to any number of first-order simultaneous equations.

In a similar way Runge–Kutta methods for the solution of a first-order differential equation may be applied to the solution of a number of simultaneous first-order equations. For the particular equations 4.112 Runge and Kutta have given formulae that incorporate an estimate of y in the interval of integration that uses the estimates of the second derivative of y generated in the course of computation. The formulae are shown as equations 4.114:

$$m_0 = hf(x_0, y_0, t_0)$$

$$m_1 = hf\left(x_0 + \frac{h}{2}, y_0 + \frac{ht_0}{2} + \frac{hm_0}{8}, t_0 + \frac{m_0}{2}\right)$$

$$m_2 = hf\left(x_0 + \frac{h}{2}, y_0 + \frac{ht_0}{2} + \frac{hm_1}{8}, t_0 + \frac{m_1}{2}\right) \qquad \cdots (4.114)$$

$$m_3 = hf\left(x_0 + h, y_0 + ht_0 + \frac{hm_2}{2}, t_0 + m_2\right)$$

$$t_1 = t_0 + \tfrac{1}{6}(m_0 + 2m_1 + 2m_2 + m_3)$$

$$y_1 = y_0 + h[t_0 + \tfrac{1}{6}(m_0 + m_1 + m_2)]$$

The more refined estimate of y that is possible because of the use of the second derivative should give a small increase in the accuracy of the method.

Ordinary differential equations of second and higher orders may be defined either by initial conditions, or by boundary-value conditions. Boundary-value problems may also be solved by the trial and iteration procedure described for simultaneous first-order equations.

A comparison between Runge–Kutta and Hamming's methods. In testing methods it is advantageous to compare attempts to solve equations for which the analytical solution is also known. The second-order differential equation:

$$\frac{d^2y}{dx^2} = y \qquad \cdots (4.115)$$

when defined by the initial values:

$$y = 1, \quad \frac{dy}{dx} = -1 \text{ at } x = 0 \qquad \cdots (4.116)$$

has the solution:

$$y = e^{-x} \qquad \cdots (4.117)$$

If a numerical solution is started at $x = 0$ and integrated in the direction of increasing x, instabilities are introduced because a portion of the unwanted solution to equation 4.115, e^x, is gradually introduced so that the integration may not converge to the correct solution. Table 4.8 shows the application of both the Runge–Kutta method and Hamming's method to the solution compared with the analytical results at different values of the increment of integration h. The starting values for Hamming's method were taken from the analytical solution.

These calculations show the greater tendency of Runge–Kutta methods to instability, because it is clear that even when the interval is as small as 0·01 the integration by Runge–Kutta does not converge to zero as $x \to \infty$. This instability is even more marked when the interval is larger and therefore Hamming's method shows greater stability in this example.

A second interesting point is brought out by the greater accuracy of the Runge–Kutta integration in the range $0 < x < 3$ when the interval is 0.1. These first few points suggest that, in this application at least, there is a higher intrinsic accuracy for the method of Runge–Kutta when the range of integration is small, although the instability of the method can be clearly seen when $x = 6$.

TABLE 4.8. *A comparison of Runge–Kutta and Hamming's methods in the solution to equation 4.115*

	x	y (Runge–Kutta)	y (Hamming)	$y = e^{-x}$
$h = 0.1$	1.0	0.367871	0.365162	0.367879
	2.0	0.135301	0.134338	0.135335
	3.0	0.049687	0.049420	0.049787
	4.0	0.018039	0.018181	0.018316
	5.0	0.005985	0.006688	0.006738
	6.0	0.000431	0.002460	0.002479
	7.0	−0.004655	0.000952	0.000912
	8.0	−0.014796	0.000330	0.000335
	9.0	−0.041010	0.000123	0.000123
	10.0	−0.111768	0.000045	0.000045
$h = 0.01$	1.0	0.367879	0.367879	0.367879
	2.0	0.135335	0.135335	0.135335
	3.0	0.049787	0.049787	0.049787
	4.0	0.018315	0.018315	0.018315
	5.0	0.006737	0.006738	0.006738
	6.0	0.002477	0.002479	0.002479
	7.0	0.000906	0.000912	0.000912
	8.0	0.000320	0.000335	0.000335
	9.0	0.000081	0.000123	0.000123
	10.0	−0.000069	0.000045	0.000045

These results are in general accord with the findings of others on these methods. Although a starter formula is required, the method of Hamming has good stability, but because the integration is based upon an extrapolation weighted by four preceding points, accuracy can be bettered in some applications by methods such as Runge–Kutta that are not so strongly bound by previous computed values. On the other hand, because previously computed values do not guide the extrapolation when Runge–Kutta is used, there is a tendency to instability and, if possible, equations should be integrated in a stable direction. The Runge–Kutta formulae are self-starting.

The numerical solution of partial differential equations

Classification of partial differential equations

Many partial differential equations of practical importance may be written in the general form:

$$a \frac{\partial^2 u}{\partial x^2} + b \frac{\partial^2 u}{\partial x \, \partial y} + c \frac{\partial^2 u}{\partial y^2} = g \qquad \qquad \dots . (4.118)$$

where a, b, c and g are functions of x, y, u, $\partial u/\partial x$, $\partial u/\partial y$ only and not of the second derivatives. This particular general form, even though non-linear, is sometimes known as a *quasi-linear* equation. When the coefficients a, b, c and g are constants or functions of x and y only, the equation is linear.

The classification of equations of the form of equation 4.118 is based upon the value of $b^2 - 4ac$. When $b^2 - 4ac$ is negative, the equation is said to be *elliptic*, when $b^2 - 4ac$ is zero the equation is *parabolic*, and when $b^2 - 4ac$ is positive the equation is said to be *hyperbolic*. In the general case it may be difficult to determine the classification because a, b and c depend upon u, the derivatives of u, and x and y. It is even possible for the equation to be parabolic in one part of the domain of integration, and to be hyperbolic or elliptic in another. However, when a, b and c are constants, or functions of x and y, the classification is known in advance and the solution of the equation can be more direct. The most effective method of numerical solution of a given equation depends upon the type of equation, and for this reason methods for the solution of parabolic and elliptic equations are discussed separately. Hyperbolic equations are not discussed in detail here and the interested reader is referred to treatments of this topic elsewhere[2].

Boundary and initial conditions for partial differential equations

The solution to a partial differential equation is defined on the boundaries of the region of integration. If the region of integration is open and the conditions are specified at the start of the region only, the differential equation is said to be defined by *initial conditions*, while if the conditions are completely specified on the boundaries of the region of integration, the equation is said to be defined by *boundary conditions*.

The defining conditions for hyperbolic equations are of the initial-value type. Integration starts from the initial curve and proceeds outwards by means of successive steps of integration. Elliptic equations are defined by boundary conditions over the extremes of the region of integration, so that the boundary conditions and the differential equation over the region have to be simultaneously satisfied. Parabolic equations are defined in part by boundary conditions, and in part by initial conditions. The solution to a parabolic equation is advanced outwards towards the open boundary, but the advance is constrained at the extremes where the solution is defined by boundary conditions at the closed portions of the boundaries. Figure 4.13 illustrates the regions of integration for an elliptic and for a parabolic equation.

In physical applications the initial and boundary conditions are determined by the constraints of the physical problem. For a typical problem the function, or a derivative of the function, or a relationship between function and derivative, is defined at the boundary.

The region of integration is first divided by means of a mesh. The function is to be evaluated at the nodes of the mesh, and the intervals of the mesh represent increments of the independent variables. When the mesh over the region of integration is fine, the increments are small and therefore the truncation error for a given finite difference approximation is small. When the mesh is coarse the probability of significant error is greater.

It is most convenient when the area of integration can be covered by a rectangular mesh with the exterior nodes lying on the boundaries of the region.

Sometimes this is not possible, and it may be more convenient to use a triangular mesh or a radial mesh if either can be fitted to the area. For irregular areas of integration, however, the nodes of the mesh cannot be arranged to coincide with the boundary and it is then necessary to use an interpolation formula (usually linear) to relate the fixed value at the boundary to the value of the function at the node that lies outside the boundary.

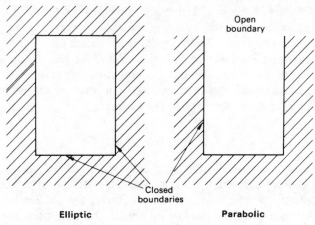

Fig. 4.13. Regions of integration for elliptic and parabolic equations

When the derivative of a function is defined at a boundary, the value of the function at the boundary, the value of the function at a neighbouring node and the mesh increment are related to the derivative by a simple finite difference formula such as equation 4.80 when the first order difference only is considered.

Partial differential equations in dimensionless form

It is often helpful to write an ordinary differential equation in dimensionless form before attempting a solution because the physical situation is clarified, and the solution, when found, has a greater generality. Partial differential equations contain at least one more independent variable, and therefore the benefits of first writing the equations in dimensionless form are even greater. Indeed, in all but the simplest of cases it will be found that a sound knowledge of the dependence of the solution upon the physical variables cannot be constructed unless the solution is presented in terms of dimensionless groups. Some examples of the transformation of dimensional equations into dimensionless form will be given.

Parabolic equations

Parabolic equations are defined by a combination of boundary values and initial conditions. The way in which defining conditions arise may be illustrated by considering a parabolic equation that describes transient conduction or diffusion. The equation written for one space dimension only is:

$$K_1 \frac{\partial^2 V}{\partial x^2} = K_2 \frac{\partial V}{\partial t}$$

.... (4.119)

where K_1 is a diffusion coefficient, K_2 is a capacity coefficient, V is concentration or temperature and t is time.

Suppose that the medium into which diffusion or conduction is taking place is a flat slab of thickness L so that $0 < x < L$. The initial conditions describe the distribution of V in the slab at a time that is taken as the start of the experiment. The conditions at the faces $x = 0$ and $x = L$ are specified as functions of time and are known as boundary conditions. The boundary conditions are completely specified for times later than the start of the experiment, while the initial conditions define V at the start.

An integration of this partial differential equation will carry the solution forward in time from the initial conditions while constrained by the boundary conditions. The scheme of integration is shown diagrammatically in Fig. 4.14.

The partial differential equation 4.119 may now be expressed in dimensionless form by introducing the dimensionless variables u, Z and θ where:

$$u = \frac{V}{V_0}, \quad Z = \frac{x}{L}, \quad \theta = \frac{K_1 t}{K_2 L^2}$$

and V_0 is a convenient reference temperature or composition. The differential equation is now:

$$\frac{\partial^2 u}{\partial Z^2} = \frac{\partial u}{\partial \theta}$$

.... (4.120)

and the boundary and initial conditions are expressed in terms of the dimensionless variables u, Z and θ. In this example it is convenient to take the group $K_2 L^2 / K_1$ as a characteristic time.

Several methods are available for the solution of parabolic equations, but with some there may be pronounced instability. The differential equation 4.120 is replaced by finite difference approximations to the derivatives and the simultaneous equations so formed are solved by appropriate methods. The methods of replacing the differentials by finite difference approximations lead either to *explicit* or to *implicit* methods of solution.

Explicit methods of solution. The domain of integration is divided into intervals of width ΔZ and $\Delta \theta$ as shown in Fig. 4.14 (page 348), and the points subdividing each interval are numbered. These points are known as nodes.

A simple expression for the first derivative is obtained by truncating equation 4.78 after the first forward difference. When expressed in terms of forward differences of all orders the first derivative at the rth node is:

$$Du_r = \frac{1}{\Delta \theta} \left[\Delta - \tfrac{1}{2}\Delta^2 + \tfrac{1}{3}\Delta^3 \ldots \right] u_r \qquad \text{(equation 4.78)}$$

The first derivative at time θ_0 when approximated is:

$$\Delta\theta \left(\frac{\partial u_r}{\partial \theta}\right)_{\theta_0} = u_{r,\,\theta_0+\Delta\theta} - u_{r,\,\theta_0} \qquad \dots (4.121)$$

This is a convenient, although rough, approximation to the first derivative. An expression for the second derivative is conveniently given in terms of central differences by equation 4.91:

$$\mathbf{D}^2 u_r = \frac{1}{(\Delta Z)^2}(\delta^2 - \tfrac{1}{12}\delta^4 + \tfrac{1}{90}\delta^6 \dots)u_r \qquad \text{(equation 4.91)}$$

When differences of orders higher than the second are neglected, the approximation to the second derivative is:

$$(\Delta Z)^2 \left(\frac{\partial^2 u_r}{\partial Z^2}\right)_{\theta_0} = (u_{r-1} - 2u_r + u_{r+1})_{\theta_0} \qquad \dots (4.122)$$

The finite difference approximation to equation 4.120 is thus:

$$u_{r,\,\theta_0+\Delta\theta} = u_{r,\,\theta_0} + \frac{\Delta\theta}{(\Delta Z)^2}(u_{r-1} - 2u_r + u_{r+1})_{\theta_0} \qquad r = 1, 2, \dots n \qquad \dots (4.123)$$

Difference corrections may be included, but the usual practice with an automatic digital computer is to choose an interval that is small enough to give the required accuracy when high-order difference corrections are neglected. The set of equations 4.123 corresponds to one equation for each of the internal nodes as shown in Fig. 4.14 (page 348). The solution to this set of linear simultaneous equations is simple.

The initial conditions define the right-hand side of equations 4.123 at $\theta = 0$. The values of u at all of the interior nodes at $\theta = \Delta\theta$ are found by straight-forward substitution of the initial values into equation 4.123. For a typical set of boundary conditions the values of u at the two exterior nodes are defined by the boundary conditions, and the solution is carried forward in time by successive substitutions. Because the values of u at a later interval of time are found directly from the values at the end of the previous interval, the method is described as *explicit*.

Although convenient for computation, it has been found that explicit methods are not stable unless $\Delta\theta < 0.5\,(\Delta Z)^2$. For large time intervals this condition is not met and the numerical solution will then show a considerable scatter about the true values. The scatter is not found when $\Delta\theta < 0.5\,(\Delta Z)^2$ and the precision of computation is adequate, but the restriction upon the length of the time interval usually requires very small intervals, and therefore a considerable amount of computation may be necessary.

Implicit methods of solution. Implicit methods of solving parabolic equations are stable for all values of the time interval, and this is often an outstanding consideration. One simple implicit method differs from the explicit method just described in that the second derivative of equation 4.120 is defined at time

$\theta_0 + \Delta\theta$ instead of time θ_0. The set of implicit equations corresponding to the set 4.123 is:

$$\left(u_{r+1} + u_{r-1} - \left(2 + \frac{(\Delta Z)^2}{\Delta\theta}\right)u_r\right)_{\theta_0+\Delta\theta} = -\frac{(\Delta Z)^2}{\Delta\theta}u_{r,\theta_0} \quad r = 1,2,...n \quad ....(4.124)$$

A difference correction can be included when required, but when the chosen intervals are small enough the correction may be neglected.

The solving of these equations is more complex than the set 4.123, but there are still no computational difficulties. When written out in matrix form it can be seen that the set 4.124 consists of a tridiagonal matrix of coefficients and a right-hand side that has in part been calculated at the end of the previous time interval, and in part is determined by the boundary values. The tridiagonal matrix may be brought into upper triangular form by means of a simple recurrence formula and the back substitution can then be accomplished by means of a second recurrence formula. For each time interval the values of u at the nodes are calculated at the stage of back-substitution.

When the set of equations 4.124 is put into matrix form the values at the nodes of Fig. 4.14 are linked by the following matrix equation:

$$\begin{bmatrix} -(\phi+2) & 1 & 0 & 0 \cdots & & 0 & 0 \\ 1 & -(\phi+2) & 1 & 0\ 0 \cdots & & & 0 \\ 0 & 1 & -(\phi+2) & 1\ 0\ 0 \cdots & & & \\ & \cdot\ \cdot\ \cdot & & & & & \\ & \cdot\ \cdot\ \cdot\ \cdot & & & & & \\ 0 & \cdots & & & 0\ 1 & -(\phi+2) & 1 \\ 0 & 0 & \cdots & & 0\ 0 & 1 & -(\phi+2) \end{bmatrix}$$

$$\times \begin{bmatrix} u_1 \\ u_2 \\ u_3 \\ \\ \vdots \\ \\ u_{n-1} \\ u_n \end{bmatrix}_{\theta_0+\Delta\theta} = - \begin{bmatrix} \phi u_{1,\theta_0} + u_{0,\theta+\Delta\theta} \\ \phi u_{2,\theta_0} \\ \phi u_{3,\theta_0} \\ \\ \vdots \\ \\ \phi u_{n,\theta_0} \\ \phi u_{n,\theta_0} + u_{n+1,\theta_0+\Delta\theta} \end{bmatrix} \quad ....(4.125)$$

where

$$\phi = \frac{(\Delta Z)^2}{\Delta\theta}$$

The values of $u_{0,\theta_0+\Delta\theta}$ and $u_{n+1,\theta_0+\Delta\theta}$ are known because they are boundary values.

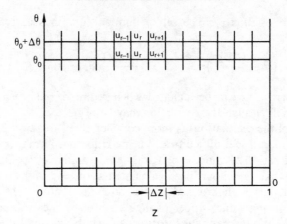

FIG. 4.14. Integration of a parabolic partial
differential equation

This matrix equation may be solved by a simple, direct method. If the first row of the augmented matrix composed of the matrix of coefficients and the right-hand side are divided by first non-zero coefficient ($\phi + 2$) and added to the second row, the coefficient of u_1 is eliminated from that row. The same procedure is repeated for the second row so that the coefficient of u_2 is eliminated from the third row, and if the elimination is carried right through the matrix, the matrix of coefficients will be brought into upper-triangular form. Systematic back-substitution will then carry the matrix into diagonal form thus giving the values $u_1, u_2, \ldots$ at the end of the time interval $\theta_0 + \Delta\theta$.

The recurrence formulae of elimination and back-substitution may be programmed for a digital computer. This direct method may be applied to make very light demands upon both computer storage and computer time, showing great savings over an automatic application of the method of pivotal condensation, as described on page 315.

In a numerical example, both the implicit and the explicit method have been used to solve equation 4.119 for the slab of thickness L subject to the boundary conditions:

$$u = 1 \quad \text{at} \quad z = 0 \quad \text{and at} \quad z = 1\cdot 0$$

and the initial condition:

$$u = 0, \quad 0 < z < 1\cdot 0$$

The value of Δz has been set at $0\cdot 1$, and $\Delta\theta$ at $0\cdot 01$ so that the value of ϕ is 1. The numerical results and the corresponding analytical solution are shown in Table 4.9.

The superiority of the implicit over the explicit method is very clear in this example. The condition of stability for the explicit method $\Delta\theta < 0\cdot 5(\Delta z)^2$ is not met and therefore the solution given by this method is obviously unstable. On the other hand the implicit method is obviously stable, and the solution given by this method compares very favourably with the analytical solution.

TABLE 4.9. *Comparison of implicit, explicit and analytical solutions to equation* 4.119

	y	Explicit	Implicit	Analytical
$\theta = 0.07$	0·0	1·00	1·000	1·000
	0·1	36·99	0·792	0·802
	0·2	−53·99	0·608	0·624
	0·3	51·99	0·467	0·484
	0·4	−37·99	0·379	0·394
	0·5	31·99	0·349	0·363
	0·6	−37·99	0·379	0·394
	0·7	51·99	0·467	0·484
	0·8	−53·99	0·608	0·624
	0·9	36·99	0·792	0·802
	1·0	1·00	1·000	1·000
$\theta = 0.08$	0·0	1·00	1·000	1·000
	0·1	−89·99	0·813	0·821
	0·2	142·99	0·645	0·660
	0·3	−143·99	0·515	0·532
	0·4	121·99	0·433	0·450
	0·5	−107·99	0·405	0·422
	0·6	121·99	0·433	0·450
	0·7	−143·99	0·515	0·532
	0·8	142·99	0·645	0·660
	0·9	−89·99	0·813	0·821
	1·0	1·00	1·000	1·000

Parabolic equations in two dimensions. The two-dimensional generalisation of equation 4.119 is:

$$K_1 \frac{\partial^2 V}{\partial x_1^2} + K_1 \frac{\partial^2 V}{\partial x_2^2} = K_2 \frac{\partial V}{\partial t} \qquad \dots (4.126)$$

where K_1 is a diffusion coefficient, K_2 is a capacity coefficient, x_1 and x_2 are coordinates, V is concentration or temperature, and t is time. Suppose that the system to be considered is a rectangular element such that $0 < x_1 < L_1$ and $0 < x_2 < L_2$. The initial conditions define the distribution of V in the element at the start of the experiment. The boundary conditions at the faces of the element are specified as functions of time. This partial differential equation may be placed in dimensionless form by the substitutions:

$$u = \frac{V}{V_0}, \quad y = \frac{x_1}{L_1}, \quad z = \frac{x_2}{L_1}, \quad \theta = \frac{K_1 t}{K_2 L_1^2}$$

where V_0 is a convenient reference temperature or composition. The boundary and initial conditions are expressed in terms of the dimensionless variables u, y, z and θ. Following the substitutions, the differential equation becomes:

$$\frac{\partial^2 u}{\partial y^2} + \frac{\partial^2 u}{\partial z^2} = \frac{\partial u}{\partial \theta} \qquad \dots (4.127)$$

An explicit method of solution similar to that described for the single-dimensional equation 4.119 may be described for this equation also, but the method is unstable under the same conditions so that it is necessary to use very small time intervals in the solution. The implicit method described for the single dimensional equation may be extended to two dimensions, and the finite difference equations then take the form of linear simultaneous equations characterised by a banded matrix of coefficients. This implicit method also is stable for all values of the time interval, but the dimensions of the matrix equation may be large, and the elimination and back-substitution procedures are more time-consuming than those for equations containing a tridiagonal matrix of coefficients.

PEACEMAN and RACHFORD[9], and DOUGLAS[10] have proposed an alternating-direction implicit method that is stable for all values of the time interval, but which requires the solution of linear simultaneous equations containing a tridiagonal matrix of coefficients. Two equations are solved alternately. The first equation expresses $\partial^2 u/\partial y^2$ by an implicit finite difference formula, and $\partial^2 u/\partial z^2$ by an explicit finite difference formula, while the second equation expresses $\partial^2 u/\partial y^2$ by an explicit formula, and $\partial^2 u/\partial z^2$ by an implicit formula. The first equation is solved for one time interval, the second equation is solved for the next time interval, and the pattern is repeated until the range of integration is completed.

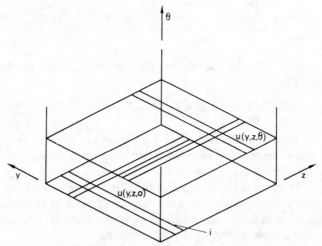

FIG. 4.15. Arrangement of nodes for the solution of a parabolic equation in two dimensions

Figure 4.15 shows the arrangement of nodes for the rectangular element. The first set of equations to be solved is implicit in the y direction and explicit in the z direction. The equations have the form:

$$-u(y - \Delta y, z, \theta + \Delta\theta) + (2 + \rho)\,u(y, z, \theta + \Delta\theta) - u(y + \Delta y, z, \theta + \Delta\theta)$$
$$= u(y, z - \Delta z, \theta) + (\rho - 2)\,u(y, z, \theta) + u(y, z + \Delta z, \theta) \quad \ldots \ldots (4.128)$$

where $\rho = (\Delta y)^2/\Delta\theta$, and Δy has been chosen equal to Δz. One equation of this type is written for every internal node. The equations are bound together in the y direction so that there are n systems each containing m equations. Thus line i shown on Fig. 4.15 contains m nodes, and one equation is written for each node. The unknowns of the m equations are the values of u at time $\theta + \Delta\theta$ and appear on the left-hand side of the equations for which equations 4.128 form a typical example. These equations may be arranged in the matrix form of equation 4.125 to give a tridiagonal matrix of coefficients. The solution is found as described for equation 4.125, and the procedure is repeated for each of the n sets of m equations.

For the next time step the sets of equations are implicit in the z direction and explicit in the y direction:

$$-u(y, z - \Delta z, \theta + \Delta\theta) + (2 + \rho)\,u(y, z, \theta + \Delta\theta) - u(y, z + \Delta z, \theta + \Delta\theta)$$
$$= u(y - \Delta y, z, \theta) + (\rho - 2)\,u(y, z, \theta) + u(y + \Delta y, z, \theta) \quad \ldots(4.129)$$

These equations are bound together in the z direction so that there are m systems each containing n equations. Each set of n equations may be arranged in the form of equation 4.125 and the presence of a tridiagonal matrix of coefficients allows a rapid and direct solution.

For the next time step the computation returns to equation 4.128, and the computation progresses by the alternate use of equations 4.128 and 4.129.

It has been shown[9] that for linear equations of the type 4.126 the method offers considerable savings in computer time over other methods of solution, such as a completely implicit scheme or explicit methods.

For non-linear partial differential equations the method suffers from the disadvantage that systems of non-linear simultaneous equations must be solved. The problem of solving non-linear simultaneous equations is by no means easy but some successful methods have been reported[2] in which the integration is advanced one step by means of an explicit method, and the values so obtained are improved by successive approximations. In many instances the type of equation to be solved may have a considerable influence upon the methods to be adopted so that useful generalities are hard to find. Some further discussion is given elsewhere[2].

Elliptic equations

A simple example of an elliptic equation is Laplace's equation in two dimensions:

$$\frac{\partial^2 u}{\partial y^2} + \frac{\partial^2 u}{\partial z^2} = 0 \qquad \ldots(4.130)$$

The finite difference equation for the second derivative has already been given as equation 4.91, and the way in which the linear simultaneous equations are formulated may be shown by referring again to the example of the rectangular element. The arrangement of the nodes is shown in Fig. 4.16.

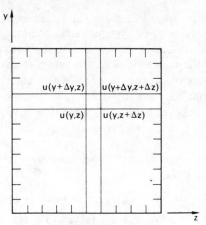

FIG. 4.16. Arrangement of nodes
for the solution of an elliptic
equation in two dimensions

Each of the set of equations to be solved has the form:

$$4u(y, z) - u(y + \Delta y, z) - u(y - \Delta y, z) - u(y, z - \Delta z) - u(y, z + \Delta z) = 0$$

$$\dots (4.131)$$

and one equation of this form is written for each internal node. The structure of the complete system of equations may be seen by writing them in matrix form:

$$\begin{bmatrix} A & I & O & O & O & \dots \\ I & A & I & O & O & \dots \\ O & I & A & I & O & \dots \\ \dots & & & & & \\ \dots & & & & & \\ \dots & & & & & \end{bmatrix} \mathbf{u}_{ij} = \mathbf{b}_{ij} \qquad \dots (4.132)$$

If each line contains m internal nodes, and there are n internal nodes altogether, the matrices $\mathbf{I}$ and $\mathbf{O}$ are the identity and null matrices each of dimension $m \times m$ while the $m \times m$ matrix $\mathbf{A}$ has the form:

$$\mathbf{A} = \begin{bmatrix} 4 & -1 & 0 & \dots & & \\ -1 & 4 & -1 & 0 & \dots & \\ 0 & -1 & 4 & -1 & 0 & \dots \\ \dots & & & & & \\ \dots & & & & & \end{bmatrix} \qquad \dots (4.133)$$

The column vector $\mathbf{u}_{ij}$ represents the n unknowns, while $\mathbf{b}_{ij}$ represents the n values of the right-hand vector that introduce the boundary values fixed on the perimeter of the region.

Methods for the solution of equation 4.132 include direct methods such as pivotal condensation if n is not too large. For many problems however n is so large that the augmented matrix cannot be held in computer storage and indirect methods such as that of Gauss and Seidel must be used.

One further method that is sometimes of interest arises from the observations that as $\theta \to \infty$ solutions to equation 4.127 approach the solution to equation 4.130. For example, the alternating direction implicit method already described may be applied to an appropriate initial condition in the region defined by equation 4.127. The solution is then carried forward in time until a steady state is reached. Some computational advantages over other direct and indirect methods of solution have been claimed for this approach[9].

Methods for the optimisation of linear and non-linear objective functions

Linear objective functions

On many occasions the resources of power, labour and machinery are combined to make a product, or to accomplish some task of economic significance. Some combinations are more wasteful of resources than others, and when the resources are scarce it is important that the most efficient combinations should be used. The efficiency of manufacture or of the methods used in a task is measured by the *objective function*, sometimes known as the cost function.

The most efficient method for a particular task is known as the *optimum method.* The optimum method is always associated with a maximum or minimum value of the objective function. When the dependence of the objective function upon the resources can be expressed by a mathematical equation, or by a set of equations when constraints upon the resources are also defined, it is possible to search for the optimum by mathematical procedures.

The methods of search depend upon the mathematical nature of the objective function and constraints. If the objective function is linear in the variables measuring the resources, and when the variables are bounded by linear constraints, the task of optimisation is known as a *linear-programming* problem. Successful methods for the solution of linear programmes were first formulated about 1950 by Dantzig and his associates (see GASS[1]). Their method of solution known as the *simplex* method has since been applied to many different linear programmes, and many modern computer programs for linear programming are based upon this procedure.

Non-linear objective functions are also important, but here no corresponding degree of success has been achieved. Nevertheless a great variety of problems in non-linear optimisation has been solved, and a small number of the more useful methods will be described.

Linear programming. The fundamental problem in linear programming is to maximise or minimise the linear objective function:

$$c_1 x_1 + c_2 x_2 + c_3 x_3 \ldots + c_n x_n = M \qquad \ldots . (4.134)$$

when the objective function is subject to the m linear constraints:

$$\left.\begin{array}{l} a_{11}x_1 + a_{12}x_2 \quad \ldots \; + a_{1n}x_n = b_1 \\ a_{21}x_1 + a_{22}x_2 \quad \ldots \; + a_{2n}x_n = b_2 \\ \ldots \\ \ldots \\ a_{m1}x_1 + a_{m2}x_2 \quad \ldots \; + a_{mn}x_n = b_m \end{array}\right\} \qquad \ldots (4.135)$$

The variables x_i are always positive, the c_i are known as cost coefficients and some of the a_{ij} and the c_i may be zero. The value of n is greater than m.

Because the number of variables n is greater than the number of equations m there is no unique solution to the set of linear equations because any number of solutions can be generated by first choosing n–m variables and then solving the m equations to find the remaining m variables. Of the infinity of solutions it is necessary to find those that maximise or minimise the objective function, and in many practical problems this condition is met by a single solution, or by a limited range of multiple solutions.

The simplex method defines a systematic search for the optimum. If the linear programming problem is a valid one the operating solution is found after a maximum of $2m$ eliminations in the set of equations 4.135. The method is readily adapted to automatic digital computation. However, before describing the simplex method in detail, there are some preliminary mathematical matters to be discussed.

Vectors and vector bases. Equation 4.135 may be written in matrix form:

$$\mathbf{Ax} = \mathbf{b} \qquad \ldots (4.136)$$

where $\mathbf{A}$ is the $m \times n$ matrix of coefficients, $\mathbf{x}$ is the $n \times 1$ column vector $(x_1, x_2, \ldots x_n)$ and $\mathbf{b}$ is the $m \times 1$ column vector $(b_1, b_2, b_3, \ldots b_m)$. This equation differs from the earlier examples in that the matrix of coefficients is not square. An alternative formulation of the equation is:

$$x_1\mathbf{p}_1 + x_2\mathbf{p}_2 + x_3\mathbf{p}_3 \, \ldots + x_n\mathbf{p}_n = \mathbf{p}_0 \qquad \ldots (4.137)$$

where $\mathbf{p}_1, \mathbf{p}_2, \mathbf{p}_3, \ldots$ are column vectors defined by:

$$\mathbf{p}_1 = \begin{bmatrix} a_{11} \\ a_{21} \\ . \\ . \\ . \\ a_{m1} \end{bmatrix}, \; \mathbf{p}_2 = \begin{bmatrix} a_{12} \\ a_{22} \\ . \\ . \\ . \\ a_{m2} \end{bmatrix}, \; \ldots \mathbf{p}_n = \begin{bmatrix} a_{1n} \\ a_{2n} \\ . \\ . \\ . \\ a_{mn} \end{bmatrix}, \; \mathbf{p}_0 = \begin{bmatrix} b_1 \\ b_2 \\ . \\ . \\ . \\ b_m \end{bmatrix}$$

Equation 4.137 defines the vector $\mathbf{p}_0$ in terms of the vectors $\mathbf{p}_1, \dots \mathbf{p}_n$. An important property of the vectors $\mathbf{p}_j$ is their dependence or independence. The vectors are said to be linearly independent if the condition:

$$\alpha_1 \mathbf{p}_1 + \alpha_2 \mathbf{p}_2 + \alpha_3 \mathbf{p}_3 \dots \alpha_n \mathbf{p}_n = \mathbf{o} \qquad \dots (4.138)$$

is met only when $\alpha_1 = \alpha_2 = \alpha_3 \dots = \alpha_n = 0$. Here $\mathbf{o}$ is the $n \times 1$ null column vector. If the condition for independence is not met the vectors constitute a dependent set.

If a problem is concerned with vectors all of which may be expressed by relations among n linearly independent vectors, then the vectors constitute an *n-dimensional space* and the n linearly independent vectors form a *basis* for the space. Every set of $n + 1$ or greater number of vectors is linearly dependent.

Convex sets. A convex combination of points in n-dimensional space $\mathbf{u}_1, \mathbf{u}_2, \dots$ defines a point:

$$\mathbf{u} = \beta_1 \mathbf{u}_1 + \beta_2 \mathbf{u}_2 \dots \qquad \dots (4.139)$$

where the β_i are scalars satisfying the conditions $\beta_i \geqslant 0$, and $\Sigma \beta_i = 1$. Any point on a line segment in n-dimensional space joining two points can be expressed as a convex combination of the two points. The n-dimensional space itself is a convex set because any point in the space can be expressed as a convex combination of other points. More familiar examples of convex sets in two dimensions are shown in Fig. 4.17, a figure that also includes a set that is not convex. The polygon is known as a convex polygon.

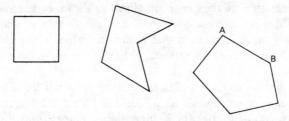

FIG. 4.17. Convex and non-convex sets

The property of convexity is an important one for linear-programming problems, because the solutions to equation 4.137 constitute a convex set. To describe these solutions a terminology has evolved that must first be introduced.

A *feasible solution* to a linear programming problem is a vector $\mathbf{x} = (x_1, \dots x_n)$ where the x_i are positive and equation 4.135 is satisfied.

A *basic feasible solution* is a feasible solution with no more than m positive x_i while the remainder are zero.

A *non-degenerate basic feasible solution* is a basic feasible solution with exactly m positive x_i, while the remainder are zero.

A *minimum feasible solution* is a feasible solution that minimises the objective function M of equation 4.134. If the problem is formulated as the maximisation of M, this object is the same as the minimisation of $-M$, so that no generality is lost by considering only problems in minimisation.

It may be proved that the set of all feasible solutions is a convex set. Suppose that there are two solutions x_1 and x_2 that satisfy both equation 4.136 and the condition of non-negativity. If β satisfies the condition $0 < \beta < 1$, then x is a convex combination of x_1 and x_2 when the relationship between them is:

$$x = \beta x_1 + (1 - \beta) x_2 \qquad \ldots (4.140)$$

As the groups A, x_1, b and A, x_2, b are each related by equation 4.136:

$$Ax = A[\beta x_1 + (1 - \beta) x_2] = \beta A x_1 + (1 - \beta) A x_2$$
$$= \beta b + (1 - \beta) b = b$$

so that x also satisfies equation 4.136. As β and $(1 - \beta)$ are both positive, as are x_1 and x_2, it follows that x satisfies the condition of non-negativity. Therefore x or any convex combination of the feasible solutions x_1 and x_2 are also feasible solutions.

One further definition is now useful. A *linear functional* $f(x)$ is a real valued function defined so that for every vector x in the n-dimensional space that may be represented as the convex combination of two other vectors $\beta x_1 + (1 - \beta) x_2$ then:

$$f(x) = f(\beta x_1 + (1 - \beta) x_2) = \beta f(x_1) + (1 - \beta) f(x_2) \qquad \ldots (4.141)$$

The objective function 4.134 is a linear functional of the vectors x that satisfy equation 4.135.

From the properties of the linear functional as shown by equation 4.141, it is clear that when x is a convex combination of the vectors x_1 and x_2, then the functional takes on its maximum and minimum values at the extremes of the combination. Thus in the example above either $f(x_1)$ or $f(x_2)$ is the minimum value of $f(x)$.

It has been shown that the set of all feasible solutions to a linear programming problem is a convex set, and as all of the constraints are linear, the extreme surfaces of the set form lines (in 2 dimensions), planes (in 3 dimensions), or *hyperplanes* (in n dimensions). The points in the exterior surfaces of the set lie in hyperplanes, and some are extreme points lying at the intersections of planes. If the set is two-dimensional, it is a convex polygon as in Fig. 4.17, with the edges of the polygon as constraints, and the extreme points at the intersection of the constraints.

As the objective function 4.134 is a linear functional, its minimum value is found at an extreme point. If the minimum is found at more than one extreme point, then from the property of the functional 4.141, the functional takes on the minimum value at every point that is a convex combination of these extreme points. Only when $f(x_1)$ and $f(x_2)$ are equal can the minimum value be found at a point that is not extreme. When this happens there exist multiple solutions to a linear programming problem.

These important properties of the solution to the linear programming problem may be summed up by the theorem: the minimum of the objective function 4.134 is found at an extreme point of the convex set of all feasible solutions. If the function takes on the minimum value at more than one extreme point, then it takes on the minimum value at every convex combination of the extreme points. Thus in the example of the convex polygon shown in Fig. 4.17, if the minimum value of the function is found at both the points A and B, this value is also found at every point along the line AB. Otherwise the minimum value is found only at one of the corners.

Obviously the point of great importance is that the minimum value of the objective function may be found by inspecting only the extreme points of the region of feasible solutions. The feasible solutions of equation 4.135 that are found at the extreme points of the convex region are basic feasible solutions as shown by the next theorem.

Although there are n variables x_i, they are related by m equations only where $m < n$. A feasible solution is therefore found by choosing $n - m$ variables and solving the m equations for the remaining m variables. A feasible solution may be defined by $k \leqslant m$ linearly independent vectors $\mathbf{p}_1, \dots \mathbf{p}_k$ so that:

$$x_1\mathbf{p}_1 + x_2\mathbf{p}_2 \dots + x_k\mathbf{p}_k = \mathbf{p}_0 \qquad \dots (4.142)$$

and the point $\mathbf{x} = (x_1, x_2, \dots x_k, 0, \dots 0)$. As $\mathbf{x}$ is a feasible solution it may be expressed as a convex combination of two other points $\mathbf{x}_1$ and $\mathbf{x}_2$. Thus $\mathbf{x} = \beta\mathbf{x}_1 + (1 - \beta)\mathbf{x}_2$ for non-negative β. Therefore, if the last $n - k$ elements x_i are zero for $\mathbf{x}$, they must also be zero for $\mathbf{x}_1$ and $\mathbf{x}_2$ because all x_i are non-negative. Hence as $\mathbf{x}_1$ and $\mathbf{x}_2$ both satisfy equation 4.136, with the same k positive x_i, it follows that $\mathbf{x}_1 = \mathbf{x}_2$; $\mathbf{x}$ is a convex combination of itself only and therefore is an extreme point.

In choosing $n - m$ variables it is obviously permissible to set them to zero, leaving the remaining m variables to be found from the m equations. The basic feasible solution so found may have exactly m non-zero x_i in which case it is a non-degenerate basic feasible solution, or it may have less than m non-zero x_i in which case it is a degenerate basic feasible solution.

Thus far it has been established that the minimum feasible solution is also basic, so that if all basic feasible solutions are examined the minimum will be found. The method of proceeding from one basic solution to the next is defined by the simplex procedure.

The simplex method. It is supposed that a basic feasible solution (non-degenerate) is available. Then:

$$x_1\mathbf{p}_1 + x_2\mathbf{p}_2 \dots + x_m\mathbf{p}_m = \mathbf{p}_0 \qquad \dots (4.143)$$

and the objective function is:

$$x_1c_1 + x_2c_2 \dots + x_mc_m = M_0 \qquad \dots (4.144)$$

As the set of vectors $\mathbf{p}_1, \dots \mathbf{p}_m$ is linearly independent any vector from the set

$\mathbf{p}_1, \dots \mathbf{p}_n$, say $\mathbf{p}_j$, can be expressed in terms of $\mathbf{p}_1, \dots \mathbf{p}_m$:

$$x_{1j}\mathbf{p}_1 + x_{2j}\mathbf{p}_2 \dots + x_{mj}\mathbf{p}_m = \mathbf{p}_j; \quad j = 1, \dots n \qquad \dots (4.145)$$

and the corresponding increment in the objective function due to $\mathbf{p}_j$ is:

$$x_{1j}c_1 + x_{2j}c_2 \dots + x_{mj}c_m = M_j; \quad j = 1, \dots n \qquad \dots (4.146)$$

Equation 4.145 is now multiplied by a positive constant θ and subtracted from 4.143 to give:

$$(x_1 - \theta x_{1j})\mathbf{p}_1 + (x_2 - \theta x_{2j})\mathbf{p}_2 \dots + (x_m - \theta x_{mj})\mathbf{p}_m + \theta\mathbf{p}_j = \mathbf{p}_0 \dots (4.147)$$

and if equation 4.146 is multiplied by θ and subtracted from 4.144, and θc_j added to both sides of the resulting equation, it is found that:

$$(x_1 - \theta x_{1j})c_1 + (x_2 - \theta x_{2j})c_2 \dots + (x_m - \theta x_{mj})c_m + \theta c_j$$
$$= M_0 - \theta(M_j - c_j) \qquad \dots (4.148)$$

On comparing equations 4.143 and 4.144 with 4.147 and 4.148, it is clear that the right-hand side of 4.148 is the objective function. Further, the inclusion of the vector $\mathbf{p}_j$ into the solution has reduced the value of the objective function provided that $(M_j - c_j) > 0$, and provided that all of the coefficients of the vectors $\mathbf{p}_i$ are non-negative. Equation 4.147 shows that provided θ does not increase beyond the minimum value of the ratio (x_i/x_{ij}) then all of the coefficients of the vectors will remain positive. In fact if:

$$\theta = \min (x_i/x_{ij}) = x_l/x_{lj} \qquad \dots (4.149)$$

then the vector $\mathbf{p}_l$ disappears from the basis and $\mathbf{p}_j$ is introduced. As the new solution is also basic, and the objective function has been reduced provided $(M_j - c_j) > 0$, the procedure can be repeated until all of the $(M_j - c_j) < 0$ when no further reduction in the objective function can be made and the minimum feasible solution has been found.

Independent vectors and the artificial basis. Although any m independent vectors may be chosen from the set $\mathbf{p}_1, \dots \mathbf{p}_n$, it is a great convenience if the unit matrix $\mathbf{I}_m$ can be found amongst the n vectors. Any independent set of m vectors may be brought into the form of a unit matrix by row operations upon the equations in the same way in which row operations upon m independent linear simultaneous equations will bring the matrix of coefficients into the form of the unit matrix $\mathbf{I}_m$. But it may be difficult to recognise and choose m independent vectors; if the unit matrix $\mathbf{I}_m$ cannot be found, then it is convenient to add sufficient independent vectors so that the unit matrix is formed.

This procedure is known as the formation of an *artificial basis.* For example, an artificial basis may be introduced into equation 4.135 by adding variables $x_{n+1}, \dots x_{n+m}$ while at the same time the objective function is also modified.

Equation 4.135 becomes:

$$
\left.
\begin{aligned}
a_{11}x_1 \ldots + a_{1k}x_k \ldots + a_{1n}x_n + x_{n+1} &= b_1 \\
a_{21}x_1 \ldots + a_{2k}x_k \ldots + a_{2n}x_n \quad + x_{n+2} &= b_2 \\
\ldots \\
a_{l1}x_1 \ldots + a_{lk}x_k \ldots + a_{ln}x_n \quad + x_{n+l} &= b_l \\
\ldots \\
a_{m1}x_1 \ldots + a_{mk}x_k \ldots + a_{mn}x_n \quad + x_{n+m} &= b_m
\end{aligned}
\right\} \quad \ldots (4.150)
$$

while the objective function is modified to give:

$$
c_1 x_1 + c_2 x_2 \ldots + c_n x_n + H x_{n+1} + H x_{n+2} \ldots + H x_{n+m} = M \quad \ldots (4.151)
$$

Note that all of the equations in 4.150 should be arranged so that all of the b_i are positive. The vector $\mathbf{x} = (0, 0, 0, \ldots x_{n+1}, \ldots x_{n+m})$ is a basic feasible solution that contains the unit matrix $\mathbf{I}_m$. The coefficient H is some very large coefficient. As M is to be minimised, the large magnitude of H ensures that the artificial variables are quickly replaced in the basis, and none of the artificial variables appears in the final solution.

The simplex algorithm. The simplex method may be described by referring to equation 4.150. For each column $1, \ldots n$ the quantities M_j are calculated by means of equation 4.146. The values of each $(M_j - c_j)$ are now inspected. In theory any $(M_j - c_j) > 0$ will improve the solution, but it is usually best to reduce the objective function as much as possible, and the choice of the vector associated with the largest $(M_j - c_j)$ as the vector to place into the basis has been found to be satisfactory in practice.

The first basic feasible solution chosen is $\mathbf{x} = (0, 0, \ldots x_{n+1}, \ldots x_{n+m})$, and it is supposed that the inspection of the $(M_j - c_j)$ has led to the choice of a new basic feasible solution that contains the column vector $\mathbf{p}_k = [a_{1k}, a_{2k} \ldots a_{mk}]$. As the initial basis contains the unit matrix $\mathbf{I}_m$, the initial values of $x_{n+1}, \ldots x_{n+m}$ are $b_1, \ldots b_m$, so that the minimum value of (x_i/x_{ik}) can be found. Suppose that this minimum value is (x_l/x_{lk}), then the task is to replace the vector $\mathbf{p}_l$ in the basis by $\mathbf{p}_k$.

The coefficient of x_k in the first row of equation 4.150 may be eliminated by subtracting an appropriate multiple (a_{1k}/a_{lk}) of row l and the coefficients of x_k in the succeeding rows except the lth can be eliminated in the same way. Equation 4.150 is brought into the form of 4.152 by these successive eliminations:

$$
\left.
\begin{aligned}
u_{11}x_1 \ldots \quad \ldots + u_{1n}x_n + x_{n+1} \ldots \quad + u_{1,n+l}x_{n+l} &= d_1 \\
u_{21}x_1 \ldots \quad \ldots + u_{2n}x_n \quad + x_{n+2} \ldots + u_{2,n+l}x_{n+l} &= d_2 \\
u_{l1}x_1 \ldots + x_k \ldots + u_{ln}x_n \quad + u_{l,n+l}x_{n+l} &= d_l \\
u_{m1}x_1 \quad + u_{mn}x_n \quad + u_{m,n+l}x_{n+l} \ldots + x_{n+m} &= d_m
\end{aligned}
\right\} \quad (4.152)
$$

A new basic feasible solution $\mathbf{x} = (0, 0, \ldots x_k, 0, \ldots x_{n+1}, \ldots x_{n+l-1}, 0, x_{n+l+1},$ $\ldots x_{n+m})$ has then been found. New values of $(M_j - c_j)$ can then be calculated and inspected and the whole procedure repeated until no $(M_j - c_j) > 0$, when the minimum feasible solution has been found.

In this description the simplex method is defined only by equality constraints. In practice constraints may be given in the form of inequalities, so that before giving a numerical example, this complication will be considered.

Inequality constraints. If the linear constraints of the problem are given in terms of inequalities instead of the equalities of equation 4.135, they may take the form of 4.153 where all b_i are positive:

$$
\left.
\begin{aligned}
a_{11}x_1 + a_{12}x_2 \ldots + a_{1n}x_n &< b_1 \\
&\ldots \\
a_{m1}x_1 + a_{m2}x_2 \ldots + a_{mn}x_n &< b_m
\end{aligned}
\right\} \quad \ldots (4.153)
$$

This form may be changed to the form of equation 4.135 by adding *slack variables* $x_{n+1}, \ldots x_{n+m}$ to successive equations:

$$
\left.
\begin{aligned}
a_{11}x_1 + a_{12}x_2 \ldots + a_{1n}x_n + x_{n+1} \quad\quad &= b_1 \\
a_{21}x_1 + a_{22}x_2 \ldots + a_{2n}x_n \quad + x_{n+2} \quad &= b_2 \\
&\ldots \\
a_{m1}x_1 + a_{m2}x_2 \quad + a_{mn}x_n \quad\quad + x_{n+m} &= b_m
\end{aligned}
\right\} \quad \ldots (4.154)
$$

As all of the b_i are positive this set of equations is the same as the set 4.150, and the initial basic feasible solution $\mathbf{x} = (0, 0, \ldots x_{n+1}, \ldots x_{n+m})$ can be used to start the simplex procedure.

Alternatively, the constraints may take the form:

$$
\left.
\begin{aligned}
a_{11}x_1 + a_{12}x_2 \ldots + a_{1n}x_n &> b_1 \\
&\ldots \\
a_{m1}x_1 + a_{m2}x_2 \ldots + a_{mn}x_n &> b_m
\end{aligned}
\right\} \quad \ldots (4.155)
$$

where all of the b_i are positive. Slack variables may be subtracted here to give:

$$
\left.
\begin{aligned}
a_{11}x_1 + a_{12}x_2 \ldots + a_{1n}x_n - x_{n+1} \quad\quad &= b_1 \\
&\ldots \\
a_{m1}x_1 + a_{m2}x_2 \quad + a_{mn}x_n \quad - x_{n+m} &= b_m
\end{aligned}
\right\} \quad \ldots (4.156)
$$

As all of the $b_1, \ldots b_m$ are positive a basic feasible solution cannot be found on multiplying each equation by -1, since $-b_1, -b_2, \ldots$ do not fulfil the condition of non-negativity and, in general, it is necessary to add an artificial basis.

In these ways both types of constraint may be brought to the form of equation 4.150.

A numerical example of a linear programme. To minimise:

$$x_1 + x_2 + x_3$$

subject to the constraints:

$$x_1 \quad + x_4 - 3x_5 = 1$$

$$x_2 \quad + 2x_4 \quad = 2$$

$$x_3 + x_4 - 2x_5 = 3$$

where all of the x_i are positive. The progress towards the solution may be shown by arranging the key figures in the form of a *simplex tableau.* The tableau has the form:

TABLEAU 4.1

i	basis	C	p_0	1	1	1	0	0
				p_1	p_2	p_3	p_4	p_5
1	p_1	1	1	1	0	0	①	−3
2	p_2	1	2	0	1	0	2	0
3	p_3	1	3	0	0	1	1	−2
4			6	0	0	0	4	−5

The column i states the row index, the column *basis* lists vectors in the basis, the column C gives the *cost coefficients* associated with the vectors, the columns $p_0, \ldots p_5$ state the current values of the vectors and the associated cost coefficients are placed at the head of each column. The bottom line contains the value of M_0 in column p_0, and the values of $(M_j - c_j)$ under the appropriate columns.

The constraints already contain the unit matrix I_3 so that the initial basic feasible solution is chosen $(x_1, x_2, x_3, 0, 0) = (1, 2, 3, 0, 0)$.

Of the $(M_j - c_j)$, $(M_4 - c_4)$ is the largest. Of the x_i/x_{i4}, both x_1/x_{14} and x_2/x_{24} are equal to 1. A method of resolving the tie that has been found satisfactory in practice is to choose the lowest index i. Hence vector p_1 is to be removed from the basis, and vector p_4 is to be placed in the basis. The coefficients of x_{i4}, $i \neq 1$, are cleared by elimination using row 1 and the results are shown in Tableau 4.2:

TABLEAU 4.2

i	basis	C	p_0	1	1	1	0	0
				p_1	p_2	p_3	p_4	p_5
1	p_4	0	1	1	0	0	1	−3
2	p_2	1	0	−2	1	0	0	⑥
3	p_3	1	2	−1	0	1	0	1
4			2	−4	0	0	0	7

The maximum $(M_j - c_j)$ is $(M_5 - c_5)$ and the minimum x_i/x_{ik} that is non-negative is x_2/x_{25}, but this has a value of 0. The third and final tableau is shown below:

TABLEAU 4.3

				1	1	1	0	0
i	basis	C	p_0	P_1	P_2	P_3	P_4	P_5
1	P_4	0	1	0	$\frac{1}{2}$	0	1	0
2	P_5	0	0	$-\frac{1}{3}$	$\frac{1}{6}$	0	0	1
3	P_3	1	2	$-\frac{2}{3}$	$-\frac{1}{6}$	1	0	0
4			2	$-\frac{5}{3}$	$-\frac{7}{6}$	0	0	0

In the final tableau all of the $(M_j - c_j) < 0$ so that the minimum feasible solution has been found. The final solution is the vector $\mathbf{x} = (0, 0, x_3, x_4, x_5)$ and therefore $\mathbf{x} = (0, 0, 2, 1, 0)$. It is interesting to note that although there are three constraints, there are only two non-zero x_i. The solution is therefore a degenerate basic feasible solution, as also is the solution given by Tableau 4.2.

The method obviously lends itself to automatic computation, so that most computer manufacturers have standard computer programs for linear programming problems. In some instances problems containing hundreds of variables may be treated.

Optimisation of non-linear objective functions

If an n-dimensional non-linear objective function is defined in terms of variables x_i:

$$M = f(x_1, x_2, \ldots x_n) \qquad \ldots (4.157)$$

subject to m constraints that will usually take the form of upper and lower bounds on the x_i, so that:

$$f_{li}(x_1, \ldots x_n) < x_i < f_{ni}(x_1, \ldots x_n) \qquad \ldots (4.158)$$

where f_{li} is the lower allowed limit, and f_{ni} is the upper allowed limit, of x_i. There may be one equation 4.158 for each variable, two or more if there are two or more expressions for the upper and lower bounds to be satisfied, or none if the variable is not constrained. In addition there may be further constraints of the type:

$$f_{lj}(x_1, \ldots x_n) < g_j(x_1, \ldots x_n) < f_{nj}(x_1, \ldots x_n) \qquad \ldots (4.159)$$

where the g_j are functions defined in terms of the x_i.

For non-linear functions it is possible that the maximum or minimum of the objective function will be found at a point that is not on the boundaries of the constrained region. For this instance the condition:

$$\frac{\partial M}{\partial x_i} = 0; \quad i = 1, \dots n \qquad \qquad \dots (4.160)$$

is satisfied, but the condition may also be satisfied at other points. However, if the maximum or minimum occurs at a boundary of the constrained region the condition 4.160 is not satisfied in general.

The methods employed to find the optimum condition may be classified as methods of direct search or of indirect search. Methods of indirect search are based upon analytical studies of the optimum condition, while methods of direct search are essentially experimental in that the local values of variables to be optimised are changed, and the effect of the change upon the value of the objective function is used as a guide to further changes in the value of the variables. Only methods of direct search will be discussed here.

Direct search in one dimension, or vector search. Suppose that the minimum value of the function:

$$M = f(x_1) \qquad \qquad \dots (4.161)$$

is to be found subject to the condition:

$$a < x_1 < b \qquad \qquad \dots (4.162)$$

where a and b are constants.

One simple method starts with the value of the objective function at one end of the range, and examines changes in the objective function as the value of x_1 is incremented. Thus x_1 is incremented until the value of M is found to increase at $a + jh$, for example, where h is the increment. New constraints are then set on the variable x_1:

$$a + (j - 2)h < x_1 < a + jh \qquad \qquad \dots (4.163)$$

and the search is repeated. If a minimum is not found within the interval, the search is extended outside the interval.

This method, although simple and effective, is not the most economical method of search and a greatly improved performance may be given by a *golden-section search*.

Suppose that a search is to be carried out within the interval AB of Fig. 4.18 which is known to contain a single minimum. Two estimates of the objective

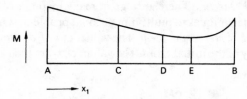

FIG. 4.18. Search by golden section

function are made within the interval at the points C and D chosen so that BC = AC(1 + $\sqrt{5}$)/2 and AC = CD(1 + $\sqrt{5}$)/2. If D yields a lower value of the objective function than C, then AC cannot contain the minimum. The value of the objective is then estimated at the point E, chosen so that BC = EC(1 + $\sqrt{5}$)/2 and the process continued until the range of uncertainty is sufficiently small.

The golden-section search may start from a point on a vector. An estimate is made at an arbitrary point on the vector and if the objective function is smaller at this point, a new trial is made at a point on the vector that is a distance of (1 + $\sqrt{5}$)/2 times the original interval from the first trial point. This process is continued until the objective function fails to decrease, when the pattern of search adopted is changed to the closed interval search of the last paragraph.

As an example of an application of the golden-section search consider the finding of the minimum of:

$$M = (x_1 - 3)^2 \qquad \qquad \dots (4.164)$$

where the constraint is:

$$0 < x_1 < 10 \qquad \qquad \dots (4.165)$$

TABLE 4.10. *Values of* x_1, *M for equation* 4.163

x_1	M
0·000	9
6·18	10·11
3·82	0·672
2·36	0·410
1·46	2·37
2·92	0·00673
3·26	0·0688
2·70	0·0869
3·05	0·00245
3·131	0·0171
2·999	$5·38 \times 10^{-8}$

The progress of the computation is shown in Table 4.10 in which the successive values of x_1 and the corresponding values of M are shown. When no information other than the value of the objective function is available, the golden-section search performs almost as well as the best that can theoretically be attained, and has two further advantages in that it is easy to program for an automatic digital computer, and has low storage requirements during machine computation.

Multi-dimensional searches. There are a great many methods of linking simple dimensional search techniques to multi-dimensional problems. One of the better known is the method of *steepest descent.* The values of $\partial M/\partial x_i$ are calculated for each x_i in the vicinity of the initial point. The values of x_i are then simultaneously changed by:

$$\Delta x_i = C \frac{\partial M}{\partial x_i}; \quad C < 0, \quad i = 1, \dots n \qquad \qquad \dots (4.166)$$

where C is some appropriate negative number. The value of the objective function may then be calculated at the point $(x_1 + \Delta x_1, \ldots x_i + \Delta x_i, \ldots)$, further derivatives calculated and the procedure repeated. Alternatively, a single-dimension search may be made in the direction of 4.166 (the *gradient vector*), for a limited number of applications. Values of Δx_i are then calculated and the procedure is repeated.

Although straightforward in application, the method of steepest descent has disadvantages when the surface is steeply ridged. Other methods that have some advantage over steepest descent include methods due to BUEHLER, SHAH and KEMPTHORNE[14], HARKINS[15], ROSENBROCK[12], and FLETCHER and POWELL[13, 16, 17]. The original papers should be consulted for details.

FURTHER READING

Modern Computing Methods (H.M.S.O., London 1961).
BICKELY, W. G. and THOMPSON, R. S. H. G.: *Matrices* (English University Press 1964).
FADDEEVA, V. N.: *Computational Methods of Linear Algebra* (Dover 1959).
INCE, E. L.: *Ordinary Differential Equations* (Dover 1956).
LAPIDUS, L. and SEINFELD, J. H.: *Numerical Solution of Ordinary Differential Equation* (Academic Press 1971).
AMES, W. F.: *Numerical Methods for Partial Differential Equations* (Nelson 1969).
Non-Linear Optimization Techniques, I.C.I. Monograph No. 5 (Oliver & Boyd 1969).

REFERENCES TO CHAPTER 4

[1] GASS, S. I.: *Linear Programming, Methods and Applications* (McGraw-Hill 1958).
[2] *Modern Computing Methods* (H.M.S.O. 1961).
[3] BUCKINGHAM, R. A.: *Numerical Methods* (Pitman 1962).
[4] WILKINSON, J. H.: *Rounding Errors in Algebraic Processes* (H.M.S.O. 1963).
[5] CLULEY, J. C.: *Electronic Computers* (Oliver & Boyd 1967).
[6] FORSYTHE, G. and MOLER, C. B.: *Computer Solution of Linear Algebraic Systems* (Prentice-Hall 1967).
[7] RUNGE, C.: *Math. Annln* **46** (1895) 167. Ueber die numerische Auflösung von Differentialgleichungen.
[8] KUTTA, W.: *Z. Math. Phys.* **46** (1901) 435. Beitrag zur näherungsweisen Integration totaler Differentialgleichungen.
[9] PEACEMAN, D. W. and RACHFORD, H. H.: *J. Soc. ind. appl. Math.* **3** (1955) 28. The numerical solution of parabolic and elliptic differential equations.
[10] DOUGLAS, J.: *J. Soc. ind. appl. Math.* **4** (1956) 20. On the relation between stability and convergence in the numerical solution of linear parabolic and hyperbolic differential equations.
[11] HAMMING, R. W.: *J. Ass. comput. Mach.* **6** (1959) 37. Stable predictor-corrector methods for ordinary differential equations.
[12] ROSENBROCK, H. H.: *Comput. J.* **3** (1960) 175. An automatic method for finding the greatest or least value of a function.
[13] FLETCHER, R. and POWELL, M. J. D.: *Comput. J.* **6** (1963) 163. A rapidly convergent descent method for minimization.
[14] BUEHLER, R. J., SHAH, B. V., and KEMPTHORNE, O.: *Chem. Eng. Prog.* Symp. Ser. No. 50, **60** (1964) 1. Methods of parallel tangents.
[15] HARKINS, A.: *Chem. Eng. Prog.* Symp. Ser. No. 50, **60** (1964) 35. The use of parallel tangents in optimization.
[16] POWELL, M. J. D.: *Comput. J.* **7** (1964) 155. An efficient method for finding the minimum of a function of several variables without calculating derivatives.
[17] POWELL, M. J. D.: *Comput. J.* **7** (1965) 303. A method for minimizing a sum of squares of non-linear functions without calculating derivatives.

LIST OF SYMBOLS USED IN CHAPTER 4

D	Differential operator
E	Displacement operator (equation 4.63)
h, H	Interval
δ	Central difference operator (equation 4.61)
Δ	Forward difference operator (equation 4.61)
∇	Backward difference operator (equation 4.61)
μ	Averaging operator (equation 4.66)

CHAPTER 5

Biochemical reaction engineering

THE PRINCIPLES OF MICROBIAL REACTORS

Micro-organisms are exploited industrially to produce additional micro-organisms and to achieve biochemical conversions of water-soluble organic compounds. The microbial cells are used as animal foodstuffs, for seeding batch processes, and as a source of enzymes for both research and commercial applications. The biochemical products, on the other hand, may be beverages like beers and wines, medicinal compounds such as antibiotics and steroids, and industrial chemicals, for instance, solvents or organic acids.

A biochemical conversion with the aid of a micro-organism differs from the purely chemical process in a number of ways, particularly:

(a) the complexity of the reactant mixture,

(b) the increase in the mass of micro-organisms simultaneously with the accomplishment of the biochemical transformation,

(c) the ability of micro-organisms to synthesise their own catalysts (enzymes),

(d) the mild conditions of temperature and pH involved, and its greater sensitivity to changes in these conditions,

(e) the difficulty in maintaining the required biochemical transformation (stability), and

(f) the restriction to the aqueous phase.

Microbiological processes also play an important part in the control of water pollution. Organic impurities of low concentration occur in vast amounts of water. These are converted to a solid phase, i.e. micro-organisms, and innocuous carbon dioxide and water. The former are removed by mechanical means.

The physiology of micro-organisms

Micro-organisms are forms of life of microscopic dimensions[9]. They may be either *saprophytes* or *parasites**. Normally it is the former that are concerned in industrial microbiological processes.

A unicellular micro-organism, as indicated in Fig. 5.1, is a highly ordered and complex entity consisting of sub-cellular units known as *organelles*, which can only be detected with the aid of an electron microscope at a magnification of 30,000 times or greater. The organelles, as with organs in the human

---

* Biological terms are defined in the glossary on pages 423–4.

body, each have a specific function in the overall life processes of the micro-organism.

Micro-organisms of industrial importance take in "food" from solution by a diffusion mechanism rather than by ingestion of "solid" particles. Like all living creatures, they are concerned with the life processes of growth and reproduction, which they accomplish primarily as a result of biochemical transformations.

The basic requirements for microbiological processes are energy, carbon, nitrogen and minerals. Frequently the source of energy and carbon is an organic compound which is oxidised to release energy and which also provides the structural carbon for the synthesis of new cellular material.

Micro-organisms which use carbon in this way are referred to as *heterotrophs*[3]. In contrast *autotrophs* are independent of external supplies of previously elaborated organic compounds and normally synthesise carbohydrates from carbon dioxide and water. Their energy requirements are met either by light (*photosynthesis*) or by the oxidation of inorganic compounds such as nitrogen, sulphur and certain metals. Structural nitrogen, however, may be required in either an inorganic or organic form.

The resulting cellular material, in either case, consists of approximately 50 per cent dry weight carbon and 10 per cent dry weight nitrogen and is de-scribed[5] by the empirical chemical formula $C_5H_7NO_2$.

The overall biochemical reaction or metabolism is on balance exothermic. The energy released by the oxidative reactions is consumed in synthetic reactions, and is dissipated as heat to the external environment and as mechanical and osmotic work[6].

Microbial metabolism may be thought of as a series of interconnecting reaction loops or metabolic pathways (Fig. 5.2) arranged spatially throughout the cell. The basic unit in a metabolic pathway is a reaction catalysed by an enzyme. The overall reaction path is controlled by the micro-organism itself, largely by adjustment of the rate of enzyme synthesis, or alternatively by inhibition of the enzymes by the product itself.

The industrial objective, at least where biochemical products are involved, is to use a part of the overall metabolism for a particular biochemical conversion. This is achieved in an advantageous manner by the supply of a primary organic raw material, or substrate, in addition to those metabolites required by the micro-organism for survival. If possible, an attempt is made to enhance the amounts and activity of the enzymes involved in the conversion, while those enzymes which act on the desired product are inhibited or repressed. These objectives are attained by the addition of appropriate chemical constituents to the reactant mixture or nutrient medium. By and large, the composition of this medium is arrived at on an empirical basis and consequently experience plays an extremely important part.

A nutrient solution is frequently of considerable chemical complexity,

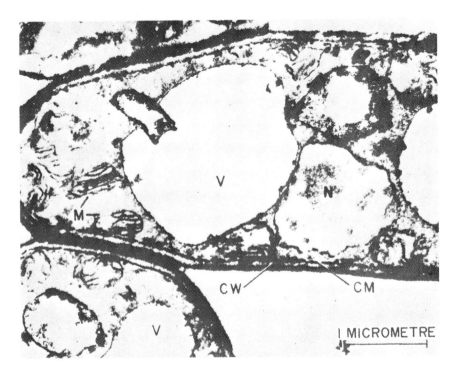

Fig. 5.1. Section through aerobic cells of the yeast *Torulopsis utilis* showing the cell wall (CW), cytoplasmic membrane (CM), nucleus (N), mitochondria (M), and vacuole (V)

(Courtesy of A. W. Linnane *et al.*: *J. Cell Biol.* **13** (1962) 345.)

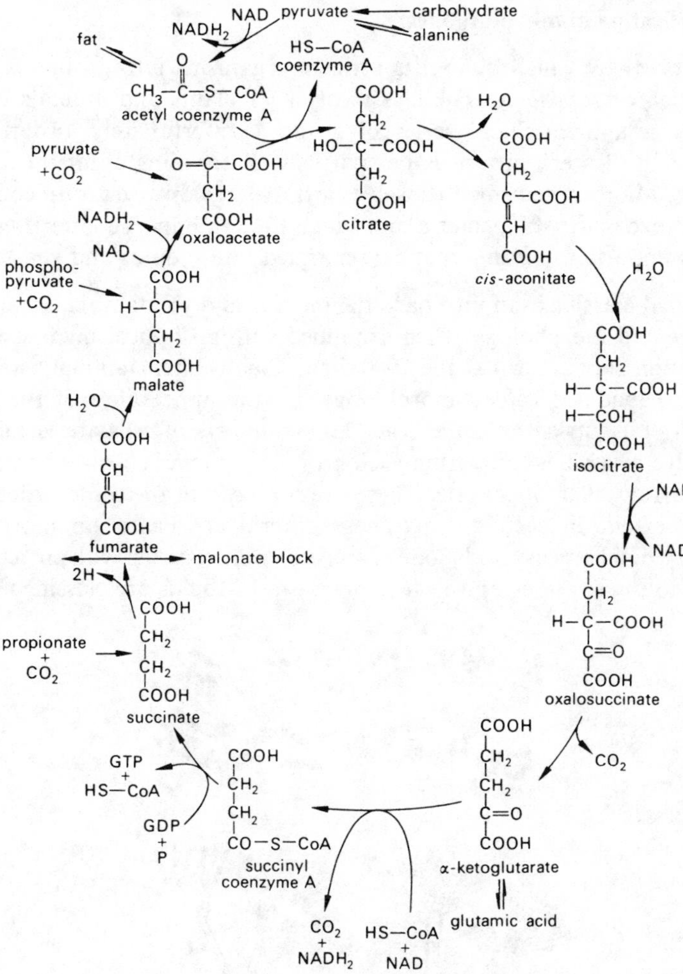

NAD pyruvate ← carbohydrate

NADH$_2$ ⇌ alanine

fat ⇌

HS—CoA
coenzyme A

O
‖
CH_3—C—S—CoA
acetyl coenzyme A

COOH
|
CH$_2$
|
HO—C—COOH
|
CH$_2$
|
COOH
citrate

H$_2$O

COOH
|
CH$_2$
|
C—COOH
‖
CH
|
COOH
cis-aconitate

H$_2$O

pyruvate
+CO$_2$

O=C—COOH
|
CH$_2$
|
COOH
oxaloacetate

NADH$_2$
NAD

phospho-
pyruvate
+CO$_2$

COOH
|
H—COH
|
CH$_2$
|
COOH
malate

H$_2$O

COOH
|
CH
‖
CH
|
COOH
fumarate — malonate block

2H

COOH
|
CH$_2$
|
H—C—COOH
|
H—COH
|
COOH
isocitrate

NAD
NADH$_2$

COOH
|
CH$_2$
|
H—C—COOH
|
C=O
|
COOH
oxalosuccinate

CO$_2$

propionate
+
CO$_2$

COOH
|
CH$_2$
|
CH$_2$
|
COOH
succinate

GTP
+
HS—CoA

GDP
+
P

COOH
|
CH$_2$
|
CH$_2$
|
CO—S—CoA
succinyl
coenzyme A

COOH
|
CH$_2$
|
CH$_2$
|
C=O
|
COOH
α-ketoglutarate

glutamic acid

CO$_2$ HS—CoA
+   +
NADH$_2$ NAD

FIG. 5.2. The Krebs cycle

though the mineral content, e.g. phosphorus, magnesium, calcium, potassium, sulphur, sodium, iron, cobalt, copper and zinc, is usually to be found in sufficient quantities simply as impurities in the materials from which the solution is prepared. The major constituents of a typical synthetic nutrient medium are given in the following table.

TABLE 5.1. A typical nutrient medium

	Concn. kg/m^3
Dextrose	100
MgSO$_4$7H$_2$O	0·2
KH$_2$PO$_4$	0·2
(NH$_4$)$_2$HPO$_4$	0·4
Tap water	

N

The classification of micro-organisms

The process by which biologists arrange organisms into groups is termed *classification* or *taxonomy*. The classification of plants and animals is based primarily on a hierarchical series of groups. Unfortunately, ordering into "higher" and "lower" groups is not particularly meaningful for most micro-organisms. Micro-organisms themselves form a large diverse group containing algae, protozoa, viruses, bacteria and fungi, though only the latter two are of industrial importance. Fungi may be separated into moulds and yeasts.

The broad classification into bacteria, moulds and yeasts is made on a basis of cell size and morphology when examined with an optical microscope at a magnification of between 400 and 1000 times, usually after staining[12] (Fig. 5.3). Use is also made of *colony* morphology, i.e. the appearance of the growth produced on a nutrient medium when large numbers of cells are visible to the unaided eye. Classification within each group, i.e. bacteria, yeasts and moulds, involves many different criteria. These are not used in the same order nor to the same extent in each group. These criteria are based on morphology, physiology, biochemistry and mode of reproduction, together with an occasional reference to the relationship to other organisms. Moulds are classified by their

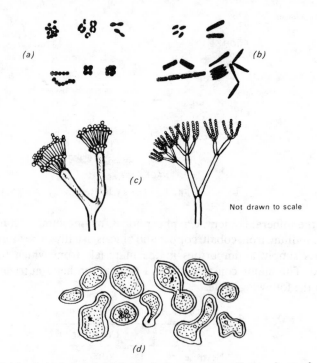

FIG. 5.3. Morphological characteristics of micro-organisms. Various forms of:
(a) spherical bacteria
(b) rod-shaped bacteria
(c) moulds
(d) yeasts

morphology and in particular the production and characteristics of their reproductive spores. For yeasts, these same criteria, together with their detailed physiology and biochemistry, are used. For bacterial classification all available criteria are used, with physiological and biochemical criteria predominant.

Bacteria are unicellular micro-organisms with their smallest dimension characteristically in the range 0·5 to 2μm and they reproduce asexually by binary fission. Yeasts, on the other hand, while also unicellular, have a size range from 5 to 10 μm and reproduce either by asexual budding or by sexual processes. Moulds may also reproduce by sexual or asexual means, but have a multicellular structure, and may be 5 μm or considerably larger in size.

The temperature range in which a given micro-organism is found to grow at finite rates leads to the classifications:
  (a) Psychrophiles—temperatures below 20°C
  (b) Mesophiles—temperatures between 20° and 45°C,
  (c) Thermophiles—temperature range 45° to 60°C.

Bacteria have an optimum pH range for growth between 6·5 and 7·5 and they display a particularly wide variety of patterns of response to free oxygen. Some have an absolute oxygen requirement while others can grow only in its complete absence, and intermediate species exist that develop with or without oxygen, though not necessarily at the same rates. The principal types of bacteria are:
  (a) aerobic bacteria which grow only in the presence of free oxygen,
  (b) anaerobic bacteria which grow only in the absence of free oxygen,
  (c) facultative anaerobic bacteria which grow in either the absence or presence of free oxygen, and
  (d) microaerophilic bacteria which grow in the presence of minute quantities of free oxygen.

Moulds, however, grow most rapidly under aerobic conditions, but generally more slowly than bacteria. They can therefore be overgrown by bacterial contaminants, though the media used are usually quite acid (say pH 5·6, which is outside the range for rapid bacterial growth) and this fact provides a degree of compensation.

An abundant supply of oxygen is required for the growth of yeasts. Nevertheless, in industrial fermentations using yeasts (e.g. beer production), it is frequently necessary to exclude oxygen and therefore reduce the growth rate of the micro-organisms, to obtain the most satisfactory yield of biochemical end products. The acid tolerance of yeast fermentations ranges from pH 2·2 to pH 8.

It must be remembered that the foregoing represent broad guide-lines and many exceptions occur. Some micro-organisms can tolerate and often thrive in what usually are very unfavourable environments (e.g. 25 per cent copper sulphate solution, 2·5N acid (=2·5 kmol/m^3); others can withstand exposure to extremes of acidity, desiccation and temperature for long periods and remain viable, i.e. retain their potential for reproduction and growth. In particular, bacterial spores are the most resistant form of life known and often cause infections in industrial fermentations.

The role of biochemical engineering and biochemical reaction engineering

Because of the problems associated with maintaining a given strain of micro-organisms, of sterility and (to some degree) of scale, microbiological processes have largely been developed on a batch or semi-continuous basis. These aspects, which have in the past tended to inhibit the progress of reactor design technology for such processes, are slowly being overcome as the tendency to large scale continuous processing inevitably continues[13]. This situation is illustrated with developments in continuous beer production, the waste treatment industry, and the production of cellular material for use as protein or for enzyme recovery.

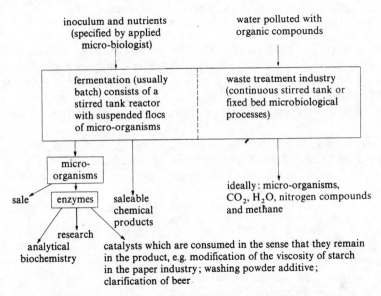

FIG. 5.4. The present pattern of the industrial exploitation of micro-organisms

The pattern of exploitation of micro-organisms is outlined on Fig. 5.4. Once the composition of nutrient solution and the species and strain of micro-organisms to be used for a particular objective have been determined from the microbiological viewpoint, it is the task of the biochemical engineer to devise an economic process involving media preparation, sterilisation, reactor design, product recovery and isolation. The design basis of such operations must inevitably be founded upon the rate processes of mass, heat and momentum transfer, together with reaction kinetics. While those aspects of the separation processes and reactor design which involve transport phenomena are moderately well understood[7, 8], it is in the areas of reaction kinetics and of the avoidance of contamination that microbiological systems pose new problems for the designer. It is thus the purpose of biochemical reaction engineering to provide appropriate kinetic rate equations to describe microbiological systems, to establish the utility of such equations for design purposes, and to devise laboratory experiments by which the semi-empirical rate coefficients inevitably involved in kinetic equations may be determined. For this endeavour, encouragement may be taken from the area of chemical reaction engineering[8] (see Chapter 1) where it has been found that complex and often unknown kinetics can frequently be described on a first or higher order basis, and that initially considerable advances were made by taking diffusional effects into consideration by the use of pseudo-orders of reaction.

The principal types of microbiological reactor

The basic reactor arrangement for the accomplishment of microbiological transformations is the *deep tank fermenter* (Fig. 5.5). The stages of innovation associated with this reactor, particularly for aerobic processes, provide interesting

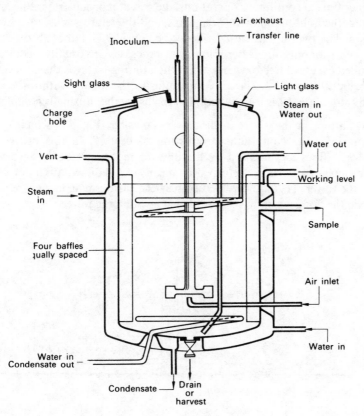

FIG. 5.5. The deep tank fermenter

technological history and may be followed through the development of the gluconic acid process as described by PRESCOTT and DUNN[1]. The basis of this reactor is the fact that micro-organisms in their normal state contain a considerable amount of water and, in consequence, have a density that differs only slightly from that of water. Very little hydrodynamic drag is required to maintain them in suspension, with the result that if the fluid surrounding them is in a mild state of motion, the micro-organisms will be suspended. The logical arrangement then becomes an essentially completely mixed reactor in which the motion is induced either by mechanical stirring, by the evolution of gas as a biochemical product, or by air bubbling through the medium. The last-mentioned system provides the free oxygen demanded by aerobic processes. Such a reactor arrangement may be used either on a batch or continuous basis as required, and has the basic advantages of being of a simple form for sterilisation purposes and of general applicability, giving the flexibility demanded by small scale batch processing. The most widely used continuous process based on this reactor type is the *activated sludge process* in the waste treatment industry[5].

The design problem associated with the deep tank fermenter lies in the specification of the size of the vessel, the process time in batch operations, the residence time in continuous operations, the reactant or substrate concentration required, the hold-up of micro-organisms, and the requirements for power and aeration. The two latter factors will not be considered, although they may be important economically. Microbiological processes are usually operated with controlled excess of free oxygen so that the limiting factor is the concentration of organic substrate[10]. The air contributes to the mixing within the reactor, but additional power is usually required to ensure that mixing is complete.

A variation on the above is the *tower fermenter* (Fig. 5.6) which has been developed for the continuous production of beer[29]. In this process, yeast particles or flocs are fluidised by the upward movement of the medium and any entrained particles are returned by means of a sedimentation device at the top of the tower. The liquid velocities are necessarily small, in view of the need to retain the flocs, which have a low density, within the tower.

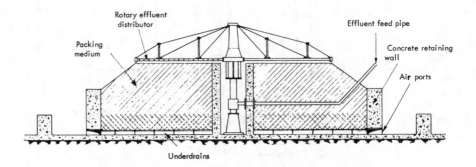

FIG. 5.7. Trickling filter

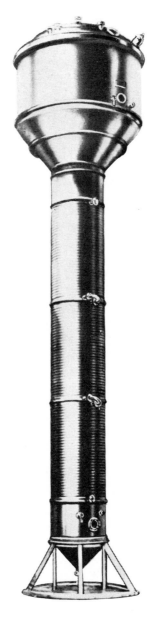

FIG. 5.6.   The tower fermenter

In *fixed bed* processes the micro-organisms adhere to a support surface. These processes are chiefly used in the waste treatment industry, usually in a *trickling filter*[5] (Fig. 5.7). Here a film or layer of micro-organisms is built up, usually on crushed stone, although plastic support surfaces are gaining popularity. The organic-bearing waste waters flow under gravity as a liquid film over the support surface. As the liquid passes through the "filter" the organic impurities are removed by the microbial mass and oxidised to compounds containing smaller carbon chains and to additional micro-organisms.

Biological rate equations

Industrial microbiological manufacturing processes invariably use organic compounds, as a source both of chemical energy and of structural carbon, in order to attain biochemical transformations. Matters are usually arranged so that the components of the nutrient mixture which are required in relatively low concentrations are provided in excess, with the result that the basic energy and carbon source is the rate controlling reactant.

The consumption of this reactant in metabolic processes results in additional micro-organisms plus any number of biochemical products. Thus:

$$K_o + K_{p1} + K_{p2} + K_{p3} + \ldots = 1 \qquad \ldots\ldots(5.1)$$

where K_o refers to the fractional conversion of substrate to additional micro-
organisms, and
K_{p1} etc. are the fractional conversions to various biochemical products.

The total yield of any product, e.g. new micro-organisms, may be obtained from the consumption of substrate according to the linear relation:

$$\text{micro-organisms produced } = S_o K_o \text{ (total substrate consumed)} \qquad \ldots\ldots(5.2)$$

where S_o is a stoichiometric coefficient, and the coefficient $S_o K_o$ is referred to as a *yield coefficient*[10].

MONOD[15] has confirmed that the yield of new microbial mass is simply related to the consumption of substrate, as indicated by equation 5.2, and that the yield coefficient $S_o K_o$ is in fact a constant, and specific to a given nutrient solution and micro-organism system; its value is affected by various physical parameters, such as temperature, pH and concentration of trace components. Correspondingly, for the biochemical products, the yield coefficients $S_{p1} K_{p1}$, $S_{p2} K_{p2}$ etc. are involved. It follows that knowledge of the yield coefficients as system constants and the total substrate consumption leads to the total amount of any given product that has been formed. A similar argument, of course, applies to the rate of production, in which case equation 5.2 becomes:

$$\text{Rate of production of micro-organisms} = S_0 K_0 \text{ (rate of substrate consumption)} \qquad \ldots\ldots(5.3)$$

It is sufficient, therefore, for design purposes, to consider only the rate of consumption of the limiting substrate providing the yield coefficients are known.

The yield of a desired component can be improved by enhancing the corresponding yield coefficient relative to those of the less desirable products. This can be achieved by using an alternative strain of micro-organisms or by exploiting the biochemistry of the process in blocking undesirable reactions. Obviously any advantageous modification to the yield coefficients produces a further benefit in the product recovery and isolation stages of the process.

The rate of substrate consumption depends, in general, on the metabolic processes in the microbial mass, as well as on diffusional effects within the liquid phase. The former has been related empirically to the substrate concentration, at the interface of the micro-organisms and the liquid phase, by the equations:

$$n = b'C^* \qquad \qquad \dots(5.4)$$

$$n = a' \qquad \qquad \dots(5.5)$$

$$n = \frac{fC^*}{g + C^*} \qquad \qquad \dots(5.6)$$

where n is the rate of removal of substrate in appropriate units,
   C^* is the substrate concentration at the interface with the microbial mass, and
   a', b', f, g may be termed biological rate coefficients with units corresponding to those of n and C^*.

Equation 5.4 is a first order rate equation which has found particular application in reactors involving microbial films[5], whereas the zero order equation[29] and the Langmuir adsorption isotherm type of equation[15] (as exemplified by equations 5.5 and 5.6) have been used successfully with microbial floc. However, experience of chemical processes suggests that it would be unlikely for completely different equations to apply to a given system as a result of a change simply in the geometrical arrangement. Thus, while sufficient evidence exists in support of equations 5.4, 5.5 and 5.6 to indicate that they do indeed describe substrate consumption, closer study of the problem might provide an interpretation in which these equations were seen simply to be parts of a more comprehensive picture in which the coefficients a', b', f, g were related.

The advantage of such knowledge is that it enables the results of experiments carried out using, say, microbial flocs to be applied to the design of reactors involving microbial films; to quantify, correlate and classify biological rate coefficients, and to understand better the problems associated with the scale-up of microbial processes.

As with the yield coefficients, the possibility exists for the improvement of the biological rate coefficients. It may be that an improvement in one set of these system constants results in an adverse effect in the other. In such a case the economic route to choose must inevitably be coupled with the costs both of the raw materials and of the recovery of the product.

Many attempts have been made to classify fermentation processes using, as a basis, the complex manner in which the concentrations of the substrate, of intermediates and of product vary over the period of a batch fermentation[4, 20]. It is apparent from the data given in support of these classification procedures that both the yield coefficients and the biological rate coefficients are themselves concentration dependent. This dependence, however, may not necessarily involve either the limiting substrate or the desired products.

Features of microbiological and chemical reactors

The chief contrast between these types of reactor lies in the fact that all microbiological reactions can be described as "autocatalytic" in the sense that the formation of one of the products, new micro-organisms, enhances the overall rate of the reaction. Conversely, in the absence of micro-organisms no reaction can take place and it is essential to retain at least a portion within the reactor. An important characteristic of this product is that it is a solid phase. Thus, microbiological reactors involve both solid and aqueous phases and can be classified as heterogeneous.

The stirred tank reactor lends itself to these systems since micro-organisms are readily suspended. When used on a continuous basis such *chemical* reactors, in order to achieve high conversions, usually require a number of units in series and the overall reactor volume involved is rather greater than for an equivalent tubular reactor (see Chapter 1). In contrast, a *single* stirred tank biochemical reactor achieves a high substrate conversion, but the use of a multi-stage system has its advantages when the yield coefficients and biological rate coefficients exhibit a dependence on concentration, or can be improved by exposing the organisms to a series of different environments. Thus multi-stage continuous biochemical reactors often require added substrate at each stage. The difference in reactor volume between a multi-stage system and a tubular reactor is usually marginal since, in general, the *biological rate equation* has a pseudo-order of between zero and unity with respect to concentration.

A tubular reactor can, of course, involve either microbial flocs, as in the tower fermenter, or microbial film, as with the trickling filter. It is immediately apparent that the former must be arranged with a vertical orientation and can only be operated as a fluidised bed. This arises because microbial flocs are deformable and any attempt to operate on a fixed bed basis would simply result in a most unsatisfactory liquid flow pattern. The alternative of using a microbial film adhering to a support surface has the attraction of retaining, by physical means, a known and roughly constant amount of micro-organisms within the reactor. The complication here lies in the difficulties in removing the growth from the tower and thus preventing blockage. In the trickling filter this removal is achieved in an uncontrolled, and generally unsatisfactory, manner by relying upon the overall ecology; e.g. the activities of worms and flies, together with the action of the liquid flow in washing micro-organisms from the filter.

A further advantage of the continuous stirred tank fermenter lies in its attenuation of any accidental shock load that may be imposed on the system. This is due to the effect of good mixing which causes the concentration levels within the reactor to approximate to the outlet rather than the inlet conditions. This flexibility is particularly important in the waste treatment industry. The improved potential for temperature control of the stirred tank fermenter, as compared with a tubular reactor (see Chapter 1), does not provide any over-riding advantages. Although the reactions involved are exothermic, the total heat released can be easily removed in either configuration.

A MODEL FOR A SINGLE MICRO-ORGANISM

Micro-organisms, while varying in size and form as indicated in Fig. 5.3 (page 370), basically have a similar cellular organisation. The intricacy of their structure is shown diagrammatically by the transverse section of yeast cells in Fig. 5.1 (facing page 368).

Providing that a given micro-organism can survive in its local environment of pH, temperature and nutrient solution, it is concerned primarily with metabolism and reproduction. Each step in the process of metabolism involves a different reaction pathway, similar to that shown in Fig. 5.2 (page 369), and each reaction is individually catalysed by a specific enzyme. The whole of this process is controlled by the micro-organism and this control, in turn, is genetic-ally determined.

The description of a micro-organism resulting from a detailed physico-chemical analysis is so complex that, if complete cognisance of all its features had to be taken into account to achieve a mathematical model, the problem would be beyond comprehension. However, experience indicates that such an interpretation would be unnecessarily complicated and that a model based on considerable simplifications would lead to a meaningful result. Such a procedure applied to the physical structure suggests that a micro-organism may be regarded as a combination of a substrate "sink" (i.e. a region where the substrate is consumed by the processes of metabolism) and a boundary region across which the substrate has first to be transported[16].

It is generally accepted[19] that the boundary region, which consists of the extramural layers, the cell wall and the cytoplasmic membrane, both preserves the integrity of the micro-organism and limits the entry and exit of solutes. Movement of molecules in this region is accomplished by either of two mechanisms, depending upon the substrate involved. One is a conventional molecular diffusion process that obeys Fick's Law, while the other is more complex, and is held by many workers[15,21] to involve stereo-specific carrier molecules. These molecules are proteins and have been given the generic name *permeases* to distinguish them from metabolic enzymes[11]. On the basis of a theoretical analysis, in conjunction with appropriate experiments, COHEN and MONOD[18] have demonstrated that this transport mechanism obeys the equation:

$$r_1 = \frac{\alpha_1 C^*}{\beta_1 + C^*} \qquad \ldots (5.7)$$

where r_1 is the transport rate per unit area of micro-organism,

C^* is the substrate concentration external to the boundary region, and

α_1 and β_1 are rate coefficients in appropriate units.

Equation 5.7 is of the same form as the Michaelis–Menten equation (5.8) for simple enzyme reactions, the Langmuir adsorption isotherm equation, and the equation describing saturation kinetics.

In contrast to the molecular diffusion mechanism, equation 5.7 is not influenced by the concentration in the "sink". This mechanism results in "up-hill" or *active* transport where the transport mechanism is working against a concentration gradient. In these circumstances the energy for transport comes from the reserves of energy within the cell. Alternatively, equation 5.7 also describes a transport mechanism that relies on thermal energy, known as *passive* transport, which occurs when the concentration within the "sink" is lower than that in the external environment.

Nevertheless, for present purposes it is sufficient that equation 5.7 describes both active and passive transport and that Fick's law describes molecular diffusion.

The individual reactions that compose the metabolic pathways are controlled by different enzymes and, from classical enzyme theory, it is known[2] that such reactions follow the Michaelis–Menten equation (5.8) given below. Similarly, the probable kinetic models of complex enzymatic systems invariably reduce to this form when the rate coefficients for individual steps in the reaction pathway are compounded into two coefficients α_2 and β_2. Equation 5.8 therefore provides a reasonable representation for the substrate consumption in the inner region or sink:

$$r_2 = \frac{\alpha_2 C}{\beta_2 + C} \qquad \ldots (5.8)$$

where r_2 is rate of consumption of substrate per unit volume of micro-organism, and

C represents the substrate concentration in the inner region.

The final physico-chemical model for a microbiological process thus emerges in the form of a cell assumed to consist of an outer transport zone and an inner metabolic region (Fig. 5.8). The passage of substrate through the outer zone takes place at a rate governed by either Fick's law or equation 5.7.

Under steady conditions:

$$r_1 = r_2 \frac{V_m}{a_m} \qquad \ldots (5.9)$$

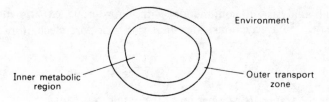

FIG. 5.8. A model for a single micro-organism

where a_m and V_m are the area and volume of a single micro-organism, and either:

$$\frac{\alpha_1 C^*}{\beta_1 + C^*} = \frac{V_m}{a_m} \frac{\alpha_2 C}{\beta_2 + C} \quad \dots (5.10)$$

or

$$-D\frac{dC}{dx} = \frac{V_m}{a_m} \frac{\alpha_2 C}{\beta_2 + C} \quad \dots (5.11)$$

The cell is capable of metabolising the substrate at a maximum rate of $(V_m/a_m)\alpha_2$. If, in the case of equation 5.10, this rate is assumed to be greater than the maximum rate of entry α_1, then both forms of transport together with metabolism consume the substrate at a rate governed by the general equation:

$$r = \frac{\alpha C}{\beta + C} \quad \dots (5.12)$$

Equation 5.12 thus describes the rate at which a substrate is removed by unit surface area of viable micro-organisms. Experimental evidence in support of this equation has been given by both CIRILLO[21] and MONOD[15].

The enzymes concerned with metabolism are known as *intracellular* enzymes. However, some enzymes are located outside the transport zone. These *extracellular* enzymes are primarily concerned with the hydrolysis of poly-saccharides to form mono-saccharides prior to their assimilation by the cell. If these reactions are assumed to take place at a rate given by the Michaelis–Menten equation, it follows that the above argument still holds in principle, if not in detail, and that the conclusions associated with equation 5.12 are applicable in these circumstances.

A MODEL FOR A MICROBIAL MASS

In industrial processes micro-organisms are employed in one of two forms, namely as biological flocs freely suspended in the fluid, or as a biological film adhering to a mechanical support surface. In practice, the characteristic size of such flocs and films is many orders of magnitude greater than the size of an individual micro-organism[24, 25]. This situation arises as a result of the gelatinous nature of micro-organisms and the fact that, in the main, they are exposed to relatively low rates of hydrodynamic shear.

A composite model of micro-organisms, arranged in the form of either flocs or film, is represented diagrammatically in Fig. 5.9. In this model, the micro-organisms are assumed to be distributed uniformly throughout a biochemically inert intercellular gel. The volume fraction of this region has been variously reported for centrifuged micro-organisms as 23–38 per cent, the variation being attributable to the different experimental procedures adopted[14, 16].

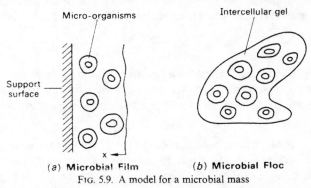

(a) **Microbial Film**   (b) **Microbial Floc**
FIG. 5.9. A model for a microbial mass

Since the microbial mass retains its cellular fluid (mainly water) when normally employed, it might be anticipated that these estimates of the "free space" are, if anything, lower than those pertaining in practice. It is of some interest to compare this situation with a random bed of packings where similar values of the "free space" occur (see Volume 2, Chapter 4).

This physical model for a microbial mass, i.e. that of micro-organisms dispersed in a biochemically inert gel, has considerable similarity to a porous catalyst. Both systems consist of inter-connected channels through which the reactants must diffuse before reaching the "active" surface.

In order to describe the problem in mathematical terms, it is inevitably necessary to make a number of further assumptions with regard to the characteristics of the microbial mass. These are as follows:

(a) A steady state in the biological sense exists in that the composition of the total microbial mass remains invariant with respect to time, in spite of the fact that the growth of each micro-organism does not fit this criterion.

(b) The cell is a unit whose functions are invariant with time and which has average gross properties that are functions of the local environment.

(c) For the microbial mass as a whole, the age distribution of the population and such biological properties as viability, mutation, selection, adaption and competition, are time invariant.

A differential mass balance, applied to the microbial mass in Fig. 5.9a (above), in which the molecular diffusion of the substrate is equated to its removal by the micro-organisms, leads to:

$$D_e \frac{d^2C}{dx^2} - ar = 0 \qquad \qquad \ldots . (5.13)$$

where r is the rate of substrate removal per unit area of micro-organism,
$\quad$ a is the active surface area per unit volume of microbial mass, and
$\quad$ D_e is an effective molecular diffusion coefficient.

Implicit in this equation are several additional assumptions which are also common to the problem of the porous catalyst particle. These are:

(a) that the molecular diffusion process within the tortuous intercellular passages can be described in terms of an effective diffusion coefficient, defined by:

$$N' = -D_e \frac{dC}{dx} \qquad \ldots.(5.14)$$

where N' is the flux of substrate within the microbial mass, and
$\quad$ C is the local concentration of substrate within the intercellular gel,

(b) that the effective diffusivity and the area a are constant in the direction of diffusion.

Clearly, in view of the fact that the microbial mass is heterogeneous in nature, equations 5.13 and 5.14, which are valid only for a continuum, are in conflict with the physical model. Thus the symbols r and N' cannot be regarded as meaningful at a point, but must be taken as quantities averaged over a neighbourhood.

Since the rate of removal of substrate per unit area of viable micro-organisms is given by equation 5.12, this may be combined with equation 5.13, with the result that:

$$D_e \frac{d^2C}{dx^2} - \frac{a\alpha C}{\beta + C} = 0 \qquad \ldots.(5.15)$$

For the situation depicted in Fig. 5.9a (page 381), i.e. the microbial film, the appropriate boundary conditions to be applied to this equation are:

$$x = L, \quad \frac{dC}{dx} = 0 \qquad \ldots.(5.16)$$

$$x = 0, \quad C = C^* \qquad \ldots.(5.17)$$

where L is the thickness of the microbial film.

The first of these conditions implies a zero flux of substrate at the interface between the micro-organisms and the support surface, while the second specifies the substrate concentration C^* imposed adjacent to the microbial mass.

The mathematical solution to equations 5.15, 5.16 and 5.17 takes the form:

$$C = f(x, D_e, a\alpha, \beta, L, C^*) \qquad \ldots.(5.18)$$

However, a solution in terms of concentration profiles is not of any particular utility; it is more convenient to express equation 5.18 in terms of a flux at the interface of the microbial mass with the nutrient solution; thus, at $x = 0$:

$$N = -D_e \frac{dC}{dx} = f(D_e, a\alpha, \beta, L, C^*) \qquad \ldots.(5.19)$$

The usefulness of this equation lies in the fact that any substrate that crosses the interface is, at steady state, ultimately consumed by the micro-organisms. Thus equation 5.19 provides the relationship for use in association with equation 5.3 from which the rate of production of both microbial and biochemical products can be obtained.

A further advantage of equation 5.19 lies in the separation of the biological parameters, namely D_e, $a\alpha$ and β, from those of a purely physical nature, namely, the biological film thickness L and the external substrate concentration C^*. The latter may be controlled with relative ease as design parameters. The former are largely determined by the micro-organism and nutrient solution involved, except in so far as they depend upon such environmental factors as temperature and pH.

THE BIOLOGICAL RATE EQUATION

An appropriate name for equation 5.19 is the *biological rate equation*. Unfortunately, because of problems associated with the process of integration, it is not possible to obtain this analytically from equations 5.15, 5.16 and 5.17, and recourse has to be made to numerical procedures. ATKINSON and DAVIES[28] have shown, on the basis of the analogy between the present problem and heterogeneous catalysis, that equation 5.19 may be written in the form:

$$N = \lambda(Lar^*) \qquad \ldots\ldots(5.20)$$

where λ is an effectiveness factor, or the ratio of the actual removal rate to the

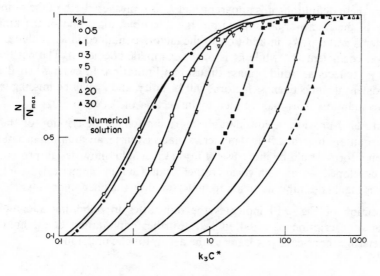

FIG. 5.10. The biological rate equation for microbial films (Table 5.2)—comparison with numerical solution

removal rate that would occur if the whole of the internal active surface were at the external concentration C^*, i.e. that given by (Lar^*). Thus, from equation 5.12:

$$N = \lambda La \frac{\alpha C^*}{\beta + C^*}$$

or

$$N = \lambda \frac{k_1 L C^*}{1 + k_3 C^*} \qquad \ldots .(5.21)$$

The data resulting from the numerical solution have been expressed in terms of equation 5.21, as is shown in Fig. 5.10, using the expression:

$$\frac{N}{N_{max}} = \lambda \frac{k_3 C^*}{1 + k_3 C^*} \qquad \ldots .(5.22)$$

where $N_{max} = \dfrac{k_1 L}{k_3}$ and λ is a function of $k_2 L$ and of $k_3 C^*$.

The coefficients k_1, k_2 and k_3 that occur in equations 5.21 and 5.22 are termed *biological rate coefficients* and are defined as:

$$\left. \begin{aligned} k_1 &= \frac{a\alpha}{\beta} \\[2mm] k_2 &= \sqrt{\frac{a\alpha}{\beta D_e}} \\[2mm] k_3 &= \frac{1}{\beta} \end{aligned} \right\} \qquad \ldots .(5.23)$$

From these definitions it is apparent that k_1 is a measure both of the number density of the micro-organisms within the microbial mass and of the kinetic coefficients α and β; k_2, in addition to the factors contained in k_1, includes the diffusion coefficient D_e, while k_3 is purely a kinetic coefficient. The net result is that k_2 reflects a "solid" phase diffusional limitation. In cases when this is non-existent the effectiveness factor λ will be unity, and k_2 will be missing from the rate equation. This may be seen by inspection of equation 5.22.

Figure 5.10 provides a considerably more convenient description of the required solution than the original numerical data. Even so, an equation in a purely functional form would be more useful in that it would allow design procedures to be developed on an analytical, rather than a purely numerical, basis; the former is highly desirable as an aid to understanding the phenomena involved.

Inspection of Fig. 5.10 indicates the difficulties in obtaining a complete functional description of equation 5.22. Further exploitation of the analogy with heterogeneous catalysis leads to the asymptotic expressions:

(a) for $\phi < 1$

$$\lambda = 1 - \frac{\tanh k_2 L}{k_2 L} \left[\frac{\phi}{\tanh \phi} - 1 \right] \qquad \ldots .(5.24)$$

(b) for $\phi > 1$

$$\lambda = \frac{1}{\phi} - \frac{\tanh k_2 L}{k_2 L}\left[\frac{1}{\tanh \phi} - 1\right] \qquad \ldots(5.25)$$

where
$$\phi = \frac{k_2 L}{(1 + 2k_3 C^*)^{\frac{1}{2}}}$$

The combination of equations 5.21, 5.24 and 5.25 provides biological rate equations for microbial films and these are given in Table 5.2 (below).

Strictly, equations 5.15, 5.16 and 5.17 apply to systems in rectangular co-ordinates—in this case, biological films. However PETERSEN[8] has shown, for problems involving diffusion with chemical reaction, how the mathematical solutions for both "slabs" and particles are essentially numerically identical if the characteristic length is selected judiciously. Thus, for the biological flocs shown in Fig. 5.9b (page 381) it is appropriate to define the characteristic length by:

$$L = \frac{V_p}{A_p} \qquad \ldots(5.26)$$

where V_p and A_p are the floc volume and external surface area.

In these circumstances equations 5.21 and 5.22 may be written:

$$N = \lambda \frac{(V_p/A_p) k_1 C^*}{1 + k_3 C^*} \qquad \ldots(5.27)$$

and
$$\frac{R}{R_{max}} = \lambda \frac{k_3 C^*}{1 + k_3 C^*} \qquad \ldots(5.28)$$

where λ is a function of $k_3 C^*$ and $k_2 V_p/A_p$.

TABLE 5.2. *The biological rate equation for microbial films*

(a) for $\phi < 1$
$$N = \left\{1 - \frac{\tanh k_2 L}{k_2 L}\left[\frac{\phi}{\tanh \phi} - 1\right]\right\}\left[\frac{k_1 L}{k_3}\right]\frac{k_3 C^*}{1 + k_3 C^*}$$

(b) for $\phi > 1$
$$N = \left\{\frac{1}{\phi} - \frac{\tanh k_2 L}{k_2 L}\left[\frac{1}{\tanh \phi} - 1\right]\right\}\left[\frac{k_1 L}{k_3}\right]\frac{k_3 C^*}{1 + k_3 C^*}$$

where $\phi = \dfrac{k_2 L}{(1 + 2k_3 C^*)^{\frac{1}{2}}}$

and the symbols have the following meanings and dimensions:

C^*	concentration of substrate at interface between micro-organisms and nutrient solution	ML^{-3}
k_1	biological rate coefficient	T^{-1}
k_2	biological rate coefficient	L^{-1}
k_3	biological rate coefficient	L^3M^{-1}
L	biological film thickness	L
N	flux substrate at interface between micro-organisms and nutrient solution	$ML^{-2}T^{-1}$

While it is appropriate to express the flux on a basis of unit external surface area for films, this is not particularly convenient in the case of flocs and a relationship for the substrate removal on a basis of unit mass of particles is to be preferred. Equation 5.28 is expressed in these terms by:

$$R = N \frac{A_p}{V_p \rho_o} \quad \text{and} \quad R_{max} = \frac{k_1}{k_3 \rho_o} \qquad \qquad \dots (5.29)$$

where R is the rate of substrate removal per unit mass of micro-organisms, and ρ_o is the density of the microbial mass.

Combination of equations 5.24, 5.25, 5.26 and 5.28 leads to the *biological rate equation* for microbial flocs in terms of the biological rate coefficients k_1, k_2 and k_3, together with the physical variables V_p/A_p and C^*. The resulting equations are given in Table 5.3 (opposite).

The biological rate equations may more readily be appreciated with reference to Figs. 5.11 and 5.12, where they are expressed diagrammatically for a given set of k_1, k_2 and k_3 in terms of the removal flux as a function of the external substrate concentration, with the characteristic size as a parameter. From

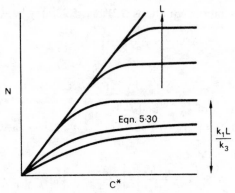

FIG. 5.11. The *biological rate equation* for microbial films

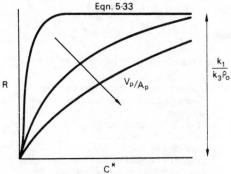

FIG. 5.12. The *biological rate equation* for microbial flocs

TABLE 5.3. *The biological rate equation for microbial flocs*

(a)   for $\phi < 1$

$$R = \left\{1 - \frac{\tanh(k_2 V_p/A_p)}{k_2 V_p/A_p}\left[\frac{\phi}{\tanh \phi} - 1\right]\right\}\left[\frac{k_1}{k_3 \rho_o}\right]\frac{k_3 C^*}{1 + k_3 C^*}$$

(b)   for $\phi > 1$

$$R = \left\{\frac{1}{\phi} - \frac{\tanh(k_2 V_p/A_p)}{k_2 V_p/A_p}\left[\frac{1}{\tanh \phi} - 1\right]\right\}\left[\frac{k_1}{k_3 \rho_o}\right]\frac{k_3 C^*}{1 + k_3 C^*}$$

where $\phi = \dfrac{k_2 V_p/A_p}{(1 + 2k_3 C^*)^{\frac{1}{2}}}$

these figures a number of asymptotes are evident, in particular at high and low substrate concentrations. This whole family of asymptotes can be obtained readily from the equations given in Tables 5.2 and 5.3. Thus for microbial films:

(a) in the absence of any "solid" phase diffusional limitation, i.e. for suffici-ently small L, equation 5.21 reduces to:

$$N = \frac{k_1 L C^*}{1 + k_3 C^*} \qquad \qquad \dots (5.30)$$

(b) in the presence of a "solid" phase diffusional limitation, i.e. sufficiently large L, equation 5.21 reduces to:

$$N = \frac{k_1}{k_2} C^* \qquad \text{(for small } C^*\text{)} \qquad \dots (5.31)$$

$$N = \frac{k_1}{k_3} L \qquad \text{(for large } C^*\text{)} \qquad \dots (5.32)$$

While for microbial flocs:

(a) with no diffusional limitation, i.e. for small V_p/A_p, equation 5.29 becomes:

$$R = \frac{k_1 C^*}{\rho_o(1 + k_3 C^*)} \qquad \dots (5.33)$$

(b) with a diffusional limitation, i.e. for large V_p/A_p:

$$R = \frac{k_1}{k_2} \frac{A_p}{V_p \rho_o} C^* \qquad \text{(for small } C^*\text{)} \qquad \dots (5.34)$$

$$R = \frac{k_1}{k_3 \rho_o} \qquad \text{(for large } C^*\text{)} \qquad \dots (5.35)$$

It is of interest to compare these asymptotes with the empirical rate equations 5.4, 5.5 and 5.6. In all cases they fall into first order, zero order and Langmuir adsorption isotherm types of equation, and this helps to confirm that the general rate equations are applicable. In Table 5.4 are given the algebraic definitions of the empirical rate coefficients (a', b', f and g) in terms of the biological rate co-efficients and the appropriate characteristic size.

TABLE 5.4. *The biological rate coefficients*

Empirical coefficient	Microbial films	Microbial floc
a'	$\dfrac{k_1 L}{k_3}$	$\dfrac{k_1}{k_3 \rho_o}$
b'	$\dfrac{k_1}{k_2}$	$\dfrac{k_1}{k_2} \dfrac{A_p}{V_p \rho_o}$
f	$\dfrac{k_1 L}{k_3}$	$\dfrac{k_1}{k_3 \rho_o}$
g	$\dfrac{1}{k_3}$	$\dfrac{1}{k_3}$

DETERMINATION OF THE BIOLOGICAL RATE COEFFICIENTS

Since the *biological rate equation*, describing substrate removal by a microbial mass, is of a general nature, i.e. it can be applied to either biological flocs or films, the opportunity arises to use experiments involving either of these geometries in determining the biological rate coefficients k_1, k_2 and k_3. It follows that the geometry selected for the microbial mass in the laboratory studies need not be the same as that in the industrial process envisaged. Indeed, it must be appreciated that a superficial similarity between the laboratory model and the full-size plant can easily lead to overconfidence in the application of the experimental results. The danger arises because the degree of mixing, the environmental conditions, and the characteristic size of the microbial mass may not be accurately reproduced in the full-size plant.

The principal difficulty associated with the determination of the biological rate coefficients arises in the experimental determination of the appropriate characteristic size, i.e. L in the case of films, and V_p/A_p for flocs. This is particularly true for small sizes and also for the ranges of particle size which might be anticipated to occur in experiments incorporating flocs.

Experiments involving biological flocs

The classical experiment involved in the study of microbiological rate processes is a batch experiment. In this experiment the microbial and substrate concentrations are monitored over a period of time and provide the characteristic *growth curve* indicated in Fig. 5.13. Initially, a nutrient solution is prepared and inoculated with a small quantity of the micro-organism under consideration; there follows an acclimatisation period known as the *lag phase*, after which additional micro-organisms and biochemical products are formed at a changing rate, as the substrate is depleted progressively. Final depletion results in a situation, the *death phase*, where the mortality rate is in excess of the rate of creation of viable new cells.

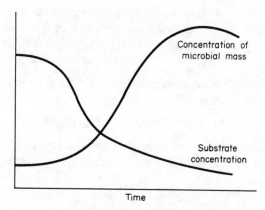

FIG. 5.13. The growth curve

In the case of aerobic reactions, the experiment is followed most conveniently on a basis of oxygen consumption using a Warburg respirometer[10] (Fig. 5.14), or one of the more recently developed automatic respirometers[26].

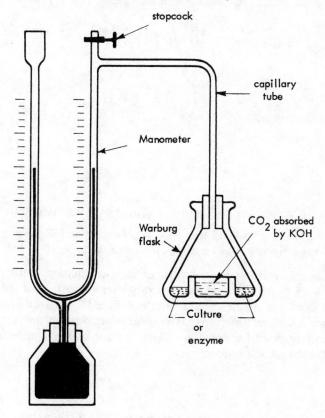

FIG. 5.14. Warburg apparatus

The *biological rate equation* is applicable to the growth phase, although the death phase could be interpreted as a variation in the biological rate coefficients k_1 and k_2, since, as discussed previously, they include in their definitions (equation 5.23) the number of viable micro-organisms through the area a.

During the growth phase a mass balance, on the microbial mass over the incremental time, Δt, yields:

$$M(t + \Delta t) - M(t) = (S_o K_o) R' M \Delta t \qquad \ldots (5.36)$$

where $M(t)$ is the mass of micro-organisms per unit volume of reactor at time t,
      R' is the rate of removal of substrate per unit mass of micro-organisms,
      and
$S_o K_o$ is the yield coefficient.

In equation 5.36, R' includes a liquid phase diffusional resistance. For reasons to be discussed subsequently (cf. equations 5.80 and 5.104) this resistance may be neglected. The rate of substrate removal R' is given therefore by the *biological rate equation* (Table 5.3, page 387).

Taking the limit of equation 5.36 leads to:

$$\frac{1}{M} \frac{dM}{dt} = (S_o K_o) R \qquad \ldots (5.37)$$

or

$$\ln \left(\frac{M_2}{M_1} \right) = (S_o K_o) R (t_2 - t_1) \qquad \ldots (5.38a)$$

or

$$\frac{M_2}{M_1} = e^{S_o K_o R (t_2 - t_1)} \qquad \ldots (5.38b)$$

providing R remains approximately constant over the interval from t_1 to t_2.

Because of the semi-logarithmic nature of equation 5.38 the growth phase of a batch experiment is also known as the *logarithmic phase*.

Equation 5.37 demands an experimental determination of the characteristic size V_p/A_p, a parameter that may (a) vary over the course of the experiment, and (b) exhibit a distribution at any one time[25]. These difficulties can be overcome if sufficient hydrodynamic shear is provided so that the particle size is sufficiently small for equation 5.33 to be applicable. Under these circumstances equation 5.37 reduces to:

$$\frac{1}{M} \frac{dM}{dt} = S_o K_o \frac{k_1 C^*}{\rho_o (1 + k_3 C^*)} \qquad \ldots (5.39)$$

Since the characteristic size is not contained in this equation, it follows that an experimental determination is unnecessary and that the complexities inherent in a distribution of characteristic sizes are removed, even though such a distribution still occurs within the experiment.

Thus equation 5.39 in conjunction with an experimentally determined "growth curve" leads to the biological rate coefficients k_1 and k_3.

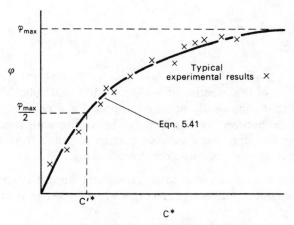

FIG. 5.15. A method for obtaining approximate values of k_1 and k_3

The experimental data are usually expressed graphically (Fig. 5.15) in terms of:

$$\phi = f(C^*) \qquad \ldots (5.40)$$

where

$$\phi = \frac{\rho_o}{(S_o K_o) M} \frac{dM}{dt}$$

The resulting plot may be curve-fitted by equation 5.39; i.e. equation 5.40 may be written:

$$\phi = \frac{k_1 C^*}{1 + k_3 C^*} \qquad \ldots (5.41)$$

The most convenient manual way to determine approximate numerical values of k_1 and k_3 is to sketch a curve through the experimental data on Fig. 5.15 and to note from equation 5.41 that:

$$\phi_{max} = \frac{k_1}{k_3} \qquad \ldots (5.42a)$$

and

$$k_3 = \frac{1}{C'^*} \qquad \ldots (5.42b)$$

where C'^* is the value of C^* when ϕ is one-half of ϕ_{max}.

This may easily be seen from equation 5.41:

$$\phi = \frac{\phi_{max}}{2} = \frac{k_1 C'^*}{1 + k_3 C'^*}$$

therefore:

$$C'^* = \frac{1}{\dfrac{2k_1}{\phi_{max}} - k_3}$$

or from equation 5.42a:

$$C'^* = \frac{1}{k_3}$$

The limitation of this experiment lies in the additional difficulties associated with the determination of k_2, namely the need for a detailed experimental knowledge of the particle size distribution. The requirements regarding hydrodynamic shear for achieving a sufficiently large particle size to determine k_2 are in conflict with those leading to complete mixing.

A further problem arises in a batch experiment since it is known that the biochemical characteristics of micro-organisms[22] change over the growth phase.

A parameter used in microbiology[9], for purposes of comparison and classification, is the *doubling time*, i.e. the time required to double the mass of a given strain of micro-organism. This may be related to the *biological rate equation* by a rearrangement of equation 5.38:

$$t_D = \frac{\ln 2}{(S_o K_o) R} \qquad \qquad \ldots (5.43)$$

The doubling time then depends both on the characteristic size and on the substrate concentration. However, at sufficiently high concentrations equation 5.35 is applicable, and then:

$$t_D = \frac{k_3 \rho_o \ln 2}{(S_o K_o) k_1} \qquad \qquad \ldots (5.44)$$

In these circumstances, the doubling time becomes a meaningful parameter and is based simply upon the maximum production rate of microbial mass.

Experiments involving biological films

The apparatus that has been developed for the study of micro-organisms which are caused to grow as a layer adhering to a mechanical support surface is referred to as the *biological film reactor*[27, 30]. In this reactor, a liquid film in laminar flow is brought into contact with a microbiological film of given thickness L (Figs. 5.16 and 5.17). This thickness is maintained essentially constant by arranging the support surface in the form of a grid of depth L and

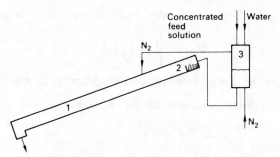

FIG. 5.16. The biological film reactor (arranged
for anaerobic studies):
  1 reaction surface
  2 wooden weir
  3 mixer and de-aerator

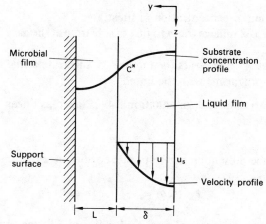

FIG. 5.17. The model of the biological film reactor

allowing the micro-organisms to develop within the grid (Fig. 5.18). As a result
of growth, the microbial mass overgrows the grid and the excess is removed
periodically by mechanical means. It is essential that in the reactor the liquid
flow be perfectly distributed across the biological film and that it takes place
rectilinearly over the surface. The achievement of a nearly perfect flow pattern
is necessary so that the liquid phase diffusional resistance can be described in
mathematical terms with accuracy. To achieve this necessitates a number of
experimental innovations, and these are described by DAOUD[31].

A differential mass balance (Fig. 5.17), equating transverse diffusion with
axial convection in the liquid film, leads to[30]:

$$u_z \frac{\partial C}{\partial z} = D \frac{\partial^2 C}{\partial y^2}$$

$$\ldots\ldots(5.45)$$

where C is the local substrate concentration,
  u_z is the local velocity of the liquid film, and
  D is the diffusion coefficient of the substrate in the liquid phase.

Since the liquid film is in laminar flow (see Volume 1, page 217):

$$u = u_s \left[1 - \left(\frac{y}{\delta} \right)^2 \right] \qquad \qquad \ldots(5.46)$$

where u_s and δ are the surface velocity and the thickness of the liquid film.

The appropriate restrictions to be applied to equation 5.45 are:

$$C(0, y) = C_I \qquad \qquad \ldots(5.47)$$

$$\frac{\partial C}{\partial y}(z, 0) = 0 \qquad \qquad \ldots(5.48)$$

$$N = -D \frac{\partial C}{\partial y}(z, \delta) \qquad \qquad \ldots(5.49)$$

where C_I is the initial concentration of substrate,

   equation 5.48 reflects the zero flux of substrate at the gas–liquid interface, and

   equation 5.49 represents the removal of substrate at the interface of the micro-organisms with the liquid.

Now, if the variation of N in equation 5.49 is taken as linear with respect to concentration, i.e. if:

$$N = k_s C \qquad \qquad \ldots(5.50)$$

where k_s is for the present an arbitrary rate coefficient, equation 5.49 becomes:

$$k_s C(z, \delta) = -D \frac{\partial C}{\partial y}(z, \delta) \qquad \qquad \ldots(5.51)$$

The solution to equations 5.45, 5.46, 5.47, 5.48 and 5.51 may be written in the form:

$$C = f(z, y, D, \delta, u_s, C_I, k_s) \qquad \qquad \ldots(5.52)$$

where z and y are independent variables, and

   D, δ, u_s, C_I, k_s are parameters.

Equation 5.52 demands an experimental determination of the substrate concentration C at a position z, y within the liquid film. This is hardly an attractive proposition from the experimental viewpoint. However the difficulties are eased considerably if equation 5.52 is expressed in terms of a mixed mean concentration C_O. This is the concentration of the fluid collected in a receptacle at the outlet of the apparatus. The mixed mean concentration is given by a velocity-weighted average over y:

$$C_O = \frac{3}{2\delta u_s} \int_0^\delta u(y) \, C(Z, y) \, dy \qquad \qquad \ldots(5.53)$$

or

$$C_O = f(Z, \delta, D, u_s, C_I, k_s) \qquad \qquad \ldots(5.54)$$

FIG. 5.18. Photograph of a supported microbial film

The solution to the complete mathematical problem has been given by ATKINSON and SWILLEY[23] in the form:

$$\frac{C_O}{C_I} = F(\eta, K) \qquad \dots(5.55)$$

where:

$$\eta = \frac{k_s\delta}{D} \qquad \dots(5.56a)$$

and

$$K = \frac{DZ}{\delta^2 u_s} \qquad \dots(5.56b)$$

Equation 5.55 has the advantage of containing only three dimensionless variables as compared with the seven dimensional variables in equation 5.54.

For convenience, equation 5.55 has been given on Fig. 5.19 in graphical form.

Now if an experiment is carried out such that the biological film thickness L is sufficiently large, the *biological rate equation* for films reduces to equation 5.31, and a comparison between this asymptote and equation 5.50 gives:

$$k_s = \frac{k_1}{k_2} \qquad \dots(5.57)$$

Thus an experiment using the biological film reactor in which C_I, C_O, u_s, δ, Z and D are determined for a sufficiently large and known L will, by means of Fig. 5.19, lead to the ratio of the rate coefficients k_1 and k_2.

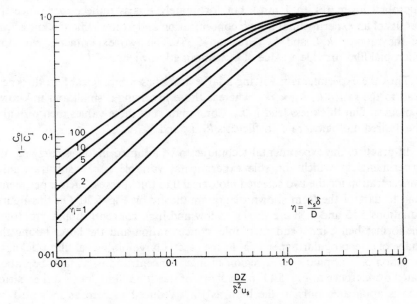

FIG. 5.19. Data corresponding to equation 5.55

If then, at the same biological film thickness L, a further experiment is carried out using a very high initial concentration C_I, the other asymptote of the *biological rate equation* will apply, i.e. equation 5.32. This equation is of zero order with respect to substrate concentration and consequently, if the concentration is sufficiently high for it to apply, then no liquid phase diffusional limitation can exist. Under these conditions a mass balance over the reactor leads to:

$$Q(C_I - C_o) = \left(\frac{k_1}{k_3} L\right) Z \qquad \qquad \ldots\ldots(5.58)$$

where Q is the peripheral volumetric flowrate.

Since L is known, equation 5.58 leads to a ratio of k_1 and k_3.

A further asymptote of the *biological rate equation* exists and this occurs at small biological film thicknesses; it corresponds to:

$$N = \frac{k_1 L C^*}{1 + k_3 C^*} \qquad \qquad \text{(equation 5.30)}$$

or, as C^* tends to zero:

$$N \rightarrow (k_1 L)\, C^* \qquad \qquad \ldots\ldots(5.59)$$

Equation 5.59 is linear with respect to substrate concentration and for such a case the theory for the biological film reactor, expressed in equations 5.45 to 5.56, applies. It follows that an experiment carried out at a very low substrate concentration and a very small biological film thickness leads to the coefficient $k_1 L$. It is difficult to obtain an accurate value for the thickness of a very thin microbial layer (0.1 to 1 mm), but fortunately this is unnecessary since the results of an experiment at a high concentration and at the same thickness lead to the ratio of $k_1 L$ and k_3 (equation 5.58). The two experiments with this microbial film provide a value of k_3 without a knowledge of L.

Thus the experiments involving a large and known biological film thickness lead to the ratios k_1/k_2, k_1/k_3, whereas those involving a small and unknown biological film thickness lead to k_3. Combination of these values then provides the desired biological rate coefficients k_1, k_2 and k_3.

In practice, the experimental technique to be adopted involves carrying out experiments in which the sole experimental variable is the substrate inlet concentration for the two selected biological film thicknesses. Such experiments lead to data of the form shown diagrammatically on Fig. 5.20. On this figure, equations 5.55 and 5.58 are shown as low and high concentration asymptotes, the former being independent of inlet concentration and the latter taking the shape of a rectangular hyperbola. In the case of large biological film thickness, equation 5.55 applies over a region of inlet concentrations, as may be anticipated by reference to Fig. 5.11 (page 386); for the "thin film", however, equation 5.55 is applicable only in the limit as C_I is reduced to zero as indicated by equation 5.59.

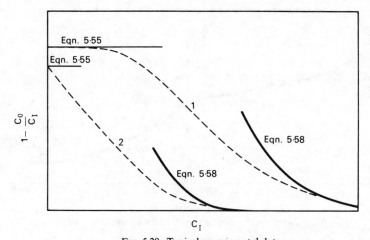

FIG. 5.20. Typical experimental data

(1) Experimental results ('solid" phase diffusional limitation)
(2) Experimental results (no "solid" phase diffusional limitation)

A complete set of experimental data for a given biological film thickness enables the two asymptotes to be joined in the manner indicated on Fig. 5.20. This may be deduced from the *biological rate equation*. A complete mathematical theory for the whole range of C_I has been provided by ATKINSON, DAOUD and WILLIAMS[30].

Calculation of biological rate coefficients from data obtained using the biological film reactor

On Figures 5.21 and 5.22 are recorded data for the efficiency of substrate removal $1 - C_O/C_I$ as a function of the inlet concentration C_I, obtained using a biological film reactor. In these experiments the micro-organisms were predominantly a mixed culture of bacteria derived from soil, and the nutrient

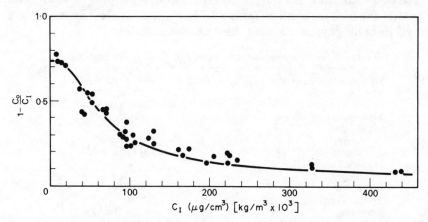

FIG. 5.21. Experimental data obtained using a "thin" microbial film[31]

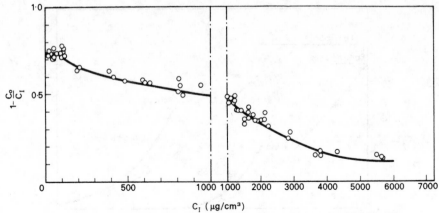

FIG. 5.22. Experimental data obtained using a "thick" microbial film[31]

solution consisted of glucose as the substrate, ammonium hydrogen phosphate as the nitrogen source, together with potassium nitrate as the oxidant. Tap water was used to supply the essential trace components and the apparatus was operated under oxygen-free conditions, i.e. anaerobically.

The first set of data was obtained using a biological film thickness which, while unknown with any degree of accuracy, was of the order of 0·25 mm. For the second set, the thickness was adjusted using the procedure indicated by Fig. 5.18 (facing page 394) to a value of 2·032 mm. A comparison between these data and Fig. 5.20 (page 397) suggests that the former refers to a situation in which the "solid" phase diffusional limitation is minimal.

As would be expected by reference to Fig. 5.11 (page 386) a vastly greater range of substrate concentration is necessary in the case of the thicker film in order to cover the range of interest. This range is dictated for the thicker film by the asymptotic equations 5.31 and 5.32 and for the thin film by equation 5.30.

The data recorded at the time of the experiments, other than those given on Figs. 5.21 and 5.22, are given in Table 5.5. The dimensions of the apparatus and the values of the physical constants used are also included.

TABLE 5.5. *Experimental data associated with the thick microbial film*

Diffusion coefficient of glucose in water (25°C)	$6·9 \times 10^{-10}$ m^2 s^{-1}
Viscosity of water	10^{-3} kg m^{-1} s^{-1}
Density of water	10^3 kg m^{-3}
Microbial film thickness	2·032 mm
Liquid film:	
length	2·11 m
width	182 mm
Microbial surface:	
length	1·83 m
width	165 mm
Temperature	22°C
Flowrate	$8·2 \times 10^{-4}$ kg s^{-1}
Surface velocity (experimental)	18·6 mm s^{-1}

The use of a grid, to support the "thick" microbial film, results in a situation where the area for substrate removal differs from that of the liquid film, and this is reflected in the data of Table 5.5. In the absence of a supporting grid, as in the case of "thin" films, the areas are the same.

Calculation of k_1/k_2 from the data for the "thick" microbial film

The average velocity $\bar{u}$ of a liquid film in laminar flow is given by (see Volume 1, page 217):

$$\bar{u} = \tfrac{2}{3}u_s \qquad\qquad ....(5.60)$$

where u_s is the surface velocity of the liquid film; therefore:

$$\bar{u} = 12.4 \text{ mm s}^{-1}$$

The thickness δ of the liquid film is given by:

$$\delta = \frac{\text{Peripheral volumetric flowrate } Q}{\text{Average velocity } \bar{u}} \qquad\qquad ....(5.61)$$

$$= 0.3619 \text{ mm}$$

The coefficient K of equations 5.55 and 5.56 is given by:

$$K = \frac{DZ}{\delta^2 u_s} \qquad\qquad \text{(equation 5.56}b\text{)}$$

The difficulty that arises with the evaluation of K is in the selection of an appropriate value for the length Z. The problem involved may be readily appreciated by reference to Fig. 5.18 (facing page 394). From this figure it is apparent that the area of the interface of the micro-organisms with the liquid film differs from the area of the liquid film itself, due to the existence of the supporting grid. This factor would be of little consequence but for the liquid phase diffusional resistance. The effect of this diffusional phenomenon is a continued rearrangement of the concentration profile when the liquid film is in contact with the grid rather than the micro-organisms. Two possible values are available for the length Z, in equation 5.56b, and neither value can exactly satisfy the problem. In view of the large liquid phase diffusional limitation the area of the liquid film has been used. Therefore:

$$K = 0.5980$$

The limiting efficiency taken from Fig. 5.22 is:

$$1 - \frac{C_o}{C_I} = 0.733 \text{ for small } C_I$$

On Fig. 5.23 are plotted data taken from Fig. 5.19 (page 395) of C_o/C_I as a function of the parameter η when K has a value of 0.5980. Interpolation leads to a value for η of 5.85.

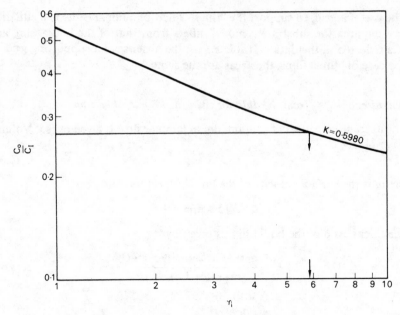

FIG. 5.23. Determination of the parameter η

Now η is given by equation 5.56a as:

$$\eta = \frac{k_s \delta}{D} \qquad \text{(equation 5.56a)}$$

and for "thick" films at low concentrations:

$$k_s = \frac{k_1}{k_2} \qquad \text{(equation 5.57)}$$

Therefore:

$$\frac{k_1}{k_2} = \frac{\eta D}{\delta} = 1 \cdot 117 \times 10^{-5} \text{ ms}^{-1} \qquad \ldots \text{(5.62)}$$

Calculation of k_1/k_3 from the data for the "thick" microbial film

At high substrate concentrations equation 5.58 applies and may be rearranged:

$$1 - \frac{C_O}{C_I} = \frac{k_1 L}{k_3} \frac{Z}{Q} \frac{1}{C_I} \qquad \ldots \text{(5.63)}$$

The data given on Fig. 5.22 have accordingly been replotted on Fig. 5.24 as:

$$1 - \frac{C_O}{C_I} = f\left(\frac{1}{C_I}\right) \qquad \qquad \dots(5.64)$$

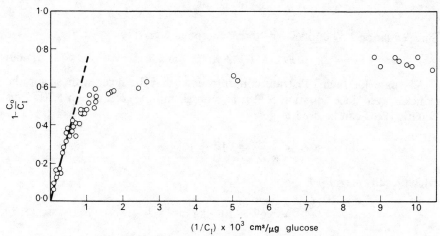

FIG. 5.24. Determination of $(k_1 L/k_3)$, "thick layer" (22°C)

The result is a linear relationship at high concentrations, the slope of which is given by:

$$\text{Slope} = \frac{k_1 L}{k_3} \frac{Z}{Q} = 0.713 \text{ kg m}^{-3} \qquad \dots(5.65)$$

Once more the problem of a judicious selection of the length Z arises. However, under conditions of "reaction" control the microbial area in contact with the liquid is dominant and this length is used accordingly.

Therefore:

$$\frac{k_1 L}{k_3} = 1.933 \times 10^{-6} \text{ kg m}^{-2} \text{ s}^{-1}$$

and from a knowledge of the microbial film thickness L:

$$\frac{k_1}{k_3} = 9.530 \times 10^{-4} \text{ kg m}^{-3} \text{ s}^{-1} \qquad \dots(5.66)$$

Calculation of k_3 from the data for the "thin" microbial layer

The liquid film average velocity and thickness, together with the parameter K, are calculated as previously:

$$\left.\begin{array}{l} \bar{u} = 12.40 \text{ mm s}^{-1} \\ \delta = 0.3622 \text{ mm} \\ K = 0.5960 \end{array}\right\} \qquad \dots(5.67)$$

o

The limiting efficiency is given from the experimental data (Fig. 5.21) as 0·735, and this leads, in the same manner as before, to:

$$\eta = \frac{k_s \delta}{D} = 6\cdot0 \qquad \qquad \ldots.(5.68)$$

Since equation 5.59 applies for "thin" microbial layers:

$$k_s = k_1 L = 1\cdot142 \times 10^{-5} \text{ m s}^{-1} \qquad \ldots.(5.69)$$

The data for high substrate concentrations may be dealt with as for the "thick" layer since equation 5.32 is also an asymptote of equation 5.30. Thus equation 5.63 can be used to give:

$$\text{Slope} = \frac{k_1 L Z}{k_3 Q} = 3\cdot14 \times 10^{-2} \text{ kg m}^{-3}$$

which results in:

$$\frac{k_1 L}{k_3} = 6\cdot69 \times 10^{-8} \text{ kg m}^{-2} \text{ s}^{-1} \qquad \ldots.(5.70)$$

However, since the microbial thickness L is an unknown in this experiment, equations 5.69 and 5.70 may only be used to determine the coefficient k_3:

$$k_3 = 1\cdot706 \times 10^2 \text{ m}^3 \text{ kg}^{-1} \qquad \ldots.(5.71)$$

Calculation of k_1, k_2 and k_3

Combination of equations 5.66 and 5.71 yields:

$$k_1 = 1\cdot624 \times 10^{-1} \text{ s}^{-1} \qquad \ldots.(5.72)$$

and combination of equations 5.62 and 5.72 yields:

$$k_2 = 1\cdot455 \times 10^4 \text{ m}^{-1} \qquad \ldots.(5.73)$$

DESIGN OF PROCESSES INVOLVING MICROBIAL FLOCS

As mentioned earlier, processes involving microbial floc are inevitably carried out in such a way that the particles are suspended. This is so because of the similarity of the specific gravity of floc (say 1·1) and of the aqueous media. Attempts to arrange the particles in a stationary manner in the form of a packed bed unfortunately result in an essentially continuous spongelike mass, due to the deformation of the flocs under their own weight and to the amalgamation of individual particles as growth takes place. The effect of this situation on the liquid flow pattern, the effective characteristic size and the pressure drop through such a bed is particularly unsatisfactory from the point of view both of reactor efficiency and of the development of satisfactory design procedures.

Batch processes

The information necessary to carry out the design of a batch reactor, for a given system of micro-organisms and nutrient solution, involves a knowledge of the biological rate coefficients k_1, k_2 and k_3, of the characteristic size V_p/A_p of the particles involved in the process, and of the liquid phase mass transfer coefficient. Even when complete mixing is achieved, these parameters are dependent on the conditions produced by the mixing device in stirred tank reactors. This is because the internal hydrodynamics influence both the effective particle size V_p/A_p and the liquid-phase mass transfer coefficient.

The variation of the concentration of the microbial mass with time in a batch reactor is given by:

$$\frac{1}{M}\frac{dM}{dt} = (S_o K_o) R' \qquad \text{(from equation 5.36)}$$

In this equation the rate of substrate removal per unit mass of micro-organisms per unit time R' is given by:

$$R' = h'(C - C^*) = R \qquad \ldots.(5.74)$$

where h' is a mass transfer coefficient based on unit mass of micro-organisms,
  C is the concentration of substrate in the bulk liquid, and
  C^* is the interfacial concentration of substrate.
The removal rate R is given by the *biological rate equation* for floc and depends upon both C^* and the particle size V_p/A_p.

Similar equations to 5.36 may be derived for both the product and the substrate, namely:

$$\frac{dp}{dt} = (S_p K_p) R'M \qquad \ldots.(5.75)$$

and

$$\frac{dC}{dt} = -R'M \qquad \ldots.(5.76)$$

where $S_p K_p$ is the yield coefficient, and
  p is the product concentration.

In addition to these equations, the fractional conversions are related by equation 5.1, i.e.:

$$K_o + K_p = 1 \qquad \ldots.(5.77)$$

and R is given in Table 5.3.

For a given set of initial conditions, namely $M(0)$, $C(0)$, $p(0)$, the objective is to describe the variation of the dependent variables M, C and p as a function of the process time, assuming that $k_1, k_2, k_3, h', V_p/A_p, S_o K_o$, and $S_p K_p$ are given.

The sequential solution of equations 5.36, 5.75 and 5.76 inevitably demands a numerical procedure because of the complexity inherent in equation 5.74, and of the fact that the equations are coupled through the concentration of both the substrate and the microbial mass. On Fig. 5.25 is given a convenient iterative procedure for the calculation of $M(t)$, $p(t)$ and $C(t)$. The procedure is presented in the form of a block diagram appropriate to a computer programme so as to illustrate clearly the iterative stages involved. The most rudimentary march into the time domain is invoked, a method which, when using finite incremental times, by no means guarantees convergence or the "true" mathematical solution for all values of the parameters. For present purposes, however,

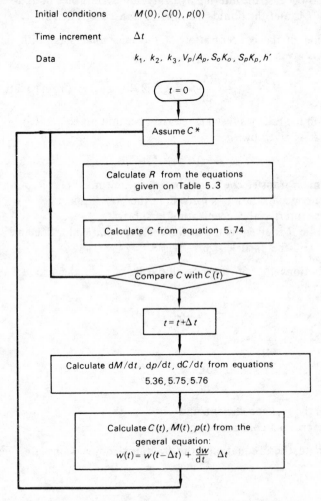

Initial conditions $M(0), C(0), p(0)$

Time increment Δt

Data $k_1, k_2, k_3, V_p/A_p, S_oK_o, S_pK_p, h'$

$t = 0$

Assume $C*$

Calculate R from the equations given on Table 5.3

Calculate C from equation 5.74

Compare C with $C(t)$

$t = t + \Delta t$

Calculate dM/dt, dp/dt, dC/dt from equations 5.36, 5.75, 5.76

Calculate $C(t), M(t), p(t)$ from the general equation:
$$w(t) = w(t - \Delta t) + \frac{dw}{dt} \Delta t$$

FIG. 5.25. Calculation procedure for the design of batch processes

the inclusion of a more sophisticated marching procedure would serve only to cloud the basic algorithm, and the reader is referred to the previous chapter where the numerical solution of ordinary differential equations is discussed.

In the algorithm on Fig. 5.25 the interfacial concentration C^* is first assumed and the rate of substrate removal R calculated. From these values and with the aid of equation 5.74 the substrate concentration C can be obtained for comparison with the substrate concentration computed from the previous time stage $C(t)$. An iterative loop is provided to allow acceptable convergence and leads to the value $C^*(t)$. Following this stage, equations 5.36, 5.75 and 5.76 are used to calculate the derivatives $\dfrac{dM}{dt}, \dfrac{dp}{dt}$ and $\dfrac{dC}{dt}$ from which, by use of a suitable incremental time, values of $M(t)$, $p(t)$ and $C(t)$ can be computed. A further assumption regarding C^* at the new time level then allows the solution to proceed further into the time domain.

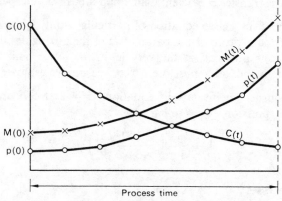

FIG. 5.26. Computed results for the path of batch
reactor performance (diagrammatic)

A diagrammatic representation of the results of such a computation is given on Fig. 5.26. The average rate of the formation of products per unit volume of the reactor is readily obtained from these data by a series of overall mass balances, as:

$$\frac{M(t) - M(0)}{t} = f(t)$$

$$\frac{p(t) - p(0)}{t} = f(t)$$

$$\left.\begin{array}{r}\end{array}\right\} \quad \dots (5.78)$$

where t is the elapsed time.

Of the data required by the design procedure, the biological rate coefficients have to be determined as described previously and appropriate experimentation is necessary using similar hydrodynamic conditions in order to determine the characteristic size V_p/A_p. As far as the liquid phase mass transfer coefficient h' is concerned, little work has been carried out that is immediately relevant as bubbles and droplets have commanded most attention. However, some conclusions can be reached if it is noted that the biological flocs will tend to follow the fluid flow pattern relatively closely, since the density difference between particle and fluid is very small. This factor, together with the assumption that the flocs are rather smaller than a turbulent "eddy", suggests that the mass transfer process might be approximated using as a basis simply molecular diffusion into an infinite region of fluid that is stationary relative to the particle. This situation is described (in Volume 1, page 177) in mathematical terms for heat transfer; by analogy for mass transfer:

$$\frac{hd}{D} = 2 \qquad \qquad \dots (5.79)$$

where d is the sphere diameter,
$\qquad D$ is the liquid phase diffusion coefficient, and
$\qquad h$ is a mass transfer coefficient defined on a basis of unit particle area.

The property of the above equation of particular utility is the fact that it is independent of the hydrodynamic parameters of the mixing device. The use of equation 5.79 thus provides, at the very least, the lower limit for the mass transfer coefficient and, quite possibly, a very reasonable estimate.

The mass transfer coefficients in equations 5.74 and 5.79 may easily be related by the expression:

$$h' = \frac{hA_p}{V_p \rho_o} = \frac{1}{3} \frac{D}{\rho_o} \left(\frac{A_p}{V_p}\right)^2 \qquad \dots (5.80)$$

Continuous processes

A homogeneously stirred tank reactor (as shown in Fig. 5.5) may be operated in a continuous manner. The basic arrangement for a single reactor unit consists of a nutrient feed with a substrate concentration C_I to a vessel in which the substrate, microbial mass and product concentration levels are C_R, M_R and p_R.

In such a system the substrate removal per unit volume is proportional to the concentration of the microbial mass M_R. It is therefore advantageous to enhance the concentration of micro-organisms in the reactor over that leaving the system M_O, by means of a centrifuge and re-cycle, or alternatively by a sedimentation procedure as adopted in the tower fermenter (Fig. 5.6).

A series of mass balances over the reactor (Fig. 5.5) leads to:

(a) micro-organisms

$$\varepsilon F M_C + (S_o K_o) M_R V R' = (1 + \varepsilon) F M_R \qquad \dots (5.81)$$

(b) substrate

$$F C_I + \varepsilon F C_R = (1 + \varepsilon) F C_R + M_R V R' \qquad \dots (5.82)$$

(c) product

$$F p_I + \varepsilon F p_R + (S_p K_p) M_R V R' = (1 + \varepsilon) F p_R \qquad \dots (5.83)$$

where ε is the re-cycle ratio,
V is the volume of the reactor,
F is the volumetric flowrate, and
M_C is the concentration of micro-organisms returned from the centrifuge.

It is convenient algebraically to relate the concentration of micro-organisms leaving the centrifuge M_C for re-cycling to the concentration leaving the reactor M_R, by a separation coefficient γ, i.e.:

$$M_C = \gamma M_R \qquad \qquad \dots\dots(5.84)$$

where $\gamma > 1$.

Combination of equations 5.81 and 5.84 leads to:

$$R' = \frac{G}{(S_o K_o)\, t_R} \qquad \qquad \dots\dots(5.85)$$

where the mean residence time t_R is given by:

$$t_R = \frac{V}{F} \qquad \qquad \dots\dots(5.86)$$

and

$$G = 1 - \varepsilon(\gamma - 1) \qquad \qquad \dots\dots(5.87)$$

The parameter G is restricted to the region $0 < G < 1$. Equations 5.82 and 5.83 then become:

$$C_I - C_R = M_R t_R R' = \frac{M_R G}{(S_o K_o)} \qquad \qquad \dots\dots(5.88)$$

$$p_R - p_I = (S_p K_p)\, M_R t_R R' = (S_p K_p)(C_I - C_R) \qquad \qquad \dots\dots(5.89)$$

In equations 5.81 to 5.89 the removal rate R' is given by equation 5.74:

$$R' = h'(C_R - C^*) = R \qquad \qquad \text{(equation 5.74)}$$

where $R = f(C^*, V_p/A_p)$.

The concentration of micro-organisms leaving the system M_O is related to the concentration within the reactor M_R by a mass balance around the centrifuge:

$$(1 + \varepsilon)\, F M_R = \varepsilon F M_C + F M_O \qquad \qquad \dots\dots(5.90)$$

or

$$M_O = G M_R \qquad \qquad \dots\dots(5.91)$$

It follows from equation 5.85 that the removal rate R' is independent of the inlet concentration to the reactor (C_I). This factor together with equation 5.74 indicates that the substrate concentration within the reactor (C_R) is independent both of C_I and of the cell concentration M_R, but is influenced by the re-circulation ratio ε in such a way that as ε is increased C_R is reduced. These comments are restricted to a reactor unit with no micro-organisms in the feed.

Since the separation factor γ is greater than unity, ε must be selected judiciously for G to remain positive, i.e. $\varepsilon < 1/(\gamma - 1)$. However, G is necessarily less than one and, as a consequence, the removal rate R' and the substrate concentration C_R are reduced by re-circulation. This effect is more than compensated by an increase in M_R and p_R as may be deduced from equations 5.85, 5.88 and 5.89. Combination of equations 5.85, 5.88 and 5.91 gives the consequent increase in M_O.

The overall effect of re-circulation, for a fixed throughput, is an advantageous change in the outlet concentrations both of micro-organisms and of bio-chemical product.

In general, simple analytical expressions relating the various concentration levels explicitly to the variables associated with the problem cannot be obtained, due to the complexity of equation 5.74. However, this does not cause too much difficulty for design purposes if the calculation procedure outlined on Fig. 5.27 is adopted.

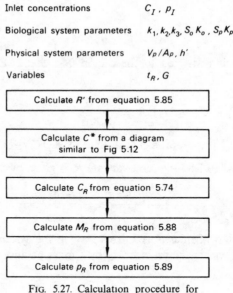

Inlet concentrations	C_I , p_I
Biological system parameters	$k_1, k_2, k_3, S_o\,K_o$, $S_p K_p$
Physical system parameters	$V_P/A_P, h'$
Variables	t_R , G

Calculate R' from equation 5.85

Calculate C^* from a diagram
similar to Fig 5.12

Calculate C_R from equation 5.74

Calculate M_R from equation 5.88

Calculate p_R from equation 5.89

Fig. 5.27. Calculation procedure for the design of a continuous stirred tank fermenter

A study of the limiting cases of equation 5.74 does assist in understanding the response of the system to changes in the independent variables, i.e. the mean residence time t_R and the re-circulation ratio ε.

Liquid phase diffusion controlled

In these circumstances the substrate concentration C^* at the interface of the microbial mass with the liquid phase is zero, and equation 5.74 reduces to:

$$R' = h'C_R \qquad \qquad ....(5.92)$$

Combination of equations 5.85 and 5.92 provides an expression for the substrate concentration within the reactor:

$$C_R = \frac{G}{(S_oK_o)\,h't_R} \qquad \qquad ....(5.93)$$

This equation confirms the statement made previously that C_R is independent both of the substrate concentration in the feed and also of the microbial concentration within the reactor M_R.

Combination of equations 5.88 and 5.93 provides the expressions for the concentration of micro-organisms

$$M_R = \frac{(S_oK_o)}{G}C_I - \frac{1}{h't_R} \qquad \qquad ....(5.94)$$

and (from equation 5.91):

$$M_O = (S_oK_o)\,C_I - \frac{G}{h't_R} \qquad \qquad ....(5.95)$$

The *biochemical product* p_R is given by:

$$p_R = p_I + \frac{(S_pK_p)}{(S_oK_o)}M_O \qquad \qquad ....(5.96)$$

On Fig. 5.28 are plotted equations 5.93, 5.94, 5.95 and 5.96 as functions of $1/t_R$. An increase in this quantity $1/t_R$ corresponds to an increase in the flowrate

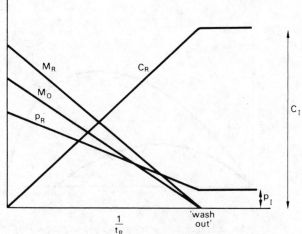

FIG. 5.28. The variation of concentration in a continuous stirred tank fermenter with liquid phase diffusion control

F to a reactor of volume V. It is evident from equation 5.93 that C_R increases linearly with F until the inlet concentration C_I is reached. Since C_I cannot be exceeded, any further increase in the flowrate simply means that the substrate passes through the reactor unchanged. This situation arises, as may be seen by inspection of equation 5.88, simply because there are no micro-organisms retained within the reactor. This phenomenon has been given the appropriate name *wash-out*. The residence time at which wash-out occurs may be calculated from equation 5.94:

$$t_{R(\text{wash-out})} = \frac{G}{(S_o K_o)\, C_I h'} \qquad \ldots(5.97)$$

Since the products are simply related to the substrate consumed, the productivity of the reactor may be expressed as:

$$\text{Productivity} = \frac{F}{V}(C_I - C_R) = \left(C_I - \frac{G}{(S_o K_o)\, h' t_R}\right)\frac{1}{t_R} \qquad \ldots(5.98)$$

This equation is quadratic in $1/t_R$ and has roots when $1/t_R$ is zero and at the position of wash-out, as would be expected. However, more important than these roots is the maximum that occurs in the equations as this refers to the maximum productivity.

The effect of the re-circulation may be appreciated from equations 5.85, 5.97 and 5.98. An increase in the re-circulation ratio reduces the parameter G which has the effect of lowering the limit on the mean residence time, i.e. it allows an increased throughput F for a given reactor volume V, thus providing a greater range of operation. From the equation for the productivity it follows that the overall effect is a general improvement in productivity for all flow rates below wash-out. These phenomena are shown diagrammatically in Fig. 5.29.

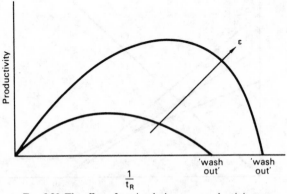

FIG. 5.29. The effect of re-circulation ε on productivity and "wash-out"

All the above comments are applicable to the general relationship given by equation 5.74, but the foregoing provides a simple vehicle for illustrating the implications involved.

Liquid phase diffusion limitation in the absence of "solid" phase diffusion limitation

Under these conditions the removal rate R in equation 5.74 is given by equation 5.33; therefore:

$$R' = h'(C_R - C^*) = \frac{(k_1/\rho_o)\, C^*}{1 + k_3 C^*} \qquad \ldots (5.99)$$

Rearranging this equation:

$$C^* = \frac{R'}{k_1/\rho_o - k_3 R'} \qquad \ldots (5.100)$$

and

$$C_R = \frac{R'}{h'} + C^* \qquad \ldots (5.101)$$

Combining equations 5.100 and 5.101:

$$C_R = \frac{R'}{h'} + \frac{R'}{k_1/\rho_o - k_3 R'} \qquad \ldots (5.102)$$

where

$$R' = \frac{G}{(S_o K_o)\, t_R} \qquad \text{(equation 5.85)}$$

From equations 5.85 and 5.102 it follows that the substrate concentration depends on the flow rate parameter t_R and the re-circulation parameter G, but not upon the microbial concentration M_R or the initial concentration C_I.

The liquid phase diffusional resistance occurs only in the first term of equation 5.102 and this term becomes progressively less important as the mass transfer coefficient is increased.

The product concentrations M_R and p_R follow from equations 5.88 and 5.89.

Combination of equations 5.85 and 5.102 leads to:

$$C_R = \frac{G}{(S_o K_o)\, t_R} \left\{ \frac{1}{h'} + \frac{1}{\dfrac{k_1}{\rho_o} - \dfrac{k_3 G}{(S_o K_o)} \dfrac{1}{t_R}} \right\} \qquad \ldots (5.103)$$

Inspection of this equation shows that as $\dfrac{1}{t_R}$ is increased from zero a discontinuity occurs, with the result that the substrate concentration C_R exceeds the inlet concentration C_I. The point of wash-out can be obtained from equation 5.103 by setting C_R equal to C_I and solving for $\dfrac{1}{t_R}$.

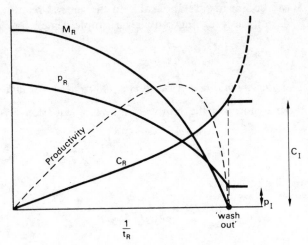

FIG. 5.30. Concentration levels in a continuous stirred
tank fermenter with a liquid phase diffusional resistance

On Fig. 5.30 are given typical curves for the concentrations M_R, C_R, p_R and
the productivity as a function of the flowrate parameter. A comparison between
this figure and Figs. 5.28 and 5.29 indicates that they differ only in the detail of
the shape of the various curves.

A further simplification follows if it is recalled that equation 5.33 is, in fact,
an asymptote applicable to a small particle size and that this in the limit refers
to the size of individual micro-organisms. Reference to equation 5.80 for the
mass transfer coefficient based on unit mass of particles shows that it is inversely proportional to the square of the characteristic size V_p/A_p:

$$h' \propto \left(\frac{A_p}{V_p}\right)^2 \qquad\qquad \ldots.(5.104)$$

The result is that experiments carried out with the microbial mass in the form
of single micro-organisms will show little liquid phase diffusional resistance.
In these circumstances equation 5.103 reduces to:

$$C_R = \frac{G}{(S_o K_o)\, t_R} \left\{ \frac{1}{\dfrac{k_1}{\rho_o} - \dfrac{k_3 G}{(S_o K_o)\, t_R}\dfrac{1}{}} \right\} \qquad \ldots.(5.105)$$

The point of "wash-out" is given by equation 5.105 when C_R is equal to C_I:

$$\frac{1}{t_{R(\text{wash-out})}} = \frac{(S_o K_o)\, k_1 C_I}{G\rho_o (1 + k_3 C_I)} \qquad \ldots.(5.106)$$

The applicability of equation 5.106 has been adequately substantiated by
HERBERT[17].

Increased re-circulation enhances the range of operation by reducing the residence time at which "wash-out" occurs and, as may easily be shown, gives rise to increased productivity.

A further factor is the effect of the liquid phase diffusional resistance. Not only does this reduce the productivity, but it also reduces the range of operation, as may be seen by comparing equations 5.97 and 5.106 (see Fig. 5.31), from which:

$$\frac{\text{Wash-out for reaction control}}{\text{Wash-out for diffusion control}} = \frac{\dfrac{k_1 C_I}{1 + k_3 C_I}}{h' C_I} \quad \ldots (5.107)$$

By definition the numerator of equation 5.107 must be greater than the denominator.

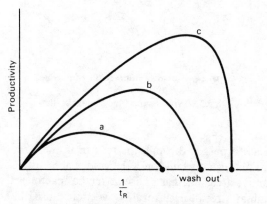

FIG. 5.31. The effect of the liquid phase mass transfer coefficient on productivity and "wash-out"

    (a) liquid phase diffusional control
    (b) liquid phase diffusional limitation
    (c) reaction control

DESIGN OF FIXED BED PROCESSES

Two arrangements are potentially useful when a microbial layer adhering to a particulate surface is used in a reactor (Fig. 5.32). The choice between the two is largely dictated by the manner in which the accumulated growth of micro-organisms is to be removed from the system.

In aseptic processes this must be achieved by hydrodynamic shear, possibly aided by a partial fluidisation of the support particles. Liquid fills the void space as a continuous phase together with, for aerobic processes, air in the form of bubbles. Alternatively, as in the non-aseptic waste treatment industry, the micro-organisms either fall from the reactor or are removed by the overall ecology, that is by worms and flies that live on the micro-organisms. In this

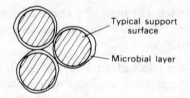

(a) **Controlled biological film thickness**

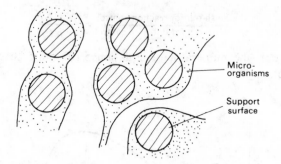

(b) **Uncontrolled biological film thickness**

FIG. 5.32. Typical supported biological films

case the liquid phase is passed through the reactor in the form of a film falling under gravity. This film usually fails to fill the void spaces or to cover completely the biologically active area. The gas phase in such a reactor is then continuous and passes through as a result of natural convection caused by the exothermic nature of the microbiological reactions.

The effect of growth on the performance of a reactor, in circumstances when the input variables are maintained constant, is to render the output time-dependent. This situation arises from the dependence of the *biological rate equation* on the growth parameter L. If, after an acceptable accumulation of micro-organisms, the microbial film thickness is maintained constant by either of the above-mentioned techniques, the output variables might be expected to attain a steady state. A further complication arises in the film flow reactor in that, while the hold-up of micro-organisms is self-regulating, a situation arises in which the microbial mass interacts with the liquid phase due to growth. The consequence of this interaction is that the extent of the interface between the microbial mass and the liquid phase becomes something of an uncontrolled dependent variable. The performance is obviously closely linked with this interfacial area, and it follows that this form of reactor is, in fact, unstable. Oscillation in output concentration is liable to occur, and the frequency and magnitude of these oscillations are intimately connected with the size of the reactor, the packing size and shape, and the average hold-up of micro-organisms under pseudo-steady conditions.

Void space filled with liquid phase

The thickness of the microbial layer within a given configuration of reactor can only be determined as a result of experimentation. However, a knowledge of this, together with the system parameters k_1, k_2, k_3, $(S_o K_o)$, $(S_p K_p)$ and the physical parameters, namely the biologically active area per unit volume a_w and the mass transfer coefficient, leads to a design procedure.

A mass balance over a differential length of reactor for a simple plug flow model, neglecting dispersion effects, gives:

$$Q' \, dC = N a_w \, dz = h_1 a_w \, dz(C - C^*) \qquad \ldots.(5.108)$$

where a_w is the biologically active area per unit volume,
$\quad C^*$ is the substrate concentration adjacent to the microbial layer,
$\quad h_1$ is the mass transfer coefficient per unit area, and
$\quad Q'$ is the volumetric flowrate per unit cross-sectional area.

The *biological rate equation* for microbial films is non-linear with respect to substrate concentration (cf. Table 5.2), and it is convenient for algebraic reasons to use a linearised form, namely:

$$N = \alpha' + \beta' C^* \qquad \ldots.(5.109)$$

Inserting 5.109 into 5.108 yields:

$$C^* = \frac{h_1 C - \alpha'}{h_1 + \beta'} \qquad \ldots.(5.110)$$

and

$$N = \frac{h_1}{h_1 + \beta'}(\alpha' + \beta' C) \qquad \ldots.(5.111)$$

Combining 5.111 and 5.108 leads to:

$$Q' \, dC = \frac{h_1}{h_1 + \beta'}(\alpha' + \beta' C) a_w \, dz \qquad \ldots.(5.112)$$

which on integration becomes:

$$C_O = \frac{\alpha'}{\beta'}\left[\exp\left(-\frac{\beta' a_w h_1 Z}{(\beta' + h_1) Q'}\right) - 1\right] + C_I \left[\exp\left(-\frac{\beta' a_w h_1 Z}{(\beta' + h_1) Q'}\right)\right]$$

$$\ldots.(5.113)$$

where C_I and C_O are the inlet and outlet substrate concentrations.

Inspection of this equation suggests the presence of an overall mass transfer coefficient h_O which is defined as:

$$h_O = \frac{1}{\dfrac{1}{\beta'} + \dfrac{1}{h_1}} \qquad \ldots.(5.114)$$

Combining equations 5.113 and 5.114 yields:

$$\frac{C_O}{C_I} = \frac{\alpha'}{\beta'C_I}\left[\exp\left(-\frac{h_O a_w Z}{Q'}\right) - 1\right] + \exp\left(-\frac{h_O a_w Z}{Q'}\right) \quad \ldots(5.115)$$

This equation may be rearranged conveniently by defining:

$$\tau = \frac{\alpha'}{\beta'C_I} \quad \ldots(5.116)$$

Thus:

$$\frac{C_O}{C_I} = (1 + \tau)\exp\left(-\frac{h_O a_w Z}{Q'}\right) - \tau \quad \ldots(5.117)$$

The equations 5.114, 5.116 and 5.117 provide the basic design equations for this reactor configuration. The product formation, both biochemical and microbiological, is related stoichiometrically to equation 5.117 through the yield coefficients $S_o K_o$ and $S_p K_p$.

The coefficients α' and β' of equation 5.109 have to be related to the biological rate equation. For present purposes this will be carried out for the reduced form of the equation, i.e. when no diffusional limitation occurs within the microbial layer. The principles are the same for the full equation though necessarily more complex from the algebraic point of view.

With this limitation, equations 5.109 and 5.30 may be equated:

$$N = \alpha' + \beta'C^* = \frac{k_1 L C^*}{1 + k_3 C^*} \quad \ldots(5.118)$$

Differentiation of equation 5.118 with respect to C^* leads to:

$$\alpha' = \frac{(k_1 L)\, k_3 C^{*2}}{(1 + k_3 C^*)^2} \quad \ldots(5.119a)$$

$$\beta' = \frac{k_1 L}{(1 + k_3 C^*)^2} \quad \ldots(5.119b)$$

Therefore, from equation 5.116:

$$\tau = \frac{\alpha'}{\beta'C_I} = \frac{k_3 C^{*2}}{C_I} \quad \ldots(5.120)$$

In order to evaluate h_O and τ, it is necessary to select an appropriate value of C^*. The particular value that recommends itself is the mean value within the reactor, namely:

$$\overline{C^*} = \frac{1}{Z}\int_0^Z C^*(z)\,dz \quad \ldots(5.121)$$

Algebraic combination of equations 5.110 and 5.115 followed by insertion into 5.121 and then integration, leads to:

$$\frac{\bar{C}^*}{C_I} = \left(\frac{Q'}{a_w Z}\right)\left(\frac{1+\tau}{\beta'}\right)\left[1 - \exp\left(-\frac{h_0 a_w Z}{Q'}\right)\right] - \tau \qquad \ldots(5.122)$$

This equation may then be rearranged as:

$$C_I = \frac{\dfrac{a_w Z \beta' \bar{C}^*}{Q'} - k_3 \bar{C}^{*2}\left[1 - \exp\left(-\dfrac{h_0 a_w Z}{Q'}\right) - \dfrac{a_w Z \beta'}{Q'}\right]}{\left[1 - \exp\left(-\dfrac{h_0 a_w Z}{Q'}\right)\right]} \qquad \ldots(5.123)$$

where the coefficients β' and τ are:

$$\beta' = \frac{k_1 L}{(1 + k_3 \bar{C}^*)^2} \qquad \ldots(5.124a)$$

$$\tau = \frac{k_3 \bar{C}^{*2}}{C_I} \qquad \ldots(5.124b)$$

Equations 5.114, 5.117, 5.123 and 5.124 provide the necessary design procedure and a suitable block diagram is given on Fig. 5.33. This is based on an initial assumption of $\bar{C}^*$ since it is more convenient to iterate than to attempt to solve for $\bar{C}^*$ for a given inlet concentration.

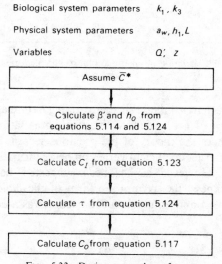

Biological system parameters $\quad k_1, k_3$

Physical system parameters $\quad a_w, h_1, L$

Variables $\quad Q', z$

| Assume $\bar{C}^*$ |
| Calculate β' and h_0 from equations 5.114 and 5.124 |
| Calculate C_I from equation 5.123 |
| Calculate τ from equation 5.124 |
| Calculate C_0 from equation 5.117 |

FIG. 5.33. Design procedure for fixed bed reactor

The mass transfer coefficient h_1 must necessarily be obtained by experiments involving a similar geometric configuration, though not necessarily in a biologically active system. The correlation of such data has received considerable attention in the past and a discussion is given in Volume 2, Chapter 6.

The film flow reactor

This reactor is important primarily in the water pollution industry where it is known as the *trickling filter*. Thick microbial layers are invariably present and, as a result, it is often found that an asymptotic form, equation 5.31, of the *biological rate equation* applies. When the reactor is started up, accumulation of micro-organisms takes place until a situation arises where performance is essentially independent of additional accumulation[27, 32]. However, even under these circumstances and with constant input characteristics, it is found that the performance of the reactor oscillates with time as the new growth interacts with the liquid flow pattern, as shown in Fig. 5.34. It is therefore an inbuilt characteristic of the reactor that the output varies with time about a mean. Inevitably such a variation is unpredictable, and the objective must be to provide a useful approximation to the mean performance.

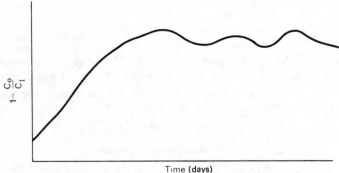

FIG. 5.34. Variation of microbial film reactor performance on start-up

The liquid flow pattern is inherently of a variable nature and indeed probably is more characteristic of a cascade flow than truly gravitational film flow[27, 32]. In these circumstances the theory of the previous section holds, although the liquid phase mass transfer coefficient verges on being incapable of interpretation and determination. However, as has been suggested, because of the large microbial film thickness, the asymptotic form of the *biological rate equation* (5.31) finds application and equation 5.109 becomes as a consequence:

$$N = \frac{k_1}{k_2} C^* \qquad \dots(5.125)$$

i.e.

$$\alpha' = 0, \qquad \beta' = \frac{k_1}{k_2}$$

In these circumstances the design equation (5.115) reduces to:

$$\frac{C_o}{C_I} = \exp\left[-\frac{h_o a_w Z}{Q'}\right] \qquad \dots(5.126)$$

Because the liquid flow pattern is strongly influenced by the presence of the micro-organisms, it is only possible to obtain realistic values of the mass transfer coefficient h_1, which is contained in h_0, in the presence of micro-organisms. The logical procedure to follow is to determine experimentally the "overall volumetric mass transfer coefficient" $h_0 a_w$ in a pilot apparatus of the process geometry and to use these data for design purposes in conjunction with equation 5.126. This removes the necessity for separate determinations of h_1, a_w and the biological rate coefficients k_1 and k_2. It is of course to be expected that $h_0 a_w$ will vary in a complex manner with flowrate due to changes both in the mass transfer coefficient h_1 and the interfacial area a_w.

IN CONCLUSION

The material that has been presented is intended to provide a basis for the design of microbiological reactors. The approach has been based on the laws of conservation, the rate processes and the kinetics of physical biocnemistry. This has led to a kinetic rate equation for microbiological processes that has been termed, for reasons of convenience, the *biological rate equation*. This equation has been used as a basis for a simple analysis of the usual reactor configurations in an attempt to illustrate the design procedures and the important physical variables involved. Relatively unsophisticated models for the hydro-dynamic phase have been adopted for reasons of clarity, and such important phenomena as incomplete mixing in stirred tank reactors and dispersion effects in packed beds have been omitted.

Of necessity, various assumptions have been made both in the derivation of the *biological rate equation* and in the procedures developed for reactor design. The experimental justification for these assumptions is, in some cases, of an unequivocal nature, while in others, necessarily circumstantial. Some sections, particularly those dealing with design procedures, represent logical mathematical extrapolations of the basic theory. The application of the *biological rate equation* to microbial flocs comes into this category.

The complete *biological rate equation* for microbial films has been confirmed by DAOUD[31] using a biological film reactor with glucose as the substrate and a mixed microbial culture derived from soil. For many years in the waste-treatment industry, when using trickling filters, a rate equation of the form of equation 5.4 or 5.31 has been generally accepted for all substrates with mixed microbial cultures. Alternatively, with microbial flocs a zero order equation has found favour both with the waste-treatment industry and in the beer industry. For many pure cultures equation 5.6, or 5.33, has been used successfully to correlate experimental data.

In addition, substantial evidence suggests that for small microbial film thicknesses the performance of a biological film reactor is, in fact, dependent on the accumulated growth[24, 27].

The above evidence, together with the fundamental basis of the derivation of the *biological rate equation*, suggests that this formulation represents a useful working hypothesis of some generality. Its advantages lie in the fact that it allows design procedures for at least a number of microbiological systems to be founded upon the same principles. As a result, data for different microbial systems in different reactor configurations can be related to a limited number of process and microbiological variables.

As far as the design procedures are concerned, most of these are based on mathematical extrapolation, particularly in the use of the *biological rate equation* for flocs and in the incorporation of liquid phase diffusional resistances. This extension of the rate equation for films is based on similar studies in the area of heterogeneous catalysis and is still in need of comprehensive experimental substantiation.

Experimental confirmation of the design procedures is limited in the case of stirred reactors to the work of HERBERT[17] who investigated a system in which sufficiently small flocs were used so that neither liquid nor solid phase diffusional limitations were in evidence. In this way the theory as described by equation 5.105 and the principles of "wash-out" have been adequately substantiated.

Considerable effort has been expended[5] in the study of the trickling filter problem, and in this work equations basically of the form of equation 5.126 have been found to correlate the data in a satisfactory way.

Many questions remain to be answered; for instance:

(a) is the *biological rate equation* of universal or only limited application?

(b) are the biological rate coefficients for a given system independent of the geometry of the microbial mass?

(c) how are the biological rate coefficients affected by pH and temperature?

(d) do thermal effects within the microbial mass influence the effectiveness factor λ?

(e) what factors determine the floc size and floc size distribution?

(f) how are the yield coefficients affected by the process variables, e.g. pH, temperature and concentration levels?

(g) to what extent can biological rate coefficients and yield coefficients determined in a batch reactor be applied to a continuous process?

(h) what is the relative importance of the liquid phase mass transfer resistance in microbiological reactors operating on a commercial basis?

Experimental methods are also needed for determining the thickness of thin biological films and for controlling the thickness in reactors of industrial dimensions.

These and many other factors of an engineering, as opposed to a microbiological, nature have to be studied and understood before the importance of the interaction of the microbiological and engineering aspects can be fully appreciated. That such effort is both justifiable and rewarding can be seen if the inherent potential of micro-organisms is considered. These living creatures are basically capable of accomplishing any reaction involving water soluble organic compounds. In fact, they might be looked upon as micro-miniaturised biochemical reactors, each of which is capable of synthesising its own catalysts and ultimately replacing itself.

ACKNOWLEDGMENT

The author wishes to express his grateful thanks to Mr. D. R. Trollope, Lecturer in Microbiology at Swansea, for helpful advice and comment on this chapter.

FURTHER READING

AIBA, S., HUMPHREY, A. E. and MILLIS, N. F.: *Biochemical Engineering*, 2nd edn. (Academic Press 1973).
ATKINSON, B.: *Biochemical Reactors* (Pion Ltd. 1974).
BAILEY, J. E. and OLLIS, D. F.: *Biochemical Engineering Fundamentals* (McGraw-Hill 1977).
BLAKEBROUGH, N.: *Biochemical and Biological Engineering Science*, Vol. 1 (Academic Press 1967).
PIRT, S. J.: *Principles of Microbe and Cell Cultivations* (Blackwell Scientific Publications 1975).
RHODES, A. and FLETCHER, D. L.: *Principles of Industrial Microbiology* (Pergamon Press 1966).

REFERENCES TO CHAPTER 5

[1] PRESCOTT, S. C. and DUNN, C. G. D.: *Industrial Microbiology,* 2nd ed. (McGraw-Hill 1949).
[2] LAIDLER, K. J.: *The Chemical Kinetics of Enzyme Action* (Oxford University Press 1958).
[3] HAWKER, L. E., LINTON, A. H., FOLKES, B. F. and CARLILE, M. J.: *An Introduction to the Biology of Micro-organisms* (Arnold 1960).
[4] DEINDORFER, F. H.: *Advances in Applied Microbiology*, Volume 2 (ed. UMBREIT, W. W.) (Academic Press 1960).
[5] ECKENFELDER, W. W. and O'CONNOR, D. J.: *Biological Waste Treatment* (Pergamon Press 1961).
[6] LOEWY, A. G. and SIEKEVITZ, P.: *Cell Structure and Function* (Holt, Rineholt & Winston 1963).
[7] BIRD, R. B., STEWART, W. E. and LIGHTFOOT, E. N.: *Transport Phenomena* (Wiley 1960).
[8] PETERSEN, E. E.: *Chemical Reaction Analysis* (Prentice-Hall 1965).
[9] PELCZAR, M. J. and REID, R. D.: *Microbiology* (McGraw-Hill 1965).
[10] AIBA, S., HUMPHREY, A. E. and MILLIS, N. F.: *Biochemical Engineering* (Academic Press 1965).
[11] ROSE, A. H.: *Chemical Microbiology* (Butterworth 1965).
[12] CRUICKSHANK, R.: *Medical Microbiology* (Livingstone 1965).
[13] MALEK, I. and FENCL, Z.: *Theoretical and Methodological Basis of Continuous Culture of Micro-organisms* (Academic Press 1966).
[14] BEETLESTONE, H. C. J.: *J. Inst. Brewing* **36** (1930) 483. Osmosis and fermentation. Part I.
[15] MONOD, J.: *Ann. Rev. Microbiol.* **3** (1949) 371. The growth of bacterial cultures.
[16] CONWAY, E. J. and DOWNEY, M.: *Biochem. J.* **47** (1950) 347. An outer metabolic region of the yeast cell.
[17] HERBERT, D.: *Soc. Chem. Ind.* Monograph No. 12 (1956) 21. A theoretical analysis of continuous culture systems.
[18] COHEN, G. N. and MONOD, J.: *Bact. Rev.* **21** (1957) 169. Bacterial permeases.
[19] MITCHELL, P.: *Ann. Rev. Microbiol.* **13** (1959) 407. Biochemical cytology of micro-organisms.

[20] GADEN, E. L.: *J. Biochem. Microbiol. Tech. Eng.* **1** (1959) 63. Fermentation process kinetics.

[21] CIRILLO, V. P.: *Ann. Rev. Microbiol.* **15** (1961) 197. Sugar transport in micro-organisms.

[22] HERBERT, D.: 11*th Symp. of the Soc. of Gen. Microbiol.* **11** (1961) 395. The chemical composition of micro-organisms as a function of their environment.

[23] ATKINSON, B and SWILLEY, E. L.: *Proc. 18th Ind. Waste Conf. Purdue Univ. Eng. Ext. Series* No. 115 (1963) 706. A mathematical model for the trickling filter.

[24] GREEN, M. B., COOPER, B. E. and JENKINS, S. H.: *Int. J. Air Wat. Pollut.* **9** (1965) 807. The growth of microbial film on vertical screens dosed with settled sewage.

[25] MUELLER, J. A., BOYLE, W. C. and LIGHTFOOT, E. N.: *Proc. 21st Ind. Waste Conf. Purdue Univ. Eng. Ext. Series* No. 121 Part 2 (1966) 964. Oxygen diffusion through a pure culture of Zoogloea Ramigera.

[26] ABSON, J. W., FURNESS, C. D. and HOWE, C.: *Wat. Pollut. Con.* (formerly *Proc. Inst. Sewage Purif.*) **66** (1967) 607. Development of the Simcar respirometer and its applications to waste treatment.

[27] ATKINSON, B., SWILLEY, E. L., BUSCH, A. W. and WILLIAMS, D. A.: *Trans. Inst. Chem. Eng.* **45** (1967) 257. Kinetics, mass transfer and organism growth in a biological film reactor.

[28] ATKINSON, B. and DAVIES, I. J.: *Trans. Inst. Chem. Eng.* **52** (1974) 248. The overall rate of substrate uptake (reaction) by microbial films—Part I: A biological rate equation.

[29] SHORE, D. T. and ROYSTON, M. G.: *Chem. Engr, London* No. 218 (May 1968) CE99. Chemical engineering of the continuous brewing process.

[30] ATKINSON, B., DAOUD, I. S. and WILLIAMS, D. A.: *Trans. Inst. Chem. Eng.* **46** (1968) 245. A theory for the biological film reactor.

[31] DAOUD, I. S.: University of Wales, Ph.D. thesis (1969). Studies on a biological film reactor.

[32] WILLIAMS, D. A.: University of Wales, Ph.D. thesis (1969). An optimisation study on an industrial trickling filter.

LIST OF SYMBOLS USED IN CHAPTER 5

A	Cross-sectional area of a packed bed	L^2
A_p	External area of a "floc"	L^2
a	Biologically active surface area per unit volume of microbial mass	L^{-1}
a'	Empirical rate coefficient	$ML^{-2}T^{-1}$ or T^{-1}
a_w	Biologically active area per unit volume of packed bed	L^{-1}
b'	Empirical rate coefficient	LT^{-1} or $L^3M^{-1}T^{-1}$
C	Substrate concentration	ML^{-3}
C^*	Concentration of substrate at interface between micro-organisms and nutrient solution	ML^{-3}
C_I	Inlet concentration of substrate	ML^{-3}
C_O	Outlet concentration of substrate	ML^{-3}
C_R	Concentration of substrate in reactor	ML^{-3}
$\bar{C}^*$	Mean of C^* defined by equation 5.121	ML^{-3}
D	Diffusion coefficient	L^2T^{-1}
D_e	Effective diffusion coefficient	L^2T^{-1}
d	Particle diameter	L
F	Volumetric flow rate	L^3T^{-1}
f	Empirical rate coefficient	$ML^{-2}T^{-1}$ or T^{-1}
G	Re-cycle parameter defined by equation 5.87	—
g	Empirical rate coefficient	ML^{-3}
h	Mass transfer coefficient	LT^{-1}
h'	Mass transfer coefficient	$M^{-1}L^3T^{-1}$
h_1	Mass transfer coefficient	LT^{-1}
h_O	Overall mass transfer coefficient defined by equation 5.114	LT^{-1}
K	Dimensionless parameter defined by equation 5.56	—
K_o	Fractional conversion of substrate to micro-organisms	—
K_p	Fractional conversion of substrate to biochemical product	—
k_1	Biological rate coefficient	T^{-1}
k_2	Biological rate coefficient	L^{-1}

k_3	Biological rate coefficient	$M^{-1}L^3$
k_s	Empirical rate coefficient	LT^{-1}
L	Biological film thickness	L
M	Concentration of micro-organisms	ML^{-3}
M_C	Concentration of micro-organisms leaving centrifuge	ML^{-3}
M_O	Concentration of micro-organisms leaving system	ML^{-3}
M_R	Concentration of micro-organisms leaving reactor	ML^{-3}
N	Flux of substrate at interface between micro-organisms and nutrient solution	$ML^{-2}T^{-1}$
N'	Flux of substrate within the microbial mass	$ML^{-2}T^{-1}$
n	Rate of substrate removal	$ML^{-2}T^{-1}$ or T^{-1}
p	Product concentration	ML^{-3}
p_I	Concentration of product entering reactor	ML^{-3}
p_R	Concentration of product in reactor	ML^{-3}
Q	Peripheral volumetric flowrate	L^2T^{-1}
Q'	Volumetric flowrate per unit area	LT^{-1}
R	Rate of substrate removal at interface between microbial mass and nutrient solution	T^{-1}
R'	Rate of substrate removal	T^{-1}
r	Rate of substrate removal	$ML^{-2}T^{-1}$
r_1	Rate of substrate removal	$ML^{-2}T^{-1}$
r_2	Rate of substrate removal	$ML^{-3}T^{-1}$
S_o	Stoichiometric coefficient relating substrate to micro-organisms	—
S_p	Stoichiometric coefficient relating substrate to biochemical product	—
t	Time	T
t_D	Doubling time	T
t_R	Residence time	T
u_s	Surface velocity of liquid film	LT^{-1}
u_z	Liquid film velocity	LT^{-1}
$\bar{u}$	Average velocity of liquid film	LT^{-1}
V	Reactor volume	L^3
V_p	Volume of a "floc"	L^3
x	Space coordinate	L
y	Space coordinate	L
Z	Length of reactor	L
z	Space coordinate	L
α	Rate coefficient	$ML^{-2}T^{-1}$
α'	Coefficient defined in equation 5.109	$ML^{-2}T^{-1}$
α_1	Transport coefficient	$ML^{-2}T^{-1}$
β	Rate coefficient	ML^{-3}
β'	Coefficient defined in equation 5.109	LT^{-1}
β_1	Transport coefficient	ML^{-3}
β_2	Biochemical rate coefficient	ML^{-3}
γ	Separation coefficient	—
δ	Liquid film thickness	L
ε	Re-cycle ratio	—
η	Dimensionless parameter.defined in equation 5.56	—
θ	Mean residence time	T
λ	Effectiveness factor	—
ρ_o	Density of microbial mass	ML^{-3}
τ	Parameter defined in equation 5.116	—
ϕ	Parameter defined in equation 5.40	$ML^{-3}T^{-1}$

GLOSSARY

Activated sludge	Material containing a very large and active microbial population. Used in one method of purifying waste waters.
Adaptation	The ability to exist in a changed environment.
Aerobe	A micro-organism that requires molecular oxygen.

Algae	A group of plants capable of photosynthesis.
Anaerobe	A micro-organism that grows in the absence of molecular oxygen.
Bacillus	A rod-shaped bacterium.
Budding	A form of asexual reproduction.
Capsule	A gelatinous layer surrounding the cell wall of many bacteria.
Coccus	A spherical bacterium.
Culture	A population of micro-organisms.
Cytoplasm	Living matter between the cell membrance and the nucleus.
Ecology	The study of micro-organisms in relation to their environment.
Enzyme	An organic catalyst produced within a micro-organism.
Facultative anaerobe	A bacterium that grows under either aerobic or anaerobic conditions.
Fermentation	Common usage: any industrial microbiological process. Specifically: anaerobic microbiological processes.
Fermenter	An industrial microbiological reactor.
Growth curve	Graphical representation of the growth of micro-organisms in a nutrient medium in a batch reactor.
Heterotroph	A micro-organism which uses organic compounds as a carbon source.
Medium	A mixture of nutrient substances.
Metabolism	The overall process by which a micro-organism uses nutrients.
Metabolite	An essential growth factor for micro-organisms.
Microaerophile	A bacterium which grows most rapidly in the presence of small amounts of molecular oxygen.
Microbial film	An adherent aggregate of micro-organisms attached to a supporting surface.
Microbial floc	An adherent aggregate of micro-organisms in suspension
Micro-organism	A form of life of microscopic dimensions.
Mixed culture	Two or more species of micro-organisms living in the same medium.
Morphology	The structure and form of micro-organisms.
Mutation	A stable change of gene inherited on reproduction.
Nutrient	A substance used as food.
Obligate	Necessary or required.
Organism	A living biological specimen.
Parasite	A living organism deriving its nutrition from another living organism.
Physiology	The study of the function of organisms.
Protein	A class of organic compounds associated with living matter. Based on combinations of amino acids.
Protozoa	Uni-cellular animals.
Pure culture	A culture containing only one species of micro-organism.
Respiration	Any chemical reaction whereby energy is released for use by the micro-organism.
Saprophytes	Micro-organisms that utilise non-living organic matter.
Slime layer	A gelatinous covering of the cell wall, used synonymously with capsule.
Species	One kind of micro-organism.
Spore	A minute, thick-walled, resistant body that forms within the cell and is considered as the resting stage.
Sterile	Free of viable micro-organisms.
Stock cultures	Known species of micro-organisms.
Strain	A pure culture of micro-organisms composed of the descendants of a single organism.
Substrate	The substance acted on by an enzyme.
Tissue	A collection of cells forming a structure.
Viable	Capable of growth.
Virus	A parasitic micro-organism smaller than a bacterium.
Zoogloeal masses	Clumps of micro-organisms in the form of á film or floc.

CHAPTER 6

Non-Newtonian technology

MANY industrial processes involve fluids or fluid mixtures which behave in various complex ways, and in consequence specialised sets of operating and design rules have been developed. One particular group of fluids of interest is that in which the effective viscosity varies with shear rate, the so-called non-Newtonian fluids, which were briefly mentioned in Volume 1, Chapter 3. Most slurries, suspensions and dispersions are non-Newtonian, as are melts of long chain polymers and solutions of polymers or other large molecules. Liquids of this type are encountered in almost all sections of the chemical industry. Indeed some types of product involve a great variety of these complex fluids: the food, paint, polymer and pharmaceutical industries draw heavily on the technology that has been evolved to handle them. Until recently the approximations provided by assuming typical average viscosities have had to suffice for design purposes, largely because the mathematical complexity of the subject has precluded a more exact analysis. However, increasing success has followed attempts to adopt a rational engineering approach to the problems. Extensive treatment of this subject has been given in several books [1, 3, 5, 6, 8, 9, 10, 12, 13].

The extent to which non-Newtonian properties as such influence design varies with the application under consideration. One area of major importance is the specification of pumping requirements for flow through pipeline systems: indeed not only in terms of the calculation of pressure drop but also in the design of the pumps themselves. Certainly, the complex behaviour shown by fluids which have some elastic properties in addition to their shear dependent viscosities has a profound influence on the design of such items as the dies used in the spinning of synthetic textile filaments or in shaping extruded polymer sections. Any other standard chemical engineering operation (like heat transfer) or process (reaction in a stirred tank, say) may be affected by the non-Newtonian properties of the liquid. Many features in the practical design of plant for these fluids are also common to those in which high viscosity Newtonian fluids (say of viscosities 10 Ns/m^2 or above) are used.*

In many instances the departure from constant viscosity that is caused by variable local shear rates is minimal, and the additional complication of allowing for it in design may not be worthwhile. This is especially true when large concentration or temperature gradients occur within the fluid.

* Viscosity in SI units may be expressed in kg/ms, Ns/m^2 or Pa s. Note: 1 centipoise ≡ 1 mN s/m^2.

In this chapter various analytical approaches to non-Newtonian fluid mechanics will be given, together with some simple examples of current practice. Before starting to consider these approaches in detail, a broad distinction will be drawn between those fluids in which the viscous properties are determined— from instant to instant—by the local shear rate, and those in which the prior shear history of the fluid sample is also important. This latter group should be further subdivided, first into those in which the viscosity is affected simply by the duration of shearing (called *time-dependent*, *thixotropic* or *rheopectic* fluids), and secondly into those in which deformations are resisted by elastic stresses as well as the usual viscous ones (called *viscoelastic* fluids). Initially, consideration will be limited to the relatively simple case of shear dependent viscosity: time dependence and viscoelasticity will be dealt with subsequently. A brief tabulation of fluids of experimental and technical interest is given in Table 6.1.

TABLE 6.1. *Some common non-Newtonian characteristics*

Practical fluid	Characteristics	Consequence of non-Newtonian behaviour
Toothpaste	Bingham plastic	Stays on brush
Drilling muds	Bingham plastic	Good lubrication properties and ability to convey detritus
Non-drip paints	Thixotropic	Thick in the tin, thin on the brush
Wallpaper paste	Pseudoplastic, often viscoelastic	Good spreading, good adhesion
Egg-white	Viscoelastic	Easy air incorporation (whipping)
Molten polymers	Viscoelastic	Thread forming properties
"Bouncing putty"	Viscoelastic	Will flow if stretched slowly, but will bounce (or shatter) if hit sharply
Wet cement aggregates	Dilatant	Permit tamping operations in which small impulses produce most complete settlement
Printing inks	Pseudoplastic	Spread easily in high speed machines yet do not run excessively at low speeds

Since many of these fluids produce effects which are functions of local velocity gradients and which vary with direction (i.e. they are anisotropic), it is convenient to start with the basic equations for three-dimensional fluid mechanics that involve viscosity, and to consider the deviations that may occur with more complex liquid behaviour in various simple situations.

THE DIFFERENTIAL EQUATIONS OF MOTION OF A VISCOUS FLUID

One of the fundamental factors in fluid mechanics is the existence of a relationship between a velocity gradient and the corresponding resisting force. In its most usual form, this is expressed as the Newtonian viscosity equation that was given as equation 3.2 of Volume 1. This, in slightly modified form, is:

$$R_{yx} = -\mu \frac{du_x}{dy} \qquad \qquad \ldots\ldots(6.1)$$

In this formulation x, y are orthogonal axes, u_x is a velocity in the x-direction and R_{yx} is the shear stress within the fluid, expressed as a resisting (negative) force per unit area for a positive velocity gradient. The parameter μ which occurs in this equation is referred to as the *viscosity* of the fluid. The viscosity of a Newtonian fluid is, by definition, independent of the velocity gradient, but will always be sensitive to variations in temperature, and to a lesser extent to changes in the applied pressure.

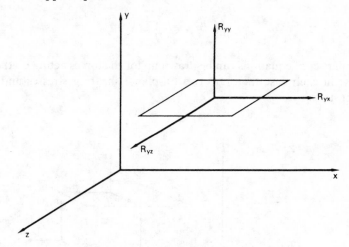

FIG. 6.1. Resolved force acting on a plane normal to the y-axis

In equation 6.1 there are terms relating to two directions in space, whereas all three directions should be considered. Figure 6.1 represents a plane perpendicular to the y-axis (a y-plane) in an x, y, z cartesian coordinate system. A stress (force per unit area) acting on such a plane may be resolved in three directions, and the components associated with this are represented by $R_{yx}R_{yz}$ and R_{yy}. In each case the first suffix relates the force to the y-plane, (i.e. of constant y) and the second to the direction of resolution. Equation 6.1 is concerned only with the x-direction—both the force and the velocity gradient are so resolved. In fact there is another limitation; the equation is strictly valid only for simple shear—that is when all the flow is in a direction parallel to the y-plane. In consequence another similar equation could be formulated if the flow were in the z-direction, in the terms:

$$R_{yz} = -\mu \frac{du_z}{dy} \qquad \dots (6.2)$$

but the third possibility concerning R_{yy} is excluded since u_y must be uniform for all values of y in order to meet the simple shear condition. It follows that the three terms R_{xx}, R_{yy} and R_{zz} are all zero for simple shear.

Naturally not all fluid motion is limited to the one directional simple shear mode: more complex partial differential equations have to be used to define other situations. The more general case may be expressed—for the y-plane—as follows (see BIRD et al.[2]):

$$R_{yy} = -2\mu \frac{\partial u_y}{\partial z} + \frac{2}{3}\mu \left(\frac{\partial u_x}{\partial x} + \frac{\partial u_y}{\partial y} + \frac{\partial u_z}{\partial z} \right) \qquad \dots (6.3)$$

$$R_{yx} = -\mu \left(\frac{\partial u_x}{\partial y} + \frac{\partial u_y}{\partial x} \right) \qquad \dots (6.4)$$

$$R_{yz} = -\mu \left(\frac{\partial u_z}{\partial y} + \frac{\partial u_y}{\partial z} \right) \qquad \dots (6.5)$$

A similar set of equations can be drawn up for the forces acting on the x- and z-planes; in each case there are two (in-plane) shearing stresses and a third (normal) stress.

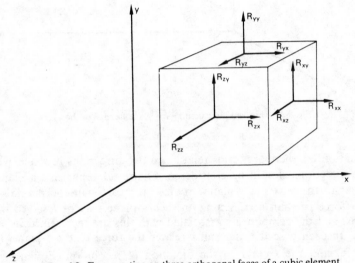

FIG. 6.2. Forces acting on three orthogonal faces of a cubic element of fluid

In Fig. 6.2 a cubic element is drawn with each of these resolved forces illustrated. From the way in which R_{yx} is formulated in equation 6.4, it is evident that it is identical to R_{xy}, and there is the same equivalence between R_{yz} and R_{zy}. This fact may be clearer if the couples acting about one corner of the cubic element shown in the figure are considered. As the size of the cube is reduced, the steady state can exist only if couples R_{xy} and R_{yx} etc. are equal and opposite.

The equations all imply the constancy of the coefficient μ, the Newtonian viscosity of the fluid, but the relationships for non-Newtonian fluids are not of such a straightforward nature. Indeed, even the simplicity of the relationship implied for simple shear breaks down, since the effective viscosity (i.e. the ratio between the shear stress and the velocity gradient) becomes a function of that velocity gradient. Thus:

$$R_{yx} = f\left(\frac{du_x}{dy}\right) \qquad \ldots(6.6)$$

In many practical cases even this type of equation is inadequate, as it is found that the fluid properties change with time t, the liquid becoming effectively more or less viscous as the shearing continues. For such a situation time must also be involved in the equation:

$$R_{yx} = f\left(\frac{du_x}{dy}, t\right) \qquad \ldots(6.7)$$

In many publications dealing with fluids of these types the symbols τ_{yx}, τ_{yz} etc. are used for shear stresses instead of R_{yx}, R_{yz} etc. They are in all respects equivalent.

This discussion generally uses velocity gradients as variables, and these have been written as $\frac{du_x}{dy}$ etc. The dimensions of velocity gradient are T^{-1} and, as such, the gradients represent shear rate. If the ratio of the linear upper plane displacement to the plane separation distance is denoted by γ, then the shear rate is $\frac{d\gamma}{dt}$, often expressed as $\dot{\gamma}$.

The presentation that will be given here is necessarily simplified. The nine components of stress represented in Fig. 6.2 are the components of a tensor, which can be written as a 3×3 matrix. Full mathematical analysis is best presented in tensor and vector notation. A proper understanding of modern publications in this field requires familiarity with the appropriate mathematics: most of the more recent books cited above give at least a summary of the necessary knowledge.

THE SHEAR BEHAVIOUR OF INELASTIC, NON-TIME-DEPENDENT FLUIDS

In most equipment used for the determination of viscosity, the shear rate varies throughout the sample of fluid. This is of no consequence with Newtonian liquids when viscosity is not a function of shear rate. However, with non-Newtonian materials the apparent viscosity will vary throughout the sample in such an experiment; in consequence the characterisation of non-Newtonian fluids is often carried out in apparatus which is designed with a view to limiting, or at least controlling, the variation of shear rate with position.

Details of different types of apparatus and some discussion of the interpretation of experiments carried out with them will follow later, but it may be mentioned here that the only design in common use that maintains a uniform shear rate throughout the specimen is that in which the fluid is confined between a cone and a plate rotating relative to one another as illustrated in Fig. 6.17. The concentric cylinder arrangement (Fig. 6.15) does not give rise to an exactly uniform shear rate distribution, but the stress and velocity field can easily be calculated. Various patterns of rotating bob, cylinder and disc are also widely used. Flow in static apparatus is the basis of the capillary tube viscometer (Figs. 6.13 and 6.14), and the picture is completed by the use of falling sphere and allied viscometers. Although the analytical simulation of some of these situations will be considered later, some description of the types of behaviour that are found, and the methods of modelling these characteristics, is appropriate first.

In Fig. 6.3 the relationship between shear stress and shear rate for two idealised fluids (*Newtonian* and *pseudoplastic*) is shown. A plot of this type is called a *rheogram*. For the Newtonian fluid the constant slope corresponds to the uniform viscosity indicated by equation 6.1. For the pseudoplastic material the slope decreases with increase of shear rate showing that the apparent viscosity is becoming less. For a given shear rate (velocity gradient) the apparent or effective viscosity of the non-Newtonian fluid is given by the slope of the line drawn directly to the origin (not by the tangent to the flow curve). At the shear rate marked S the apparent viscosity of this pseudoplastic is the same as that of the Newtonian liquid. If the apparent viscosity μ_a is plotted as a function of shear rate then the behaviour represented in Fig. 6.3 may be shown as in Fig. 6.4. In many cases it is useful to present this information as double logarithmic plots which are shown as Figs. 6.5 and 6.6.

Other types of behaviour that were represented in Fig. 3.4 of Volume 1, including two forms of *viscoplastic* behaviour (*Bingham plastic* and *false body*), are shown in Fig. 6.7, with the corresponding effective viscosities in Fig. 6.8; and the log/log plots of the same data are given in Figs. 6.9 and 6.10. The corresponding curves for *shear thickening* or *dilatant* materials are also shown in the same figures.

It should be noted that any fluid that needs a finite stress to produce flow—e.g. the Bingham plastics—is not really a liquid according to the strict physical definition. In particular, such a fluid will not level out under gravity to form an absolutely flat free surface. Irregularities in free surfaces are very evident if viewed with specularly reflected light at a low angle of incidence. Very small finite yield stresses of some materials can be detected in this way.

Equations that have been used to describe these various types of fluid behaviour are given below. They are all approximations and appreciable deviations may occur in practice, particularly if a wide range of shear rate is considered. If an experiment is carried out over several decades of shear rate, a pattern of behaviour of a real pseudoplastic fluid might be of the type illustrated in Fig. 6.11, with the corresponding apparent viscosity as shown in the log/log plot of Fig. 6.12. These curves can frequently be conveniently regarded as being approximated by the three linear regions A–B, C–D and E–F and their extrapolations.

At the ends of the shear rate range there are two regions in which the material is essentially Newtonian in character, with an effective viscosity that is uniform. These are termed the lower and upper Newtonian regions. It is difficult to

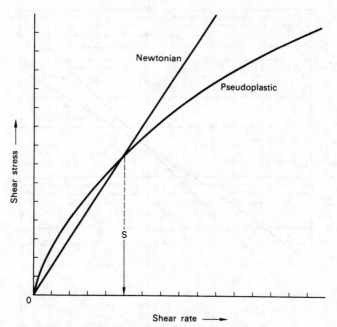

FIG. 6.3. Shear stress–shear rate behaviour of Newtonian and pseudoplastic fluids plotted using arithmetic coordinates

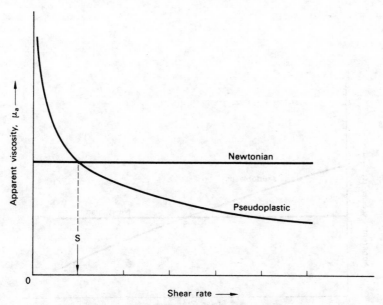

FIG. 6.4. Shear rate dependence of apparent viscosity for Newtonian and pseudoplastic fluids plotted arithmetically

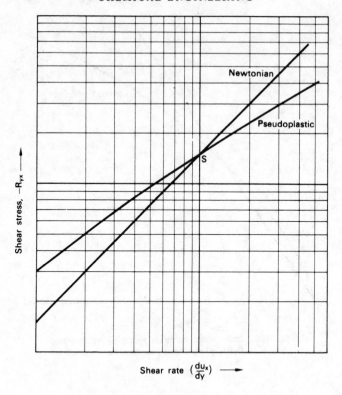

FIG. 6.5. Log/log plot of the data in Fig. 6.3

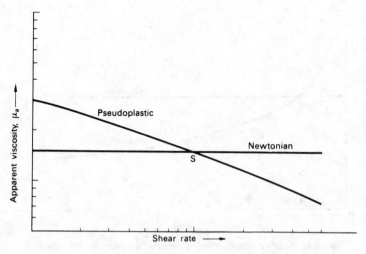

FIG. 6.6. Log/log plot of the data in Fig. 6.4

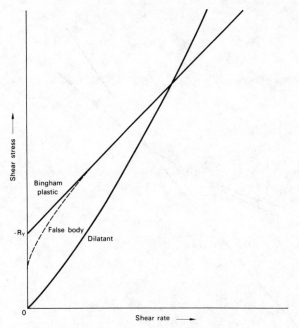

FIG. 6.7. Shear stress–shear rate data for Bingham plastic and
dilatant fluids (arithmetic)

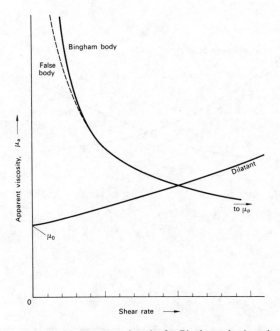

FIG. 6.8. Apparent viscosity for Bingham plastic and
dilatant fluids

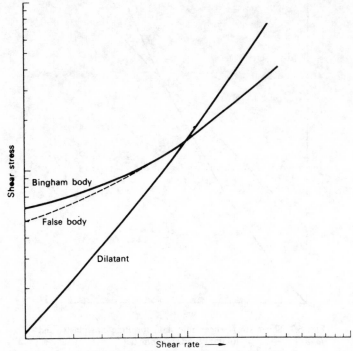

FIG. 6.9. Log/log plot of the data in Fig. 6.7

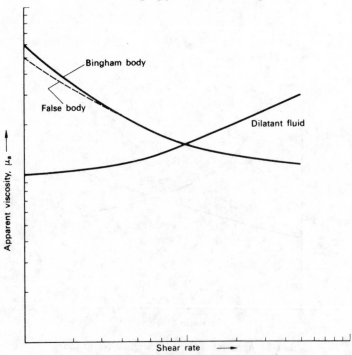

FIG. 6.10. Log/log plot of the data in Fig. 6.8.

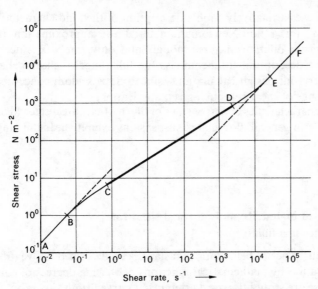

FIG. 6.11. Behaviour of a typical shear-thinning fluid
plotted logarithmically for several orders of shear rate

provide valid generalisations, but many materials meet these conditions for
shear rates below about $0{\cdot}1\ \text{s}^{-1}$ and above $10^5\ \text{s}^{-1}$. In the intermediate region
(C–D on the shear rate plot), the slope is less than unity, reflecting shear-
thinning or pseudoplastic behaviour. Data such as these are extremely difficult
to obtain satisfactorily, as it is almost impossible to arrange for a single instrument to have both the sensitivity required for the measurements at low shear
rates and the robustness for those at high rates.

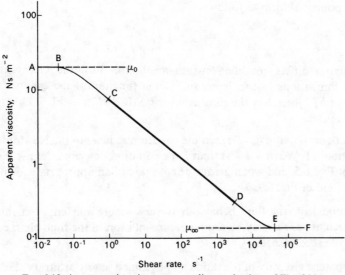

FIG. 6.12. Apparent viscosity corresponding to the data of Fig. 6.11

In many cases a relatively simple description of the fluid characteristics will be sufficient; for a non-Newtonian material the approximation that it is pseudoplastic or dilatant may be enough, and only rarely is it necessary to elaborate the description to encompass the full range of shear behaviour in order to carry out design calculations, since one is seldom concerned with a very wide range of shear rates in a given application.

The flow characteristics of suspensions of fine particles are discussed in Volume 2, Chapter 7; in general, flocculated suspensions exhibit shear thinning characteristics.

Equations used to describe an isothermal shear rate dependence in a fluid

The various types of behaviour that have been described above are capable of representation by mathematical functions of varying degrees of complexity. Some of these are straightforward attempts at curve fitting, giving an empirical relationship for the shape of the shear stress–shear rate curve for example, while others have some theoretical foundation in the theories of statistical mechanics—as an extension of the application of kinetic theory to the liquid state.

1. *The power law*

The relationships between shear stress and shear rate (or velocity gradient) that are illustrated in Fig. 6.5 for a pseudoplastic material, and in Fig. 6.9 for a dilatant fluid, are nearly straight lines on a log/log plot and may be approximated by an exponential form as follows:

$$R_{yx} = -k \left(\frac{du_x}{dy}\right)^n \qquad \ldots .(6.8)$$

This equation differs from the Newtonian equation 6.1 in the important respect that the dimensions of the "constant" are a function of the exponent on the second term; k here has the dimensions of $\mathbf{ML^{-1}T^{n-2}}$ or $\mathbf{FL^{-2}T^n}$, whilst n is dimensionless.

It can be seen that if $n = 1$, then the equation reduces to the Newtonian form of equation 6.1. When n is less than 1 the equation describes the pseudoplastic model of Fig. 6.5, and when greater than 1 it is of an appropriate form for the dilatant fluid of Fig. 6.9.

This model for the fluid behaviour is very widely applied, even though in any particular case it may be a poor representation of the fluid characteristics over a wide shear range.

An apparent viscosity of the fluid can be defined at any arbitrary shear stress as the viscosity of a Newtonian fluid which would flow at the same rate. Thus:

$$R_{yx} = -k \left(\frac{du_x}{dy}\right)^n = -\mu_a \left(\frac{du_x}{dy}\right)$$

Therefore:

$$\mu_a = k \left(\frac{du_x}{dy}\right)^{n-1} \qquad \ldots (6.9)$$

For all values of n other than unity the apparent viscosity would approach an unrealistic limit for creeping flow, i.e. at very low rates of shear; the pseudoplastic materials becoming very highly viscous and the dilatant fluids very mobile. In practice, shear rates of less than $1 \cdot 0 \text{ s}^{-1}$ are seldom significant and this limiation is often not serious.

The power-law equations 6.8 and 6.9 use two parameters n and k in order to provide approximations to the real behaviour shown graphically in Figs. 6.3 to 6.12. Such a curve fit is of necessity only reasonable when the fluid behaviour is more or less linear when drawn on log/log graph paper. Naturally this approach breaks down when ranges of shear rate over several orders of magnitude are considered and the fluid has non-Newtonian properties, say like those possessed by the pseudoplastic liquid represented in Figs. 6.11 and 6.12. Under such conditions it may well be useful to use three or more constants in the descriptive analytical equation (the "equation of state"), and with more complex formulations than the simple exponential dependence embodied in the power law.

2. The Ellis model

Of particular value for describing the behaviour at relatively low shear rate ranges is the Ellis equation, which is written as follows:

$$R_{yx} = -\frac{1}{\beta + \Gamma(-R_{yx})^{\alpha-1}} \left(\frac{du_x}{dy}\right) \qquad \ldots (6.10)$$

It will be noticed that this equation is expressed in terms of a non-linear function of the stress, and the shear rate raised to an exponent of unity. This form of equation has advantages in that it permits easy calculation of velocity profiles from stress distributions, but renders the reverse operation complicated. The dimensions of the constants are as follows:

$$\alpha \quad \text{dimensionless}$$
$$\beta \quad \mathbf{M^{-1}LT}$$
$$\Gamma \quad \mathbf{M^{-\alpha}L^{\alpha}T^{2\alpha-1}}$$

3. The Prandtl–Eyring model

This two-parameter model which has some quasi-theoretical justification from the theory of rate processes gives the following equation:

$$R_{yx} = -R_{PE} \sinh^{-1} \left(\theta_{PE} \frac{du_x}{dy}\right) \qquad \ldots (6.11)$$

with the dimensions of the constants being:

$$R_{PE} = \mathbf{ML^{-1}T^{-2}}, \quad \theta_{PE} = \mathbf{T}$$

Unfortunately the equation does nothing to provide any real advantage over the simple power-law equation in terms of the accuracy with which the fluid characteristics are represented, and it is evidently less tractable as a model for calculations.

4. The Reiner–Philippoff equation

This equation is a three-parameter curve fit which has been suggested in an attempt to describe the entire range of shear rates in terms of the upper and lower Newtonian viscosities (see Fig. 6.11):

$$R_{yx} = - \left[\mu_\infty + \frac{\mu_0 - \mu_\infty}{1 + \left(\dfrac{R_{yx}}{R_{RP}} \right)^2} \right] \frac{du_x}{dy} \qquad \ldots (6.12)$$

The dimensions of the viscosities are, as usual, $\mathbf{ML^{-1}T^{-1}}$. The constant R_{RP} has the dimensions of a stress, i.e. $\mathbf{ML^{-1}T^{-2}}$. The fitting of this equation is empirical, the value of R_{RP} determining a point of transition between the upper and lower limiting viscosities. It should be noted that the algebraic form of this equation describes 90 per cent of the entire transition from one limiting viscosity to the other within only two decades of shear stress variation, even when there is a thousandfold difference in the two limiting viscosities.

5. The Sutterby equation

An empirical three-parameter equation that has been of value in describing behaviour at low shear rates is due to SUTTERBY[27]. This is expressed in terms of the ratio of the effective viscosity μ_a to that at extremely slow rates of shear (μ_0):

$$\frac{\mu_a}{\mu_0} = \left[\frac{\sinh^{-1} \theta_S \left(\dfrac{du_x}{dy} \right)}{\theta_S \left(\dfrac{du_x}{dy} \right)} \right]^{\alpha_S} \qquad \ldots (6.13)$$

In this expression the viscosities are, as usual, of the dimensions $\mathbf{ML^{-1}T^{-1}}$. The constant α_S is dimensionless and θ_S has the dimensions of time.

An equation of this form may be used to fit data of the type illustrated in Fig. 6.29 with a reasonable degree of accuracy; the asymptotic slope is described by the α_S term, and the point where the extrapolations of the limiting values would intersect is defined by θ_S. Again, there is no way of allowing for differences in the range of shearing rates over which the Newtonian behaviour is replaced by a power-law behaviour for different liquids. In that region (often of order 0.1 s^{-1}) the value of viscosity given by equation 6.13 may be in error by 20 per cent or so, but away from that change-over region the predictions are very good. The data of Fig. 6.29 have been fitted by lines of this general form.

6. The Bingham plastic equation

This equation is the simplest that may be used to describe the behaviour of a fluid with a yield stress, as illustrated in Fig. 6.7. The equation is:

$$R_{yx} - R_Y = -\mu_p \left(\frac{du_x}{dy}\right) \qquad \ldots(6.14)$$

In this equation the term R_Y is used for the yield stress of the fluid, and $|R_{yx}|$ must exceed $|R_Y|$ in order that any flow may take place; i.e. below that value of stress $\frac{du_x}{dy}$ is always zero.

7. The Casson equation

Many foodstuffs and biological materials are well described by the empirical Casson equation that can fit a "false body" material:

$$|R_{yx}|^{\frac{1}{2}} - |R_Y|^{\frac{1}{2}} = \left\{\mu_p \left|\frac{du_x}{dy}\right|\right\}^{\frac{1}{2}} \qquad \ldots(6.15)$$

Blood, yoghurt, yeast suspensions and molten chocolate are among the materials for which this equation has been used.

FLUID CHARACTERISATION

In order to determine fully the flow properties of a fluid, careful experimentation is required. With a Newtonian fluid it is strictly necessary to define only the temperature and pressure at which the viscosity is determined, and to measure the shear stress at one known shear rate. For all non-Newtonian fluids some variant of the viscosity determination must be carried out at more than one shear rate. The further complications that arise when viscous behaviour is accompanied by elastic will be considered later.

It is appropriate now to consider the general features of apparatus that is commonly used to determine flow characteristics. Detailed treatment of the various instruments will be found in specialised books, e.g. WILKINSON[5], VAN WAZER et al.,[7] SKELLAND[9] and especially WALTERS[11]. There are two main groups of characterising apparatus: that in which fluid flows through a tube, channel or orifice, and that in which it is sheared between moving surfaces.

Tube viscometry

Several forms of apparatus have been constructed for the determination of viscosity from knowledge of the pressure drop–flowrate relationship for a tube. The flow may be achieved simply as a result of gravitational head, as in the standard U-tube viscometers that are described in B.S. 188 (1977)[41]. They are not very suitable for measurements with non-Newtonian fluids, not only because of the non-uniform shear stress across the tube section, but also because the rate of flow through them varies with time as the level falls. Even so, they may be used in some circumstances. More usually the flowrate is maintained by pressure from a gas reservoir and the fluid is forced at constant velocity through the calibrated tube. It is necessary to allow for the non-linearities that occur at inlet and exit, and various empirical and semi-theoretical methods are available that permit prediction of these corrections. The most usual, and probably the most satisfactory, approach is to examine the pressure drop–flow relationships in two tubes that are identical in all respects except that they differ in length, though even in that case it is necessary to ensure that the entry lengths for fully developed flow are exceeded in all the measurements. For work of this type the length to diameter ratios of the tubes normally exceed 50, and are often of the order of 500. These viscometers are illustrated diagrammatically in Figs. 6.13 and 6.14.

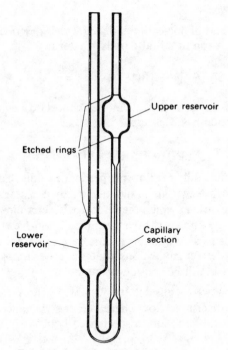

FIG. 6.13. A standard U-tube viscometer

Rotational viscometry

Several patterns of rotational viscometer are in common use. These are only briefly described here; reference should be made to the specialised sources[5, 7, 9, 11] for fuller details.

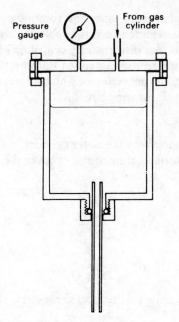

FIG. 6.14. A pressurised capillary viscometer

Concentric cylinders

Concentric-cylinder viscometers are in widespread use. Figure 6.15 represents a partial section through such an instrument in which liquid is contained and

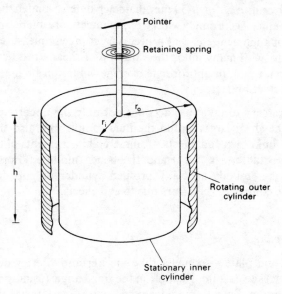

FIG. 6.15. Partial section of a concentric-cylinder
viscometer

sheared between the stationary inner and rotating outer cylinders. Either may
be driven, but the flow regime which is established with the outer rotating and
the inner stationary is less disturbed by centrifugal forces. The couple transmitted through the fluid to the suspended stationary inner cylinder is resisted
by a calibrated spring, the deflection of which allows calculation of the torque,
and hence the inner wall shearing stress R_i:

$$C = -R_i \times r_i \times 2\pi r_i h \qquad \qquad \dots\text{(6.16}a\text{)}$$

This couple C originates from the outer cylinder which is driven at a uniform
speed. On the inner surface of the outer cylinder the shear stress is R_o and we
can write:

$$C = -R_o \times r_o \times 2\pi r_o h \qquad \qquad \dots\text{(6.16}b\text{)}$$

therefore:
$$R_i = \frac{-C}{2\pi r_i^2 h}$$

and
$$R_o = \frac{-C}{2\pi r_o^2 h} \qquad \qquad \left.\rule{0pt}{45pt}\right\} \dots\text{(6.16}c\text{)}$$

For any intermediate radius r, the local shear stress is:

$$R_r = \frac{-C}{2\pi r^2 h} = R_o\left(\frac{r_o^2}{r^2}\right) = R_i\left(\frac{r_i^2}{r^2}\right) \qquad \qquad \dots\text{(6.17)}$$

If the gap between the cylinders is small relative to their radii, the minor
difference in shear stress across the gap can be neglected. However, it should
be noted that as the stress variation is proportional to the square of the radius
the value of r_o needs to be not more than $1\cdot024 \times r_i$ for the variation to be less
than 5 per cent, e.g. for a 50 mm diameter outer cylinder the gap can be only
about 0·6 mm. Accordingly, strict requirements are therefore imposed on the
accuracy of alignment etc. that is necessary if precise measurements are needed.
Fortunately, with many fluids the variation of shear stress across the gap is not
too important and an appropriate average value can be used with analytically
derived correction factors[5].

It is necessary to take into account not only the uncertainty of the average
value of the shear stress within the fluid sample, but also those errors which
arise from unknown end effects. Commercial instruments of this type—like the
Ferranti portable (Fig. 6.16) and the bench mounted Epprecht and Haake
instruments—embody specially profiled cylinder or cup designs to provide
reasonable control of the errors due to end effects.

Cone and plate

A cone and plate viscometer (like the Ferranti–Shirley or the Weissenberg
instruments) shears a fluid sample in the small angle (usually 4° or less) between
a flat surface and a rotating cone whose apex just touches the surface. Figure
6.17 illustrates one such arrangement. This geometry has the advantage that

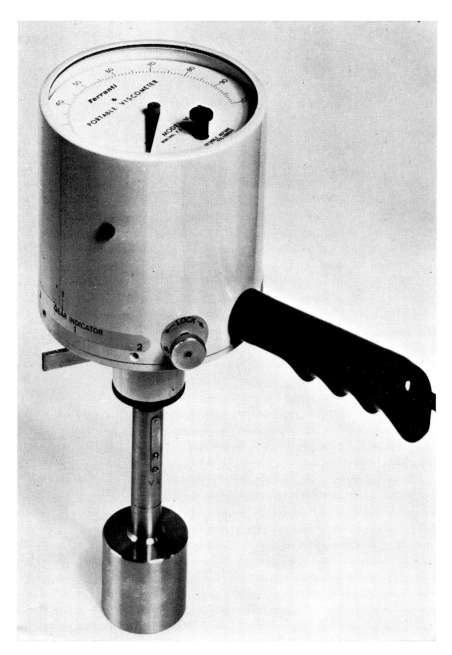

FIG. 6.16. Ferranti portable (concentric-cylinder) viscometer

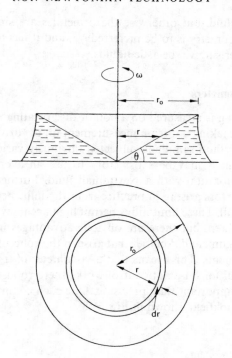

FIG. 6.17. Cone and plate viscometer

the shear rate is everywhere uniform and equal to $\omega/\sin\theta$, since the local cone velocity is ωr and the separation between the solid surfaces at that radius is $r\sin\theta$; ω is the angular velocity of rotation. The shear stress R_r acting on a small element or area dr wide will produce a couple $-2\pi r\,dr\,R_r r$ about the axis of rotation. With a uniform velocity gradient at all points in contact with the cone surface, the surface stress R will also be uniform, so we can omit the suffix and express the total couple about the axis as:

$$-C = \int_0^{r_o} 2\pi r^2\,R\,dr = \frac{2}{3}\pi r_o^3 R \qquad\qquad \ldots(6.18)$$

The shear stress within the fluid can therefore be evaluated immediately from equation 6.18.

This type of apparatus needs to be extremely carefully manufactured; any eccentricity in the orientation of the cone or in the drive results in serious errors in the readings. This is a more serious problem than with concentric cylinders which, to some extent, are self-centring.

Parallel plate viscometers

Parallel plate instruments are not a limiting case of the cone and plate geometry since there is no central contact and the shear rate is not uniform throughout the fluid, but otherwise the principles are similar. From certain viewpoints this geometry is to be preferred[11], and it has been more especially useful in characterising viscoelastic liquids.

Infinite fluid viscometers

The viscous drag is measured on a bob or disc rotating in a large volume of fluid, usually by making a torsion measurement on the driving shaft. The shear rate to which the fluid sample is subjected obviously varies with the distance from the bob, but it may be satisfactory to use an average value, possibly derived from calibration with a Newtonian fluid. Further, it is necessary to ensure that the errors arising in practice from the influence of the walls of the container are small. These limitations permit high accuracy to be achieved only if great care is taken, but there are obvious advantages in avoiding the con-centricity and alignment difficulties that arise in the concentric cylinder or cone and plate instruments. This means that a viscometer of this general pattern is particularly useful for on-plant control work where robustness and simplicity of use are more important than precision. One such commercial instrument is that made by Brookfield, shown in Fig. 6.18.

Falling ball measurements

One other method that may be used for characterisation, which is particularly useful at low rates of shear, is the falling ball viscometer. The terminal velocity of a spherical particle is determined and the effective viscosity of the fluid is calculated by applying Stokes' Law (see Volume 2, Chapter 3). For accurate work it is necessary to correct for the fact that the ball is continuously displacing its own volume of fluid. This technique was used effectively by SUTTERBY[27].

The falling sphere method differs from the rotational methods in the im-portant respect that the fluid is not in a steady state of flow during the determina-tion, and results have therefore to be treated with reserve for viscoelastic fluids. Fortunately, the importance of the elastic effects becomes less as the shear rates are reduced and, since a common objective is to determine the lower Newtonian viscosity, the limitation is often acceptable.

ANALYTICAL EQUATIONS FOR FLOW IN SIMPLE GEOMETRIES

With a Newtonian fluid flowing through a pipe or channel it is possible to calculate a velocity profile as a result of an application of the basic viscous flow equation to the stress field. This is of value for laminar flow situations, which are of only limited importance in Newtonian fluid mechanics. With non-Newtonian systems, however, the problem is rather different since the great majority of flow situations are laminar in character, partly because of the

FIG. 6.18. Brookfield portable (infinite fluid) viscometer

intrinsically high viscosity of most practical non-Newtonian fluids, and partly—as will be seen later—because there is some difficulty in inducing a fully turbulent flow mode in such fluids.

Laminar flow in a tube

Fully developed laminar flow in a circular pipe was considered in Volume 1, Chapter 3. The shell balance for such a section of pipe (Fig. 6.19) leads us to

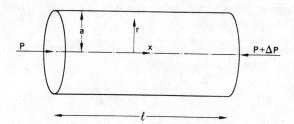

FIG. 6.19. Coordinate system for a pipe section

equate the force attributable to the pressure drop over the length of the pipe to that arising from the total shearing stress in the fluid at the wall. For the whole pipe section:

$$-\Delta P\pi a^2 = -2\pi a l R_{wx} \qquad \ldots (6.19)$$

or for some smaller radius:

$$-\Delta P\pi r^2 = -2\pi r l R_{rx} \qquad \ldots (6.20)$$

Confusion occasionally occurs from the signs arising in these equations, which are a consequence of defining ΔP in the sense used in Fig. 6.19. For normal uniform pipeline flow ΔP would be negative, but in the more general case (when dealing with changes in cross-section for example) this is not necessarily so.

The relations 6.19 and 6.20 lead us immediately to the conclusion that the shear stress distribution is a linear function of radius. Thus:

$$\frac{R_{rx}}{R_{wx}} = \frac{r}{a}$$

i.e.

$$R_r = R_w \frac{r}{a} \qquad \ldots (6.21)$$

It is important to notice that the shear distribution calculated here is a consequence of the macroscopic or shell balance, and involves no assumptions about the nature of the flow or the fluid except in so far as the pipe is of uniform circular cross-section and there is fully developed flow. The shear stress profile is shown in Fig. 6.20; the fluid is in simple shear (there is no radial flow). In

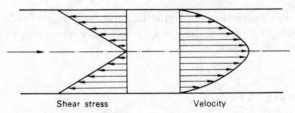

FIG. 6.20. Shear stress and Newtonian velocity
profiles for laminar flow in a pipe

cylindrical coordinates this means we can ignore terms in x and ψ and restrict consideration to those involving r.

For the Newtonian case we can apply a form of equation 6.1, together with the shear stress distribution, and write:

$$\frac{-\Delta P r}{2l} = -\mu \frac{du_x}{dr} \qquad \ldots(6.22)$$

Separating the variables and integrating we obtain immediately:

$$\frac{-\Delta P}{2l} \left[\frac{r^2}{2}\right]_0^r = -\mu \left[u_x\right]_{u_{CL}}^{u_x}$$

Therefore:

$$\frac{-\Delta P r^2}{4l} = \mu \left[u_{CL} - u_x\right] \qquad \ldots(6.23)$$

where u_{CL} is the velocity at the centre of the pipe and u_x is the velocity at radius r, both in the x-direction. As a boundary condition, we know that at the wall the velocity is zero (assuming there is no slip). It follows that:

$$u_{CL} = \frac{-\Delta P a^2}{4\mu l}$$

Substituting this into equation 6.23 we have then:

$$\frac{-\Delta P r^2}{4l} = \mu \left(\frac{-\Delta P a^2}{4\mu l} - u_x\right)$$

whence:

$$u_x = u_{CL} \left(1 - \frac{r^2}{a^2}\right) \qquad \ldots(6.24)$$

A similar technique is followed in the non-Newtonian case using the appropriate velocity gradient/stress relationship. For example, for a fluid described by the power-law equation (6.8) we have, instead of equation 6.22:

$$\frac{-\Delta P r}{2l} = k \left(-\frac{du_x}{dr}\right)^n \qquad \ldots(6.25)$$

therefore:

$$\left(\frac{-\Delta P r}{2lk}\right)^{1/n} dr = -du_x$$

$$\frac{n}{n+1}\left(\frac{-\Delta P}{2lk}\right)^{\frac{1}{n}} r^{\frac{n+1}{n}} = u_{CL} - u_x \qquad \ldots (6.26)$$

The boundary condition is that when $r = a$, $u_x = 0$ and therefore:

$$u_{CL} = \frac{n}{n+1}\left(\frac{-\Delta P}{2lk}\right)^{\frac{1}{n}} a^{\frac{n+1}{n}}$$

from which it follows that:

$$u_x = u_{CL}\left[1 - \left(\frac{r}{a}\right)^{\frac{n+1}{n}}\right] \qquad \ldots (6.27)$$

which reduces to the Newtonian result of equation 6.24 when $n = 1$.

It is perhaps more useful to calculate the velocity profile in terms of the average velocity of the fluid u, since commonly we have knowledge of the total volumetric flowrate rather than of the centre-line velocity. To do this we simply calculate the total flow Q from equation 6.27 by integrating over the whole pipe area and equate the result to that calculated from the defined average velocity u, as follows:

$$Q = \pi a^2 u = \int_0^a 2\pi r u_x \, dr \qquad \ldots (6.28)$$

$$= \int_0^a \frac{2\pi u_{CL}}{a^{\frac{n+1}{n}}}\left[ra^{\frac{n+1}{n}} - r^{\frac{2n+1}{n}}\right] dr$$

$$= \frac{2\pi u_{CL}}{a^{\frac{n+1}{n}}}\left[\frac{r^2}{2}a^{\frac{n+1}{n}} - \frac{nr^{\frac{3n+1}{n}}}{3n+1}\right]_0^a$$

$$= \frac{2\pi u_{CL}}{a^{\frac{n+1}{n}}}\left[\frac{a^{\frac{3n+1}{n}}}{2} - \frac{na^{\frac{3n+1}{n}}}{3n+1}\right]$$

whence

$$\frac{u}{u_{CL}} = \frac{2}{a^{\frac{3n+1}{n}}}\left[\frac{a^{\frac{3n+1}{n}}}{2} - \frac{na^{\frac{3n+1}{n}}}{3n+1}\right] = \frac{n+1}{3n+1} \qquad \ldots (6.29)$$

Substituting from 6.29 into 6.27 therefore, the dimensionless velocity profile is:

$$\frac{u_x}{u} = \frac{3n+1}{n+1}\left[1 - \left(\frac{r}{a}\right)^{\frac{n+1}{n}}\right] \qquad \ldots (6.30)$$

The velocity profiles calculated from equation 6.30 for various values of n are shown in Fig. 6.21.

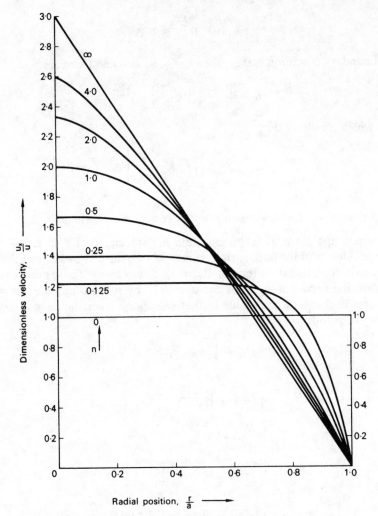

FIG. 6.21. Fully developed laminar velocity profiles for power-law
fluids in a pipe

Example

Calculate the fully developed volumetric flowrate for an Ellis fluid (equation 6.10) flowing down
a wide vertical surface in a film 3 mm thick if the fluid constants are as follows:

$$\text{Density } 1200 \text{ kg/m}^3, \quad \beta = 3\cdot 0 \text{ m s/kg}$$
$$\Gamma = 4\cdot 0 \text{ m}^{1\cdot 3} \text{ s}^{1\cdot 6} \text{ kg}^{-1\cdot 3}, \quad \alpha = 1\cdot 3$$

Solution

A general equation will be derived first.

The weight of fluid outside the plane (Fig. 6.22) distant y from the surface is balanced by the shear
stress at that plane; i.e. for unit width and unit height:

$$-R_{yx} = \rho g(s - y)$$

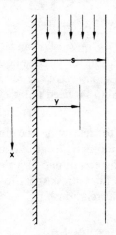

FIG. 6.22. Coordinates
for film flow example

The equation of state (equation 6.10 above) may be rearranged:

$$-R_{yx}\beta + (-R_{yx})^{\alpha}\Gamma = \frac{du_x}{dy}.$$

Substituting, it follows that:

$$du_x = \{\rho g(s - y)\beta + [\rho g(s - y)]^{\alpha}\Gamma\}\,dy$$

and integrating:

$$u_x = \rho g\left(sy - \frac{y^2}{2}\right)\beta - \frac{(\rho g)^{\alpha}}{\alpha + 1}(s - y)^{\alpha + 1}\Gamma + \text{constant}.$$

Applying the condition of no slip at the wall $(u_x)_{y=0} = 0$, the constant is evaluated as $\dfrac{(\rho g)^{\alpha}}{\alpha + 1}\Gamma s^{\alpha + 1}$, and:

$$u_x = \rho g\left(sy - \frac{y^2}{2}\right)\beta - \frac{(\rho g)^{\alpha}}{\alpha + 1}\Gamma\{(s - y)^{\alpha + 1} - s^{\alpha + 1}\}$$

The total volumetric flow Q is given by $\int_0^s u_x\,dy$, i.e.:

$$Q = \rho g\beta\left[\frac{sy^2}{2} - \frac{y^3}{6}\right]_0^s + \frac{(\rho g)^{\alpha}\Gamma}{\alpha + 1}\left[\frac{(s - y)^{\alpha + 2}}{\alpha + 2} + s^{\alpha + 1}y\right]_0^s$$

$$= \frac{\rho g\beta s^3}{3} + \frac{(\rho g)^{\alpha}\Gamma}{\alpha + 1}\left[s^{\alpha + 2} - \frac{s^{\alpha + 2}}{\alpha + 2}\right]$$

$$= \frac{\rho g s^3}{3}\beta + \frac{(\rho g)^{\alpha}}{\alpha + 2}s^{\alpha + 2}\Gamma$$

For this example:

$$Q = \frac{1200 \times 9{\cdot}81 \times 27 \times 10^{-9} \times 3}{3} + \frac{(1200 \times 9{\cdot}81)^{1{\cdot}3} \times (3 \times 10^{-3})^{3{\cdot}3} \times 4}{3{\cdot}3}$$

$$= 1{\cdot}44 \times 10^{-3}\ \text{m}^3\text{/s per metre width.}$$

General equations for flow in tube geometries

The analysis above is valid for power-law and Newtonian fluids, and can be extended to any given formulation for viscous liquid properties. However, it is usually the case that the precise nature of the relationship between shear stress and shear rate is not known, but that need not prevent a worthwhile analysis of the situation being undertaken.

A common experiment is to determine the flowrate through a tube as a function of the pressure drop along its length. Knowledge of the linear form of the stress distribution (given by equation 6.21 above) combined with a general functional relationship for the fluid stress/shear rate relationship (equation 6.6) allows formulation of a general relationship:

$$\frac{du_x}{dr} = f(R_{rx}) = f\left(R_{wx}\frac{r}{a}\right) \qquad \ldots\ldots(6.31)$$

which leads to:

$$u_x - u_w = \int_a^r f\left(R_{wx}\frac{r}{a}\right) dr$$

or

$$u_x = \int_a^r f\left(R_{wx}\frac{r}{a}\right) dr \qquad \ldots\ldots(6.32)$$

providing there is no slip at the wall, i.e. $u_w = 0$.

Again Q can be found by integrating over the pipe section:

$$Q = \int_0^a 2\pi r u_x \, dr = \pi \int_0^{a2} u_x \, d(r^2) \qquad \ldots\ldots(6.33)$$

$$= \pi \left[r^2 u_x - \int_a^r r^2 \, du_x\right]_{r=0}^{r=a}$$

$$= -\pi \int_0^a r^2 f\left(R_{wx}\frac{r}{a}\right) dr \qquad \ldots\ldots(6.34)$$

(since $u_w = 0$)

Substituting $r = a\dfrac{R_{rx}}{R_{wx}}$ it follows that:

$$\frac{Q}{\pi a^3} = -\frac{1}{R_{wx}^3}\int_0^{R_{wx}} R_{rx}^2 \, f(R_{rx}) \, d(R_{rx}) \qquad \ldots\ldots(6.35)$$

It will be realised that R_{rx} is the shear stress at radius r and that $f(R_{rx})$ is the functional relationship between such a stress and the local velocity gradient.

Equation 6.35 embodies a finite integral, the value of which depends only on the values of the integral function at the limits, and not on the nature of the (continuous) function that is integrated. For this reason it is necessary to evaluate only the wall shear stress and the associated velocity gradient at the wall. This may be accomplished by the use of Leibniz' rule which allows a differential of an integral of the form $\dfrac{d}{dz'}\left\{\displaystyle\int_0^{z'} z^2\, f(z)\, dz\right\}$ to be written as $z'^2\, f(z')$. Multiplying equation 6.35 first by R_{wx}^3 and then differentiating:

$$R_{wx}^3 \frac{d(Q/\pi a^3)}{d(R_{wx})} + 3R_{wx}^2 \frac{Q}{\pi a^3} = -R_{wx}^2\, f(R_{wx})$$

$$\therefore \quad \frac{-\Delta Pa}{2l} \frac{d(Q/\pi a^3)}{d(-\Delta Pa/2l)} + \frac{3Q}{\pi a^3} = -f(R_{wx}) = -\left(\frac{du_x}{dr}\right)_w$$

$$\therefore \quad \frac{4Q}{\pi a^3} \frac{d\ln(Q/\pi a^3)}{4d\ln(-\Delta Pa/2l)} + \frac{3Q}{\pi a^3} = -\left(\frac{du_x}{dr}\right)_w \qquad \ldots(6.36)$$

$$\therefore \quad \frac{4Q}{\pi a^3}\left[\frac{1}{4n'} + \frac{3}{4}\right] = \frac{4Q}{\pi a^3}\left[\frac{3n'+1}{4n'}\right] = -\left(\frac{du_x}{dr}\right)_w \qquad \ldots(6.37)$$

where n' is the slope of the log/log plot of $-\Delta Pa/2l$ versus $Q/\pi a^3$ for the particular values of Q and $-\Delta P$ in question. For any small region on that graph an equation can be written in the form:

$$-R_{wx} = \frac{-\Delta Pa}{2l} = \text{constant}\left(\frac{Q}{\pi a^3}\right)^{n'}$$

It is convenient to express this relationship in the slightly different form (which does not alter the value of n'):

$$-R_{wx} = \frac{-\Delta Pa}{2l} = k'\left(\frac{4Q}{\pi a^3}\right)^{n'} \qquad \ldots(6.38)$$

When n' is unity this equation reduces to the Poiseuille–Newtonian relationship with k' equal to μ:

$$Q = \frac{\pi a^4(-\Delta P)}{8l\mu} \qquad \text{(see Volume 1, Chapter 3)}$$

It is important to note that n' and k' are parameters determined from pressure drop and volumetric flowrate data, and they can be defined *whatever the characteristics of the fluid*. There are however simple relationships between n' and k' and the rather similar-looking power-law parameters n and k.

The relation between n and n'

From equation 6.37:

$$\ln\frac{4Q}{\pi a^3} + \ln\frac{3n'+1}{4n'} = \ln\left(-\frac{du_x}{dr}\right)_w \qquad \ldots(6.39)$$

and therefore:

$$\frac{d\ln\dfrac{4Q}{\pi a^3}}{d\ln(-R_{wx})} + \frac{d\ln\dfrac{3n'+1}{4n'}}{d\ln(-R_{wx})} = \frac{d\ln\left(-\dfrac{du_x}{dr}\right)_w}{d\ln(-R_{wx})} \qquad \ldots(6.40)$$

Further, from the definition used in equation 6.8:

$$\ln\left(-R_{wx}\right) = \ln k + n \ln \left(-\frac{du_x}{dr}\right)_w$$

Since k is a constant it follows that:

$$n = \frac{d\ln\left(-R_{wx}\right)}{d\ln\left(-\dfrac{du_x}{dr}\right)_w} \qquad\qquad \ldots\text{(6.41)}$$

Considering the terms of equation 6.40 therefore, the first is by definition $\dfrac{1}{n'}$ and the last $\dfrac{1}{n}$, and we have the result:

$$\frac{1}{n} = \frac{1}{n'} + \frac{d\ln\dfrac{3n'+1}{4n'}}{d\ln\left(-R_{wx}\right)} \qquad\qquad \ldots\text{(6.42)}$$

In this function, if n' is not dependent on R_{wx} (i.e. the plot of $\log\left(\dfrac{-\Delta Pa}{2l}\right)$ versus $\log\left(\dfrac{Q}{\pi a^3}\right)$ is a straight line), the second term on the right becomes zero. For that case $n' = n$ and the real liquid is properly and fully described by the model power-law of equation 6.8.

The relation between k and k'

As will be shown later (equation 6.45 leads straight to this result):

$$\frac{-\Delta Pa}{2l} = k\left(\frac{3n+1}{4n}\right)^n\left(\frac{4Q}{\pi a^3}\right)^n$$

By using this together with equation 6.38 it follows that, when $n = n'$:

$$k' = k\left(\frac{3n+1}{4n}\right)^n \qquad\qquad \ldots\text{(6.43)}$$

Analytical equations for the laminar pressure drop–flowrate relation with specific fluid models

(1) Power-law liquid

If the properties of the fluid are defined by the power law (equation 6.8) the value of $f(R_{rx})$ as required by equation 6.31 becomes:

$$f(R_{rx}) = -\left(\frac{-R_{rx}}{k}\right)^{1/n}$$

Therefore, from equation 6.35:

$$\frac{Q}{\pi a^3} = \frac{1}{k^{1/n}(-R_{wx})^3} \int_0^{R_{wx}} (-R_{rx})^{(2+\frac{1}{n})} d(-R_{rx}) \qquad \ldots\ldots(6.44)$$

$$= \frac{1}{k^{1/n}(-R_{wx})^3} \left[\frac{n}{3n+1}\right] (-R_{wx})^{\frac{3n+1}{n}}$$

Now $-R_{wx} = \dfrac{-\Delta Pa}{2l}$, therefore:

$$\frac{Q}{\pi a^3} = \frac{1}{k^{1/n}} \frac{n}{3n+1} \left(\frac{-\Delta Pa}{2l}\right)^{1/n}$$

or

$$Q = \frac{n}{3n+1} \pi a^3 \left(\frac{-\Delta Pa}{2lk}\right)^{1/n} \qquad \ldots\ldots(6.45)$$

In applying equation 6.45 it will be noted that $-\Delta P$ varies as $a^{-(3n+1)}$. In other words, for a given volumetric flowrate in laminar flow the pressure drop is inversely proportional to the fourth power of the pipe diameter with a Newtonian fluid ($n = 1$), but for a very pseudoplastic liquid (as $n \to 0$) the pressure drop becomes more nearly inversely proportional to the first power of the diameter.

(2) Bingham plastic fluid

In this case the value of $f(R_{rx})$ depends on whether the yield stress is being exceeded. The relationships derived from the model equation 6.14 are as follows:

$$|R_{rx}| > |R_Y|, \quad f(R_{rx}) = (R_{rx} - R_Y)\mu_p^{-1}$$
$$|R_{rx}| < |R_Y|, \quad f(R_{rx}) = 0$$

From equation 6.35:

$$\frac{Q}{\pi a^3} = \frac{-1}{\mu_p R_{wx}^3} \int_{R_Y}^{R_{wx}} R_{rx}^2 (R_{rx} - R_Y) d R_{rx} \qquad \ldots\ldots(6.46)$$

$$= \frac{-1}{\mu_p R_{wx}^3} \left[\frac{R_{rx}^4}{4} - \frac{R_{rx}^3 R_Y}{3}\right]_{R_Y}^{R_{wx}}$$

$$= \frac{-R_{wx}}{\mu_p} \left[\frac{R_Y^4}{12R_{wx}^4} + \frac{1}{4} - \frac{R_Y}{3R_{wx}}\right]$$

Now $-R_{wx} = \dfrac{-\Delta Pa}{2l}$, therefore:

$$\frac{Q}{\pi a^3} = \frac{-\Delta Pa}{2l\mu_p} \left[\frac{R_Y^4}{12}\left(\frac{2l}{-\Delta Pa}\right)^4 + \frac{1}{4} - \frac{2}{3}\left(\frac{R_Y l}{-\Delta Pa}\right)\right] \qquad \ldots\ldots(6.47)$$

This equation does not lead to an explicit solution for $-\Delta P$ at a given Q, but for a fixed pressure difference there will be a single value for the volumetric flowrate.

Generalised Reynolds numbers for non-Newtonian fluids

Flow properties have been defined on the basis of the pressure drop in a pipe, thus:

$$\frac{-\Delta Pd}{4l} = \frac{-\Delta Pa}{2l} = k'\left(\frac{4Q}{\pi a^3}\right)^{n'} \qquad \text{(equation 6.38)}$$

A *friction factor* may be defined as $\dfrac{-R_{wx}}{\rho u^2} = \dfrac{-\Delta Pa}{2l}\dfrac{1}{\rho u^2} = \dfrac{f}{2} = \phi$ (see Volume 1, Chapter 3).

If a *Reynolds group* is defined so that in the laminar flow regime the same relationship applies as for a Newtonian fluid, then:

$$\phi = \frac{f}{2} = \frac{-\Delta Pa}{2l\rho u^2} = \frac{8}{Re.^*} = \frac{k'}{\rho u^2}\left(\frac{4u}{a}\right)^{n'}$$

from which:

$$Re.^* = \frac{8^{1-n'}\rho u^{2-n'}d^{n'}}{k'} \qquad \qquad \dots(6.48)$$

For the special case of power-law fluid described by equation 6.43, $n' = n$ and $k' = k\left(\dfrac{3n+1}{4n}\right)^n$, and the Reynolds number becomes:

$$Re.^* = \frac{\rho u^{2-n}d^n}{\dfrac{k}{8}\left(\dfrac{6n+2}{n}\right)^n} \qquad \qquad \dots(6.49)$$

It will be realised that by defining the generalised Reynolds numbers in this way, the same friction factor data can be used for Newtonian and non-Newtonian fluids in the laminar region. We are, by definition, writing:

$$Re.^* = \frac{ud\rho}{\mu_{\text{effective}}} \qquad \qquad \dots(6.50)$$

and using the flow curve to decide the effective viscosity.

Transitional and turbulent flow

As with all other fluids, the nature of movement in non-Newtonian materials is governed by inertial and viscous forces. The balance between these forces is characterised by the Reynolds number with Newtonian liquids: the breakdown of laminar (viscosity controlled) flow into the turbulent (inertia dominated) mode in pipes and over surfaces has already been discussed in Volume 1 (see Chapters 3 and 9). With materials possessing a variable viscosity it may appear only of limited value to use a Reynolds number. Although a generalised group like that defined in equation 6.49 is of considerable use under laminar flow conditions, its application to turbulent flow must be carefully considered. The use of shear stress/shear rate information obtained or defined under rigorous laminar flow conditions may be unrealistic. Fortunately, even with turbulent Newtonian fluids, pressure drop correlations and boundary layer velocity profiles (Volume 1, Chapter 9) are not greatly affected by the viscosity variable. In the case of a power-law fluid this would imply a lack of sensitivity at least to the consistency k, though of course it is not possible to assume that the exponent n may be of similarly little account. Indeed, as the other variable (n) may be accommodated in the dimensionless equations relating pressure drop to the generalised Reynolds number, it is reasonable to expect that the form of these equations will be:

$$\frac{f}{2} = \phi = \frac{R}{\rho u^2} = f(Re.^*, n) \qquad \ldots.(6.51)$$

An empirical equation for the friction factor/Reynolds number relationship in Newtonian flow (Chapter 3 of Volume 1) was:

$$\phi^{-\frac{1}{2}} = 2 \cdot 5 \ln (Re. \, \phi^{\frac{1}{2}}) + 0 \cdot 3 \qquad \ldots.(6.52)$$

The analogous form for a non-Newtonian fluid would be the empirical equation:

$$\phi^{-\frac{1}{2}} = A \ln \{Re.^* \phi^{1-n'/2}\} + B \qquad \ldots.(6.53)$$

where A and B are constants.

The Reynolds number was defined in such terms as to give a fit to the experimental data in the laminar region (equation 6.48).

An empirical equation by DODGE and METZNER[19], justified on semi-theoretical grounds, is as follows:

$$\phi^{-\frac{1}{2}} = 2 \cdot 5 \ln \left[Re.^* \phi^{1-n'/2}\right] + 0 \cdot 3 \qquad \ldots.(6.54)$$

which reduces to the Newtonian form when $n' = 1 \cdot 0$. The form of this relationship is shown in Fig. 6.23.

Materials with a finite yield stress present special problems. If we apply equation, 6.21, we have to assume an unsheared core to the fluid in the pipe. In fact, because of the velocity fluctuations momentum is transported at right angles to the flow direction in turbulent flow and the stress distribution is therefore modified (e.g. as assumed by Prandtl—see Volume 1, Chapter 9). For laminar flow, however, an average velocity can be derived:

$$\bar{u} = \frac{a|R_Y|}{\mu_p} \left(\frac{b^4 - 4b + 3}{12b}\right) \qquad \ldots.(6.55)$$

where

$$b = \frac{|R_Y|}{|R_{wx}|} = \frac{r_P}{a} \qquad \ldots.(6.56)$$

in which r_p = radius of the unsheared plug of material.

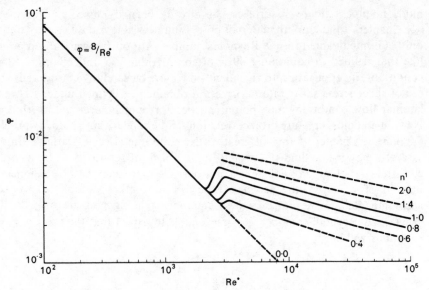

FIG. 6.23. Friction factor–generalised Reynolds number plot for power-law type fluids

Use of an equation of this type in the turbulent region requires definition of the friction factor and Reynolds number as follows:

$$\phi_B = \frac{-\Delta Pa}{2\rho lu^2(1 - b)} \qquad \ldots\ldots(6.57)$$

and

$$Re._B = \frac{du\rho\,(1 - b)\,(b^4 - 4b + 3)}{\mu_p \qquad 3} \qquad \ldots\ldots(6.58)$$

This relationship, which has been used successfully by TOMITA[18], is shown in Fig. 6.24 and has the analytical form:

$$\phi_B^{-\frac{1}{2}} = 2\cdot5 \ln Re._B\phi_B^{\frac{1}{2}} + 0\cdot3 \qquad \ldots\ldots(6.59)$$

Although equations 6.54 and 6.59 apply reasonably well to polymer solutions, they may underpredict the pressure drops for suspensions of fine solids, for which the normal Newtonian friction factor can be used, provided that the density of the suspension is used in place of that of the fluid.

Turbulence damping and drag reduction (the TOMS[17] effect)

Considerable interest has been generated in recent years in the discovery that dilute solutions of polymer molecules often give far lower friction factors than might be expected at the Reynolds numbers obtaining. The concentrations of polymer required to produce this reduction are so low (of the order of a few parts per million) that the liquids are to all intents Newtonian, and the conventional friction factor–Reynolds number plots can be drawn. Figure 6.25 shows some results for a dilute solution of polyethylene oxide in water[31]. It will be seen that there are apparently two effects: a lowering of the friction factor at high Reynolds numbers, and a slight increase in the instability at the

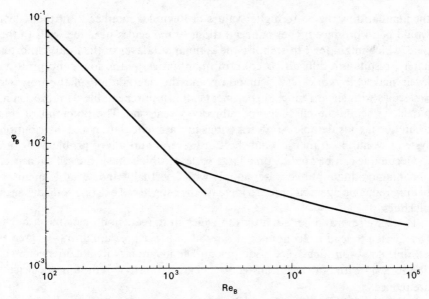

FIG. 6.24. Friction factor–generalised Reynolds number plot for Bingham fluids

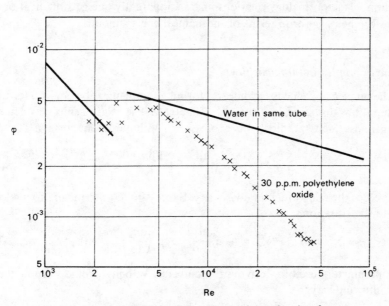

FIG. 6.25. Typical friction factor—Reynolds number data for very dilute aqueous polymer solutions

lower boundary of the transition region. The mechanisms producing this behaviour are far from clear. It is certain that the type of random rotational motion characteristic of free turbulence will repeatedly deform any long chain materials that happen to be present, and that such molecules will therefore tend to smooth out instabilities in the flow. This might be expected to prolong the laminar flow region to higher values of Reynolds number. A further effect would be to produce higher concentrations of molecules near the walls of the pipe, which might tend to stabilise the laminar sub-layer in the fluid. Unfortunately, results are difficult to confirm quantitatively—the low concentration levels make the work very demanding and the degradation of the polymer molecules with shear renders experiments almost unrepeatable with precision, but nevertheless the effect is undoubtedly a real one. The potential of this technique is considerable—power savings in water circulation and distribution systems would be dramatic—but there are several unresolved problems before the technique can be applied on a large scale, of which one is the ethical aspect of contaminating a public water supply with a virtually undetectable quantity of drag reducing agent in order to increase the capacity of existing long distance pipelines.

There are several other situations in which drag reduction phenomena might be expected to lead to economies as a result of lower pressure drop—or greater possible flow—through process equipment. The advantages are not often realised, however, since for both heat transfer[33] and mass transfer[39], the film coefficients are reduced.

The reduction is approximately that which would be expected from the friction $-j$ factor analogy relationships, so apparently one gets the heat or mass transfer one pays for in terms of mechanical energy input.

Velocity profiles in turbulent flow

The universal velocity profile in turbulent pipeline flow of a Newtonian material was discussed in Volume 1, Chapter 9. The relationship between shear stress and velocity gradient was expressed in Volume 1, Chapter 9 as:

$$R_{yx} = -\left(\frac{\mu}{\rho} + E\right)\frac{d(\rho u_x)}{dy} \qquad \dots (6.60)$$

where E is the eddy kinematic viscosity. Expressing E in terms of mixing length led to the equation:

$$R_{yx} = -\mu\frac{du_x}{dy} - \lambda_E^2 \rho \left(\frac{du_x}{dy}\right)^2 \qquad \dots (6.61)$$

The parameters used in deriving the universal velocity profile were defined in the following way:

$$u^* = \sqrt{\frac{-R_{wx}}{\rho}} = \phi^{\frac{1}{2}} u \qquad \dots (6.62)$$

$$u^+ = \frac{u_x}{u^*} = \frac{u_x}{u}\phi^{-\frac{1}{2}} \qquad \dots (6.63)$$

$$y^+ = \frac{yu^*\rho}{\mu} = \frac{y}{d}\phi^{\frac{1}{2}}Re. \qquad \dots (6.64)$$

The shear stress was assumed uniform across the pipe section, and hence R_{rx} was put equal to R_{wx}.

The basic velocity profile equations derived were then, for the laminar sub-layer (Volume 1, Chapter 9):

$$u^+ = y^+ \qquad \qquad \ldots\ldots(6.65)$$

and for the turbulent core:

$$u^+ = 2\cdot5 \ln y^+ + \text{Constant} \qquad \qquad \ldots\ldots(6.66)$$

where the constant has the value 5·5 with a smooth pipe and some rather smaller value when the pipe is rough.

Application to power-law fluids

With non-Newtonian fluids the equations 6.62 and 6.63 defining u^* and u^+ are again valid, but equation 6.64 has to be modified to accommodate the variations in viscosity. The generalised Reynolds number defined in equation 6.49 above can conveniently be modified by dividing by $8\left(\dfrac{n}{6n+2}\right)^n$ when (for a power-law fluid) the new form becomes:

$$Re._1^* = \frac{u^{2-n}d^n\rho}{k} \qquad \qquad \ldots\ldots(6.67)$$

This leads to an equation for y^+:

$$y^+ = \frac{(u^*)^{2-n}y^n\rho}{k} = \frac{u^{2-n}d^n\rho}{k}\left(\frac{u^*}{u}\right)^{2-n}\left(\frac{y}{d}\right)^n$$

$$= Re._1^*\phi^{1-n/2}\left(\frac{y}{d}\right)^n \qquad \qquad \ldots\ldots(6.68)$$

For the laminar sub-layer:

$$-R_{wx} = k\left(\frac{u_x}{y}\right)^n$$

$$\rho u^{*2} = ky^{-n}(u^*u^+)^n$$

$$(u^+)^n = \frac{\rho(u^*)^{2-n}y^n}{k} = y^+ \qquad \qquad \ldots\ldots(6.69)$$

For the turbulent core:

$$u^+ = C_1 \ln y^+ + C_2.$$

where C_1 and C_2 are constants.

Using n' to characterise the fluid BOGUE and METZNER[24] suggest:

$$u^+ = 2\cdot42 \ln (y^+)^{1/n} - 1\cdot21\left(\frac{d^n(u^*)^{2-n}\rho}{k}\right)^{1/n} + 0\cdot984\sqrt{\phi} + 3\cdot63 \qquad \ldots\ldots(6.70)$$

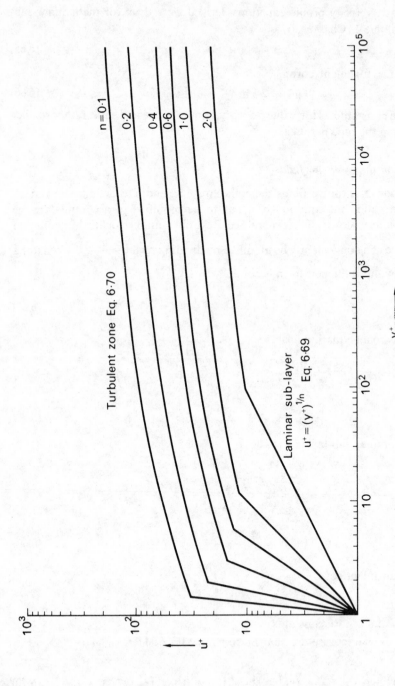

FIG. 6.26. Universal velocity profiles for turbulent power-law fluids

with the addition of a small correction factor ($<$1·5) which is a function of position and friction factor ϕ. Details of this treatment are given by SKELLAND[9]. Figure 6.26 shows the universal velocity profiles for power-law fluids.

Extensive experimental work indicates that the form of the turbulent velocity profiles established in non-elastic fluids without yield stress is very similar to that found with Newtonian fluids. This supports the contention, implicit in the discussion on friction factors above, that it is the properties of the fluid in the region immediately adjacent to the wall which are most important. This is particularly important with pseudoplastic materials, in which this region, which is subject to the higher shear rates, therefore has a lower effective viscosity than the bulk.

A recent application of this principle has been the investigation of the possibility of pulsing flow in pipelines carrying pseudoplastic materials. The superimposition of a small oscillating component to the bulk velocity has the effect of raising the average shear rate appreciably. It appears advantageous to put in some of the pumping energy in this way thereby using smaller pumps than would otherwise be required.

Equations that allow for temperature and pressure dependence

The approach that provides the basis for the Prandtl–Eyring equation of state (equation 6.11) depends on the concept of an activation energy per mol that must be exceeded before flow can take place: this is analogous to the energy needed for an atom in a crystal lattice to move to a vacant site in an adjoining location in the crystal. The conventional form for the activation type equation, which is known as the Arrhenius relation and expresses the variation of chemical rate constant with temperature, is shown as equation 1.7. When expressed in the manner relevant to viscous behaviour it takes the form:

$$\mu_T = \eta \exp \frac{\varepsilon}{\mathbf{R} T} \qquad \qquad ....(6.71)$$

where ε is the molar activation energy for flow,
$\quad$ $\mathbf{R}$ is the gas constant, and
$\quad$ T is the absolute temperature.
The constant η is characteristic of the fluid.

This equation leads to the ratio of the viscosities at two temperatures T_1 and T_2 as:

$$\frac{\mu_{T_1}}{\mu_{T_2}} = \exp\left\{\frac{\varepsilon}{\mathbf{R}}\left(\frac{1}{T_1} - \frac{1}{T_2}\right)\right\} \qquad \qquad ....(6.72)$$

Applying the power-law relationship for apparent viscosity (equation 6.9), this leads to:

$$\ln \frac{\left[k\left(\frac{du_x}{dy}\right)^{n-1}\right]_1}{\left[k\left(\frac{du_x}{dy}\right)^{n-1}\right]_2} = \frac{\varepsilon}{\mathbf{R}}\left(\frac{1}{T_1} - \frac{1}{T_2}\right) \qquad \qquad ....(6.73)$$

Since it is known that n is much less sensitive than k to temperature variation, this equation can be reasonably approximated by:

$$\ln \frac{k_1}{k_2} = \frac{\varepsilon}{R} \left(\frac{1}{T_1} - \frac{1}{T_2} \right) \qquad \dots (6.74)$$

If values of k are known at two (absolute) temperatures T_1 and T_2, an estimate of $\frac{\varepsilon}{R}$ can be made and equation 6.74 used for interpolation.

However, a more satisfactory approach was used by HANKS and CHRISTIANSEN[21] who reformulated equation 6.71 in stress terms, viz.:

$$R_{yx} = -\eta \left(\frac{du_x}{dy} \right) \exp \frac{\varepsilon}{RT} \qquad \dots (6.75)$$

and then modified this for a power-law fluid:

$$R_{yx} = -k_1 \left[\frac{du_x}{dy} \exp \frac{\varepsilon}{RT} \right]^n \qquad \dots (6.76)$$

It will be noted that in this case the temperature is used to generate a correction factor applied to the shear rate and the consistency or exponent parameters are not modified; and while this approach provides a means of accommodating temperature variations within a particular apparatus, it does not convey information of the physical changes in the liquid under consideration.

As stated above, measurements show that for a power-law fluid the variation of k is much greater than that of n. With a Newtonian fluid n is always 1, whatever the temperature range; the variation in viscosity ($k = \mu$) can often be expressed as:

$$\mu = 10^{\alpha + \beta T} \qquad \dots (6.77)$$

where α and β are empirically determined constants. In the absence of data for a non-Newtonian fluid it is probably reasonable to apply equations of the form:

$$k' = 10^{\alpha + \beta T}$$
$$\qquad \dots (6.78)$$
$$n' = \text{Constant}$$

The pressure dependence of viscosity is not significant in most chemical processing operations, but in two particular fields it has considerable importance, viz., polymer processing (especially extrusion) and lubrication. Under high pressures (up to 1500 bar $= 1 \cdot 5 \times 10^8$ N/m² in polymer processes, for example) there are dramatic increases in viscosity which can be explained, qualitatively at least, in terms of the increased molecular interaction as the fluid is compressed. The extent of the viscosity change is a function of both the compressibility of the fluid and the complexity of the molecular species, and becomes much greater under extreme conditions. If it is necessary to interpolate or extrapolate viscosity data, it is generally reasonable to use a logarithmic relationship of a form:

$$\ln \frac{\mu_P}{\mu_{P_0}} = \frac{P}{\psi} \qquad \dots (6.79)$$

where μ_P is the viscosity at a pressure P,

    μ_{P_0} is the viscosity at low pressure P_0 (accurately enough 1 atmosphere),

    and

    ψ is a constant dependent on temperature and material properties.

For non-Newtonian power-law materials it should be adequate to use the apparent viscosity relation of equation 6.9 to give:

$$\ln \frac{\left[k \left(\frac{du_x}{dy} \right)^{n-1} \right]_P}{\left[k \left(\frac{du_x}{dy} \right)^{n-1} \right]_{P_0}} = \frac{P}{\psi} \qquad \dots (6.80)$$

If k is much more sensitive than n to pressure variations, equation 6.80 reduces to:

$$\ln \frac{k_P}{k_{P_0}} = \frac{P}{\psi} \qquad \dots (6.81)$$

Typical values of ψ are given in Table 6.2.

TABLE 6.2. *Pressure dependence of apparent viscosity*

Material	Temperature		ψ (0–3000 bar)
	K	°C	bar
Polystyrene	455	~180	280
Polyethylene	475	~200	760
Polydimethylsiloxane	375	~100	440
	495	~220	830
Polyisobutylene	375	~100	490
	495	~220	830
Polybutene	400	~25	230

1 bar = 10^5 N/m^2 (10^5 Pa)

Time dependence

There are two distinct ways in which time may enter the consideration of fluid behaviour. Firstly, there is the whole family of elastic effects which become important in any accelerating or non-uniform flow, and which will be considered in detail later. There is also a large number of liquids whose properties are influenced by the sample history. Many non-Newtonian fluids degrade (i.e. usually become less viscous and/or less elastic) as a result of prolonged shearing, though some become apparently more viscous under similar conditions. In either case, some of the liquids recover their earlier behaviour if allowed to rest. Others appear to possess a "memory"—an effect that usually originates in an orientation at some stage of the flow process.

Materials that show a fall in viscosity with shearing are called *thixotropic*, and those that become more viscous are *rheopectic*.

Some picture of the physical changes that can lead to thixotropy is obtained by considering a liquid containing long chain molecules in solution, in which the dynamic molecular state is tending to lead to the slow repeated formation and breakdown of gel type links. At any given shear rate there will eventually be an equilibrium amount of *structure* present, but it will take time to reach this condition. In consequence the material will tend to thicken slowly on standing and to thin out slowly during shearing. Rheopectic behaviour can be considered as a reverse procedure. For example, the development of an ordered structure of fibrous particles can be achieved in the presence of a high shear field, and this may give a higher viscosity than the less oriented equilibrium arrangement when the mixture is sheared less vigorously.

It is extremely difficult to characterise these fluids adequately; usually the best that can be achieved is the development of a standardised testing procedure, subjecting the samples to a given sequence of increasing and decreasing shear rates, for example.

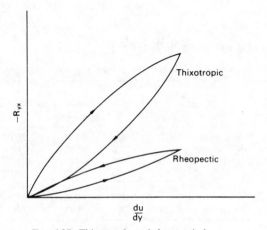

FIG. 6.27. Thixotropic and rheopectic behaviour

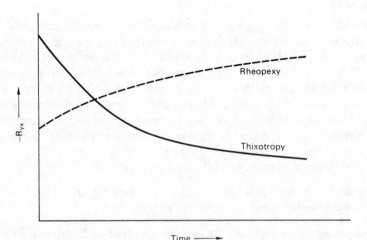

FIG. 6.28. Shear stress changes with continued uniform rate of
shear

Rheograms for such liquids are drawn in Fig. 6.27, for an experiment in which shear rates are steadily increased and then decreased. If the shear rate is kept constant at some value, then the corresponding stress measurement will fall, tending to approach a limiting value (see Fig. 6.28).

Shear is only one of several possible causes of time dependence. Experimental fluids which are solutions of certain salts may slowly hydrolyse, or be modified as a result of bacteriological action; such slow chemical changes can complicate characterisation, giving effects similar to thixotropy or rheopexy, but of course generally irreversibly.

Industrial processes involving time dependent materials cannot be designed adequately from first principles unless the condition of the fluid is known fully, and even then design may have to be based on some rather over-simplified description of the fluid properties: usually, either the initial condition or that after prolonged shearing, whichever imposes the limiting criteria.

As an example of a time dependent change, the data shown in Fig. 6.29 are of interest. This shows the apparent viscosity for a 0·6 per cent polyox solution before and after prolonged shearing. The fluid properties are described in terms of the Sutterby model (equation 6.13) which accommodates the change in limiting viscosity at low shear rates. Sutterby parameters for this fluid to be used in equation 6.13 are:

	μ_0 (Ns/m^2)	α	θ_s (s)
before shearing	0·30	0·55	1·55
after shearing	0·11	0·55	0·22

The only region in which a power-law curve fit could reasonably be applied would be above the shear rate P on the figure (region PQ), when the corresponding parameters would be:

$$k = 0\cdot4 \text{ Nm}^{-2} \text{ s}^{-0\cdot45} \qquad n = 0\cdot45$$

These values are not altered as a result of the shearing.

For many industries (notably foodstuffs) the way in which the rheology of the material affects the process is far less significant than the effects that the process might have on the rheology. Implicit here is the recognition of the importance of the time dependent properties of these materials (almost all being suspensions or dispersions) which can be profoundly influenced by mechanical working on the one hand or from an ageing process during a prolonged shelf life on the other.

VISCOELASTIC FLUIDS [10, 11, 12, 13, 14]

The elasticity of solids can be expressed as a linear relationship between stress and strain. For tension this is known as Hooke's law, with Young's modulus as the coefficient of proportionality. A similar relationship exists for distortion produced by shear, thus:

$$R_{yx} = -\,G\,\frac{dx}{dy} \qquad\qquad \dots (6.82)$$

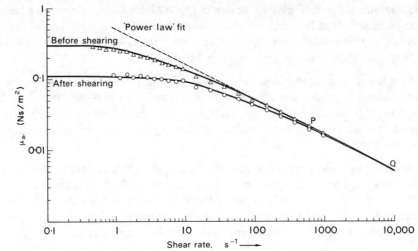

FIG. 6.29. Apparent viscosity change with time

where dx is the shear displacement of two elements separated by a distance dy, and

G is the modulus of rigidity.

With a solid deformed elastically, removal of a stress producing deformation allows the material to recover its original shape. If, however, the yield stress of the solid has been exceeded, then there will have been a certain amount of creep which is not recovered on removal of the stress—the solid has flowed.

Many liquids show effects which are elastic in nature: they have some ability to store and recover shear energy. Perhaps the most easily observed experiment is the "soup bowl" effect. If a liquid in a dish is made to rotate by means of gentle stirring with a spoon, on removing the energy source (i.e. the spoon), the inertial circulation will die away as a result of viscous forces. If the fluid is viscoelastic (and many proprietary soups are) the liquid will be seen to slow to a stop, and then unwind a little. This sort of behaviour is closely associated with the tendency for a gel structure to form within the fluid; such an element of rigidity makes simple shear less likely to occur—the shearing forces tending to act as couples to produce rotation of the fluid elements as well as pure slip. Such incipient rotation produces a stress perpendicular to the direction of shear, i.e. R_{yy} in Fig. 6.2. Measurement and analysis of these *normal* stresses provide a means of describing the degree of elasticity quantitatively. This ability to store and recover elastic energy complicates the flow behaviour of these fluids considerably. If a fluid is a true liquid (i.e. has a zero value of yield stress) but is elastic, then the imposed stresses will be dissipated by viscous flow.

The shear fields that exist within any flowing fluid produce an orientation of the constituent parts of the fluid. In particular, this happens to any long chain molecules that may be present, in a solution or melt. Such an ordered state is at a lower entropy level than a random one, and there is a tendency for the characteristic randomness to be restored when the shear field is removed. A simplified picture of the orientation effect is provided by considering the viscous forces acting on a small rod in a shear field (i.e. a region with a velocity gradient).

Local velocity profile relative to element centre

FIG. 6.30. Forces on a rigid body in a shear field

Figure 6.30 shows that the ends of the rod experience a couple tending to turn the rod into alignment with the local streamlines. In essence, each segment of a long chain molecule will tend to be affected in this way and, as a result, there will be a straightening and stretching effect along the direction of motion. Alignment of a rigid body along the streamlines is however unstable. Any further displacement leads to the rod rotating. Since the forces causing the rotation depend on orientation, rotation is far from uniform, the rod rotating rapidly while across the streamlines and slowly when nearly aligned. Flexible rods (or long molecules) tend to fold back on themselves, but the tendency towards rotation ensures that kinks can straighten out, and alignment is reasonably efficient. If the flow is at all convergent, the elongation accentuates the alignment and this can be useful in certain circumstances, as for example for incorporating a reinforcing material into a polymer[37].

As mentioned, once the molecule has been brought into alignment with the streamlines, the main forces acting on it become less effective—the couple is reduced. One result of this is that there is no longer a force tending to extend the molecule. Given time, the elongation will decrease, and the energy that is stored by virtue of the stretching is released. It follows that not only is the intensity and duration of passage of fluid elements through the shear field important, but also the rate at which the shear is developed.

When any fluid is squirted from a jet, the velocity profile within the jet is modified from the more or less paraboloid form within the tube to the uniform velocity across the jet that must obtain some distance from the point of discharge. Figure 6.31 shows the behaviour of Newtonian and viscoelastic jets. For most values of tube Reynolds number the behaviour of a Newtonian jet is that shown in (a). The process is momentum controlled and there is a contraction of the jet diameter to about 63 per cent of the tube diameter. At low velocities (b) the process is affected by surface tension and there may be a very slight enlargement of the discharging fluid, to the extent of about 15 per cent of the tube diameter possibly[22]. With a highly viscoelastic material (c) there is a spectacular increase in the jet diameter. The increase in jet cross-section leads to a lower kinetic

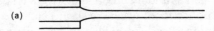

(a)

(b)

(c)

FIG. 6.31. Profiles of fluid jets

a Newtonian *Re* > 1
b Newtonian *Re* < 0·1
c viscoelastic

energy associated with the free stream compared with the usual Newtonian case. The force producing the deceleration comes from the elastic energy stored in the liquid at the point of discharge; this results from the orientation and stretching of the long molecules in the fluid. This increase is conveniently described in terms of a swelling ratio. The effect of the increasing relaxation with longer tubes

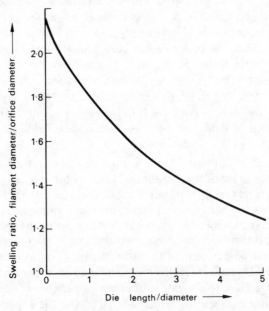

FIG. 6.32. Swelling ratio of extruded polymer fibre

is shown by the reduction in swelling ratio represented in Fig. 6.32. If the effect were simply one resulting from orientation and stretching, then the increasing of tube length or the reduction of tube diameter would be expected to increase the swelling ratio. Die swell behaviour is of very great significance in polymer processing operations; even the drawing of simple fibres requires sophisticated prediction of the appropriate orifice sizes, and the design calculations for more complex extruded sections requires a combination of theoretical understanding with the relevant know-how. Some discussion of the methods adopted is given by McKELVEY[6].

The Weissenberg effect

The physical forces that are known as Weissenberg forces arise from normal stresses. There is a variety of demonstrations which depend on the behaviour of fluid contained between surfaces rotating relative to each other. The simplest of these is the situation that is illustrated on Fig. 6.33. A Newtonian fluid will

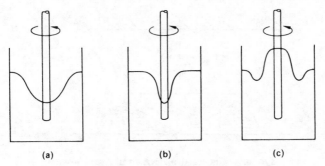

(a)   (b)   (c)

FIG. 6.33. Surfaces profiles established when a rotating rod
is immersed in a reservoir of fluid

a Newtonian
b pseudoplastic
c viscoelastic

form some approximation to a free vortex with the surface profile as shown in a. A pseudoplastic fluid will show a generally similar result, but owing to the higher viscosity at points remote from the rotating surface the surface profile is steeper, as shown in b. The viscoelastic materials behave quite differently however, with the fluid tending to climb up the rod as shown in c.

A simple intuitive picture of the mechanism that leads to this phenomenon is provided by considering the fluid as a suspension of fibrous molecules. The mechanism outlined above that produces alignment along the streamlines leads to a strangling effect in these circular shear fields. This produces a region of high pressure near to the rod, which is made evident by the resulting climbing action of the fluid. Indeed the larger scale suspensions of fibres that are encountered in the paper-making industry will produce exactly this effect. Several examples of this effect—called the Weissenberg effect—are given by REINER[1, 3]

Oscillatory experiments

The elastic forces that are established in a viscoelastic liquid flowing through a tube are significant principally in the entry and exit regions; in steady fully developed flow situations there is a dynamic equilibrium between extension and relaxation. However, if the pressure producing the flow is reduced to zero, the elastic energy stored by the enforced alignment of the fluid elements is then subsequently relaxed by viscous flow; i.e. flow continues after removal of the obvious driving force. This energy is sometimes called recoverable shear. An oscillating applied pressure will, as in a purely viscous liquid, produce an oscillating flow, but with the important difference that, as a result of the re-coverable shear, the flow will not be exactly in phase with the pressure. The calculations underlying the determination of the elastic modulus are too com-plicated to detail here. One principle may be indicated. Figure 6.34 represents a cone C suspended by a torsion bar T just in contact with a plate P. The fluid sample L fills the gap between the cone and the plate. If the cone is given an initial twist and then released, the frequency of oscillation of the whole system

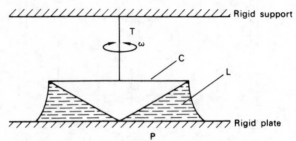

FIG. 6.34. Cone and plate geometry used to study viscous damping and elastic resonance in a fluid sample

will be a function of the spring constant, the amount of inertia of the cone and the effective inertia and springiness of the fluid brought into movement in the gap. These last two quantities will be complicated functions of the density, viscosity and elasticity of the liquid. The damping out of the oscillation will likewise be determined by the same fluid properties, but with a different sort of relationship. Similar measurements are obtainable by carrying out a forced oscillation of the plate P, and measuring the transmitted phase shift and torque (i.e. from the small movement of the cone C) at different frequencies.

Normal stress effects under steady shear conditions

The same forces that produce the Weissenberg effect shown in Fig. 6.33c will tend to force apart a cone and plate between which a viscoelastic fluid is sheared. The thrust that is exerted by the fluid in this way may be measured, and can provide quantitative information about the elasticity of the liquid. One appara-tus used in this way is the *rheogoniometer* (sketched in Fig. 6.35) in which

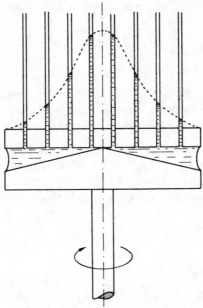

FIG. 6.35. A manometer head in a rheogoniometer

manometer tubes on the upper (stationary) plate allow determination of the pressure profile across the plate while the cone is rotating steadily. In fact, an idealised theoretical analysis of this pressure profile indicates that, at a given (slow) rotational speed, a relationship should exist:

$$\frac{dP}{d(\ln r)} = 2R_{zz} - R_{xx} - R_{yy} \qquad \qquad \ldots.(6.83)$$

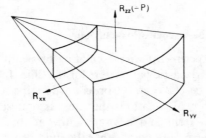

FIG. 6.36. Normal stress components on a fluid element sheared in a cone and plate geometry

The R.H.S. of this equation is sometimes expressed as $-[(R_{yy} - R_{zz}) + (R_{xx} - R_{zz})] = -(\sigma_1 + \sigma_2)$ where σ_1 and σ_2 are called the first and second normal stress (see Fig. 6.36) differences respectively[9]. The total thrust on the plate may be calculated (under certain even more restrictive conditions) as:

$$\psi = \frac{\pi a_c}{2}(\sigma_1 - \sigma_2) = \frac{\pi a_c}{2}(R_{yy} - R_{xx}) \qquad \ldots .(6.84)$$

where a_c is the cone radius.

In principle, equations 6.83 and 6.84 may be used to determine σ_1 and σ_2 separately, but in practice this is not easy since σ_2 is very small relative to σ_1. Both σ_1 and σ_2 are of course functions of the velocity gradient.

Stress differences found in this way are characteristic of the particular material, and, as such, provide ways of defining the elasticity of the fluid. For all fluids that have been examined, $\dot\sigma_1$ has been found to be greater than σ_2 (which has been assumed by many workers to be zero), but there is no known reason why this should be so.

Other types of rheometer

Elastic effects can be quantified in experiments in which the fluid experiences regular deformations superimposed on a steady flow.

Geometries that have been used are those of flow between two concentric spheres rotating around intersecting axes (the balance rheometer), between two parallel plates rotating around parallel non-colinear axes and between two cylinders (Couette style) in which the axes are parallel but not colinear. These geometries are attractive, especially since mechanical design of the equipment is simpler and the interpretation of the results is more straightforward than for those determining elastic properties from purely oscillatory experiments. A full discussion of these modern instruments is given by WALTERS[11].

Equations to describe viscoelastic behaviour

The results of experiments with oscillating or steady shear may be used to calculate viscous and elastic parameters for the fluid. In general, the equations produced need to be elaborate in order to describe a real fluid adequately. In the simplest forms of equations the stress is represented as the sum of an elastic deformation term and a viscous flow term. There are two models commonly used, depending on whether the fluid behaves more like an elastic solid or a viscous liquid. The first, known as the Voigt model equation, has the form:

$$-R_{yx} = \mathbf{G}\frac{dx}{dy} + \mu\frac{du_x}{dy} \qquad \ldots .(6.85)$$

The solid behaviour limit is seen as the flow rate, $\dfrac{du_x}{dy}$, approaches zero (i.e.

very slow deformation) when the stress is accommodated by the modulus $\mathbf{G}$ and the shear distortion by $\dfrac{dx}{dy}$.

More fluid-like behaviour may be represented by the Maxwell model equation:

$$-R_{yx} = \frac{\mu}{\mathbf{G}} \frac{d(-R_{yx})}{dt} + \mu \frac{du_x}{dy} \qquad \dots (6.86)$$

This permits slow viscous flow when the rate of stress change $\left(\dfrac{d(-R_{yx})}{dt}\right)$ is small. In the event of sudden stress, with impact for example, the elastic term is dominant.

For both Voigt and Maxwell models the ratio $\mu/\mathbf{G}$ has the dimensions of time, though its significance differs. For the Voigt solid it represents the time needed for a strain rate resulting from imposed constant stress to be reduced to one half of its initial value, and hence is called the retardation time of the solid given by:

$$t_\varepsilon = \mu/\mathbf{G} \qquad \dots (6.87)$$

For the Maxwell liquid this same ratio is the time needed for the stress produced as a result of a constant strain to decay to one half of its value; it is therefore called the relaxation time of the fluid and:

$$t_r = \mu/\mathbf{G} \qquad \dots (6.88)$$

For further discussion see WILKINSON[5].

Unfortunately a single modulus $\mathbf{G}$ is no more adequate to describe elasticity under all conditions than is a constant μ to characterise effective viscosity. For a real fluid both $\mathbf{G}$ and μ are in general functions of shear rate.

Many more elaborate theoretical and semi-theoretical constitutive equations have been formulated (see, for example, BIRD et al.[13, 14]). Unfortunately, the mathematical analysis of these systems has proved to be exceedingly complex and, to the present time at least, application to real situations has been limited to certain specialised flows in simple geometries. There has been some success in the analysis of certain extrusion die geometries, and in the description of fibre spinning and film blowing operations, in all of which the non-Newtonian behaviour of the fluid is of paramount importance. However, it must be admitted that the majority of engineering applications rest upon empirical methods. Nevertheless, determination of these fundamental physical properties often helps to elucidate which are the important variables in real situations and provide a means for the quantitative assessment of process intermediates and products.

Elongational viscosity

Flows in which material is extended in one or two dimensions occur in various processes, fibre spinning and polymer film blowing being only the most immediate examples. Coalescence of two bubbles involves a very similar stretching of the liquid film between them until it bursts. The mode of extension affects the way the fluid resists deformation, and although this resistance can be referred to loosely as being quantified in terms of an *elongational* (or *extensional*) viscosity, this parameter is in general not necessarily constant. The earliest determinations (by TROUTON[16]) were for uniaxial extension (i.e. straightforward stretching of a fibre) and he and many later workers found that, at low strain rates, the elongational viscosity is just three times the value of the shear viscosity. At higher deformation rates many materials show a larger value for this ratio. There appears to be no straightforward relationship between the elongational viscosity and the other material parameters, and to date determinations of it rest almost entirely on experiments—themselves often constrained by the difficulty of establishing and maintaining an elongational flow field for long enough for transient effects in the fluid to die out[11]. One elegant method of achieving this is the inverted syphon in which liquid is drawn upwards from a bath into a suction tube, the elasticity of the fluid allowing sufficient tension to maintain the thread[35].

Secondary flow phenomena

It is to be expected that pressure distributions giving rise to effects like the Weissenberg effect will generate flow fields within the fluid that differ from those established in Newtonian liquids. This does in fact happen, and flow patterns that are so induced are called secondary flows. Secondary flow may exist in Newtonian systems when inertial effects become significant. The best known of these is the flow pattern that is established in a helical coil. Fluid flowing down the axis of the pipe is subject to greater centrifugal forces than that flowing near the inner pipe wall. As a result, there is an unbalanced pressure that induces the flow that is illustrated in Fig. 6.37. The fluid flowing down the tube follows a helical path as it travels around the bend.

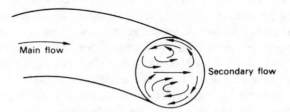

Main flow

Secondary flow

FIG. 6.37. Inertial secondary flow in a helical coil

Another form of inertial instability that leads to secondary flow is described by SCHLICHTING[4], and is illustrated in Fig. 6.38. It is relevant to the flow of fluid between two rotating cylinders. If the inner cylinder is rotating the fluid moving with it will be subjected to a centrifugal force because of the angular motion.

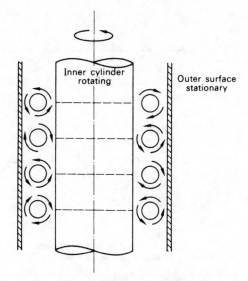

FIG. 6.38. Inertial secondary flow between rotating cylinders

Above a critical Reynolds number (defined as $\dfrac{2\pi N\rho}{\mu}\,r_i^{0\cdot5}c^{1\cdot5}$) of 41·3, the flow field breaks down into a set of stable vortex cells, called *Taylor vortices*.

Similar centrifugal effects may be observed in the rotating cone and plate geometry. The fluid that is in contact with the rotating surface will tend to flow radially outwards, to be replaced by liquid flowing inwards over the stationary surface. This is illustrated in Fig. 6.39. The analogous situation when a sphere is rotating in an "infinite" fluid is shown in Fig. 6.40.

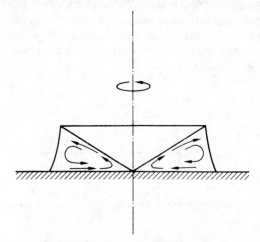

FIG. 6.39. Inertial secondary flow between
a rotating cone and a stationary plate

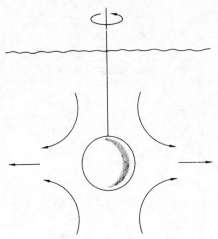

FIG. 6.40. Inertial secondary flow
near a rotating sphere

The presence of viscoelasticity in a fluid accentuates or modifies this behaviour. The work of THOMAS and WALTERS[23] is of interest; they report that considerable reductions in the pressure drop along slightly curved pipelines are achieved with very dilute polymer solutions. This appears to be a different phenomenon from that reported by WHITE[31] which was concerned with turbulence suppression, and which was discussed above. It seems certain that these results are a consequence of an enhanced circulation and increased stability of the laminar flow mode.

The transitional Reynolds number for the presence of stable Taylor vortices is affected; GIESEKUS[28] reports a reduction of a factor of about 60 for the instability. The cone and plate flow is also considerably changed. The elastic forces that produce the rod-climbing Weissenberg effect tend to give an inflow along the rotating surface towards the point of contact, while further away in the region of greater centrifugal effects the inertial flow remains dominant. Figure 6.41 shows the resulting components of flow at right angles to the main direction of shear.

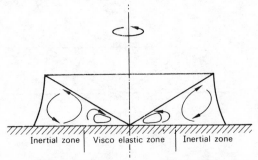

Inertial zone | Visco elastic zone | Inertial zone

FIG. 6.41. Viscoelastic secondary flow between
a rotating cone and a plate

The flow field around a sphere is modified in a rather similar way; Fig. 6.42 shows the flow patterns that are established[25]. The size of the zone within which the viscoelastic properties are dominant is naturally a function of their magnitude, and in principle this can be used as a means of characterising the fluid.

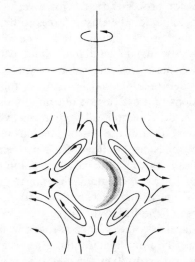

FIG. 6.42. Viscoelastic secondary flow near a rotating sphere

Secondary flows are of course of relevance to the determination of viscous flow properties, since the types of apparatus discussed above rely on the assumption that the fluid is subjected to simple shear. However, the differences in measured shear stresses that arise from such behaviour are generally minor, the angles that the streamlines make with those predicted for pure shear are generally very small—often of the order of a few degrees—and it is only for the most accurate work that these deviations must be taken into account.

Although the examples that have been presented here may appear to be rather academic, it should not be thought that these secondary flow phenomena are without practical implications. A high speed mixer of the turbine or the propeller type is liable to produce secondary flows similar to those shown in Fig. 6.42, if the fluid is elastic enough. In consequence, an impeller of such a pattern is totally unsuited for blending these solutions.

Dimensionless characterisation of viscoelastic flows

The enormous practical value obtained from the consideration of viscous flows in terms of Reynolds numbers leads to the consideration of the possibility of using analogous dimensionless groups to characterise elastic behaviour. The Reynolds group uses a ratio of inertial to viscous forces, and it might be reasonable to expect that a ratio involving elastic and inertial forces would be useful. Unfortunately, attempts to achieve meaningful correlations have not been very successful, perhaps most frequently being defeated by the complexity of natural situations and real materials. One simple parameter that may prove of value is the ratio of a characteristic time of deformation to a natural time for the fluid. The precise definition for these times is a matter for argument, but it is evident that for processes that involve very slow deformation of the fluid elements it is possible for the elastic forces to be released by the normal processes of relaxation as fast as they build up. As examples of the flow of rigid (apparently infinitely viscous) material over long periods of time, even the thickening of the lower parts of mediaeval glass windows is insignificant beside the plastic flow and deformation that lead to the folded strata of geological structures. In operations that are carried out rapidly the extent of viscous flow will be minimal and the deformation will be followed by recovery when the stress is removed. To get some idea of the possible regions in which such an analysis can provide guidance, consider the flow of a 1 per cent solution of polyacrylamide in water. Typically this might have a relaxation time (in Maxwell model terms) of the order of 10^{-2} s. If the fluid were flowing through a packed bed it would be subject to alternating acceleration and deceleration as it flowed through the interstices of the bed. With a particle size of the order of 25 mm diameter and a superficial flow rate of 0·25 m/s, one would not expect the elastic properties to influence the flow significantly. However, in a free jet discharge with a velocity of the order of 30 m/s it might be reasonable to expect some evidence of elastic behaviour near to the point of discharge.

The ratio known as the Deborah number has been defined by METZNER, WHITE and DENN[30] in these terms:

$$Db. = \frac{\text{Characteristic process time}}{\text{Characteristic fluid time}} \qquad \ldots (6.89)$$

In the example above, the packed bed, $Db.$ would be 10; for the jet and the nozzle $Db.$ would be 10^{-1}. The smaller the number the more likely is elasticity to be of practical significance.

Unfortunately, this group depends on the assignment of a single characteristic time to the fluid (perhaps a "relaxation time"). While this is better than no description at all, it appears to be inadequate for many viscoelastic materials which show different relaxation behaviour under differing conditions.

PROCESSING OPERATIONS

The shear dependent properties that we have considered in the context of various simple flows will also be significant in real processing situations. The use of a generalised Reynolds number (equation 6.48) may permit prediction of pressure requirements for pipeline flow, but difficulties of pump specification and performance remain. Similarly, equipment used for blending Newtonian liquids may not be suitable for these fluids.

Pumping

Centrifugal pumps

The commonest type of pump used in the chemical industry is the centrifugal (Volume 1, Chapter 6). The principle of operation is the conversion of kinetic energy into a static pressure head which produces the flow. For a pump of this type the distribution of shear within the pump will vary with throughput. Considering Fig. 6.43, when the discharge is completely closed off the highest degree of shearing is in the gap between the rotor and the shell, i.e. at B. Within the vanes of the rotor (region A) there will be some circulation as sketched in Fig. 6.44, but in the discharge line C the fluid will be essentially static. If fluid is flowing through the pump, the differences between these shear rates will still be present but not as extreme. If the fluid has pseudoplastic properties, then

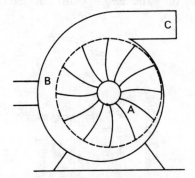

FIG. 6.43. Zones of differing shear in a centrifugal pump

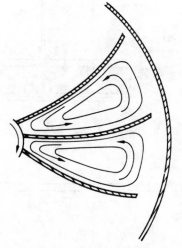

FIG. 6.44. Circulation within a centrifugal pump impeller

the effective viscosity will vary in these different regions, being less at B than at A and C. Under steady conditions the pressure developed in the rotor produces a uniform flow through the pump. However, there may be problems on starting, when the very high effective viscosities in the fluid as the system starts from rest might lead to overloading of the pump motor. At this time too the effective viscosity of the liquid in the delivery line is at its maximum value, and the pump may take an inordinately long time to establish the required flow. Many pseudoplastic materials are damaged and degraded by prolonged shearing, and such a pump would be unsuitable.

Positive displacement pumps

Difficulties experienced in getting pseudoplastic materials to flow initially are sometimes countered by the use of positive displacement pumps, of which

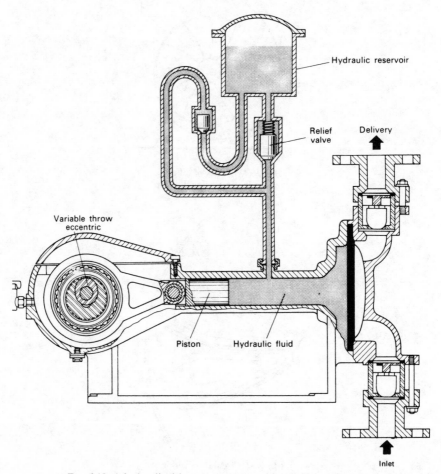

FIG. 6.45. A hydraulic drive to protect a positive displacement pump

various types were described in Volume 1. These non-Newtonian fluids which are sensitive to breakdown, particularly agglomerates in suspension, are handled with pumps which subject the liquid to a minimum of shearing. Diaphragm pumps (Volume 1, Chapter 6) are frequently used for this purpose, but care has always to be taken that the safe working pressure for the pump is not exceeded. This can be achieved conveniently by using a hydraulic drive for a diaphragm pump, equipped with a pressure limiting relief valve which ensures that no damage is done to the system (Fig. 6.45).

Screw pumps

A most important class of pump for dealing with highly viscous material is represented by the screw extruder used in the polymer industry. Extruders find their main application in the manufacture of simple and complex sections (rods, tubes, beadings, curtain rails, rainwater gutterings and a multitude of other shapes). However, the shape of section produced in a given material is dependent only on the profile of the hole through which the fluid is pushed just before it cools and solidifies. The screw pump is of more general application and will be considered first.

The principle is shown in Fig. 6.46. The fluid is sheared in the channel between the screw and the wall of the barrel. The mechanism that generates the pressure can be visualised in terms of a model consisting of an open channel covered by a moving plane surface (Fig. 6.47). This representation of a screw pump takes

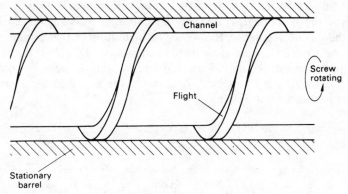

FIG. 6.46. Section of a screw pump

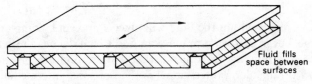

FIG. 6.47. Planar model of part of a screw pump

as the frame of reference a stationary screw with rotating barrel. The planar simplification is not unreasonable, provided that the depth of the screw channel is small with respect to the barrel diameter. It should also be recognised that the distribution of centrifugal forces would be different according to whether the rotating member is the wall or the screw; this distinction would have to be drawn if a detailed force balance were to be attempted, but in any event the centrifugal (inertial) forces are generally far smaller than the viscous forces.

If the upper plate moved along in the direction of the channel, then a velocity profile would be established that would approximate to the linear velocity gradient that exists between parallel walls moving parallel to each other. If it moved at right angles to the channel axis, however, a circulation would be developed in the gap, as drawn in Fig. 6.48. In fact, the relative movement of the barrel wall is somewhere between, and is determined by the pitch of the screw.

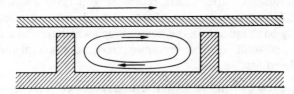

FIG. 6.48. Fluid displacement resulting from movement of plane surface

The fluid path in a screw pump is therefore of a complicated helical form within the channel section. The nature of the velocity components along the channel depends on the pressure generated and the amount of resistance at the discharge end. If there is no resistance, the velocity distribution in the channel direction will be the Couette simple shear profile shown in Fig. 6.49a. With a totally closed discharge end the net flow would be zero, but the velocity components at the walls would not be affected. As a result, the flow field necessarily would be of the form shown in Fig. 6.49b.

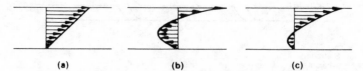

(a)   (b)   (c)

FIG. 6.49. Velocity profile produced between screw pump surfaces

    a with no restriction on fluid flow
    b with no net flow (total restriction)
    c with a partially restricted discharge

Viscous forces within the fluid will always prevent a completely unhindered discharge, but in extrusion practice an additional die head resistance is used to generate backflow and mixing, so that a more uniform product is obtained. The flow profile along the channel is then of some intermediate form such as shown in Fig. 6.49c.

It must be emphasised that flow in a screw pump is produced as a result of viscous forces. Pressures achieved with low viscosity materials are negligible. The screw pump is *not* therefore a modification of the Archimedes screw used in antiquity to raise water—that was essentially a positive displacement device using a deep cut helix mounted at an angle to the horizontal, and not running full. If a detailed analysis of the flow in a screw pump is to be carried out, then it is also necessary to consider the small but finite leakage flow that can occur between the flight and the wall. With the large pressure generation in a polymer extruder (commonly 100 bar $= 10^7$ N/m^2) the flow through this gap, which is typically about 2 per cent of the barrel internal diameter, can be significant. The pressure drop over a single pitch length may be of the order of 10 bar $= 10^6$N/m^2, and this will force fluid through the gap. Once in this region the viscous fluid is

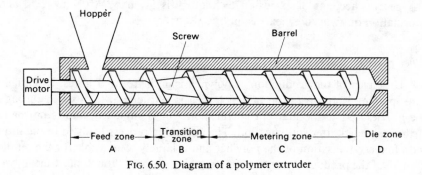

FIG. 6.50. Diagram of a polymer extruder

subject to a high rate of shear (the rotation speed of the screw is often about 2 Hz), and an appreciable part of the total viscous heat generation occurs in this region of an extruder.

Polymer extruders

A simple complete extruder has four main sections which are detailed in Fig. 6.50. The feed section A takes the raw polymer, usually in the form of small chips, from a hopper into a melting zone. In this part of the extruder the polymer tends to be conveyed forwards more rapidly than it is subsequently removed; considerable packing together and mechanical deformation is experienced by the solid in this region. The heat required for melting may be directly applied to the outside of the barrel but will be supplemented by heat generation from sheared polymer. A short transitional section B leads into the metering zone C of the extruder. This uniform screw section is intended to provide a suitable stabilisation and mixing zone in order to even out any in-homogeneities that may occur in the melt. The die section D is profiled so as to produce an extruded material of the correct dimensions after allowance has been made for die swell and similar related entrance and exit effects. The die section is almost invariably provided with the facilities for independent heating or cooling; the thermal condition of the polymer has an important effect on the quality of the product from the machine. Elaborations of this simple pattern are

484   CHEMICAL ENGINEERING

common. By appropriate choice of screw pitch and depth, the pressure development along the length of the extruder can be varied, and many extruders exploit this by having a low-pressure zone after melting is complete, in order to allow venting and removal of volatile materials.

Difficulties with polymers like PVC, which is rather easily degraded if subjected to temperatures not far above its melting range, have led to the increasing popularity of twin screw extruders. These have two intermeshing screws (either co- or counter-rotating) and convey the material in the C-shaped chambers on either side of the screws.

Twin screw extruders are dearer than single screw machines for the same output, but with a larger heating (or cooling) surface per unit length and relatively less mechanical energy dissipated as heat, they are easier to control.

A general account of extrusion technology is given by McKELVEY[6], and information on twin screw extrusion by JANSSEN[15].

Other polymer processes

A large part of the production of polymer is in the form of fibres. Essentially fibre spinning is an operation in which melt is extruded at high speed through fine nozzles. While the filament is still molten it is stretched, and is normally further extended after solidification. The stretching process leads to orientation and often crystallisation of the polymer and is largely responsible for the tensile strength of the product. A continuous stretching process, though this time in two directions, is also used in film blowing. Polymer is extruded as a tube, and through the axis of the die air is introduced at such a pressure that a bubble is blown in the melt. When the polymer has travelled a sufficient distance to have solidified, the sides of the bubble are collapsed through nip rollers, and the double thickness film is then rolled up.

Analyses of these processes are given by HAN[12] and PEARSON[8].

Mixing

The agitation and mixing of Newtonian fluids was discussed in Chapter 13 of Volume 2. With low viscosity fluids high speed agitators (usually turbines or propellers) are usually used, and in properly baffled tanks establish the flow regimes that are shown in Figs. 6.51 and 6.52.

The turbine, which produces a flow which is essentially radial in the plane of the agitator, leads to a double-celled system with two toroidal (doughnut shaped) vortices. Mixing between the upper and lower regions of the tank occurs only as a result of turbulent interchange of fluid in the discharge from the agitator. The propeller mixing impeller is an axial flow device which can induce the single large circulatory flow shown. Appreciable vertical concentration gradients are less likely to persist than with the turbine. The liquid moves through regions of varying shear conditions while circulating in the tank and, if the fluid has non-Newtonian properties, then this may be important. In particular, pseudoplastic fluids behave as if their viscosity is low when near the

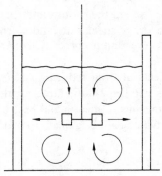

FIG. 6.51. Flow in
a baffled tank with a
turbine agitator

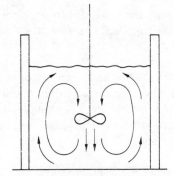

FIG. 6.52. Flow with
a propeller impeller
in a baffled tank

impeller and high when some distance from it. One result is that the impeller tends to rotate in a small circulating vortex of highly sheared fluid, while the main bulk of the liquid scarcely moves at all[20]. Effective homogenisation of the liquid inevitably will be accompanied by excessive shearing of at least some of the fluid. The situation can be improved if a propeller agitator is shrouded by a draft tube which causes larger scale movement.

The power levels required for mixing non-Newtonian systems have been extensively studied[9], and the normal correlating procedure is similar to that adopted for pipeline systems, viz. a generalised Reynolds number is formulated in terms of the drag expected in a purely viscous Newtonian system. One form of generalised Reynolds number is:

$$Re.^{+} = \frac{D^2 N^{2-n} \rho}{k} 8 \left(\frac{n}{6n+2}\right)^n \qquad \dots.(6.90)$$

which reduces to the conventional stirrer Reynolds number when $n = 1$. (See discussion in Volume 2, Chapter 13.)

In these systems which have a relatively consistent flow pattern over a large range of Reynolds numbers (particularly with most industrial equipment operating with $Re.^{+}$ over 500), it may be reasonable to attempt to describe an average shear rate as a linear function of the stirring speed. For design purposes it is reasonable to put:

$$\dot{\gamma}_{av} = KN \qquad \dots.(6.91)$$

Average shear rate = Constant × Rotational speed of agitator

The value of the constant of proportionality is about 12. The usual practice of using power input measurements to estimate this constant is open to criticism in that the relationship is defined in terms of what is happening in the region of the impeller, and this represents only a small fraction of the fluid in the tank.

However, the usual design problem is that of sizing the motor, and the use of such an empirical constant to estimate an average shear rate (and so effective viscosity) allows the use of the Newtonian power input curves (Chapter 13 of Volume 2).

Some studies have been made on the effects of drag reducing additives on power consumption in stirred vessels[38]. As with pipe flow, however, the suppression of turbulence is generally disadvantageous to mixing and heat or mass transfer and there seems to be little overall advantage to be gained from attempts to save energy in this way.

Gas dispersion and mass transfer

When non-Newtonian behaviour is involved in gas–liquid processes three effects must be considered. Because of the hydrodynamic action of the impeller, the dispersion process may be profoundly influenced[40]. Further, the flow field surrounding the bubbles is altered, so influencing their rising dynamics[34], and the distribution of the velocity field (including the scale and intensity of turbulence) will be affected. All these factors influence the mass transfer factor (coefficient multiplied by contact area), so estimates made for design purposes must be treated with some reserve.

Stirred tanks with low speed agitators

To counter the effects of the local thinning associated with high speed agitators in pseudoplastic fluids, frequent use is made of designs that sweep the volume of the vessel. The most common patterns are the gate (Fig. 6.53), anchor (Fig. 6.54) and related paddle agitators, and the more elaborate (and more efficient) helical shapes.

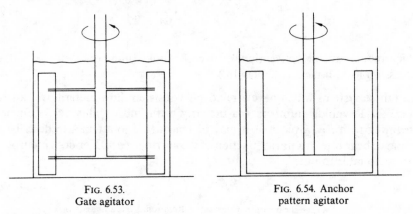

FIG. 6.53.
Gate agitator

FIG. 6.54. Anchor
pattern agitator

The flow patterns established in mixing vessels with these agitators are generally of the form illustrated in Fig. 6.55. The feature of such a streamline pattern is the very small extent of radial mixing that occurs, at least so it appears in this somewhat simplified two-dimensional picture. The extent of the asymmetry that is seen in the vicinity of the blade depends on the stirrer Reynolds

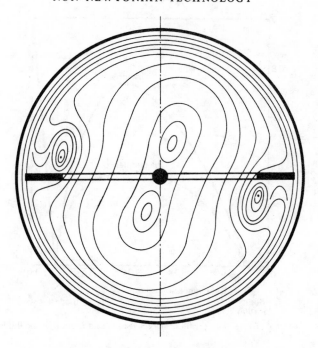

FIG. 6.55. Streamlines in a tank with a gate agitator,
drawn relative to the arms of the stirrer

number $ND^2\rho/\mu_a$. For values of Reynolds number below about 10 the streamline pattern is essentially symmetrical in front of, and behind, the blade. As the
Reynolds number is increased, circulation takes place in the vortex that travels
along with the blade (Fig. 6.55). These vortices with vertical axes are firmly
associated with the blade and only when the Reynolds number exceeds about
1000 do they break away in a vortex street; this is about two orders of magnitude
greater than for the case where the fluid is not bounded by a nearby moving
surface (see Volume 1, Chapter 9).

   With both gate and anchor agitators the principal action is the sweeping of
the vertical members close to the wall. Under "no-slip" conditions the material
between the wall and the outer edge of the blade is subjected to a high rate of
shear which leads to the effective viscosity of this layer being reduced if the
material is subject to shear thinning. (This could be expected to be advantageous
for wall heat transfer and similar processes that depend on the local effective
Reynolds number.) Such wall thinning is evident from the velocity profiles
which are shown in Fig. 6.56. The two profiles that are drawn here represent
the average fluid velocity over a sector, obtained from the time taken for fluid
at a particular initial radius to traverse an arbitrary angular sector around the
blade. For simplicity, the velocity profiles are presented relative to the blade.
The greater tendency for the shear-thinning material to rotate in bulk is evident
from this figure. The reduced velocity gradients in the inner part of the tank

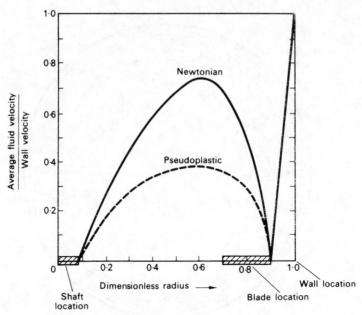

FIG. 6.56. Velocity profiles for Newtonian and pseudoplastic fluids

will mean a reduction in the rate of mixing by the deformation of elements of the fluid.

In all mixing operations that involve blending material throughout a tank, the vertical flow components are important. In the flow field illustrated in Fig. 6.55 the vertical flows are small—commonly only 5 or 10 per cent of the horizontal velocities. Any rotational motion induced within a vessel will tend to produce a secondary flow in the vertical direction; the fluid in contact with the bottom of the tank is stationary, while that which is rotating at a higher level will experience centrifugal forces. Consequently, there exist unbalanced pressure forces within the fluid which lead to the formation of a toroidal vortex. This secondary system may be single-celled (Fig. 6.57), or double-celled (Fig. 6.58); low viscosity liquids tend to give the former, while high viscosity liquids (over 1 Nsm^{-2} Newtonian) and shear-thinning materials tend to form the double-cell system.

In mixing with close clearance agitators like these, the effect of the shearing between the blade and the wall might be expected to affect the performance. In fact, for a given stirrer speed, halving the gap will double the shear rate momentarily imposed on the fluid near the wall. However, the volume that passes through the gap and is sheared will be reduced to approximately half, so that the average shear rate experienced by the fluid will not change very much.

In these systems the flow patterns change as the stirring rate is increased. The average shear rate cannot be completely described by a linear relationship like that of equation 6.91, and the criticism made above concerning high speed agitators is even more valid in this case. However, various workers have

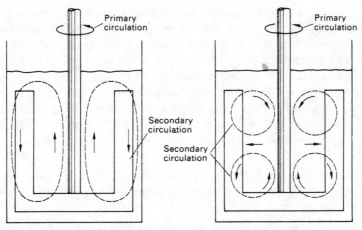

FIG. 6.57. Single-celled secondary flow with an anchor agitator

FIG. 6.58. Double-celled secondary flow

attempted such correlations, and one that is valid up to Reynolds numbers of
50 was developed by BECKNER and SMITH[29] and gives:

$$\dot{\gamma}_{av} = a'(1 - n)\,N \qquad \qquad \ldots(6.92)$$

$0.25 < n < 0.75$, where a' is a geometric constant, equal approximately to
$40(1 - 3c/D_T)$.

Helical blade mixers

Various methods have been used to improve mixing in the vertical direction.
One of the most efficient is the use of a helical ribbon blade (Fig. 6.60). An open
spiral blade passes close to the wall of the containing vessel, and the helix
induces a flow field as shown in Fig. 6.59[26, 32]. This type of agitator is much

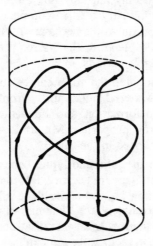

FIG. 6.59. Principal
flow field established
by a helical impeller

more complicated to make than the simple gate or anchor pattern, but there is a considerable improvement in performance. The efficiency (in terms of total power requirement to achieve a given result) is about twice that of the simple anchor in heat transfer applications, and five or ten times greater for homogenisation. The reduction in operating costs and/or process time is often adequate to compensate for the higher manufacturing costs.

An alternative way of encouraging the overall circulation in the vessel is to mount a screw impeller in a draught tube. This provides a controlled circulation, and the liquid is usually satisfactorily mixed within three or four circuits of the tank.

There is, however, evidence[36] that viscoelastic effects can slow down the blending process: this may be because of stable secondary flows encouraging regions of segregation.

A variation on the simple helix mixer is the *helicone* (Fig. 6.61), which has the additional advantage that the clearance between the blade and the wall is easily adjusted by a small axial shift of the impeller.

Mixing heavy pastes and similar materials. Many mixing operations with pastes and slurries are carried out in vessels with twin close-clearance rotating blades which are often of a skewed sigmoidal form, like those illustrated in Fig. 6.62. It is usual for the two blades to rotate at different speeds—a ratio of 3:2 is common. Various blade profiles of different helical pitch are used. The blade design differs from that of the helical ribbons considered above in that the much higher effective viscosities (of the order of 10^4 Ns/m^2) require a more solid construction; the blades consequently tend to sweep a greater quantity of fluid in front of them, and the main small-scale mixing process takes place by extrusion between the blade and the wall. Partly for this reason, mixers of this type are usually operated only partially full, though the Banbury mixer (Fig. 6.63) used in the rubber industry is filled completely and pressurised as well. The pitch of the blades produces the necessary motion along the channel, and this gives the larger scale blending that is required in order to limit batch blending times to reasonable periods.

The motion of the material in the mixer can be considered in three stages, as illustrated in Fig. 6.64. Material builds up in front of the blade in region A where it will undergo deformation and flow—the relative extent of these depending on the material properties. Materials will tend to be rolled and deformed until some is trapped in the gap B.

The difference between the solid and liquid behaviour will be evident in the regions where the shear stresses are least, i.e. less than the yield stress. Whether there is a radially inward flow in front of the blade will depend on the magnitude of this yield stress. Unfortunately, the dynamics of this region are so complex that it is not possible to quantify this generalisation.

Once the material has entered region B it will be subject to an unsteady state developing flow situation; in the fully-developed form the velocity gradient is

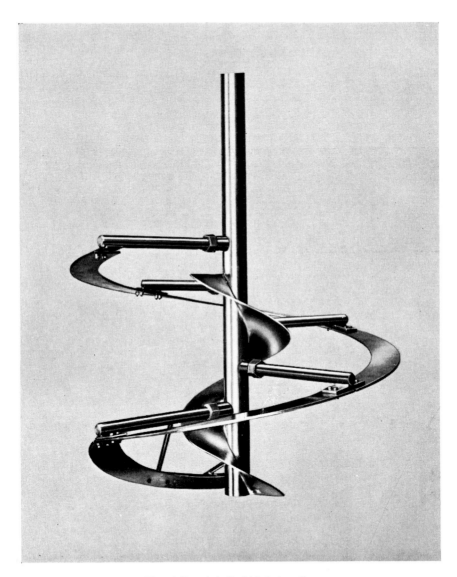

FIG. 6.60.   A helical blade impeller

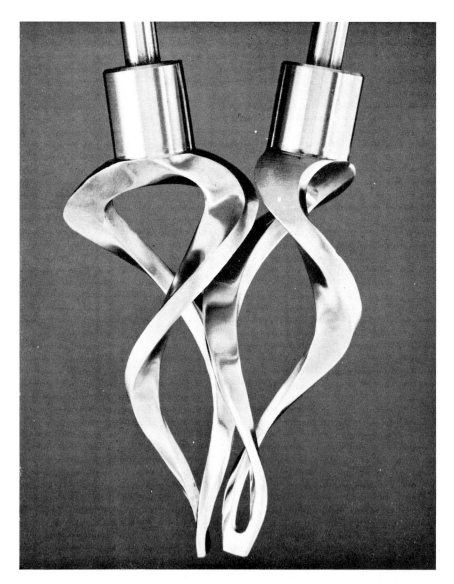

Fig. 6.61.   A double helicone impeller

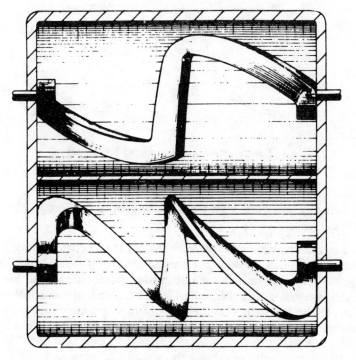

FIG. 6.62. A sigma blade mixer

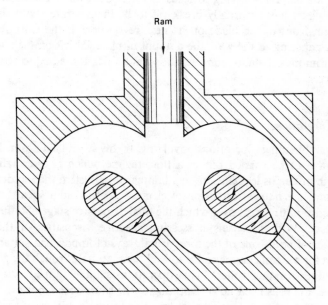

FIG. 6.63. A Banbury mixer

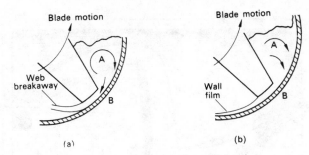

FIG. 6.64. Fluid motion in a sigma blade mixer

a non-wetting paste
b surface-wetting fluid

linear, changing uniformly from the blade velocity at the blade surface to zero at the vessel wall.

The general situation is the same as that existing between moving planes; normally the radius of the vessel is large and the shear stress is sensibly uniform throughout the gap. Under these circumstances fully developed flow will be established providing that the *channel* length (i.e. the thickness of the edge of the blade facing the wall) is of the order of two or three times the clearance between the blade and the wall. For this situation the mean velocity of the material in the gap is just half the velocity of the blade relative to the wall. A solid material may well not achieve this condition, however, since the rate at which it is drawn into the gap is largely controlled by the deformation in zone A.

On discharge from the gap the material will either break away from the wall (as in *a*) or remain adhering to it (as in *b*). In the former case, a web may be formed which will eventually break off to be incorporated with the material coming in front of the blade on its next revolution. If the extruded material remains adhering to the wall, the amount of material left on the wall is determined from mass balance considerations, at a thickness equal to about half the clearance.

Rolling operations

Several blending operations involving highly viscous or non-Newtonian fluids have to be carried out in a flow regime which is essentially laminar. Although mixing is less efficient in a laminar flow system than under turbulent conditions, the natural extension of elements of fluid in a shear field reduces the effective path length over which the final dispersive stage of diffusion has to take place. With materials possessing effective viscosities of the order of thousands of poise, one of the more usual ways of imposing a suitable velocity gradient for a significant period is to feed the material through the gap between two counter-rotating rollers. This operation is often termed *calendering*. In principle, the rolls need not be rotating at the same speed, nor even need they be the same diameter. The simplest case to consider, however, is that where the diameters and speeds are the same; this has been analysed by PEARSON[8] and McKELVEY[6]. The difficulties in the analysis lie in the fact that it is not possible to solve unambiguously for the position at which the breakaway from the roll surface will occur.

Figure 6.65 shows the situation in the "nip" region between two rolls.

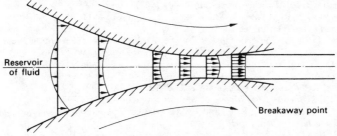

Reservoir of fluid

Breakaway point

FIG. 6.65. Flow between rolls

Rolling machines are commonly used for batch blending very viscous materials, and the method of operation is shown in Fig. 6.66. The preferential adherence of the extruded film to one of the rollers can be obtained with a small speed differential. The viscosity of the fluid is high enough to prevent centrifugal

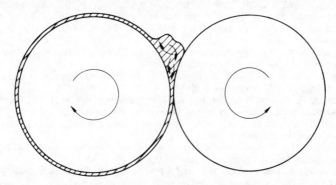

FIG. 6.66. Mixing with a rolling machine

forces throwing it clear of the roller, and it returns to the feed side of the machine where it is accumulated in the highly sheared circulating fillet of excess material. This flow is, of course, simply circumferential around the roller and through the gap, and it is necessary to ensure adequate lateral mixing along the length of the machine by separate mechanical or manual means from time to time.

FURTHER READING

BIRD, R. B., STEWART, W. E. and LIGHTFOOT, E. N.: *Transport Phenomena* (Wiley 1960).
MIDDLEMAN, S.: *Fundamentals of Polymer Processing* (McGraw-Hill 1977).
PEARSON, J. R. A.: *The Principles of Polymer Melt Processing* (Pergamon Press 1960).
SKELLAND, A. H. P.: *Non-Newtonian Flow and Heat Transfer* (Wiley 1967).
WALTERS, K.: *Rheometry* (Chapman & Hall 1975).
WILKINSON, W. L.: *Non-Newtonian Fluids* (Pergamon Press 1960).

REFERENCES TO CHAPTER 6

(1) REINER, M.: in *Rheology* (edited by EIRICH, F. R.), Volume 1 (Academic Press 1956).

(2) BIRD, R. B., STEWART, W. E., and LIGHTFOOT, E. N.: *Transport Phenomena* (Wiley 1960).

(3) REINER, M.: *Deformation, Strain and Flow,* 2nd ed. (London, H. K. Lewis 1960).

(4) SCHLICHTING, H.: *Boundary Layer Theory,* 6th ed., translated by KESTIN, J. (McGraw-Hill 1969).

(5) WILKINSON, W. L.: *Non-Newtonian Fluids* (Pergamon Press 1960).

(6) MCKELVEY, J. M.: *Polymer Processing* (Wiley 1962).

(7) VAN WAZER, J. R., LYONS, J. W., KIM, K. Y., and COLWELL, R. E.: *Viscosity and Flow Measurement* (Interscience 1963).

(8) PEARSON, J. R. A.: *The Principles of Polymer Melt Processing* (Pergamon Press 1966).

(9) SKELLAND, A. H. P.: *Non-Newtonian Flow and Heat Transfer* (Wiley 1967).

(10) ASTARITA, G. and MARRUCCI, G.: *Principles of Non-Newtonian Fluid Mechanics* (McGraw-Hill 1974).

(11) WALTERS, K.: *Rheometry* (Chapman & Hall 1975).

(12) HAN, C. D.: *Rheology in Polymer Processing* (Academic Press 1976).

(13) BIRD, R. B., ARMSTRONG, R. C. and HASSAGER, O.: *Dynamics of Polymeric Liquids,* Vol. 1 *Fluid Mechanics* (Wiley 1977).

(14) BIRD, R. B., HASSAGER, O., ARMSTRONG, R. C. and CURTISS, C. F.: *Dynamics of Polymeric Liquids,* Vol. 2 *Kinetic Theory* (Wiley 1977).

(15) JANSSEN, L. P. B. M.: *Twin Screw Extrusion* (Elsevier 1978).

(16) TROUTON, F. T.: *Proc. Roy. Soc.* **A77** (1906) 426. The coefficient of viscous traction and its relation to that of viscosity.

(17) TOMS, B. A.: *Proc. Int. Rheology Congr.* II (1948) 135. North-Holland Publ. Co. The flow of linear polymer solutions through straight tubes.

(18) TOMITA, Y.: *Bull. of Japan. Soc. Mech. Eng.* **2** (1959) 10. A study on non-Newtonian flow in pipe lines (in English).

(19) DODGE, D. W. and METZNER, A. B.: *A.I.Ch.E.Jl.* **5** (1959) 189. Turbulent flow of non-Newtonian systems. (Note: corrigenda in *A.I.Ch.E.Jl.* **8** (1962) 143.)

(20) METZNER, A. B. and TAYLOR, J. S.: *A.I.Ch.E.Jl.* **6** (1960) 109. Flow patterns in agitated vessels.

(21) HANKS, R. W. and CHRISTIANSEN, E. B.: *A.I.Ch.E.Jl.* **7** (1961) 519. The laminar non-isothermal flow of non-Newtonian fluids.

(22) MIDDLEMAN, S. and GAVIS, J.: *Phys. Fluids* **4** (1961) 355. Expansion and contraction of capillary jets of Newtonian fluids.

(23) THOMAS, R. H. and WALTERS, K.: *J. Fluid Mechanics* **16** (1963) 228. On the flow of an elastico-viscous liquid in a curved pipe under a pressure gradient.

(24) BOGUE, D. C. and METZNER, A. B.: *Ind. Eng. Chem. Fundamentals* **2** (1963) 143. Velocity profiles in turbulent pipe flow.

(25) GIESEKUS, H.: *Proc. 4th Int. Congress Rheology, Providence 1963,* **1** (1965) 249. Some secondary flow phenomena in general viscoelastic fluids.

(26) BOURNE, J. R. and BUTLER, H.: *A.I.Ch.E.–I. Chem. E. Symposium Series* No. 10 (1965), *Mixing—Theory related to Practice,* 89. Some characteristics of helical impellers in viscous liquids.

(27) SUTTERBY, J. L.: *A.I.Ch.E.Jl.* **12** (1966) 63. Laminar convergent flow of dilute polymer solutions in conical sections.

(28) GIESEKUS, H.: *Rheologica Acta* **5** (1966) 239. Zur Stabilität von Strömungen viskoelastischer Flüssigkeiten.

(29) BECKNER, J. L. and SMITH, J. M.: *Trans. Inst. Chem. Eng.* **44** (1966) T224. Anchor-agitated systems: power inputs with Newtonian and pseudoplastic fluids.

(30) METZNER, A. B., WHITE, J. L., and DENN, H. M.: *A.I.Ch.E.Jl.* **12** (1966) 863. Constitutive equations for viscoelastic fluids for short deformation periods and for rapidly changing flows: significance of the Deborah Number.

(31) WHITE, A.: *Hendon (London) Coll. of Tech. Res. Bull.* **4** (1967) 75. Turbulence and drag reduction with polymer additives.

(32) BOURNE, J. R. and BUTLER, H.: *Trans. Inst. Chem. Eng.* **47** (1969) T11. An analysis of the flow produced by helical ribbon impellers.

(33) SMITH, K. A., KEUROGHLIAN, G. H., VIRK, P. S., and MERRILL, E. W.: *A.I.Ch.E.Jl.* **15** (1969) 294. Heat transfer to drag reducing polymer solutions.

(34) CARREAU, P. J., DEVIC, M. and KAPELLAS, M.: *Rheol. Acta* **13** (1974) 477. Dynamique des boules en milieu viscoélastique.

(35) MOORE, C. A. and PEARSON, J. R. A.: *Rheol. Acta* **13** (1974) 436. Extensional rheology of Separan AP 30 solutions.

(36) CHAVAN, V. V., ARUMUGAM, M. and ULBRECHT, J.: *A.I.Ch.E.Jl.* **21** (1975) 613. On the influence of liquid elasticity on mixing in a vessel agitated by a combined ribbon–screw impeller.

(37) HARRIS, J. B. and PITTMAN, J. F. T.: *Trans. I. Chem. E.* **54** (1976) 73. Alignment of slender rod-like particles in suspension using converging flow.

(38) QURAISHI, A. Q., MARSHELKAR, R. A. and ULBRECHT, J.: *J. non-Newt. Fluid Mechanics* **1** (1976) 223. Torque suppression in mechanically stirred liquid and multiphase liquid systems.

(39) McCONAGHY, G. A. and HANRATTY, T. J.: *A.I.Ch.E.Jl.* **23** (1977) 493. Influence of drag reducing polymers on turbulent mass transfer to a pipe wall.

(40) RANADA, V. R. and ULBRECHT, J.: *Proc. 2nd Eur. Mixing Conf. BHRA* (1977), F6-83. Gas dispersion in viscous inelastic and elastic liquids.

(41) B.S. 188: British Standard 188 (1977). Methods for determination of the viscosity of liquids.

LIST OF SYMBOLS USED IN CHAPTER 6

A	Area	L^2
a	Pipe radius	L
a_c	Cone radius	L
b	Ratio $\dfrac{R_Y}{R_w}$	—
C	Couple	(FL) ML^2T^{-2}
c	Clearance	L
D	Stirrer diameter	L
D_t	Diameter of tank	L
d	Diameter	L
E	Eddy kinematic viscosity	L^2T^{-1}
f	Fanning friction factor $(=2\phi)$	—
G	Mass rate of flow	MT^{-1}
G	Dynamic rigidity	(FL^{-2}) $ML^{-1}T^{-2}$
g	Acceleration due to gravity	LT^{-2}
h	Height of concentric cylinder apparatus	L
k	Constant for power-law model (equation 6.8)	$(FL^{-2}T^n)$ $ML^{-1}T^{n-2}$
k_1	Constant in equation 6.76	$(FL^{-2}T^n)$ $ML^{-1}T^{n-2}$
k'	Constant for power-law relationship derived from pipe flow experiment (equation 6.38)	$(FL^{-2}T^n)$ $ML^{-1}T^{n'-2}$
l	Length	L
N	Stirrer speed (revolutions per unit time)	T^{-1}
n	Exponent for power-law fluid model (equation 6.8)	—
n'	Exponent for power-law relationship derived from pipe flow experiment (equation 6.38)	—
P	Pressure	(FL^{-2}) $ML^{-1}T^{-2}$
Q	Volume rate of flow	L^3T^{-1}
R	Shear stress	(FL^{-2}) $ML^{-1}T^{-2}$
R_{rx}	Shear stress in x-direction at radius r	(FL^{-2}) $ML^{-1}T^{-2}$
R_{wx}	Shear stress at wall in x-direction	(FL^{-2}) $ML^{-1}T^{-2}$
R_{xx}	Tensile stress in x-direction	(FL^{-2}) $ML^{-1}T^{-2}$
R_{yx}	Shear stress in x-direction in a plane normal to y-axis, etc.	(FL^{-2}) $ML^{-1}T^{-2}$
R_Y	Yield stress	(FL^{-2}) $ML^{-1}T^{-2}$
$\mathbf{R}$	Universal gas constant	$L^2T^{-2}\theta^{-1}$
r	Radial coordinate	L
r_i	Radius of inner cylinder	L
r_P	Radius of unsheared plug	L
s	Thickness of film flowing over an inclined plane surface	L
T	Absolute temperature	θ
t	Time	T
t_r	Relaxation time (equation 6.88)	T
t_ε	Relaxation time (equation 6.87)	T
u	Velocity—mean velocity in pipe	LT^{-1}

u_{CL}	Velocity at pipe axis	LT^{-1}
u_w	Velocity at wall	LT^{-1}
u_x, u_y, u_z	Velocity in x-, y-, z-directions (point values)	LT^{-1}
u^+	Dimensionless velocity (equation 6.63)	—
u^*	Shearing stress velocity (equation 6.62)	LT^{-1}
v	Volume per unit mass	$M^{-1}L^3$
x	Distance in x-direction	L
y	Distance in y-direction	L
y^+	Dimensionless derivative of y (equation 6.64)	—
z	Distance in z-direction (usually vertical)	L
α	Constant in equations of state	—
β	Constant in equation of state (6.10)	$(F^{-1}L^2T^{-1})\,M^{-1}LT$
Γ	Constant in equation 6.10	$(F^{-\alpha}L^{2\alpha}T^{-1})M^{-\alpha}L^\alpha T^{2\alpha-1}$
γ	Strain	—
$\dot\gamma$	Shear rate	T^{-1}
$\dot\gamma_{av}$	Mean shear rate	T^{-1}
ε	Activation energy for flow	L^2T^{-2}
λ_E	Mixing length	L
μ	Viscosity	$(FL^{-2}T)\,ML^{-1}T^{-1}$
μ_a	Apparent or effective viscosity	$(FL^{-2}T)\,ML^{-1}T^{-1}$
μ_p	Plastic viscosity	$(FL^{-2}T)\,ML^{-1}T^{-1}$
μ_0	Viscosity at zero shear rate	$(FL^{-2}T)\,ML^{-1}T^{-1}$
μ_∞	Viscosity at infinite shear rate	$(FL^{-2}T)\,ML^{-1}T^{-1}$
η	Constant in equation 6.71	$(FL^{-2}T)\,ML^{-1}T^{-1}$
ρ	Density	ML^{-3}
σ_1, σ_2	Normal stress differences	$(FL^{-2})\,ML^{-1}T^{-2}$
θ	Time constant	T
ψ	Constant in equation 6.79	$(FL^{-2})\,ML^{-1}T^{-2}$
ϕ	Friction factor $(=\tfrac{1}{2}f)$	—
ϕ_B	Friction factor for Bingham plastics (equation 6.57)	—
ω	Angular velocity	T^{-1}
$Db.$	Deborah number (equation 6.89)	—
$Re.$	Reynolds number	—
$Re.^*$	Generalised (non-Newtonian) Reynolds number (equation 6.48)	—
$Re._1{}^*$	Modified generalised Reynolds number (equation 6.67)	—
$Re._B$	Generalised Reynolds number for Bingham plastics (equation 6.58)	—
$Re.^+$	Generalised Reynolds number for mixing (equation 6.90)	—

Suffixes

i	Inner wall
o	Outer wall
r	Radius r
w	Wall
x, y, z	x-, y-, z-directions
PE	Prandtl–Eyring model (equation 6.11)
RP	Reiner–Philippoff model (equation 6.12)
S	Sutterby model

CHAPTER 7

Sorption processes

TO THE chemical engineer adsorption is a separation process and, in common with other such processes, it involves two phases between which components become differentially distributed. Except for foam separations[14] which will not be discussed, one phase is solid and the other liquid or gas. To the physical chemist, adsorption is an equilibrium phenomenon studied for the light it throws on the properties of solid surfaces and the fluids adsorbed thereon. Until recent years, the literatures relating to the interests of chemical engineers and of physical chemists were developed separately, but they are now coming together in such applications as preparative chromatography. The technology of adsorption requires some understanding of both equilibrium and rate processes. This chapter aims to develop both.

Adsorption differs from the more common process of absorption by the degree of homogeneity that exists at equilibrium in that phase to which molecules are transferred. In absorption, the molecules are uniformly mixed down to a molecular scale. In adsorption, the molecules are evenly distributed but confined to the surface of micropores which permeate the solid structure. The pore sizes could be one to four orders of magnitude greater than the sizes of the molecules, so on a molecular scale adsorption is not homogenous. The adsorbed phase is distinct by virtue of its mobility, and sometimes its density, from both the fluid and the solid phases with which it is associated.

Because adsorption is a surface phenomenon, only those solids which contain large amounts of internal surface are likely to be useful as adsorbents. Even when a solid is finely divided, its external surface is small compared with the total areas found in commercial adsorbents. Simple geometry shows that spherical particles of iron oxide with a radius of 5 μm and a specific gravity of 5 have an external surface of 12,000 m^2/kg. An average figure for commercial adsorbents would be 300,000 m^2/kg, contributed almost entirely by internal surface. Surfaces of this magnitude are found only in highly porous solids. If the pores are assumed to be cylindrical with a mean length L_p and a radius of r, the pore volume V_p and the curved surface area A_p can be written:

$$V_p = n_p \pi r^2 L_p$$

$$A_p = 2n_p \pi r L_p$$

where n_p is the number of pores.

Hence:

$$r = \frac{2V_p}{A_p}$$

If V_p is given a typical value of 3×10^{-4} m³/kg, a surface area of 3×10^5 m²/kg requires that $r = 2 \cdot 0$ nm which is a pore radius of 20 Å (Ångstrom units).

Molecules adsorbed on to an empty surface are held by forces that emanate from the surface. These can be physical, and are known as van der Waals forces[36] because they are the same forces that give rise to the a constant in the van der Waals equation. The forces can also be chemical, in which case they lead to bonds that are electrostatic or involve the sharing of an electron in the manner of a chemical compound; the process is then one of *chemisorption*[19]. Van der Waals forces are relatively weak so physical adsorption is generally more easily reversed than chemisorption. Physical adsorption can be likened to condensation; the heat released when it occurs is somewhat greater than the latent heat of condensation, 40–50 MJ/kmol, because of the heat of wetting of the surface. The heat of chemisorption, on the other hand, is more like that of a chemical reaction, about 200 MJ/kmol. The rate of chemisorption increases rapidly with temperature. In chemisorption, molecules are not attracted to all points on the surface of the solid but specifically to active centres, so that a surface which is completely chemisorbed may not be completely covered by the chemisorbed molecules. In physical adsorption, molecules are attracted to all points on the surface and are limited only by the number that can squeeze into each layer of adsorbed molecules. If the point at which the first layer is complete can be detected, then the number and size of the physically adsorbed molecules can give an indication of the surface area available.

The number of molecules that can accumulate on a surface depends on several factors. Since, in general, the process is reversible, a low concentration in the fluid will cover the surface of the adsorbent only up to the point when the pressure exerted by the adsorbed phase is equal to that in the fluid. Given a sufficient concentration in the fluid, forces of physical adsorption may continue to have influence until several layers of molecules, perhaps five or six, have accumulated on the surface. If the surface exists in narrow pores, then the maximum number of layers may be restricted by the dimensions of the pore itself. The forces in chemisorption are more specific, the chemical attraction between the solid and molecules in the fluid being saturated when each active centre is occupied. Chemisorption does not extend beyond the first layer, but it is then possible that additional physical adsorption will occur.

The structure of the solid may so limit the number of molecules that can be admitted to a single pore that it is difficult to think of any separate adsorbed phase. The spaces are then no more than vacancies in a crystal lattice, and the intimacy of the process is more akin to absorption. In *ion exchange*, for example, an ion in a liquid phase is exchanged for one in the solid resin. Though mathematically very similar, such a process does not conform to the classical idea of adsorption, so a more embracing term is required. In 1909 McBain[26, 31] used the term *sorption* to describe the transfer of molecules between solid and fluid.

The term will be used in the remainder of this chapter to describe physical adsorption, chemisorption, ion exchange, condensation in capillaries and associated phenomena. The solid will be termed the sorbent and the distributed component the sorbate whether it exists in the fluid or solid phases. This terminology is in keeping with current engineering practice[18].

COMMON SORBENTS

In spite of a long history of its application to the clarifying of liquids and to the drying of gases, sorption retains its reputation as a "newer method of separation"[14]. There have been several obstacles to its large scale adoption in competition with distillation, extraction and absorption. First, there was the difficulty of producing sorbents with properties that were consistent from one batch to the next. Sorbent manufacture was an art. Then there was the lack of theory adequate to predict the operation of large scale equipment. Sorber design was also an art. Finally, there was the need to reconcile the continuous operation of most chemical plant with the requirement that the sorbent had to be regenerated at regular intervals—a matter of stand-by equipment or a mechanism for moving the solid continuously.

But as manufacturers have become more skilled in the production of tonnage quantities of sorbents with specific properties, and as theoretical and experiment studies lead to a better understanding of the basic phenomena, so the applications of sorbents in commercial processes increase.

One of the earliest sorbents to be used on a large scale was *activated carbon*. When a carbonaceous material such as bones, coal, coconut shell or wood is carbonised at a temperature below about 870 K, the resulting carbon is either active or capable of being activated and becomes suitable as a sorbent. The process of activation is essentially a selective oxidation of residual hydrocarbons on the sorbent by steam, air or some other oxidising agent. The properties of the activated carbon will depend upon the source material and the precise conditions of activation. It can be used to recover solvent vapours, to purify water, to decolourise solutions of sugar, and for many similar applications. Internal surface areas of about 10^6 m^2/kg are commonly claimed for this material. Another common sorbent is *silica gel*. It is a granular, amorphous form of silica, made by heating to about 630 K the gel obtained when a solution of sodium silicate is acidified. The hard, glassy material is highly porous; it is used for drying both liquids and gases, also for recovering hydrocarbons[5]. There are several commercial varieties of this material, some having a better resistance to contact with liquid water; this can shatter some gels through the rapid release of the heat of adsorption. Internal surface areas in excess of 3.5×10^5 m^2/kg are commonly found for silica gels.

An adsorbent with good mechanical strength and which is popular, therefore, in moving bed applications is *activated alumina*. It is manufactured by heating to about 670 K, at a carefully controlled rate, one or other of the alumina hydrates, usually a precipitated mixture containing both mono and trihydrates. The dehydration of the alumina hydrates leads to the creation of an internal surface of about $2 \cdot 5 \times 10^5$ m^2/kg. The mechanism by which this occurs has been studied by many workers (e.g. DE BOER[8]) and it seems likely that the internal surface is the result of the structure of the solid being ruptured as the water molecules in the hydrates are expelled into the surrounding atmosphere. The micrograph of a commercial alumina reproduced in Fig. 7.1 shows a pore structure that includes a well-defined pattern of pores[89].

In the three adsorbents mentioned, there is a distribution of pore sizes from about 1 to 1,000 nm, with a mean pore diameter of between 4 and 10 nm. During the last thirty years other materials have been studied, first in their naturally occurring form, and then as synthesised products which offer exact pore sizes. These materials are mineral families, such as the zeolites, which derive their pore spaces from vacancies left in the crystal lattice when water molecules are driven off by heating. The materials that result are known by the general name of *molecular sieves*, and consist of porous aluminosilicate frameworks. A typical structure is composed of clusters of SiO$_4$ and AlO$_4$ tetrahedra arranged to form larger polyhedra. For example, twenty-four tetrahedra can form cubo-octahedra (Fig. 7.2) which can be stacked to give different frameworks. Access to the body of

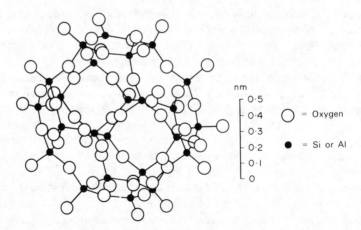

FIG. 7.2. A cubo-octahedral unit composed of SiO$_4$ and AlO$_4$ tetrahedra[70]

the framework is through *windows* which limit the size of molecules that can enter. Large cavities capable of holding a number of sorbed molecules are contained in the body of the framework. Though this sorptive capacity cannot meaningfully be related to an area in the way that was thought appropriate for the sorbents referred to previously, an equivalent area, usually about 8×10^5 m^2/kg, enables a comparison to be made. Five types of sieve are available commercially. These are shown in Table 7.1, together with some of the molecules that

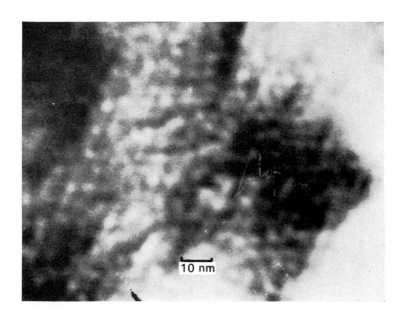

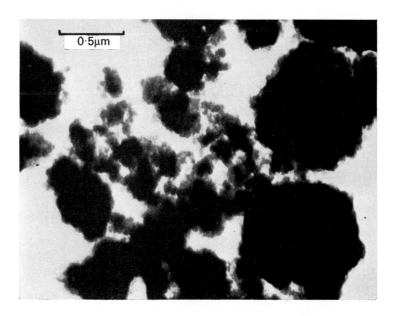

FIG. 7.1. Electron micrographs of a commercial activated alumina

TABLE 7.1. *Classification of some molecular sieves*[70]

Molecular Size Increasing ⟶

He, Ne, A, CO H_2, O_2, N_2, NH_3, H_2O	Kr, Xe CH_4 C_2H_6 CH_3OH CH_3CN CH_3NH_2 CH_3Cl CH_3Br CO_2 C_2H_2 CS_2	C_3H_8 n-C_4H_{10} n-C_7H_{16} n-$C_{14}H_{30}$ etc. C_2H_5Cl C_2H_5Br C_2H_5OH $C_2H_5NH_2$ CH_2Cl_2 CH_2Br_2 CHF_2Cl CHF_3 $(CH_3)_2NH$ CH_3I B_2H_6	CF_4 C_2F_6 CF_2Cl_2 CF_3Cl $CHFCl_2$	SF_6 iso-C_4H_{10} iso-C_5H_{12} iso-C_8H_{18} $CHCl_3$ $CHBr_3$ CHI_3 $(CH_3)_2CHOH$ $(CH_3)_2CHCl$ n-C_3F_8 n-C_4F_{10} n-C_7F_{16} B_5H_9	$(CH_3)_3N$ $(C_2H_5)_3N$ $C(CH_3)_4$ $C(CH_3)_3Cl$ $C(CH_3)_3Br$ $C(CH_3)_3OH$ CCl_4 CBr_4 $C_2F_2Cl_4$	C_6H_6 $C_6H_5CH_3$ $C_6H_4(CH_3)_2$ Cyclohexane Thiophen Furan Pyridine Dioxane $B_{10}H_{14}$	Naphthalene Quinoline, 6-decyl-1, 2, 3, 4-tetrahydro-naphthalene, 2-butyl-1-hexyl indan $C_6F_{11}CF_3$	1, 3, 5 triethyl benzene 1, 2, 3, 4, 5, 6, 7, 8, 13, 14, 15, 16-decahydro-chrysene	$(n-C_4F_9)_3N$
Size limit for Ca- and Ba-mordenites and levynite about here (~0·38 nm)	Size limit for Na-mordenite and Linde sieve 4A about here (~0·4 nm)	Size limit for Ca-rich chabazite, Linde sieve 5A, Ba-zeolite and gmelinite about here (~0·49 nm)				Size limit for Linde sieve 10X about here (~0·8 nm)	Size limit for Linde sieve 13X about here (~1·0 nm)		

Type 5

Type 4

Type 3

Type 2

Type 1

each type will admit. Molecular sieve type 3, for example, will admit CH_2Cl_2, will exclude $CHCl_3$, whilst $CHFCl_2$ is border-line.

The advantage of molecular sieves over traditional sorbents is that the sieves can be tailor-made for particular applications. In the petroleum industry, for example, sieves withstand conditions of excessive fouling because a size can be made which will admit the hydrocarbon reactants but exclude the polymers that cause fouling. The chief disadvantage of sieves is their relatively poor mechanical strength. Some uses and properties are given by COLLINS[92].

As well as aluminosilicates which start with cavities filled with water, there are others in which the cavities are filled with the positive and negative ions that make up a salt. In the dry state, the ions retain their average position in the framework, but when immersed in a polar liquid, water for example, one or both ions are free to move. It is often found that one ion is free to move through the solid whilst the other is held firmly to the framework. The mobile ion can move into surrounding liquid if the solid as a whole remains electrically neutral, that is if ions of equivalent charge pass from the liquid into the solid. The same requirement of electrical neutrality excludes, from the solid, ions of the same polarity as those held firmly by the framework. The net result of the process is to exchange mobile ions in the solid for ions of the same polarity in the liquid. This process is termed *ion exchange* and can be used to recover ions from solution or to replace those in solution by others with more suitable properties, as in water softening.

A disadvantage of the aluminosilicates is their instability in the presence of mineral acids. It was not possible to effect exchanges involving hydrogen ions until the discovery of ion exchange materials of organic origin. It was shown first that sulphonated coal[16], and later that a synthetic phenol-formaldehyde resin, could be used for cation exchange. At about the same time organic resins for anion exchange were synthesised by condensing formaldehyde and aromatic amines.

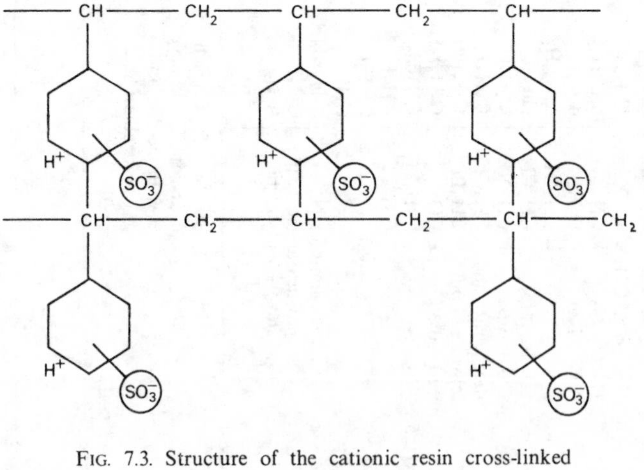

FIG. 7.3. Structure of the cationic resin cross-linked polystyrene sulphonic acid

Now cross-linked polymers are used as a basis for ion exchange resins, particularly the polystyrenes. The amount of cross-linking within the polymer must be sufficient to prevent the resin dissolving in the liquid, but it must not be so great as to prevent the resin swelling as a normal consequence of the exchange process. A certain amount of swelling will occur because, as well as the exchange of ions described above, equal numbers of positive and negative ions in the liquid can enter the solid without affecting the electroneutrality. The basic polymer framework can be made into a cation or anion exchange resin by suitable treatment. When it is sulphonated, polystyrene becomes the *cationic* exchange resin — polystyrene sulphonic acid (Fig. 7.3). An *anionic* exchange resin with an exchangeable chlorine ion can be made by treating the polystyrene with monochloroacetone and trimethylamine. Resins can be made which contain exchangeable cations and anions. These are termed *amphoteric*.

Ion exchange resins can be manufactured in sheet form, as membranes. The membranes retain the ability of the resins to admit ions of only one polarity. By siting alternate anionic and cationic membranes between electrodes to which an e.m.f. is applied, it is possible to remove both anions and cations from solution. This is the basis of the process known as *electrodialysis*.

In Fig. 7.4, the framework of the resin is assumed to have a fixed negative

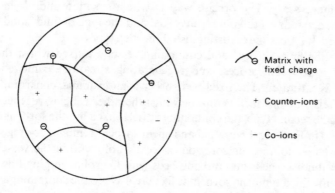

FIG. 7.4. A symbolic representation of a cationic resin and associated ions

charge. It is therefore a cationic resin. Mobile ions with the same polarity are found almost exclusively in the fluid outside the resin and are termed *co-ions*. Ions of opposite polarity are the exchangeable ions and are termed *counter-ions*. At equilibrium, some neutral solution will have entered the resin so it will contain some co-ions as well as counter-ions.

Five kinds of sorbents have been described — activated carbon, silica gel, activated alumina, molecular sieves and ion-exchange resins. Together they are characteristic of the majority of sorbents used on any scale in industry. In principle, any solid can be used, at least as a physical adsorbent. Only if it can be obtained in a porous form with a large internal surface will it be of practical interest for that purpose.

SORPTION EQUIPMENT

The scale and complexity of equipment making up a sorption unit varies from the 4 mm diameter chromatographic column used for laboratory analysis to a fluidised bed, some 12 m across, for recovering solvent vapours; from a simple container in which charcoal and a liquid to be clarified are mixed, to a highly automated moving bed with solids in plug flow, some 25 m high and 1·5 m in diameter.

All such units have this in common, that the sorbent becomes saturated as the process goes on. For continuous operation, the sorbent must be replaced and, since it is generally an expensive commodity, it must be regenerated, as far as possible being restored to its original condition.

In most systems, regeneration is carried out by heating the spent sorbent in a suitable atmosphere. The unit will, in that case, include equipment for heating and cooling. It is sometimes convenient to remove the sorbed phase without heat — by washing through with a suitable liquid as in ion-exchange, or by eluting with an inert gas as in chromatography—and in this case cooling too is unnecessary. The precise way in which sorption and regeneration are achieved depends on the phases involved and the type of fluid–solid contacting employed. It is possible to distinguish three such types.

(a) In the first, sorbent and containing vessel are fixed whilst the inlet and outlet positions for process and regenerating streams are moved when the sorbent is saturated. The fixed bed sorber is an example, consisting of at least two vessels, one of which is on-line whilst the other is being regenerated.

(b) In the second type, the containing vessel is fixed but the sorbent moves with respect to it. Fresh sorbent is fed and spent sorbent removed for regeneration at such a rate as to keep its sorption process wholly within the vessel. The type includes fluidised beds and moving beds with the solids in plug flow.

(c) In the third type, the sorbent is fixed relative to the containing vessel which moves relative to fixed inlet and outlet positions for process and regenerating fluid. The rotary bed adsorber is an example.

Fixed (or packed) bed

The sorption unit most widely used for industrial sorption processes consists of cylindrical vessels packed with graded pellets of sorbent, through which the fluid to be treated passes[82]. Figure 7.5 shows three large alumina driers. The common application of a fixed bed sorber is for drying gases. A typical unit consisting of two beds is shown diagrammatically in Fig. 7.6. While one sorber is on-line, a second is being regenerated; there is a heater to give hot regenerating gas and a cooler/condenser to cool the water-vapour laden gas that emerges from the bed being regenerated. The water that has condensed is removed from the system. If the sorbate cannot be condensed in

FIG. 7.5. Ethylene plant No. 3 at B.P. Chemicals, Grangemouth. Three alumina driers are in the centre foreground

the conditions of regeneration, it must be removed from the system in some other way. When there are no safety problems, it may be preferable to discharge the gas leaving the regenerating bed to atmosphere, dispensing with the cooler/condenser. It is normal for the regenerating gas to flow in the reverse direction

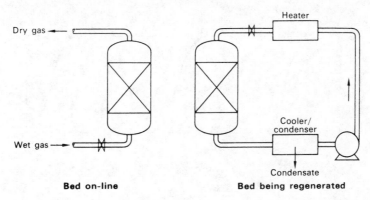

FIG. 7.6. A two-bed unit for drying gas. Regeneration is carried out
in a separate circuit using hot gas

so as to produce the maximum possible degree of regeneration at the end of the bed, which will be the exit for the drying run. In this unit it is assumed that all heat of regeneration is supplied by the hot air from the heater. It is possible to supplement this heat by electrical heaters or steam coils inserted in the bed, but the procedure is not as advantageous as it might at first appear because of the poor thermal conductivity of the sorbent. The on-line and regenerating sorbers are usually made the same size, though the latter could become a smaller stand-by unit if regeneration were carried out in a shorter time than that spent on-line. In the unit described, regeneration must include a period for cooling before replacing the unit on-line. When a fluid, containing a constant concentration C_0 of sorbate, enters a fixed bed, the concentration in the fluid falls to a low value in equilibrium with the fresh sorbent over a finite length of column. That length is termed the *sorption zone* or mass transfer zone, and is shown as Oa in Fig. 7.7a. As the sorbent at the plane OC_0 becomes saturated in equilibrium with C_0, so the *sorption wave*, which is C_0a at time t_1, moves to the position eb at t_2. After a time t_3, when the low concentration end of the sorption wave has reached the exit of the bed, it becomes necessary to take the bed off-line for regeneration. Operation beyond the time t_3 would result in the sorbate con-centration of the effluent rising rapidly until eventually it reached C_0. Figure 7.7b shows the sorbate concentration in the stream leaving the bed as a function of time. It is known as the *breakthrough curve* and the point t_b at which the concentration begins to rise rapidly is called the *breakpoint*. It can be specified as a time, or as the volume òf fluid that has entered the bed by that time.

Figure 7.7a shows the variation in fluid concentration along the column. The concentration of the sorbed phase follows a similar line, and it can be seen that some sorbent at the end of the bed is not saturated when the bed is taken off-line.

The shaded portion is the unsaturated fraction of the sorption zone. The extent
to which the sorbent is utilised depends on the shape of the sorption wave at the
breakpoint. A rough measure of the sorptive efficiency of the bed is its *capacity
at breakpoint*. This is the mass of sorbate retained by the bed at its breakpoint,

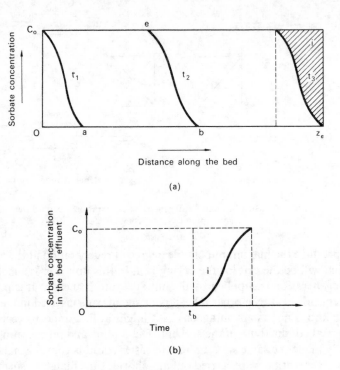

FIG. 7.7. The distribution of sorbate concentration in the
fluid phase through a bed

(a) development and progression of a sorption wave
through a length z_e
(b) a break-through curve

expressed as a fraction of the mass of sorbate-free sorbent. The length of bed
occupied by the sorption zone depends on the time and conditions of operation.
To be able to produce a sorbate-free effluent, a bed must be long enough to
contain the whole sorption zone. It must also be long enough to give the zone
a residence time at least equal to the required on-line time; 8, 12 or 24 hours is
normal in conventional designs, but periods may be as short as ten minutes in
others.

If, for reasons of safety or contamination of the process stream, air cannot be
used for regeneration, an inert gas such as nitrogen is a possible alternative.
Often, a more convenient method is to use the process flow itself for regenerating.
This leads to the integrated regeneration circuit shown in Fig. 7.8. Regenerating

flows in excess of the process flow can be obtained by including a blower in the regeneration circuit. During the cooling of the bed being regenerated, it may be necessary to by-pass the heater because of its thermal capacity.

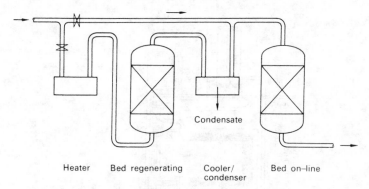

Heater   Bed regenerating   Cooler/   Bed on–line
                                  condenser

FIG. 7.8. A two-bed sorption unit with integrated regeneration
(not all pipework shown)

The criterion that has to be satisfied in any of the two-bed arrangements described is that the spent sorbent can be heated and cooled in the time that the stand-by bed is on-line. The regenerating time depends on the flow and the temperature of the regenerating stream. If the temperature is fixed at the optimum value recommended by the manufacturers, the only variable is the flow. The restriction that the integrated regeneration scheme places on the flow may result in the heat requirements not being met for a particular application. These limitations are discussed more fully on page 586.

Fixed beds are used to recover valuable solvents from gases and to remove toxic substances from effluents. These applications often involve large volumes of gases. In order to keep the pressure drop through the bed low, large cross-sectional areas are required. Since it is not easy to distribute gas evenly over a large area, it is necessary to use several beds if a conventional cylindrical design is adopted. More compact geometries have been developed to economise in flow area, such as that shown diagrammatically in Fig. 7.9. In this three-bed unit, using activated carbon as sorbent, two beds are on-line whilst the third is heated with live steam to recover the sorbate. It is assumed that the sorbate is a liquid immiscible with water, so that a separator can be used to recover pure product. A view of an actual plant of this design is shown in Fig. 7.10.

The advantage of a fixed bed is that it is simple and inexpensive to construct; it causes a minimum of attrition in the sorbent. There is some loss, however, partly attributable to slight movements of the bed when it is switched to the next phase of the cycle.

The fixed bed has a number of disadvantages which arise from that very simplicity which makes it attractive.

(a) Because it is discontinuous it has to be switched to regeneration at regular intervals. This can be done manually or automatically, the control system being

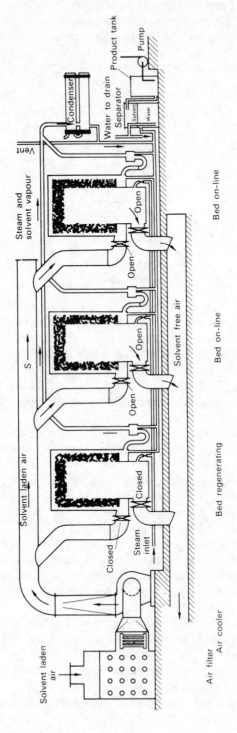

FIG. 7.9. A three-bed unit with annular sorbers

FIG. 7.10. A packaged sorber unit with a carbon bed of vertical annular design

timed or activated by the temperature changes that accompany sorption and desorption, or by the effluent concentration.

(b) In order to give the sorption zone a convenient residence time and to provide a stand-by bed, much more sorbent is tied up in the unit than is actually being used at any one time. This also leads to a higher pressure drop than is strictly necessary.

(c) Because of its poor thermal conductivity, a large volume of sorbent is difficult to heat and to cool rapidly. This has two effects; firstly it leads to longer regeneration cycles, and secondly it results in an increase in bed temperature during the sorption stage. Because it is difficult to remove the heat of sorption, the solid retains less sorbate for a given fluid concentration at the higher temperatures that result. It will be shown later (page 537 *et seq.*) that, for vapours below their critical points, it is a good approximation to say that the concentration of the sorbed phase depends on the relative saturation of the sorbate in the gas phase, rather than on its absolute concentration. An increase in bed temperature will decrease the relative saturation of the gas phase and the concentration of the sorbed phase.

(d) It is not usually possible to effect much heat economy, and therefore all the heat of regeneration is lost during each cycle. Though, in principle, the heat removed at the cooler could be used to preheat the heating stream, the problem is one of gas–gas heat exchange with poor thermal characteristics. If the process stream is at a high pressure, there is a problem of pressure loss because the pressure will be let down to atmospheric before regeneration is started. There is compensation in the latter case, however, since the pressure reduction alone will bring about some desorption.

Short cycle operation

If the disadvantages of conventional fixed bed operations are to be avoided, more complex equipment must be used. A relatively straightforward change is to decrease the time of the sorption cycle so that less sorbent is needed[76, 83]. One consequence of short cycle operation, with possibly no more than ten minutes allowed for sorption and the same for regeneration, is that switching to the next stage of the cycle must be automatic. If the sorber cannot be heated and cooled in the time available it will be necessary to use some other or supplementary method of regeneration. The sorbate can be drawn off using a vacuum, for example.

If a gas at high pressure is to be treated, short cycle operation is attractive. Desorption can be achieved by taking a fraction of the desorbed gas and passing it at atmospheric pressure through the bed to be regenerated. If the vapour pressure of the sorbed phase is P^0 at the temperature of the regenerating bed, and the regenerating air leaves in equilibrium with the sorbed phase, the rate at which water vapour leaves the bed will be $(P^0/\bar{P})\,V'$ where $\bar{P}$ is the mean total pressure and V' the flowrate of regenerating gas. As long as the air is in equilibrium with the sorbed phase, the flow of regenerating gas required to achieve a given rate of desorption is inversely proportional to its pressure. For an inlet pressure of 4 bar (4×10^5 N/m^2) about 25 per cent of the gas flow would, at atmospheric pressure, achieve the same regeneration as the total flow at the higher pressure. The technique has been described by SKARSTROM[73] in his account of a "heatless adsorption drier".

Ion exchange

Fixed beds are used in the liquid phase with ion exchange resins. The biggest single use is in the treatment of water — to soften it or for complete demineralisation. Natural water is likely to contain calcium, magnesium and sodium as cations; chloride, sulphate and bicarbonate as anions; with lower concentrations of nitrate, phosphate and silica. The water is termed hard if it contains more than 50 ppm of Ca^{++} plus Mg^{++}. If these ions exist in solution with bicarbonate ions, the hardness is referred to as temporary because boiling releases the carbonate which is precipitated as $CaCO_3$ and $MgCO_3$. The hardness is permanent if the anions in solution are chloride or sulphate.

When hard water is brought into contact with a sulphonic resin $H^+R_C^-$ (the subscript C denoting that the resin R is a cation exchanger) the ions responsible for hardness can be exchanged:

$$2H^+R_C^- + Ca^{++}Cl_2^- = Ca^{++}(R_C^-)_2 + 2H^+Cl^-$$

Though the water can be softened by this step, its mineral content has not been reduced because of the formation of the mineral acid. Further treatment with a basic resin $R_A^+OH^-$ is necessary for complete demineralisation:

$$R_A^+OH^- + H^+Cl^- = R_A^+Cl^- + H_2O$$

The total operation can be carried out in a mixed demineralising bed (Fig. 7.11). Both cationic and anionic exchangers are packed at random in the bed.

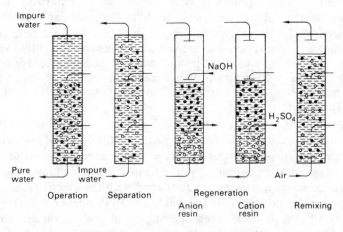

FIG 7.11. A mixed demineralising bed[24]

After either has become spent, so that the mineral concentration in the de-
mineralised water stream begins to increase, the bed is taken off-line and a
counter-current flow of untreated water used to separate the bed into an anion
layer (specific gravity 1·1) above a cation layer (specific gravity 1·4). Regeneration
can be carried out using a five per cent solution of caustic soda which is introduced at the top of the bed and removed at the bottom of the anion layer. After
rinsing to remove the greater part of the residual caustic, five per cent hydrochloric acid is introduced above the cation layer for its regeneration. When
regeneration is complete, the resins are remixed with a flow of air and the bed
is ready for re-use (Fig. 7.11). The treatment of water by ion exchange is discussed
by ARDEN[24]. A detailed discussion of all aspects of ion exchange is contained
in a book by HELFFERICH[16].

Chromatography

An example of fixed bed sorption in a different context is chromatography.
As an extremely effective analytical technique, it has inspired a considerable
volume of literature. A number of texts are available[11, 12, 13, 15]. It is also used
for measuring physical properties[93]. Only fluid–solid chromatography is a
truly adsorptive phenomenon but gas–liquid chromatography, which is absorp-
tive, is so similar that no distinction will be made in the following outline.
Columns for analytical chromatography are commonly about 4–6 mm in
diameter and are packed with 200–250 μm solid. When a non-adsorbed carrier
gas, containing a single component, is fed to a packed column of sorbent,
the concentration of sorbate falls to a low value in equilibrium with the fresh
sorbent (page 506 *et seq.*). If the gas contains two sorbates, each will have its own
sorption characteristics on the solid, one being more strongly held by it than the
other. The column becomes saturated with respect to each sorbate, the first to
emerge in the column effluent being the least strongly held. If the column is long
enough, the sorbates will emerge as separate steps in the breakthrough wave
(Fig. 7.12a). The first step corresponds to the pure lightly adsorbed component,
the second step corresponds to the emergence of the second component mixed
with the first at its inlet concentration. The process can be extended to any
number of components and the amount of each estimated if a detector is used
which responds to the total concentration in the effluent. This technique is
analogous to the fixed bed operations already described and is used for chromatographic analysis under the name of *frontal analysis*. A technique more widely
used is *elution analysis*. In this, a non-adsorbed carrier gas flows through a
packed column. The sample to be analysed is introduced into the carrier gas at
the inlet to the column. If the sample contains non-adsorbed components, these
will emerge from the column after an interval equal to the mean residence time
in the column voids and associated equipment. The sorbates in the sample are
taken up by the solid at the inlet to the column. They are then desorbed by the
carrier gas, the flow of which is continued after the sample has been introduced.
The sorbates gradually move through the column at characteristic rates, by
successive adsorption and desorption steps. The aim of the elution process is

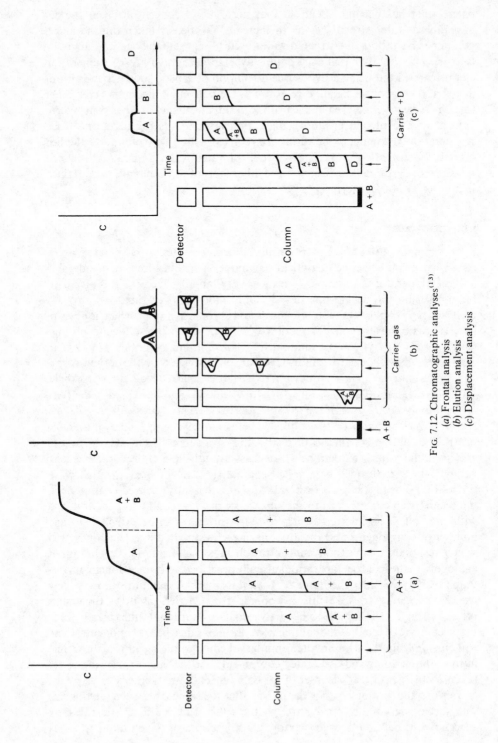

FIG. 7.12. Chromatographic analyses[13]
(a) Frontal analysis
(b) Elution analysis
(c) Displacement analysis

to *develop* the *chromatogram*; that is, to separate the components of the sample along the column so that each can be identified (Fig. 7.12*b*).

Chromatography was first developed by TSWETT[25] at the end of the last century for a liquid–solid system. His object was to isolate the components of pigments obtained from green leaves. A sample was placed at the top of a column packed with powdered calcium carbonate, and petroleum ether was poured through the column. It was found that the constituents of the sample formed bands of different colour down the column. The column could then be sectioned and each pigment analysed, or more ether added and each pigment collected as it emerged from the column.

The value of Tswett's technique as a method of analysis was not recognised until the 1930's. By 1941[39] a modification of the first procedure had been suggested, in which the active solid was replaced by an inert solid impregnated with a suitable liquid. What had been an adsorptive exchange in the earlier experiments became gas–liquid partition in the modified technique.

Gas–liquid chromatography is used more widely than gas–solid. It is easier to prepare liquids of given properties than it is solids; more is known about the equilibrium behaviour of gas–liquid systems. It can normally be assumed that the ratio of concentrations in the two phases, that is the distribution ratio or the partition coefficient, is constant.

Two modes of analysis have been referred to, frontal and elution. A third mode called *displacement analysis* combines aspects of both. A discrete sample is placed at the column inlet, as for elution analysis, and the sample is displaced by a fluid containing a more strongly sorbed component. Whereas in frontal analysis every component that emerges after the first is contaminated with those that have gone before, displacement analysis yields pure components. The sorbent in this case is difficult to regenerate. Figure 7.12*c* shows the breakthrough pattern for this mode of analysis. Elution analysis is the most common of the three; their relative advantages are discussed by PURNELL[15] and others.

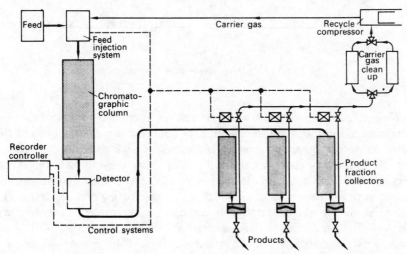

FIG. 7.13. Representation of a production scale chromatographic unit[95]

Production-scale chromatography

As well as being used for analysis, gas chromatography can be used for the purification of materials, and commercial equipment is now available for a scale of up to about 0·1 kg. In principle there is no difference between a conventional fixed bed sorber and a unit used for preparative gas chromatography. The aim in the latter unit will be to duplicate the ideal conditions that are possible in the laboratory column by careful design and choice of operating conditions. Since comparability of separation efficiencies on the laboratory and the preparative scale will entail a small particle size being used in both, pressure drops are likely to be large. Only separations difficult to achieve by other means are likely to be economic[95,96,97]. Pilot plant has been built for producing kilograms per day of material of very high purity.

Figure 7.13 shows a schematic arrangement of a production scale unit.

Chromatographic terms and definitions

The plot shown in Fig. 7.14 represents the elution analysis of a sample containing two sorbable components. Any non-sorbed components of the

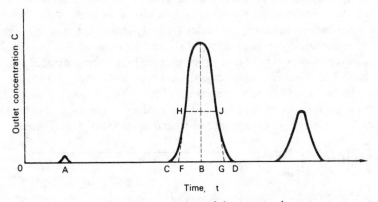

FIG. 7.14. The terminology of chromatography

sample emerge after a time OA which reflects the interstitial volume of the bed, the volume of the injection system and that of the detector. After a time OC the first of the sorbed components emerges to reach a peak concentration after a time OB. This time is characteristic of the component, the stationary phase and the conditions of operation of the column. Ideally, the peak will be symmetrical but may in practice show some *tailing*, that is, a lengthening of the distance BD. With mixtures for which components emerge in quick succession, tailing can obscure small peaks. CD is the *peak base* and FG the *peak width*. OB is the *retention time*, proportional to the *retention volume* of the eluent. The amount of component is proportional to the peak area.

As defined, the retention volume is characteristic not only of the component but also of the apparatus and the conditions of operation. The *net retention volume AB* has more fundamental significance, and even better is *the specific retention volume* which is the net retention volume, corrected for pressure and temperature, per unit mass of sorbent. The pressure correction reduces the volume from that corresponding to the mean column pressure to a value for the column inlet pressure. The temperature correction is from the column temperature to freezing point.

A suitable mean pressure can be defined using Poiseuille's law for the flow of viscous fluid:

$$u = -\frac{b'\,dP}{\mu\,dz}$$

where b' is a characteristic of the bed.

From the ideal gas law, for isothermal conditions:

$$u_0 P_0 = uP = \bar{u}\bar{P}$$

where subscript 0 denotes conditions at the column inlet and $\bar{u}$, $\bar{P}$ are mean conditions. Now:

$$dz = -\frac{b'P\,dP}{\mu P_0 u_0} \qquad\qquad \ldots\,(7.1)$$

Define the mean pressure $\bar{P}$ by:

$$\bar{P} = \int_0^{z_e} P\,dz \bigg/ \int_0^{z_e} dz \qquad\qquad \ldots\,(7.2)$$

where subscript e denotes conditions at the exit of the column. Substituting into equation 7.2 from 7.1 and integrating gives:

$$\bar{P} = \frac{2P_e}{3}\left[\frac{(P_0/P_e)^3 - 1}{(P_0/P_e)^2 - 1}\right] = \frac{P_e}{j}$$

The specific retention volume V_g can now be defined:

$$V_g = \frac{j(AB)T_0}{W_0 T}$$

where W_0 is the mass of sorbent, and AB is expressed as a volume.

Detectors

The detection of components emerging from a chromatographic column is no less important than their separation by the column. A standard work[11, 12, 13, 15] should be consulted for a fuller discussion of the characteristics and relative advantages of different detectors. Four of the most common kinds will be described briefly.

The *katharometer*, or thermal conductivity detector, compares the resistance of a heated wire exposed to the eluent gas with that of an identically heated wire in a stream of pure carrier gas. The resistances of the wires are equal until the temperature of the one exposed to eluent gas changes. The rate at which heat is conducted from the wire to the eluent gas depends on the physical properties of the gas and hence on its composition. The wires form two arms of a Wheatstone bridge, the degree of out-of-balance being a measure of the difference in gas compositions.

A *flame ionisation detector* measures the change in resistance of a flame when it contains an eluted component. A potential of several hundred volts is applied between a metal nozzle at which a hydrogen flame burns and a metal grid placed above the flame. The eluting gas may also be hydrogen or it may mix with the hydrogen before it is burned. The *flame detector* is similar. The grid is replaced by a thermocouple which measures changes in flame temperature as eluted components enter the flame. Both detectors require a separate supply of oxygen or air to burn the hydrogen.

The *gas density balance* exploits differences in density between carrier gas and eluent gas to create a gas flow which can be measured.

The katharometer and the flame ionisation detector are the most common. The latter will respond to concentrations as little as one part in 10^9. The other detectors have a sensitivity of about one part in 10^6.

Use of many small beds

It is often only a relatively small part of a fixed bed which is contributing to the sorption process at any one time. In the unit which is on-line, the inlet end of the bed makes no further contribution once it has become saturated. Similarly the inlet end of the unit being regenerated is desorbed long before desorption of the whole bed is complete. The advantage of using a moving bed, described

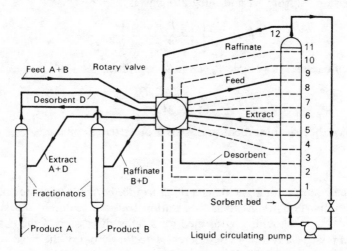

FIG. 7.15. A number of small beds used with a rotary valve to simulate a moving bed[90]

in the next section, is that the sorbent leaves the section in which it is sorbing or being regenerated as soon as its treatment is complete.

This advantage of the moving bed can be simulated in a fixed bed system if several small beds are used. The pipework connecting the beds to the process or the regenerating streams becomes more complicated but the use of multiway valves can bring about some streamlining both in appearance and operation. A system devised for removing n-hydrocarbons from others using molecular sieves has been described[90]. Sorption is effected from the liquid phase, and a desorbing liquid containing another n-hydrocarbon is used to displace the first by virtue of its higher concentration. Figure 7.15 shows a unit consisting of 12 small beds housed in one column and fed through a rotary valve. The feed consists of components **A** and **B**, the former being the more strongly adsorbed. Desorption is carried out with the most strongly adsorbed component **D**. The valve rotates at regular intervals and the positions of the inlet and outlet for the process stream and the regenerating stream move to each of the numbered positions in turn. The satisfactory operation of this kind of equipment depends on an efficient valve, on an even distribution of liquid across the column, and on the lines being flushed between their use for regeneration and for carrying the outlet streams.

Moving beds

In the system discussed in the preceding section, it was shown how a sorption process can be streamlined by increasing the number of beds. In the limit an infinite number of fluid connections and a rotary valve with a corresponding number of inlet and outlet parts would be required. A more feasible arrangement is to move the sorbent whilst keeping the containing vessels and the fluid connections stationary. Such a system can give truly continuous sorption as distinct from the quasi-continuous, unsteady state sorption described above. It also enables the inventory of sorbent, and the heat requirement in systems employing heat regeneration, to be kept to a minimum. On the other hand, moving bed equipment is more costly and results in greater attrition of the sorbent. Of the moving bed sorbers in service, many are of a type in which the solid is induced to move in plug flow down a column, and is transported from the bottom to the top of the column by fluid or mechanical means. In others, the sorbent and containing vessel move together relative to fixed inlet and outlet connections. In yet a third kind, the solid is fluidised.

Solids in consolidated plug flow

The essential features of the first kind of moving bed are shown in Fig. 7.16. It can be used for separating a multi-component feed into fractions, but will be described for a feed consisting of components **A** and **B**, with **A** the more strongly sorbed. In the top section of column the rising fluid meets sorbent containing a desorbing fluid **D**. At that stage of the process, the concentration of

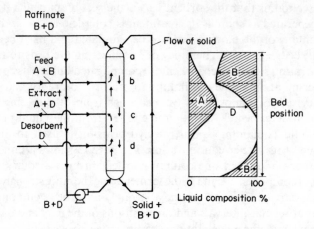

FIG. 7.16. Schematic arrangement of a moving bed[(90)]

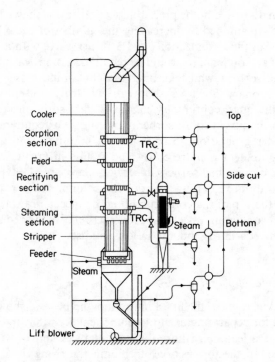

FIG. 7.17. The hypersorber

D in the fluid phase is small. Some **D** will desorb so that its concentrations in both phases approach an equilibrium condition. At the same time **A** will be sorbed so the stream leaving the top of the column will contain **B** and **D** with only traces of **A**. The length ab is the sorption section, since its purpose is to remove **A**. The section must clearly be long enough to contain the sorption zone under all permissible conditions of operation. When the sorbent leaves the sorption section, it is in contact with the feed and will contain component **A** and some component **B**, as well as a residual quantity of **D**. In the section bc, below the feed, the small amount of **B** can be recovered in the fluid phase by counter-current desorption in a stream containing **D** which comes from a point lower down the column. The sorbent entering the next section cd does so free of **B**. In cd which is a desorbing section fluid **D** displaces **A** as the sorbed phase, and a stream of **A** and **D** is recovered from the column. Some **D** with associated **A** continues up the column past the take-off point of the stream, to supply section bc with desorbing fluid.

For liquid–solid systems the desorbing fluid may well be a liquid. For gas–solid systems using hot regeneration, the desorbing stream can be a hot inert gas, possibly with supplementary heating of the desorption section; when activated carbon is the sorbent, **D** may be live steam and additional heat supplied to the solid through a heat exchanger.

One of the earliest applications of solids in plug flow was in 1946 when the *hypersorber* (Fig. 7.17) was developed for separating light hydrocarbons in refinery gas. Activated carbon was used as sorbent with steam for regeneration. The carbon was taken from the bottom to a hopper at the top of the column by an air lift pipe. The hot carbon flowed from the hopper through a cooler before being admitted to the sorption sections. A gas stream flowing counter-current to the carbon passing through the cooler removes any water retained by the carbon after it leaves the stripping section. Feed gas consisting of a mixture of hydrocarbons is fed near the centre of the sorption sections. Light components which are not sorbed are discharged just below the cooler. Inter-mediate and heavy components pass downwards as a sorbed phase. In the rectifying section, intermediates are liberated by a rising gas phase containing heavy components returned as a reflux to the column. Intermediates accumulate in the gas phase and can be removed from the top of the rectifying section as a side-cut. Some heavy components are liberated as the carbon continues down-wards through a short steaming section, and the remainder in the stripper proper. The heavy components that are not returned as reflux are removed as bottom product.

The rate at which carbon circulates is controlled by the rate at which it is allowed to enter the lift pipe. The normal conditions of regeneration in the stripping section do not maintain the high degree of carbon activity that is necessary for continuous operation. Consequently, a small proportion of the regenerated carbon is additionally steam treated in a small column to a temperature of about 870 K. A description of the *hypersorber* and its operation has been given by MANTELL[5].

A recent application on pilot plant scale is a German process for removing sulphur dioxide from flue gases. It is called the Reinluft process and uses activated carbon to remove sulphur dioxide[75, 87, 100].

Tests of the process at the Warren Spring Laboratory indicated that there was an appreciable loss of carbon through attrition and for chemical reasons. Treatment of flue gases from a large power station would require a bed of prohibitively large diameter or a large number of smaller beds (page 522).

There are two important difficulties to be overcome before this moving bed arrangement can be applied to a gas–solid system. The first is to reduce attrition of sorbent to an acceptable value; the second is to ensure a smooth flow of solids through the equipment.

For the liquid–solid contacting encountered in ion exchange, a method of moving solids discontinuously was developed by Higgins (see SETTER et al.[72]).

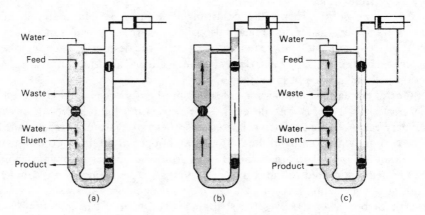

FIG. 7.18. Principle of the Higgins contactor

(a) Solution pumping (several minutes)
(b) Resin movement (3–5 seconds)
(c) Solution pumping (several minutes)

For about five seconds in a cycle time of several minutes, the exchange resin is moved by pulses applied to the liquid by a double-acting piston which simultaneously sucks liquid from the top of the column and delivers it to the bottom. The arrangement is represented in Fig. 7.18.

Moving vessel—rotary bed

If sorbent moves down a column, equipment has to be provided to raise it up to the top again. This can be an air-lift, a bucket elevator or, in liquid systems, the hydraulic ram can be used. The first two arrangements particularly will lead to an accumulation of fines in the system as a result of solid attrition. The problem can be avoided if both the sorbent and the containing vessel move together relative to fixed inlet and outlet connections (for the fluid to be processed and that used for regeneration).

Figure 7.19 shows the principle of operation of a rotary bed sorber, often used for solvent recovery from air. Solvent laden air is filtered and admitted to the drum housing. The drum contains activated carbon in a thick annular layer, divided into cells by radial partitions. Air can enter most of the drum circumference and passes through the carbon layer to emerge free of solvent.

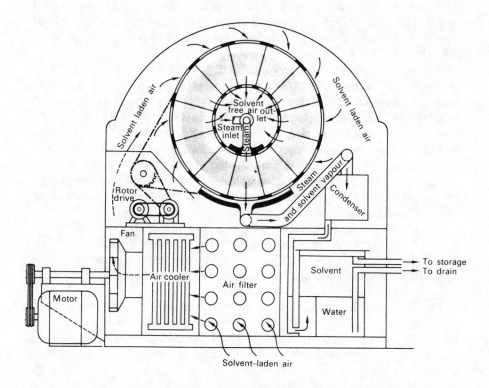

FIG. 7.19. Rotary bed sorber

Clean air leaves the equipment through a duct connected along the axis of rotation of the drum. As the drum rotates, the carbon enters a section in which it is exposed to steam. Steam flows from the inside of the annulus to the outside, so that the inner layer of carbon, which determines the solvent content of the effluent air, shall be regenerated as thoroughly as possible. The steam containing the solvent is led to condensers and the solvent recovered, by decantation if immiscible. There is no separate stage for cooling the carbon after steaming. Whilst it is cooling, the carbon will retain less solvent than when it has reached its working sorption temperature. The proportion of the carbon annulus which is cooling at any one time is small, so that its effect on the overall sorption efficiency is small. The drum should rotate at such a speed that carbon does not become saturated before it enters the regeneration section.

Fluidised bed

To maintain the lowest sorbate concentration in the effluent, a sorption operation should be carried out in equipment in which the solid is not allowed to mix. This ensures that the solid at the exit end of the bed contains the least amount of sorbate. There are other considerations, however, particularly the flowrate of fluid to be processed.

To keep the pressure drop down in fixed or moving bed equipment, cross-sectional areas have to be increased as the flow increases. When very large flow rates are met—500 m³/s of flue gas from a power station for example—one fixed bed is not a practical arrangement, and many beds in parallel would occupy a large ground area as well as requiring complicated pipework for gas distribution. In these circumstances, a fluidised bed adsorber may well be justified. Fluidisation is discussed at length in Volume 2, Chapter 6. Its advantage in the present context is that pressure drop through the bed of particles does not go on increasing once fluidisation has started and, consequently, much larger flows are possible for a given flow area.

A diagram of a plant designed to remove carbon disulphide in concentrations of over 1000 ppm v/v from ventilation air at a viscose factory is shown in Fig. 7.20[78]. Some 80 per cent of the CS_2 used in the process leaves in the ventilation air, so its recovery at a rate of some $1\frac{1}{2}$ t/h (0·417 kg/s) is an interesting economic and environmental proposition.

The CS_2 is sorbed on to activated carbon fluidised in a bed some 12 m in diameter, by an air flow of 120 m³/s. To treat the air in a conventional fixed bed would require a diameter of about 30 m. The sorber consists of five fluidised stages, one above the other. Regenerated carbon is fed to the top plate, flows across the plate in a fluidised state and overflows across a weir to the stage below. Ventilation air from the factory enters below the lowest plate and passes upwards, fluidising the solid on each plate. When it emerges from the top plate, the air contains about 100 ppm of CS_2 and is discharged to atmosphere.

Activated carbon laden with CS_2 overflows from the bottom plate through an inert-gas seal to the stripper–drier where it is met by a countercurrent flow of stripping steam. The steam and desorbed CS_2 passes to a series of condensers from which the CS_2 is recovered by decantation. Now contaminated only by steam, the carbon continues down the stripper, through a battery of heaters in which the steam is driven off the carbon. Dry carbon passes to a separate fluidised bed where it is cooled with air before being raised in an elevator to the top of the adsorber. In the sorption process some deactivation of carbon occurs due to the simultaneous adsorption of H_2S from the ventilation air. There is also deactivation due to some breakdown of CS_2, the products from which poison the carbon. To remedy this, a purge of carbon from the stripper

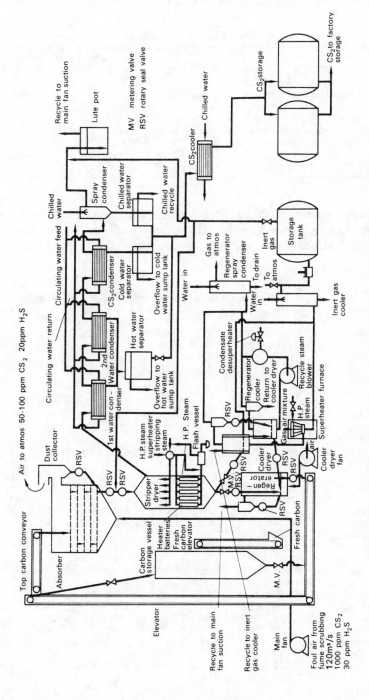

FIG. 7.20. Recovery of CS_2 by sorption on active carbon in a fluidised bed. The *Landmark process*

is regenerated in high temperature steam before being returned to the air cooler. Operating conditions for one such plant are given in Table 7.2.

TABLE 7.2. *Operating conditions on the Landmark CS$_2$ recovery plant*

Air flow	130 m³/s, 299 K, 100% R.H.
Inlet CS$_2$ concentration	1100–1200 ppm v/v
Outlet CS$_2$ concentration	100 ppm v/v
Inlet H$_2$S concentration	30 ppm v/v
Outlet H$_2$S concentration	10–15 ppm v/v
Carbon circulation rate	32 t/h (8·89 kg/s)
L.P. steam consumption	10 t/h (2·78 kg/s)
H.P. steam consumption	5·4 t/h (1·50 kg/s)
CS$_2$ recovery rate	1·5 t/h (0·42 kg/s)
Main fan power	820 kW

Essentially similar plant can be used for the recovery of many solvents, including alcohols, acetone, methyl ethyl ketone, isopropyl acetate, methylene chloride, trichloroethylene, perchloroethylene, ethyl acetate, butyl acetate and

TABLE 7.3. *Properties of some solvents recoverable by sorptive techniques*

	Mol wt	Specific heat capacity at 293K kJ/kg K	Density at 293 K kg/m³	Latent heat of evapora- tion kJ/kg	Boiling point K	Explosive limits in air at 293 K % v/v low	high	Solubility in water at 293 K kg/m³
acetone	58·08	2·211	791·1	524·6	329·2	2·15	13·0	∞
allyl alcohol	58·08	2·784	853·5	687	369·9	2·5	18	∞
n-amyl acetate	130·18	1·926	876	293	421·0	3·6	16·7	1·8
iso-amyl acetate	130·18	1·9209	876	289	415·1	3·6	—	2·5
n-amyl alcohol	88·15	2·981	817	504·9	410·9	1·2	—	68
iso-amyl alcohol	88·15	2·872	812	441·3	404·3	1·2	—	32
amyl chloride	106·6	—	883	—	381·3	—	—	Insol.
amylene	70·13	1·181	656	314·05	309·39	1·6	—	Insol.
benzene	78·11	1·720	880·9	394·8	353·1	1·4	4·7	0·8
n-butyl acetate	116·16	1·922	884	309·4	399·5	1·7	15	10
iso-butyl acetate	116·16	1·921	872	308·82	390·2	2·4	10·5	6·7
n-butyl alcohol	74·12	2·885	809·7	600·0	390·75	1·45	11·25	78
iso-butyl alcohol	74·12	2·784	805·7	578·6	381·8	1·68	—	85
carbon disulphide	76·13	1·005	1267	351·7	319·25	1·0	50	2
carbon tetrachloride	153·84	0·846	1580	194·7	349·75	Non-flam.		0·084
cellosolve	90·12	2·324	931·1	—	408·1	—	—	∞
cellosolve methyl	76·09	2·236	966·3	565	397·5	—	—	∞
cellosolve acetate	132·09	—	974·8	—	426·0	—	—	230
chloroform	119·39	0·942	1480	247	334·26	Non-flam.		8
cyclohexane	84·16	2·081	778·4	360	353·75	1·35	8·35	Insol.
cyclohexanol	100·16	1·746	960	452	433·65	—	—	60
cyclohexanone	98·14	1·813	947·8	—	429·7	3·2	9·0	50
cymene	134·21	1·666	861·2	283·07	449·7	—	—	Insol.
n-decane	142·28	2·177	730·1	252·0	446·3	0·7	—	Insol.
dichloroethylene	96·95	1·235	1291	305·68	333·0	9·7	12·8	Insol.

TABLE 7.3 (*cont.*)

	Mol wt	Specific heat capacity at 293K kJ/kg K	Density at 293 K kg/m³	Latent heat of evapora-tion kJ/kg	Boiling point K	Explosive limits in air at 293 K % v/v		Solubility in water at 293 K kg/m³
						low	high	
ether (diethyl)	74·12	2·252	713·5	360·4	307·6	1·85	36·5	69
ether (di-*n*-butyl)	130·22	—	769·4	288·1	415·4	—	—	3
ethyl acetate	88·10	2·001	902·0	366·89	350·15	2·25	11·0	79·4
ethyl alcohol	56·07	2·462	789·3	855·4	351·32	3·3	19·0	∞
ethyl bromide	108·98	0·812	1450	250·87	311·4	6·7	11·2	9·1
ethyl carbonate	118·13	1·926	975·2	306	399·0	—	—	V.sl.sol.
ethyl formate	74·08	2·135	923·6	406	327·3	2·7	16·5	100
ethyl nitrite	75·07	—	900	—	290·0	3·0	—	Insol.
ethylene dichloride	96·97	1·298	1255·0	323·6	356·7	6·2	15·6	8·7
ethylene oxide	44·05	—	882	580·12	283·5	3·0	80	∞
furfural	96·08	1·537	1161	450·12	434·8	2·1	—	83
n-heptane	100·2	2·123	683·8	318	371·4	1·0	6·0	0·052
n-hexane	86·17	2·223	659·4	343	341·7	1·25	6·9	0·14
methyl acetate	74·08	2·093	927·2	437·1	330·8	4·1	13·9	240
methyl alcohol	32·04	2·500	792	1100·3	337·7	6·72	36·5	67·2
methyl cyclohexanol	114·18	—	925	—	438·0	—	—	11
methylcyclohexanone	112·2	1·842	924	—	438·0	—	—	30
methyl cyclohexane	98·18	1·855	768	323	373·9	1·2	—	Insol.
methyl ethyl ketone	72·10	2·085	805·1	444	352·57	1·81	11·5	265
methylene chloride	84·93	1·089	1336	329·67	313·7	Non-flam.		20
monochlorobenzene	112·56	1·256	1107·4	324·9	404·8	—	—	0·49
naphthalene	128·16	1·683	1152	316·1	491·0	0·8	—	0·04
nonane	128·25	2·106	718	274·2	422·5	0·74	2·9	Insol.
octane	114·23	2·114	702	296·8	398·6	0·95	3·2	0·015
paraldehyde	132·16	1·825	904	104·75	397·0	1·3	—	120
n-pentane	72·15	2·261	626	352	309·15	1·3	8·0	0·36
iso-pentane	72·09	2·144	619	371·0	301·0	1·3	7·5	Insol.
pentachlorethane	202·33	0·900	1678	182·5	434·95	Non-flam.		0·5
perchlorethylene	165·85	0·879	1662·6	209·8	393·8	Non-flam.		0·4
n-propyl acetate	102·3	1·968	888·4	336·07	374·6	2·0	8·0	18·9
iso-propyl acetate	102·3	2·181	880	332·4	361·8	2·0	8·0	29
n-propyl alcohol	60·09	2·453	803·6	682	370·19	2·15	13·5	∞
iso-propyl alcohol	60·09	2·357	786·3	667·4	355·4	2·02	—	∞
propylene dichloride	112·99	1·398	1159·3	302·3	369·8	3·4	14·5	2·7
pyridine	79·10	1·637	978	449·49	388·3	1·8	12·4	∞
tetrachloroethane	167·86	1·130	1593	231·5	419·3	Non-flam.		3·2
tetrahydrofuran	72·10	1·964	888	410·7	339·0	1·84	11·8	∞
toluene	92·13	1·641	871	360	383·6	1·3	7·0	0·47
trichloroethylene	131·4	0·934	1465·5	239·9	359·7	Non-flam.		1·0
xylene	106·16	1·721	897	347·1	415·7	1·0	6·0	Insol.
water	18	4·183	998	2260·9	373·0	Non-flam.		∞

heptane. Because it is run in a steady state condition, the process is readily adaptable for automatic control. If deactivation of the carbon does not occur, the equipment can be simplified by the omission of the high temperature reactivation step. Properties of a wide variety of solvents which, in principle, can be recovered by the sorptive techniques described are given in Table 7.3.

EQUILIBRIUM PROPERTIES

The design of the equipment described in the previous sections depends on a knowledge of both the quantity of sorbate which can be taken up by the solid and the rate of the process. The first of these requirements is determined by the equilibrium that is set up between a fluid and a sorbent. There is a considerable body of literature devoted to this particular topic which will be examined in this section. The physical chemistry of the matter is complex, and no single theory of sorption has been put forward which satisfactorily predicts all conditions. Fortunately, for the engineer, what is needed is an accurate representation of the equilibrium, the molecular significance of which need not concern him overmuch. For that reason, some of the earliest theories of sorption are still the most useful, even though the assumptions on which they were based may now be seen to have limited validity.

When a fluid containing a sorbate, or composed wholly of sorbate, is in contact with a solid, molecules of sorbate transfer from the fluid to the surface of the solid until the concentration (pressure in the case of a gas) exerted by the sorbed phase is equal to that of the component in the fluid. A dynamic equilibrium is set up in which sorption and desorption rates are equal. The equilibrium is a function of temperature. Three quantities are involved, C (or P) the concentration (or pressure) in the fluid, C_s the concentration on the solid and T the temperature. The equilibrium may be represented graphically by keeping C constant (sorption isobars), by keeping C_s constant (sorption isosteres), or by keeping T constant (sorption isotherms). The last mentioned are by far the most common; Fig. 7.21 shows examples of all three. The concentration of the sorbed phase, usually expressed in kg (or m^3 of equivalent vapour) of sorbate per kg of sorbent, is not a true solution concentration since sorption is a surface rather than a bulk phenomenon, but the distribution of surface area is sufficiently intimate for one kilogram of sorbent to be representative of any other.

Only sorption isotherms will be discussed in any detail, the treatment being divided as shown in Fig. 7.22.

Gas–solid isotherms; single sorbate

Though the earliest applications of adsorption involved liquids and solids, most of the theories of adsorption have been developed in respect of gas–solid systems largely because, experimentally, such systems are simpler to analyse.

The sorption of pure vapours is of interest mainly in connection with the laboratory measurement of isotherms, from which sorbent or sorbate properties may be calculated. Sorption of single components from a mixture with inert gases is the most common industrial application, though the separation of multicomponent mixtures is being carried out on an increasing scale. The presence of inerts in the gas phase does not affect the shape of the isotherm which depends on the partial pressure of the sorbate.

Only sorbents such as molecular sieves which derive their pores from precise molecular vacancies in a crystal lattice can be manufactured with closely

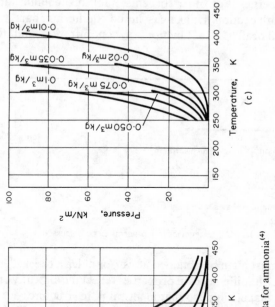

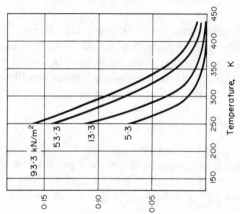

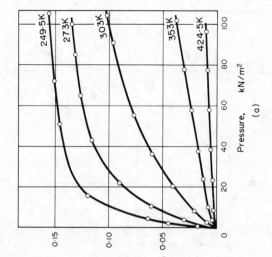

Fig. 7.21. Equilibrium sorption data for ammonia[4] on charcoal

(a) Sorption isotherms
(b) Sorption isobars
(c) Sorption isosteres

reproducible properties. For other sorbents, published equilibrium properties should be used with caution. Unlike gas–liquid equilibrium data, which can be tabulated unequivocally as a function of concentration and temperature,

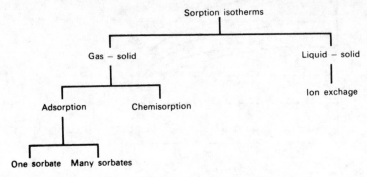

FIG. 7.22. Categories of isotherms to be discussed

fluid–solid data must often be obtained afresh, or at least checked, for each new batch of material. The differences in properties reflect the difficulty of controlling the conditions of manufacture so that identical pore structures are obtained in all batches. Experimental methods of obtaining sorption equilibrium data are well documented[17, 20, 23] and will not be treated further.

Theories of equilibrium

In 1940[37] a classification of known isotherms was proposed which consisted of the five shapes shown in Fig. 7.23. Only gas–solid systems provide examples of all the shapes, and not all occur frequently. It is not possible to predict the

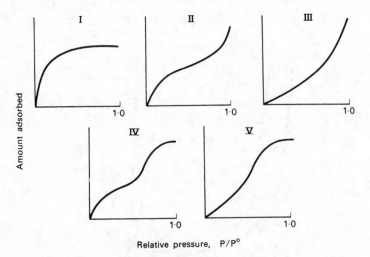

FIG. 7.23. Classification of isotherms into five types by BRUNAUER, DEMING, DEMING and TELLER[37]

shape of an isotherm for a given system, but it has been observed that some shapes are often associated with particular sorbent or sorbate properties. Charcoal, with pores only a few molecules in diameter, almost always gives a Type I isotherm. A non-porous solid is likely to give Type II. If the cohesive forces between sorbate molecules are greater than the adhesive forces between sorbate and sorbent, a Type V isotherm is likely to be obtained for a porous sorbent and a Type III for a non-porous one.

Some isotherms do not fit the above classification. If a solid surface is energetically very uniform the sorbed phase will tend to build up in well defined layers. This leads to a stepped isotherm, each step corresponding to another layer. It is possible that stepped isotherms can also be caused by phase transformation within a sorbed layer[17].

Useful as the classification of isotherms by shape is, it is usually more convenient in numerical work to have the equilibrium data expressed as an equation. If an equation can be found which describes not only the shape of the isotherm, but also the mechanisms of sorption, it may be possible to learn something about the structure of the solid.

Sorption sometimes occurs in three stages. These are illustrated in the Type IV isotherm found, for example, in the activated alumina–air–water vapour system at normal temperature. It consists of two regions which are concave to the gas concentration axis, separated by a region which is convex. The concave region that occurs at low gas concentrations is usually associated with the formation of a single layer of sorbate molecules over the surface. The convex portion coincides with the build-up of additional layers, whilst the other concave region is the result of bulk condensation of sorbate in the pores—capillary condensation. The behaviour of the sorbed molecules is likely to be different in each of the three regions. At low gas concentrations whilst the monolayer is still incomplete, the sorbed molecules will be relatively immobile. In the multilayer region the adsorbed molecules will behave more like a liquid film[7]. In the region of capillary condensation, the amount sorbed will depend on the pore sizes and their distribution, as well as the concentration in the gas phase. There is no single isotherm equation that satisfactorily describes all mechanisms and all shapes, so the equation used will be determined by the region or the range of conditions of interest.

Immobile layers

If adsorption is confined to a monolayer and the adhesive forces are considerably greater than the cohesive, or if chemisorption occurs, it is reasonable to assume that sorbed molecules will not move freely over the surface.

In 1938 BRUNAUER, EMMETT and TELLER[35, 38] developed what is now known as the BET theory, based on the concept of a sorbed molecule which was not free to move over the surface and which exerted no lateral forces on molecules sorbed on adjacent sites. When a surface is in equilibrium with a gas phase at a particular concentration, the number of molecules on each part of the surface will not, in general, be the same, but the net amount of surface associated with

S

a monomolecular layer, with a bimolecular layer, and so on, will be constant. Figure 7.24 shows a typical piece of BET surface. Consider, for example, the constancy of monolayer surface for a particular equilibrium condition. The rate at which monolayers disappear is equal to the rate at which the layers are created. Monolayers disappear as a result of further adsorption on to such layers and of desorption from them. Monolayers are created by adsorption

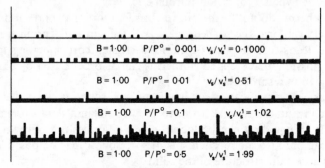

FIG. 7.24. Profiles of typical BET surfaces for different
values of parameters[17]

on to empty surface and by desorption from bilayers. The rates of adsorption will be proportional to the amount of area available for the adsorption and the frequency with which molecules strike that surface. At constant temperature, on the basis of the kinetic theory of gases, the latter is proportional to the pressure of molecules in the gas phase.

Rate of adsorption on empty surface $= k_0 a_0 \, P$

Rate of adsorption on monolayers $= k_1 a_1 \, P$

where k_0, k_1 are adsorption velocity constants for the empty surface and the monolayer area, and

a_0, a_1 are the fractions of empty surface and of surface covered by a monolayer.

Desorption is an activated process. If E_1 is the excess energy one mol in the monolayer requires to overcome the surface forces, the proportion of molecules possessing such energy is $e^{-E_1/RT}$.

Rate of desorption from a monolayer $= k_1' \, e^{-E_1/RT}$

Rate of desorption from a bilayer $= k_2' \, e^{-E_2/RT}$

where k_1', k_2' are the desorption velocity constants.

The dynamic equilibrium of the monolayer is given by:

$$k_0 a_0 P + k_2' a_2 \, e^{-E_2/RT} = k_1 a_1 P + k_1' a_1 \, e^{-E_1/RT} \qquad \ldots\ldots (7.3)$$

A similar argument can be applied to empty surface:

$$k_0 a_0 P = k_1' a_1 \, e^{-E_1/RT} \qquad \ldots\ldots (7.4)$$

Equations 7.3 and 7.4 give:

$$k_2' a_2 \, e^{-E_2/RT} = k_1 a_1 P \qquad \ldots\ldots (7.5)$$

The argument is extended to n layers. If it is assumed that each energy of activation, after the first layer, equals the latent heat of condensation:

$$E_2 = E_3 = E_4 \ldots . E_n = \lambda$$

From equation 7.4:

$$a_1 = \frac{k_0}{k_1'} e^{E_1/RT} P a_0 = \alpha a_0 \qquad \ldots . (7.6)$$

From equation 7.5:

$$a_2 = \frac{k_1}{k_2'} e^{\lambda/RT} P a_1 = \beta a_1 \qquad \ldots . (7.7)$$

β is assumed to be constant for all subsequent layers. Then:

$$a_i = \beta^{i-1} a_1 = B \beta^i a_0$$

where $B = \alpha/\beta$, and
$\quad a_i$ is the fraction of the area on which there are i layers of adsorbate.

As the a's are fractional areas, then summation over n layers gives:

$$1 = a_0 + \sum_{i=1}^{n} a_i$$

$$= a_0 + \sum_{i=1}^{n} B \beta^i a_0$$

The total volume of adsorbate associated with unit area of surface is given by:

$$v_s = v_s^1 \sum_{i=1}^{n} i\, a_i = v_s^1 \sum_{i=1}^{n} i B \beta^i a_0$$

where v_s^1 is the volume of adsorbate in a unit area of each layer.

It is implied in the last equation that unit area in the first layer corresponds to unit area in subsequent layers. This will not be true for highly convex or concave surfaces. From the two preceding equations:

$$\frac{v_s}{v_s^1} = \frac{\displaystyle\sum_{i=1}^{n} iB\beta^i a_0}{a_0 + \displaystyle\sum_{i=1}^{n} B\beta^i a_0} \qquad \ldots . (7.8)$$

The numerator of the ratio can be written as:

$$Ba_0\beta \frac{d}{d\beta} \left(\sum_{i=1}^{n} \beta^i \right)$$

$$= Ba_0\beta \frac{d}{d\beta} \left\{ \left(\frac{1 - \beta^n}{1 - \beta} \right) \beta \right\}$$

The denominator of equation 7.8 can be written as:

$$a_0 \left(1 + B\beta \left(\frac{1 - \beta^n}{1 - \beta}\right)\right)$$

By substituting these values and rearranging the ratio becomes:

$$\frac{v_s}{v_s^1} = \frac{B\beta}{1 - \beta} \frac{[1 - (n + 1)\beta^n + n\beta^{n+1}]}{[1 + (B - 1)\beta - B\beta^{n+1}]} \qquad \ldots(7.9)$$

where:

$$B = \frac{k_0 \, k_2'}{k_1' \, k_1} e^{(E_1 - \lambda)/RT}$$

On a flat, unrestricted surface there is no theoretical limit to the number of layers that can build up. When $n = \infty$ equation 7.9 becomes:

$$\frac{v_s}{v_s^1} = \frac{B\beta}{(1 - \beta)(1 - \beta + B\beta)} \qquad \ldots(7.10)$$

This is known as the infinite form of the BET equation. It enables the physical meaning of β to be established for, when the pressure of the sorbate in the gas phase has increased to the saturated vapour pressure, condensation occurs on the solid surface and v_s/v_s^1 approaches infinity. This condition can be met by equation 7.10 if $\beta = 1$ (the condition that $\beta = 1/(1 - B)$ is not helpful). From the definition of β in equation 7.7:

$$1 = \frac{\cdot \, k_1}{k_2'} e^{\lambda/RT} P^0$$

where P^0 is the saturated vapour pressure. Then $\beta = P/P^0$. Applied not to unit surface but to the total mass of adsorbent, equation 7.9 becomes:

$$\frac{V_s}{V_s^1} = B \frac{P}{P^0} \frac{[1 - (n + 1)(P/P^0)^n + n(P/P^0)^{n+1}]}{(1 - P/P^0)[1 + (B - 1)(P/P^0) - B(P/P^0)^{n+1}]} \qquad \ldots(7.11)$$

V_s^1 is the volume of adsorbate contained in a complete monolayer.

Equation 7.11 is the form of the BET equation limited to n layers of adsorbate. When sorption is confined to one layer, in chemisorption for example, $n = 1$ and equation 7.11 becomes:

$$\frac{V_s}{V_s^1} = \frac{B(P/P^0)}{1 + B(P/P^0)} \qquad \ldots(7.12)$$

Equation 7.12 is the Langmuir isotherm[29] proposed many years earlier. The condition that $n = 2$ and $B = 4$ also reduces to the Langmuir form, but no special significance can be attached to it under those conditions.

Equation 7.11 can be rearranged in a linear form:

$$\frac{\phi_1}{V_s} = \frac{1}{V_s^1 B} + \frac{\theta_1}{V_s^1} \qquad \ldots(7.13)$$

where:

$$\phi_1 = \frac{P/P^0[1 - (P/P^0)^n - n(P/P^0)^n(1 - P/P^0)]}{(1 - P/P^0)^2}$$

and

$$\theta_1 = \frac{P}{P^0}\frac{[1 - (P/P^0)^n]}{(1 - P/P^0)^n}$$

The conformity of experimental data to equation 7.11 can be tested by plotting ϕ_1/V_s against θ_1. The quantities V_s^1 and B can be found from the slope and the intercept. Tabulated values of ϕ_1 and θ_1 for different values of P/P^0 are available. At low relative pressures, when the restrictions on the number of adsorbed layers that can form are not apparent, the infinite form of the BET equation can be used to find V_s^1, or V^1 the corresponding change in gas phase volume, and B:

$$\frac{P}{V(P^0 - P)} = \frac{1}{V^1B} + \frac{B-1}{V^1B}\left(\frac{P}{P^0}\right) \qquad \ldots (7.14)$$

The sorbed volumes are conveniently expressed as changes in gas phase volume V.

Equation 7.11 has not been bettered for correlating isothermal data. When n is 3, or somewhat greater, the equation can represent all five types by using an appropriate value of B. In its simpler, infinite form equation 7.14 can describe Type II and the rarer Type III isotherm if a suitable value of B is used. As B increases, the point of inflexion or "knee" of the Type II isotherm becomes more prominent. It corresponds to an increasing tendency for the monolayer to become complete before the second layer starts forming. When B is less than 2 there is no point of inflexion and Type III isotherms are obtained. The condition $1 > B > 0$ corresponds physically to a tendency for molecules to adsorb in clusters rather than in complete layers. The monolayer form, equation 7.12, can represent Type I isotherms.

Limitations of the BET approach. The success of equations 7.11, 7.12 and 7.14 in representing experimental data is not an indication of the accuracy of the model from which the equations were derived. The main deficiencies of the model as a representation of mechanism are (a) that it does not allow for interactions between adsorbed molecules, (b) that it assumes immobile adsorption on an energetically uniform surface, and (c) that enthalpy changes for layers after the first will not, as assumed, be equal to the enthalpy of liquefaction. The coordination number of a molecule in the second layer, for example, will not be the same as that of a molecule in bulk liquid.

Various modifications have been suggested to remove some of the grosser simplifications of the model[56]. HILL[48,60] and others have incorporated terms for sorbate interactions. The effects of varying the heat of adsorption in the second and higher layers[51] and the consequence of a non-uniform surface[20, 79] have also been discussed. The matter is reviewed in some detail by YOUNG and CROWELL[7].

Mobile layers

In the middle range of fluid concentrations, sorbed molecules are unlikely to be fixed to a site on the solid surface. There is indeed experimental evidence to suggest that sorbate moves over some surfaces even at low concentrations[48]. If the sorbed layers are regarded as a mobile phase which is distinct from both gas and solid, the procedures of classical thermodynamics can be applied.

The state properties of the adsorbed phase are functions of temperature, pressure, number of molecules n_s adsorbed and the area A_s available to the adsorbed film. The Gibbs free energy G of the film can be expressed:

$$G = F(P, T, n_s, A_s)$$

A differential change in G can be expressed in terms of the partial derivatives with respect to each of the independent variables[22]:

$$dG = \left(\frac{\partial G}{\partial P}\right)_{T, n_s, A_s} dP + \left(\frac{\partial G}{\partial T}\right)_{P, n_s, A_s} dT + \left(\frac{\partial G}{\partial n_s}\right)_{T, P, A_s} dn_s + \left(\frac{\partial G}{\partial A_s}\right)_{T, P, n_s} dA_s$$

At constant temperature and pressure this becomes:

$$dG = \left(\frac{\partial G}{\partial n_s}\right) dn_s + \left(\frac{\partial G}{\partial A_s}\right) A_s$$

$$= \mu_s dn_s - \Gamma dA_s \qquad \qquad \dots (7.15)$$

where μ_s is the free energy per mol or the chemical potential of the film. Γ is defined as a two-dimensional or spreading pressure. The total Gibbs free energy can be written:

$$G = \mu_s n_s - \Gamma A_s$$

so that:

$$dG = \mu_s dn_s + n_s d\mu_s - \Gamma dA_s - A_s d\Gamma \qquad \qquad \dots (7.16)$$

A comparison of equations 7.15 and 7.16 shows:

$$d\Gamma = \frac{n_s}{A_s} d\mu_s \qquad \qquad \dots (7.17)$$

If the gas phase is perfect and equilibrium exists between it and the sorbed phase then, by definition:

$$d\mu_s = d\mu_g = \mathbf{R}T d(\ln P)$$

where μ_g is the chemical potential of the gas.

Substituting for $d\mu_s$ in equation 7.17:

$$d\Gamma = \frac{n_s}{A_s}\mathbf{R}T d(\ln P) \qquad\qquad \dots(7.18)$$

Equation 7.18 has the makings of an isotherm since it relates the amount sorbed and the pressure, but an expression is needed for Γ. Assuming an analogy between adsorbed and liquid films, HARKINS and JURA[42] proposed that:

$$\Gamma = \alpha_1 - \beta_1 a_m$$

where α_1 and β_1 are constants and $a_m = A_s/\mathbf{N}n_s$, the area per molecule of sorbate. $\mathbf{N}$ is the Avogadro number. Substituting for $d\Gamma$ in equation 7.18:

$$d\Gamma = -\beta_1 da_m = A_s\frac{\beta_1}{\mathbf{N}}\frac{dn_s}{n_s^2}$$

$$= \frac{n_s}{A_s}\mathbf{R}T d(\ln P) \qquad\qquad \dots(7.19)$$

If the area A_s available to the film is constant, equation 7.19 can be integrated simply from some condition P_1, n_{s_1} at which the sorbed film becomes mobile to an arbitrary coverage n_s at pressure P, giving:

$$\ln P - \ln P_1 = \frac{1}{\mathbf{R}T}\frac{\beta_1 A_s^2}{2\mathbf{N}}\left[\frac{1}{n_{s_1}^2} - \frac{1}{n_s^2}\right] \qquad\qquad \dots(7.20)$$

Equation 7.20 can be redrafted:

$$\ln\frac{P}{P^0} = L - \frac{M}{V^2} \qquad\qquad \dots(7.21)$$

where V is the volume occupied in the gas phase by n_s mols of sorbate at a temperature T and pressure P,

$$L = \ln\frac{P_1}{P^0} + \frac{1}{\mathbf{R}T}\frac{\beta_1 A_s^2}{2\mathbf{N}}\frac{1}{n_{s_1}^2}$$

$$M = \frac{1}{\mathbf{R}T}\frac{\beta_1 A_s^2}{2\mathbf{N}}\frac{1}{C_M^2}$$

where C_M is the molar density of the gas phase. Equation 7.21 is the Harkins–Jura (H–J) equation. It can be used to correlate sorption data and to obtain an estimate of the surface area of a sorbent (page 550).

The H–J and BET equations are found to be compatible only over certain ranges of relative pressure, indicating the limitations of one or both models. The H–J model can be criticised because the comparison between a sorbed film on a solid surface and a liquid film on a liquid surface will not stand detailed analysis. The film approach to the sorbed layer makes no allowances for the energetic heterogeneity of the solid. The matter is reviewed in some detail by YOUNG and CROWELL[17].

A comprehensive method of allowing for variations in energy over a solid surface has been described by Ross and Olivier[20]. The distribution of energy is assumed to be Gaussian and the sorbed film to be composed of small areas of homogeneity, each having its own potential energy. An equation of state relating pressure in the sorbed film to the fraction a of surface covered, obtained by Hill[45, 48, 51] and based on a two-dimensional van der Waals equation, was applied to each homogeneous area:

$$P = K' \frac{a}{1-a} \exp \left[\frac{a}{1-a} - \frac{2\alpha_2 a}{k_N T \beta_2} \right] \qquad \dots (7.22)$$

where K' is a constant depending on the energy of the area,

   k_N is the Boltzmann constant, and

   α_2, β_2 are two-dimensional van der Waals constants.

Corresponding values of a and P/K' have been tabulated as a function of a parameter:

$$\frac{2\alpha_2}{\mathbf{R} T \beta_2}$$

and standard deviations of the distribution. A comparison of tabulated and experimental results enables properties of sorbate and sorbent to be estimated.

Capillary condensation

In that region of the Type IV or Type V isotherm which is concave to the gas concentration axis at high concentrations, bulk condensation of sorbate occurs in the capillaries. An equation relating the volume condensed to the sorbate pressure in the gas phase can be found by assuming transfer of sorbate to occur in three stages:

(a) transfer from the gas to a point above the meniscus in a capillary,

(b) condensation on the meniscus, and

(c) transfer from a plane surface source of liquid sorbate to the gas phase in order to maintain stage (a). At equilibrium, the free energy changes associated with steps (b) and (c) are zero.

For step (a) at constant temperature T:

$$\Delta G = \int_{P^0}^{P_m} V \, dP = n_s \mathbf{R} T \ln \frac{P_m}{P^0}$$

where P^0 is the vapour pressure over the plane surface, and

   P_m is the vapour pressure over the meniscus.

Another expression for ΔG can be obtained from a consideration of interfacial changes that occur as the capillaries fill[1, 10]:

$$\Delta G = \Delta A_p (\sigma_{sl} - \sigma_{sv}) = - \Delta A_p \sigma_{lv} \cos \phi$$

where σ's   are the surface tensions at the three interfaces of the solid (s), liquid (l) and vapour (v),

   ϕ is the interfacial angle between liquid and solid, and

   ΔA_p is the change in interfacial area.

Then:

$$n_s \mathbf{R} T \ln \frac{P_m}{P^0} = - \Delta A_p \sigma_{lv} \cos \phi$$

If the transfer of $\mathrm{d}n_s$ mols results in a change in interfacial area $\mathrm{d}A_p$:

$$\frac{\mathrm{d}n_s}{\mathrm{d}A_p} = - \frac{\sigma_{lv} \cos \phi}{\mathbf{R}T \ln P_m/P^0}$$

$$= \frac{1}{V_M} \frac{\mathrm{d}V}{\mathrm{d}A_p}$$

where $\mathrm{d}V$ is the volume occupied by $\mathrm{d}n_s$ mols in the vapour phase, and
$\quad V_M$ is the molar volume,

$$\frac{\mathrm{d}V}{\mathrm{d}A_p} = \frac{V_M \sigma_{lv} \cos \phi}{\mathbf{R}T \ln P^0/P_m} \qquad \dots (7.23)$$

Equation 7.23 relates the pressure of adsorbate in the gas phase to

$$\frac{\mathrm{d}V}{\mathrm{d}A_p}$$

a characteristic dimension of the capillary in which condensation occurs. Such condensation will normally occur on top of existing adsorbed layers, so equation 7.23 is not a total isotherm. The equation can be used for estimating the sizes and the size distribution of pores (page 552) if an allowance is made for the thickness of layers already adsorbed.

Temperature and equilibrium

The effect of temperature is important because sorption in gas–solid systems on an industrial scale is rarely isothermal. The isotherm equations already referred to have included a temperature dependency, but a particularly convenient equation can be derived from the potential theory of adsorption[17]. It is based on a model proposed by Polanyi in 1914 who considered the adsorbed phase to be contained in adsorption space (Fig. 7.25). The potential of a point in this space is a measure of the work done by the surface forces in bringing a mol of adsorbate to that point from infinity, or a point at such a distance from the surface that those forces exert no attraction. The work depends on the phases involved. Polanyi considered three possibilities—that the temperature of the system was well below the critical temperature of the adsorbate, and the adsorbed phase could be regarded as liquid; that the temperature was just below the critical temperature, and the adsorbed phase was a mixture of vapour and liquid; that the temperature was above the critical temperature, and the adsorbed phase was a gas. Only the first possibility, the simplest and most common, will be considered.

At the limit of adsorption space the surface forces exert no attraction and the pressure is P. For conditions below the critical point of the adsorbate, the pressure at the adsorbent surface is the saturated vapour pressure of the liquid adsorbate (P^0). For an ideal gas, the work done bringing a mol from the limit of

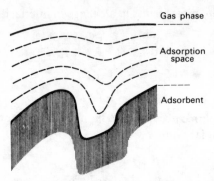

FIG. 7.25. The concept of "adsorption space" used by POLANYI[4]

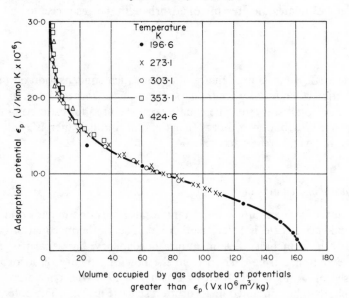

FIG. 7.26. Characteristic curve for the sorption of carbon dioxide on charcoal[4]

adsorption space to the adsorbed phase is given by:

$$\varepsilon_p = RT \ln \frac{P^0}{P} \qquad \dots (7.24)$$

The potential theory postulates a unique relationship between the adsorption potential ε_p and the volume of adsorbed phase contained between the equipotential surface and the solid. It may be preferable to express that volume as a change in volume of the gas phase. Then:

$$\varepsilon_p = f(V) \qquad \qquad \dots (7.25)$$

The function is assumed to be independent of temperature. Equations 7.24 and 7.25 can be combined to give an equilibrium equation, but one that is not explicit until the form of $f(V)$ has been specified. It is sometimes convenient to express the function graphically as a plot of ε_p against V which is called the characteristic curve (Fig. 7.26). The graph is valid for a particular adsorbate and adsorbent at any temperature. There is experimental evidence[8, 17] to suggest that the validity of such a graph can be extended to many adsorbates on a particular adsorbent by writing equation 7.25:

$$\varepsilon_p = \beta_3 f_1(V)$$

where β_3, the coefficient of affinity, has been equated to V'_M the molar volume of the liquid adsorbate at a temperature T. Values of β_3 for hydrocarbons are listed in Table 7.4.

TABLE 7.4. *Values of coefficients of affinity relative to $\beta_3 = 1$ for benzene, and of $V'_M / V'_{M \, (benzene)}$*

Vapour	β_3 expt	$V'_M / V'_{M \, (benzene)}$
C_6H_6	1	1
C_5H_{12}	1·12	1·28
C_6H_{12}	1·04	1·21
C_7H_{16}	1·50	1·65
$C_6H_5CH_3$	1·28	1·19
CH_3Cl	0·56	0·59
CH_2Cl_2	0·66	0·71
$CHCl_3$	0·88	0·90
CCl_4	1·07	1·09
C_2H_5Cl	0·78	0·80
CH_3OH	0·40	0·46
C_2H_5OH	0·61	0·65
$HCOOH$	0·60	0·63
CH_3COOH	0·97	0·96
$(C_2H_5)_2O$	1·09	1·17
CH_3COCH_3	0·88	0·82
CS_2	0·70	0·68
CCl_3NO	1·28	1·12
NH_3	0·28	0·30

The pressures exerted by a particular volume of two sorbates (subscripts 1 and 2) on the same sorbent are then related by:

$$\left[\frac{T}{V'_M} \ln \frac{P^0}{P_1} \right]_1 = \left[\frac{T}{V'_M} \ln \frac{P^0}{P} \right]_2 \qquad \dots (7.26)$$

where V'_M is the liquid molar volume at the temperature T.

LEWIS *et al.*[54] have suggested an improved form of equation 7.26 in which fugacity replaces pressure:

$$\left[\frac{T}{V'_M}\ln\left(\frac{f^0}{f}\right)\right]_1 = \left[\frac{T}{V'_M}\ln\left(\frac{f^0}{f}\right)\right]_2$$

where V'_M is measured at the temperature for which $P^0 = P$.

Several algebraic forms of the function in equation 7.25 have been proposed. If the potential at a point is inversely proportional to the square of its distance from the surface of the solid, then:

$$\varepsilon_p = \frac{\beta_4}{(\alpha_4 + V)^2}$$

where β_4 is a constant, and
α_4 is a correction for the finite size of the molecules.

Substituting for ε_p in equation 7.24 and rearranging:

$$V = \left(\frac{\beta_4}{RT}\right)^{\frac{1}{2}}\left[\ln\frac{P^0}{P}\right]^{-\frac{1}{2}} - \alpha_4 \qquad\qquad \ldots\ldots(7.27)$$

Equation 7.27 is linear in V and $[\ln P^0/P]^{-\frac{1}{2}}$ at constant temperature.
It has been suggested that V is a Gaussian function of the corresponding adsorption potential[17, 50]:

$$V = V_T \exp\left[-\alpha_5\left(\frac{\varepsilon_p}{\beta_5}\right)^2\right]$$

where β_5 is a constant,
V_T is the total volume of sorption space, assumed to be the volume of the micropores, and
α_5 is a constant characteristic of the distribution of pore sizes.

Substituting for ε_p from equation 7.24 yields, after taking logs:

$$\ln V = \ln V_T - \frac{\alpha_5}{\beta_5^2}(RT)^2\left(\ln\frac{P^0}{P}\right)^2 \qquad\qquad \ldots\ldots(7.28)$$

Data conforming to this equation are linear in $\ln V$ and $(\ln P^0/P)^2$. V_T can be obtained from the intercept.
A modified approach[71] in which the distribution of sorption sites, or the fraction of surface covered, is taken as a Gaussian function of adsorption potential leads to:

$$\frac{V}{V^1} = \exp\left[-\alpha_6\left(\frac{\varepsilon_p}{\beta_6}\right)^2\right]$$

where β_6 is a constant,
V^1 is the gas volume of adsorbate in a complete monolayer, and
α_6 is a constant characterising the distribution.

Substituting for ε_p yields an equation similar to equation 7.28:

$$\ln V = \ln V^1 - \frac{\alpha_6}{\beta_6^2}\,(\mathbf{R}T)^2 \left(\ln \frac{P^0}{P}\right)^2 \qquad \dots (7.29)$$

The intercept on the $\ln V$ axis gives the volume of the monolayer. The equation is confined to monolayer adsorption.

Clearly, because the intercepts are different, equations 7.28 and 7.29 cannot both be valid for a particular system for the same range of relative pressures. It is claimed that equation 7.29 is valid at relative pressures P/P^0 below 10^{-4}, and equation 7.28 at higher relative pressures (up to 0.2).

Gas–solid isotherms; two or more sorbates

The previous sections have brought out the difficulty of representing the sorption equilibrium of even a single component. When several components compete for surface, the difficulties are multiplied. If a fluid contain two sorbates **A** and **B**, each obeying the Langmuir equation for monolayer adsorption, a binary isotherm equation can be obtained without difficulty.

For component **A**, at equilibrium:

$$k_A(1 - a_A - a_B)\,P_A = k_A' a_A \qquad \dots (7.30)$$

where a_A, a_B are the fractions of surface occupied by **A** and **B**,

$\quad k_A$ is the adsorption velocity constant, and

$\quad k_A'$ is the desorption velocity constant, incorporating an energy of activation term that was written explicitly in equation 7.3.

For the equilibrium of **B**:

$$k_B(1 - a_A - a_B)\,P_B = k_B' a_B. \qquad \dots (7.31)$$

Equations 7.30 and 7.31 can be solved simultaneously to give:

$$a_A = \frac{(k_A/k_A')\,P_A}{1 + (k_A/k_A')\,P_A + (k_B/k_B')\,P_B}$$

and

$$a_B = \frac{(k_B/k_B')\,P_B}{1 + (k_A/k_A')\,P_A + (k_B/k_B')\,P_B}$$

It is convenient to express the fractional coverages a_A, a_B as ratios of the volume of the component sorbed to the monolayer volume of that component. There is some difficulty when more than one sorbate is involved because the monolayer volumes need not be the same for all components. A mean monolayer volume $\overline{V}_s^1$ is chosen. Then, the total-volume sorbed becomes:

$$V_s = V_{sA} + V_{sB} = \overline{V}_s^1 \, \frac{(k_A/k_A')\,P_A + (k_B/k_B')\,P_B}{1 + (k_A/k_A')\,P_A + (k_B/k_B')\,P_B}$$

By analogy, an equilibrium equation for n components can be written:

$$V_s = \sum_1^n V_{si} = \overline{V}_s^1 \sum_1^n b_i P_i \big/ \Big[1 + \sum_1^n b_i P_i\Big]$$

where $b_i = k_i/k_i'$.

Other single sorbate equations, as well as Langmuir's, have been extended to two or more components. The matter is reviewed by YOUNG and CROWELL[17].

Chemisorption

It is implied in the theories of sorption so far discussed that the forces between the solid molecules at the surface and the sorbed molecules on the surface are of a physical or van der Waals kind. In such systems, the forces between the solid and the first sorbed layer or between the first and the second sorbed layers are not very different so that several layers of molecules can build up. In other systems, bonds between the solid and the first layer can be more specific and involve the sharing of electrons. In those circumstances, the phenomenon is known as *chemisorption*; and the bonds are normally stronger than those resulting from physical forces of attraction.

There is no single criterion which will distinguish between adsorption and chemisorption in all systems, but there are a few which are generally valid. The most common is the magnitude of the heat of sorption. When the bonds are physical, the heat is little more than the latent heat of condensation of the sorbate, usually about 1·5 times the latent heat. The difference is due to a heat of wetting of the solid surface. For chemisorption, however, the heat may be as large as ten times the latent heat.

Another distinction between adsorption and chemisorption is the rate at which each occurs. The adsorption step itself, though not usually the physical diffusion that precedes it, is fast. The chemisorption process on the other hand may have an appreciable energy of activation which limits the rate at low temperatures and leads to a rapid increase in rate with temperature. Physical adsorption is unlikely to occur to any appreciable extent at temperatures above the boiling point of the sorbate.

Because electrons are shared between solid and sorbate molecules, no more chemisorption can occur when the surface of the solid has been covered with enough sorbate to satisfy the residual valency requirements of the surface atoms. Chemisorption cannot result in more than one layer of sorbate molecules, though the situation does not preclude additional layers of physically adsorbed molecules forming.

When molecules of sorbate are confined to a single sorbed layer, the derivation of an equilibrium sorption isotherm is simplified. Using the nomenclature on page 530, equilibrium can be written:

$$k_0 a_0 P = k_1' a_1 \, e^{-E_1/RT}$$

a_0 is proportional to:

$$\left(1 - \frac{V_s}{V_s^1}\right)$$

a_1 is proportional to:

$$\frac{V_s}{V_s^1}$$

Hence:

$$\frac{V_s}{V_s^1} = \frac{k''P}{1 + k''P} = \frac{V}{V^1} \qquad \dots (7.32)$$

where:

$$k'' = \frac{k_0}{k_1' \exp\left(\dfrac{-E_1}{RT}\right)}$$

Equation 7.32, like equation 7.12, is termed the Langmuir isotherm. Its derivation requires that the solid surface is energetically homogeneous, that only one molecule can be accommodated on a sorption site and that the sorbed molecules are not free to move over the surface[19]. Langmuir's is a Type I isotherm; at low pressures it becomes:

$$\frac{V_s}{V_s^1} = \alpha_3 P$$

a relationship akin to Henry's Law. At high pressures:

$$\frac{V_s}{V_s^1} \to 1$$

An alternative way of representing a Type I isotherm, but yielding a slightly different shape from that obtained from Langmuir's isotherm, was suggested by FREUNDLICH[2]:

$$V = \alpha_7 P^{1/n} \qquad \dots (7.33)$$

By choosing appropriate values of α_7 and n, a number of Type I isotherms can be fitted. Though first introduced as an empirical expression, equation 7.33 can be justified thermodynamically and statistically[19]. α_7 and n depend on temperature, usually decreasing with increasing temperature.

The Langmuir isotherm is limited by the assumption that the energy E_1 is constant for all coverage. A modification due to Temkin (see YOUNG and CROWELL[17]) is often in better agreement with experimental data than is the Langmuir equation.

Equation 7.32 can be rearranged:

$$\frac{V/V^1}{1 - V/V^1} = \frac{k_0}{k_1'} P \, e^{E_1/RT} \qquad \dots (7.34)$$

Temkin's assumption was that the energy E_1 could be expressed as a linear function of coverage:

$$E_1 = E_0(1 - \beta_7 V/V^1)$$

where β_7 is a constant, and

E_0 is the activation energy of desorption on an empty surface.

Substituting for E_1 in equation 7.34 and taking logarithms:

$$\ln P = -\ln\frac{k_0}{k_1'} e^{E_0/RT} + \frac{E_0\beta_7 \, V/V^1}{RT} + \ln\frac{V/V^1}{1 - V/V^1} \quad \ldots\ldots (7.35)$$

In the region of $V/V^1 = 0.5$, the third term on the right-hand side of equation 7.35 is small. For chemisorption E_0 is large so that the equation may be approximated to:

$$\frac{V}{V^1} = \frac{RT}{E_0\beta_7} \ln\left\{\frac{k_0}{k_1'} P \, e^{E_0/RT}\right\} \quad \ldots\ldots (7.36)$$

Equation 7.36 is the Temkin isotherm. A plot of V/V^1 against $\ln P$ should yield a straight line. The equation is not likely to be valid outside the middle range of relative pressures because the last term in equation 7.35 has been neglected.

Liquid–solid isotherms

Early work by FREUNDLICH[2] on the sorption of organic compounds from aqueous solution on to charcoal led to an empirical equilibrium equation:

$$c_s' = \alpha_8 (C^*)^{1/n} \quad \ldots\ldots (7.37)$$

where c_s' is the mass of sorbed solute per unit mass of charcoal,

α_8 and n are constants (the latter greater than unity), and

C^* is the concentration of solute in solution in equilibrium with that on the solid.

In gas–solid systems it is convenient to measure the amount sorbed by noting the changes in pressure and volume of the gas phase, or by measuring the change in the mass of the solid. Neither approach is practicable in liquid–solid systems because the volume changes in the liquid phase are small, and because of the difficulty of distinguishing between the adsorbed phase and liquid entrained by the solid without change in concentration.

Instead, the extent to which adsorption has occurred is calculated from the change in concentration of a fixed volume of liquid after it has reached equilibrium with a known amount of sorbent. Solid and liquid of a known composition are sealed into a tube, if necessary freezing the mixture first to prevent vaporisation losses. The tube is tumbled so as to keep the solid falling through the liquid[66] until equilibrium is reached. The liquid is separated from the solid and analysed. The total amount adsorbed at the temperature and equilibrium composition of the liquid can be calculated so that equation 7.37 becomes:

$$V_0(C_0 - C^*) = \alpha_9 (C^*)^{1/n} \quad \ldots\ldots (7.38)$$

where V_0 is the volume of solution of initial concentration C_0, associated with unit mass of sorbent. Equation 7.38 has been found to apply to weak binary solutions, often those in which the solubility of one component is limited. For a liquid solution the components of which are miscible over a wide range of compositions, a different method of representing equilibrium data has been devised, called the composite isotherm.

Composite isotherms

A binary liquid solution consists of components **A** and **B**. The total mass of solution is w_0 initially. After reaching equilibrium with a mass W_0 of sorbent, the components are distributed as follows:

$$\text{Mass of } \mathbf{A} \text{ adsorbed} = w_{sA}$$
$$\text{Mass of } \mathbf{A} \text{ in solution} = w_A$$
$$\text{Mass of } \mathbf{B} \text{ adsorbed} = w_{sB}$$
$$\text{Mass of } \mathbf{B} \text{ in solution} = w_B$$

The mass fraction of **A** in solution initially is:

$$x_{w_0} = \frac{w_A + w_{sA}}{w_0}$$

The mass fraction of **A** in solution at equilibrium is:

$$x_w = \frac{w_A}{w_A + w_B}$$

The change in concentration of **A** in solution is:

$$\Delta x_w = x_{w0} - x_w$$

$$\Delta x_w = \frac{w_A + w_{sA}}{w_0} - \frac{w_A}{w_A + w_B} \qquad \ldots (7.39)$$

$$= \frac{(w_A + w_B)(w_A + w_{sA}) - w_A w_0}{w_0(w_A + w_B)}$$

But $w_0 = w_A + w_B + w_{sA} + w_{sB}$.

Substituting for w_0 in the numerator of equation 7.39:

$$w_0 \Delta x_w = \frac{w_B w_{sA} - w_A w_{sB}}{w_A + w_B}$$

Dividing by W_0, the mass of fresh sorbent:

$$\frac{w_0 \Delta x_w}{W_0} = (1 - x_w) x_{rA} - x_w x_{rB} \qquad \ldots (7.40)$$

where x_{rA}, x_{rB} represent the mass of sorbed phase per unit mass of empty sorbent.

A graph of $w_0 \Delta x_w / W_0$ against x_w is called the composite isotherm for component A. When the mass fraction of A is 0 or 1, Δx_w must be zero. Between these extremes, the isotherms will go through at least one turning point. An attempt to classify composite isotherms has been made by Nagy and Schay (see KIPLING[21]) and is shown in Fig. 7.27. Coordinates are frequently measured in mols and mol fractions. Equation 7.40 then becomes:

$$\frac{n_0 \Delta x_m}{W_0} = (1 - x_m) n_{sA} - x_m n_{sB} \qquad \qquad \dots (7.41)$$

where n_{sA}, n_{sB} are the mols of sorbed component on unit mass of empty sorbent. There is evidence to suggest that sorption from the liquid phase is chiefly in the monolayer[21]. This can be tested by noting that the maximum value of the left-hand side of equation 7.41 is reached when $n_{sB} = 0$. Then:

$$\frac{n_0 \Delta x_m}{W_0} = (1 - x_m) n_{sA}^1$$

where n_{sA}^1 is the number of mols of A in a monolayer. If it can be shown experimentally that $n_0 \Delta x_m / W_0$ exceeds $(1 - x_m) n_{sA}^1$, sorption is not confined to a monolayer.

Types 4 and 5 of Fig. 7.27 show a negative portion of the composite isotherm, indicating a range over which component A is no longer preferentially sorbed.

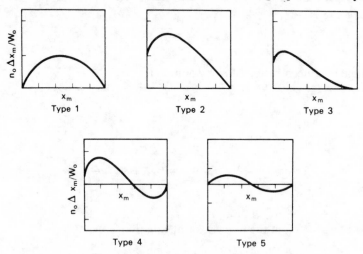

Type 1   Type 2   Type 3

Type 4   Type 5

FIG. 7.27. Classification of liquid–solid, composite isotherms[21]

Individual isotherms

Composite isotherms are the result of adding individual isotherms as shown in Fig. 7.28[21]. For a two-component solution, equation 7.41 can be used to relate n_{sA} and n_{sB} to a chosen x_m and the corresponding $n_0 \Delta x_m / W_0$. The equation can be solved for n_{sA} and n_{sB} if another relationship between the quantities is known. A suitable relationship can be found, if it is assumed that sorption is confined to a monolayer and that the monolayer is always complete. Then:

$$n_{sA}\, a_{MA} + n_{sB}\, a_{MB} = a^1 \qquad\qquad \dots (7.42)$$

where a_{MA}, a_{MB} are the areas occupied by a mol of **A** and of **B** in the sorbed
      state, and
    a^1 is the area of a monolayer.

The area of the monolayer can also be expressed in terms of the sorption of
each of the pure components:

$$a^1 = n_{sA}^1\, a_{MA} = n_{sB}^1 a_{MB}$$

where n_{sA}^1, n_{sB}^1 are the number of mols of **A** and of **B** to complete a monolayer.

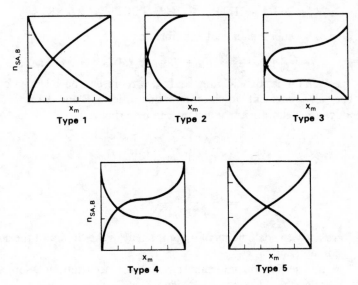

FIG. 7.28. Individual isotherms for two-component liquid–solid
systems[52]

Substituting in equation 7.42:

$$\frac{n_{sA}}{n_{sA}^1} + \frac{n_{sB}}{n_{sB}^1} = 1 \qquad\qquad \dots (7.43)$$

Equations 7.41 and 7.43 can be solved simultaneously to give n_{sA} and n_{sB} as
functions of fluid concentration. Relationships more complex than that of
equation 7.42 are discussed by KIPLING[21].

Ion exchange

An instance of sorption from the liquid phase which is of wide commercial
interest is ion exchange (page 510). Attainment of ion exchange equilibrium

is complicated by the existence of electrostatic as well as concentration differences. If, for example, a cationic resin is immersed in a dilute solution of electrolyte, the cations in the resin, which are the counter-ions in this case, will tend to diffuse out of the resin. From the solution, both the original counter-ions and the co-ions will tend to diffuse into the resin. But a condition of electrostatic neutrality has to be maintained. The net positive charge is constant if for every solution counter-ion diffusing into the resin, a resin counter-ion diffuses out into solution. Since the negative charges on the resin are fixed, no such exchange of anions can occur. Movement of small numbers of anions down the anion concentration gradient will generate an electrostatic potential between resin and solution which effectively excludes the anions. This is known as the Donnan potential[16]. Anion exchange such as that used to soften water:

$$Ca^{++} + 2Cl^- + 2H^+R_{\bar{C}} = 2H^+ + 2Cl^- + Ca^{++}(R_{\bar{C}})_2$$

can be written as a general chemical equilibrium:

$$v_B C_A + v_A C_{sB}(R_C)_{v_B} = v_A C_B + v_B C_{sA}(R_C)_{v_A}$$

where R_C is a resin for cation exchange and v_A, v_B represent the valencies of ions **A** and **B**. They also represent stoichiometric quantities of **B** and **A** respectively. The thermodynamic equilibrium constant can be written:

$$K = \frac{(\gamma_B C_B)^{v_A} (\gamma_{sA} C_{sA})^{v_B}}{(\gamma_A C_A)^{v_B} (\gamma_{sB} C_{sB})^{v_A}}$$

$$= \frac{(\gamma_B)^{v_A} (\gamma_{sA})^{v_B}}{(\gamma_A)^{v_B} (\gamma_{sB})^{v_A}} K_c$$

where γ, γ_s are activity coefficients of a component in the fluid and the sorbed phase, and

K_c is an equilibrium constant based on concentrations—sometimes called the *selectivity coefficient*.

The activity coefficients approach unity and K approaches K_c in dilute solutions.

If the total concentration of exchangeable ions in solution is C_0 and in the solid is Q_s, then for a binary system, that is one containing only two exchangeable ions:

$$x_A = \frac{C_A}{C_0} = 1 - x_B$$

$$y_A = \frac{C_{As}}{Q_s} = 1 - y_B$$

A separation factor, α_{10}, analogous to relative volatility in vapour–liquid equilibria (see Volume 2, Chapter 11), may be defined by:

$$\alpha_{10} = \frac{y_B/x_B}{y_A/x_A} = \frac{x_A(1 - y_A)}{y_A(1 - x_A)}$$

and

$$y_A = \frac{x_A}{\alpha_{10} + (1 - \alpha_{10}) \, x_A} \qquad \dots (7.44)$$

Equation 7.44 can be plotted as shown in Fig. 7.29. Now for equimolar exchange in a binary system:

$$K_x = \frac{y_A(1 - x_A)}{x_A(1 - y_A)}$$

where K_x is an equilibrium constant. Therefore:

$$\alpha_{10} = \frac{1}{K_x}$$

When the exchange is not equimolar, an average value for α_{10} can often be obtained.

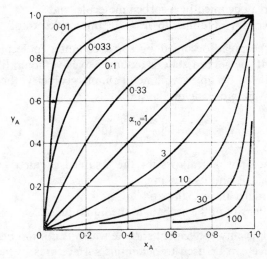

FIG. 7.29. Ion exchange isotherms expressed as dimensionless concentrations with the separation factor as parameter

An equation in the form of equation 7.44 can also be used to describe adsorption equilibrium. For Langmuir kinetics, equilibrium is expressed by equation 7.12, which can be rewritten for dimensionless concentrations:

$$y = \frac{K_{ax}x}{1 + K_{ax}x}$$

where K_{ax} is an equilibrium constant for adsorption.

Substituting for y in equation 7.44 yields:

$$\alpha_{10} = \frac{1}{1 - K_{ax}x}$$

Interpretation of equilibrium data

Gas–solid isotherms

As well as being a convenient way of expressing equilibrium data, the iso-therm, as a graph or as an equation, can provide information about the structure of the sorbent. The surface area of the sorbent, which is a measure of its capacity for taking up sorbate, is often found from an estimate of the number of mole-cules contained in a monomolecular layer, and a knowledge of the area occupied by one molecule.

If n_s^1 is the number of mols of sorbate per unit mass of sorbent corresponding to a complete monolayer:

$$a^1 = n_s^1 N a_m \qquad \dots (7.45)$$

where a_m is the area occupied by a sorbed molecule, and
         N is the Avogadro Number (6.02×10^{26} molecules/kmol).

Estimates of the areas occupied by some common molecules have been tabulated[17, 33]. If the sorbed state is assumed to have liquid-like properties, and each molecule to occupy a volume of simple geometry, then the area, in nm², projected by the molecule on the surface is:

$$a_m = J \left(\frac{M_w}{\rho N} \right)^{2/3} 10^{14} \qquad \dots (7.46)$$

where J is a factor which allows for the assumed geometry of the sorbed molecule and its packing,
       M_w is the molecular weight, and
       ρ is the density of the sorbate in the liquid phase.

Though, in principle, any convenient sorbate which is not too big to penetrate the internal surface can be used to determine surface area, it has been found that different sorbates do not necessarily predict the same surface area for a particular sorbent. Because of this uncertainty, it has become conventional to use nitrogen at 78 K for finding surface areas. The molecular area calculated from equation 7.46 becomes 0.162 nm² for liquid nitrogen.

Monolayer volume

The volume or mass of adsorbate contained by the complete monolayer can be calculated from the isotherm equation in a number of cases.

The equation most frequently used is the infinite form of the BET equation:

$$\frac{P}{V(P^0 - P)} = \frac{1}{V^1 B} + \frac{B-1}{V^1 B} \left(\frac{P}{P^0} \right) \qquad \text{(equation 7.14)}$$

From the slope and the intercept of a plot of $P/V(P^0 - P)$ against P/P^0 the volume V^1 and incidentally B can be found. Equation 7.14 assumes that the build-up of layers is not restricted by the solid structure. It is most likely to apply at low relative pressures.

The Harkins–Jura equation in its modified form:

$$\ln \frac{P}{P^0} = \ln \frac{P_1}{P^0} + \frac{1}{RT} \frac{\beta_1 A_s^2}{2N} \frac{1}{n_{s1}^2} - \frac{1}{RT} \frac{\beta_1 A_s^2}{2N} \frac{1}{C_M^2 V^2} \quad \text{(equation 7.21)}$$

can be plotted as:

$$\ln \frac{P}{P^0} \quad \text{versus} \quad \frac{1}{V^2}$$

to give a straight line, the slope of which is proportional to A_s^2. The constant of proportionality can be found using the same sorbate on a solid of known surface area. Equation 7.20 was derived assuming the film to be mobile, a condition most likely to be met in the middle ranges of relative pressure. At higher values the situation could be complicated by capillary condensation; at lower values the film is unlikely to be mobile.

An explicit form of the sorption potential equation:

$$\ln V = \ln V^1 - \frac{\alpha_6}{\beta_6^2} (RT)^2 \left(\ln \frac{P^0}{P} \right)^2 \quad \text{(equation 7.29)}$$

predicts that a plot of $\ln V$ against $(\ln P^0/P)^2$ will be linear. The intercept of such a plot will give V^1. The equation has been found to fit experimental data in the low relative pressure range (less than 10^{-4}).

Point B method. Many isotherms of Type II or IV show a straight section in the middle range of relative pressures (Fig. 7.30). The lower pressure at which

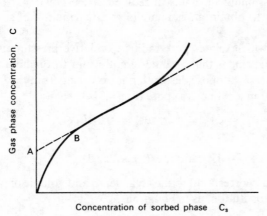

FIG. 7.30. The "point B" method for estimating surface area

the isotherm deviates from the straight section is called point B. A study of many such isotherms led EMMETT et al.[33,40] to conclude that the point corresponded to the completion of a monolayer. This interpretation is widely accepted as a convenient empiricism.

Micropore volume

A sorbent is often found to contain pores of two kinds: micropores which are less than 100 nm and may be related to the structure of the solid lattice; macropores which are generally larger with no definable origin. Micropores are those in which the bulk of the capillary condensation occurs.

The micropore volume can be an important property if the sorbent is to operate in such a condition that the pores become filled with liquid condensate. Since the volume of the micropores is presumably constant, the volume of liquid filling them will also be constant. The statement is the basis of Gurvitsch's rule[28] which predicts that the volume of adsorbate at saturation will be the same for all sorbates on the same sorbent.

Micropore volumes can sometimes be predicted from an explicit form of the sorption potential theory:

$$\ln V = \ln V_T - \frac{\alpha_5}{\beta_5^2} (RT)^2 (\ln P^0/P)^2 \qquad \text{(equation 7.28)}$$

A plot of $\ln V$ against $(\ln P^0/P)^2$ is linear with an intercept from which V_T can be calculated. Equation 7.28 is thought to be valid up to a relative pressure of about 0·2.

Pore sizes

Internal surface in a solid is only of use if the pores giving access to it are large enough to admit the sorbate molecules. A method of finding the mean size of pores by assuming a cylindrical geometry was shown on page 497, but it is often necessary to obtain an estimate of the distribution of sizes about the mean.

Equation 7.23 which relates pore radius to relative pressure, is based on a thermodynamic argument that assumes equilibrium throughout the sorption or desorption process. Suppose desorption is occurring from the free surface of sorbate condensed in a cylindrical pore (Fig. 7.31a). Then:

$$dV = dV_p = d(\pi r^2 L_p) = \pi r^2 \, dL_p$$

and

$$dA_p = d(2\pi r L_p) = 2\pi r \, dL_p$$

where A_p, L_p and V_p correspond to the area, depth and volume of pore occupied by condensate. Substitution, as in equation 7.23, gives:

$$r = \frac{2V_M}{RT} \frac{\sigma_{lv} \cos \phi}{\ln P^0/P} \qquad \dots (7.47)$$

This is known as the Kelvin equation and is thought to be valid for the desorption process[34, 101]. When adsorption occurs the term:

$$\frac{dV}{dA_p}$$

of equation 7.23 is not the same as that calculated for desorption (Fig. 7.31b). The changes in volume dV_p and in area dA_p resulting from a small change dr in the inner radius of the sorbed layer are given by:

$$dV = dV_p = d\,(\pi r^2 L_p) = 2\pi r L_p\,dr$$

$$dA_p = d\,(2\pi r L_p) = 2\pi L_p\,dr$$

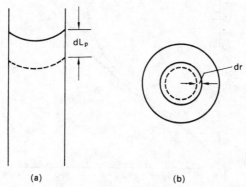

(a)   (b)

FIG. 7.31. The capillary condensation equation applied to a cylindrical pore (a) for adsorption, (b) for desorption

In equation 7.23 these values lead to

$$r = \frac{V_M}{RT}\frac{\sigma_{lv}\cos\phi}{\ln P^0/P} \qquad \ldots (7.48)$$

Equations 7.47 and 7.48 have been used to calculate a distribution of pore sizes. The equations predict that evaporation of condensate from a cylindrical pore takes place at a different relative pressure from that at which condensation in the pore occurred. For the majority of porous sorbents over the range of relative pressures in which condensation occurs, say 0·4 to 0·95, the sorption isotherm has separate desorption and adsorption branches. It exhibits a hysteresis effect[9].

The isotherm can be used to find the change in volume of sorbate corresponding to an incremental change in relative pressure. The pore radius corresponding to the mean relative pressure in the increment can be calculated using equation 7.47 or 7.48, depending on whether the desorption or adsorption branches were used to find the change in volume. Interpretation of the pore radius thus calculated is complicated by the fact that adsorption on the surface of the solid occurs simultaneously in those pores as yet too big for condensation. Since the pores in which condensation occurs already contain sorbed layers,

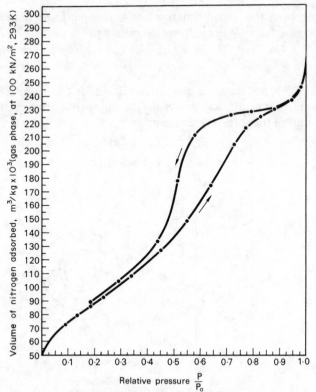

FIG. 7.32. Nitrogen isotherm for activated alumina determined at
the boiling point of liquid nitrogen

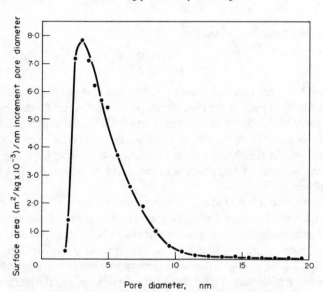

FIG. 7.33. The pore size distribution of activated alumina calculated
from the isotherm by the method of CRANSTON and INKLEY[67]

the radius calculated from equation 7.47 or 7.48 is a net radius. An allowance has to be made for the thickness of the sorbed layers if a true pore radius is to be estimated[67]. This thickness can be found experimentally as a function of relative pressure, using a non-porous solid. Values have been published for nitrogen[81].

Pore structure

Equations 7.47 and 7.48 demonstrated that the sorbate volume corresponding to a particular relative pressure could be different during adsorption and desorption. The result is a hysteresis loop in the range of medium to high relative pressures, such as that shown in Fig. 7.32. It may, conversely, be possible to make certain deductions about the shape of capillaries from a study of the hysteresis loops. The problem has been discussed by DE BOER[8].

For cylindrical pores the mean size is given (page 498) by:

$$r = 2\frac{V_p}{A_p}$$

The expression has been generalised by EVERETT[9] for any shape of capillary:

$$r = v\left(2\,\frac{V_p}{A_p}\right) \qquad\qquad \ldots (7.49)$$

where v takes a value that depends on the geometry of the capillary, and is unity for cylindrical capillaries and for parallel-sided fissures.

Methods of predicting pore structure that are based on equation 7.23 can be only as accurate as the assumptions of that equation. In pores of molecular dimensions it is difficult to picture molecules condensing in a manner analogous to condensation in capillaries. Attempts have been made to observe pores directly using an electron microscope. Electron transmission micrographs obtained for the thin edges of particles of a commercial alumina show an hexagonal array of parallel cylindrical micropores of approximately 2·7 nm in diameter (Fig. 7.1[89]). As well as well-defined micropores, there appears to be a distribution of random pores spanning the range from the micropores to macropores existing between particles that make up the granules. A pore size distribution plotted from the isotherm is shown in Fig. 7.33. There is nothing to indicate well-defined micropores and macropores, as were observed in the electron micrographs. Clearly distributions obtained from the isotherm are liable to misinterpretation.

PROCESS DESIGN OF SORPTION UNITS

Sorption units may be designed to operate with fixed or moving beds. The former are mechanically simple but operate in an unsteady state mode. This can lead to the need for complex mathematics to predict the amount of sorbent required for a given duty. Moving beds are mechanically more elaborate, and the mechanical design may be more exacting than the sorption process design. The process design of an ideal fluidised bed in which the solids are assumed to be uniformly mixed is similar to that of a stirred tank reactor. The mechanical and fluid-dynamic design of a fluidised bed can be complicated and is discussed in Chapter 6 of Volume 2. Beds in which solids move in plug flow relative to the containing vessel are designed to operate in the steady state. In that respect their process design is simpler than the design of fixed beds. In units such as the rotary sorber shown in Fig. 7.19, the sorbent and containing vessel move slowly relative to the feed point. Except for its geometry which results in a radial flow of gas, the bed has many features in common with a fixed bed. The problem of radial flow through annular sorbers has been treated by LAPIDUS and AMUNDSON[53].

Fundamental to the design of any sorption equipment is an understanding of what goes on in a fixed bed. It is the topic that has attracted most study for the reason given, and because the cylindrical fixed bed is the unit most commonly used for industrial sorption processes. Most of this section will be about predicting sorbate concentrations as a function of time and distance along a fixed bed. The problem has received increasing attention over the last twenty years, many solutions being offered for different operating conditions. The failure, in general, of those solutions to predict the behaviour of large beds with accuracy is a measure of the complexity of the phenomena being analysed.

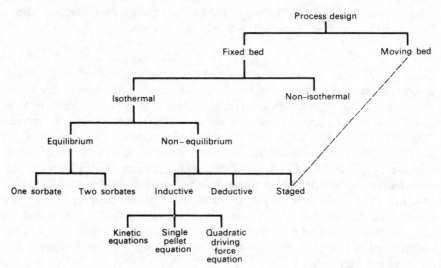

FIG. 7.34. The arrangement of the discussion on the process design of sorption equipment

Consequently, the design of fixed bed sorbers still rests heavily on empirical techniques which will be referred to at the end of the section. It is clear, however, that better design waits on a more thorough understanding of the molecular processes involved, and an appreciation of their relevance to a mathematical description of the system. Whilst it is true that the application of computers to the solution of fixed bed equations has made it possible to include more and more effects that were previously neglected in the interests of an analytical solution, computation times for all but the shortest beds can be considerable. It is still necessary therefore to examine the importance of an effect before including it.

For many of the solutions quoted, a detailed derivation is beyond the scope of the section. In those instances, equations describing the system will be set up and reference made to a published derivation. Figure 7.34 illustrates the plan which discussion in this section will follow.

Fixed bed equations

An energy or mass balance across an elementary section of a column such as that shown in Fig. 7.35 can be written generally:

$$\text{Input} - \text{Output} = \text{Accumulation} + \text{Loss by sorption} \quad \ldots. (7.50)$$

The equation can be applied to the fluid phase only, to the solid phase or to both simultaneously. The simplest situation is one in which the velocity of

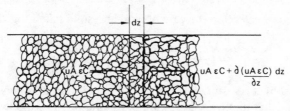

FIG. 7.35. Mass conservation of sorbate in a fixed bed

the fluid is constant across the bed and in which longitudinal diffusion can be neglected.

Fluid phase sorbate balance

$$\text{Input} - \text{Output} \quad = \quad -\frac{\partial(uA\varepsilon C)}{\partial z}\,dz$$

$$\text{Accumulation} \quad = \quad \frac{\partial(\varepsilon AC\,dz)}{\partial t}$$

$$\text{Loss by sorption} \quad = \quad N_2 A(1 - \varepsilon)dza_p = N_2 A dza_z$$

Solid phase sorbate balance

$$\text{Input} - \text{Output} \quad = \quad 0$$

$$\text{Accumulation} \quad = \quad \frac{\partial((1 - \varepsilon)AC_s\,dz)}{\partial t}$$

$$\text{Loss by sorption} \quad = \quad - N_2 A(1 - \varepsilon)\,dza_p$$

Fluid and solid phase considered together

Input – Output $=$ $-\dfrac{\partial(uA\varepsilon C)}{\partial z}\,dz$

Accumulation $=$ $\dfrac{\partial((1-\varepsilon)AC_s\,dz)}{\partial t}+\dfrac{\partial(\varepsilon AC\,dz)}{\partial t}$

Loss by sorption $=$ 0

where A is the cross-sectional area of empty column,
   a_p is the external area of granule per unit volume of granule.
   a_z is the external area of granule per unit volume of fixed bed,
   C is the molar concentration of sorbate in the fluid,
   C_s is the concentration of sorbate in the solid,
   N_2 is the flux of sorbate per unit of external area of granules,
   $m\ =\ \varepsilon/1-\varepsilon$ where ε is the intergranular voidage,
   t is time,
   u is the intergranular fluid velocity, and
   z is the distance along the column.

Applying equation 7.50 to the case of fluid and solid taken together yields:

$$u\left(\frac{\partial C}{\partial z}\right)_t+\left(\frac{\partial C}{\partial t}\right)_z=-\frac{1}{m}\left(\frac{\partial C_s}{\partial t}\right)_z \qquad \ldots\,(7.51)$$

In equation 7.51 it is assumed that the fluid velocity remains constant throughout the bed. This implies that the fluid volume of the sorbate transferred is not high.

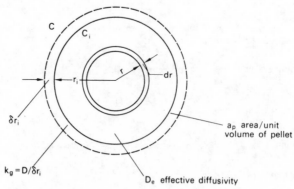

FIG. 7.36. Mass conservation of sorbate in a spherical granule

The *Loss* term for the solid phase is negative, indicating a gain by that phase. The gain can be expressed in a number of ways. Using the nomenclature of Fig. 7.36, the gain can be expressed in terms of the external film mass transfer coefficient k_g, thus:

$$N_2 = k_g(C - C_i) \qquad \qquad \dots (7.52)$$

where C_i is the sorbate concentration in the fluid at the external surface of the granule.

The flux can also be expressed in terms of a diffusivity and a concentration gradient measured either in the fluid or the sorbed phase. For diffusion through a stationary fluid:

$$N_2 = - D_e \left(\frac{\partial C_s}{\partial r}\right)_{r=r_i} = - D_c \left(\frac{\partial C}{\partial r}\right)_{r=r_i} \qquad \qquad \dots (7.53)$$

where r_i is the position of the external surface of the granule,

D_e is the effective diffusivity for a granule referred to the sorbed phase, and

D_c is the effective diffusivity for a granule referred to the fluid phase.

Equations 7.52 and 7.53 assume that the concentrations of sorbate in the fluid are small.

Effective diffusivity through a granule

The diffusivity must allow for gas phase diffusion through a fraction ε of external surface. The kind of gas phase diffusion that takes place in the pores of a granule depends on the relative magnitude of the diameter of the pores and the mean free path of the gas under the conditions that exist in the pores. At atmospheric pressure, the mean free path of a gas molecule is about 100 nm (approximately $100/P$ nm where P is the pressure in bar). In pores appreciably smaller than this, molecules will tend to collide with the pore walls rather than with other molecules and the diffusion is termed Knudsen diffusion[55]. A Knudsen diffusivity can be written:

$$D_K = \frac{2r}{3} \sqrt{\left(\frac{8k_N T}{\pi m_m}\right)} = \frac{2r}{3} u_m$$

where m_m is the mass of a molecule,

r is the radius of a pore, and

u_m is the root mean square velocity of a molecule.

If the diameter of the pores is appreciably greater than the mean free path, collisions between gas molecules are more numerous than those with the pore walls. Diffusion is then similar to that found in bulk gases. If the pore diameter is of the same order of magnitude as the mean free path, the diffusivity will have features of both Knudsen and bulk gas diffusion. A working formula to account for both has been proposed[55]:

$$D' = D(1 - e^{-D_K/D})$$

As well as moving through the gas phase, sorbate can diffuse into a granule over the internal surfaces. In general, surface diffusivities will depend on the concentration of the sorbed phase, being greater at medium to high coverage of the surface. The effects of the surface and pore diffusion are additive. For the particular condition of constant diffusivities and equilibrium between fluid and solid within a pore, pore and surface diffusivities can be combined analytically[86].

Isothermal operation

Equilibrium case

One sorbate. When resistance to the transfer of sorbate from the fluid to the sorbed state is negligible, concentrations in the bulk fluid and in the solid at a point in the bed are related[41] through the sorption isotherm $C_s = f(C)$.

The conservation equation 7.51 can be written in terms of the total volume V'' that has entered unit cross-section of empty bed by using the relation:

$$V'' = u\varepsilon t$$

Then:

$$\frac{\partial C}{\partial z} + \varepsilon \frac{\partial C}{\partial V''} = -(1 - \varepsilon)\frac{\partial C_s}{\partial V''} \qquad \dots (7.54)$$

For equilibrium operation, C is a function of the distance along the column and the volume of fluid that has entered the bed. Thus:

$$\left(\frac{\partial C}{\partial V''}\right) = -\left(\frac{\partial C}{\partial z}\right)_{V''}\left(\frac{\partial z}{\partial V''}\right)_C \qquad \dots (7.55)$$

Equations 7.54 and 7.55 can be combined and integrated at constant C to give:

$$\frac{V''}{\varepsilon + (1 - \varepsilon)f'(C)} = z - z_0 \qquad \dots (7.56)$$

where z_0 is the position at which the concentration is initially C_s (it would be zero for all C_s in a column initially free of sorbate), and

$$f'(C) = \frac{\partial C_s}{\partial C}$$

is the slope of the isotherm at the concentration C_s.

Differentiating equation 7.56 yields the rate at which a point of constant concentration moves along the bed:

$$\left(\frac{\partial z}{\partial t}\right)_C = \frac{u}{1 + \left(\frac{1}{m}\right)f'(C)} \qquad \dots (7.57)$$

Equation 7.57 is important because it illustrates, in the equilibrium case, a principle that carries over to the non-equilibrium cases that are more commonly encountered. The principle relates to the way in which the shape of the sorption wave changes as it moves along the bed. If the isotherm is concave to the fluid concentration axis, then it is termed *favourable*, and points of high concentration in the sorption wave move faster than points of low concentration.

It is physically impossible for points of high concentration to overtake points of low concentration so, in practice, the sorption zone becomes narrower as it moves down the bed, tending to a step change from inlet to outlet concentrations. In these conditions the zone is termed *self-sharpening*.

If the isotherm is convex to the fluid concentration axis, it is termed *unfavourable*; the sorption zone becomes broader as it moves along the bed and is termed *proportional*. For the intermediate case of a linear isotherm, the sorption zone goes through the bed unchanged. The development of a zone under various conditions is illustrated in Fig. 7.37.

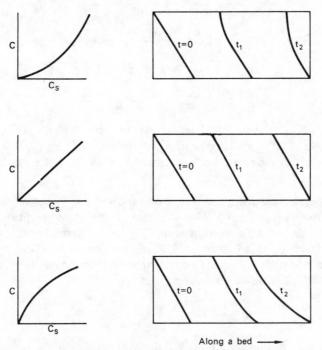

FIG. 7.37. Effect of the shape of the isotherm on the development of a sorption wave through a bed with the initial distribution of sorbate shown at $t = 0$

In non-equilibrium operation, resistance to transfer from fluid to solid results in a zone of finite width. If the isotherm is favourable, the zone is propagated through the bed as a wave of constant shape, after an initial period during which the wave becomes fully developed. The property is important because it leads to simplified methods of solution, valid at all except small times (page 570).

Two sorbates. If the fluid contains two sorbates, equation 7.56 can be applied to each. For a column initially free of sorbate:

T

$$\frac{V''}{z} = \varepsilon + (1 - \varepsilon) f'(C)$$

where V'' is the volume of liquid that has entered the column per unit cross-sectional area and must be the same for both components. At a fixed value of z, it follows that:

$$f'(C_A) = f'(C_B)$$

The conditions can be written as a ratio of finite differences:

$$\frac{(C_{sA})_2 - (C_{sA})_1}{(C_A)_2 - (C_A)_1} = \frac{(C_{sB})_2 - (C_{sB})_1}{(C_B)_2 - (C_B)_1} \qquad \ldots (7.58)$$

Suppose conditions $(C_A)_1$, and $(C_B)_1$, are known as well as the isotherm relationships:

$$C_{sA} = f_1(C_A, C_B) \qquad \ldots (7.59)$$

$$C_{sB} = f_2(C_A, C_B) \qquad \ldots (7.60)$$

Equation 7.58 can then be solved by trial and error. For a selected value of $(C_A)_2$, values of $(C_B)_2$ are tried until equations 7.58, 7.59 and 7.60 are satisfied.

In principle this method can be extended to many sorbates.

With longitudinal dispersion. Equilibrium operation is not often found, but the condition is approached at low flowrates. In the same condition, dispersion along the column could be significant[58]. For isothermal operation, mass is conserved according to equation 7.50. Dispersion in the fluid results in an additional component of the *Input* and *Output* terms over those that yielded equation 7.51. The new equation becomes:

$$u \frac{\partial C}{\partial z} - D_L \frac{\partial^2 C}{\partial z^2} + \frac{\partial C}{\partial t} = -\frac{1}{m} \frac{\partial C_s}{\partial t} \qquad \ldots (7.61)$$

where D_L, the effective longitudinal diffusivity in the column, is assumed to be constant.

The equation has been integrated[58], using the Laplace transform method, for the general case of a known initial distribution of sorbate along the column, and a sorbate concentration in the fluid entering the bed which is a known function of time. The general solution is restricted to a linear isotherm and is presented in an integral form requiring numerical solution. More immediately useful is a restricted solution applicable to long beds, initially free of sorbate and subject to a constant inlet fluid concentration C_0. Then

$$\frac{C}{C_0} = \frac{1}{2} \left\{ 1 + \mathrm{erf} \left[\left(\frac{uz}{4D_L} \right)^{\frac{1}{2}} \frac{V'' - V''_{min}}{(V'' V''_{min})^{\frac{1}{2}}} \right] \right\}$$

where V''_{min} is the minimum volume to saturate a bed of unit cross-section and length z, i.e.

$$V_{min}'' = z \left[\varepsilon + (1 - \varepsilon) \frac{C_{s\infty}}{C_0} \right]$$

where $C_{s\infty}$ is the concentration of sorbed phase in equilibrium with C_0.

The effect of longitudinal dispersion is to lengthen the sorption zone. In this respect its effect is the same as resistance to transfer within the sorbent pellets. The correct choice of D_L can simulate non-equilibrium between fluid and pellets. The effects of longitudinal diffusion and of resistance to mass transfer are discussed by VAN DEEMTER et al.[65] with particular reference to chromatography.

Non-equilibrium case for isothermal operation

Equilibrium operation implies that the fluid concentration in the pores of the granule is the same as that in the body of the fluid outside the granule; that the concentration of the sorbed phase is uniform throughout the granule and related to the fluid concentration through the sorption isotherm. These conditions are not achieved in large scale plant and seldom in the laboratory.

In non-equilibrium conditions, concentration gradients develop across the boundary film outside the granule and through the granule itself. Four stages of transfer can be distinguished: through the boundary film; through the pores and over the internal surface; sorption on to internal surface; possibly diffusion through the crystallites of the solid. Not all stages need offer the same resistance to the transfer of sorbate, but the stages occur in series so the rate of transfer through the first three will be virtually the same. The greater the resistance offered by a stage, the greater the concentration gradient across it. If diffusion through the crystallite occurs, this need not be at the same rate because molecules can accumulate in the sorbed phase.

In physical adsorption, it is usually the diffusional processes through the boundary film, through the granule pores and over the granule internal surface that are important; the first is significant only initially. In many of the developments that follow, use will be made of the idea of a "controlling mechanism" in order to simplify the mathematics. This refers to the stage of transfer that offers the greatest resistance. If, by comparison, the other stages offer little resistance, the rate equation can be written in terms of the controlling mechanism only. For a more rigorous approach to the interaction of physical diffusion and non-catalytic reaction within a solid, recent literature such as that by WEN[98] should be consulted.

There is a formidable body of literature dealing with the solution of fixed bed equations for non-equilibrium operation. For the isothermal problem, three general approaches can be discerned. The first is *inductive* in the sense that a solution is used which has been found for a particular kind of rate expression. The actual conditions of operation for which the solution is required are then arranged in the form of this rate expression. In the *deductive* approach, the relevant mechanisms of transfer are picked out and combined to give a rate equation. This equation is combined with the conservation equation to yield whatever solutions are possible, usually by numerical procedures. A third approach uses the concept of *staged separations*, sometimes in a graphical method.

Inductive approach

(a) *Kinetic rate equation*

In 1944, in a paper that was a model of brevity, THOMAS[43] published a theory of fixed bed performance for application to ion exchange columns. It was assumed that the rate was controlled by the ion exchange step, the rate equation being:

$$\frac{\partial C_s}{\partial t} = k\left[C(C_{s\infty} - C_s) - \frac{1}{K_i}C_s(C_0 - C)\right] \qquad \dots (7.62)$$

where k is the forward velocity constant of the exchange,

$C_{s\infty}$ is the concentration equivalent to total quantity of exchangeable ion in the exchange resin,

C, C_s are fluid and solid concentrations of an exchange component, and

K_i is an equilibrium constant for ion exchange.

Equation 7.62 can be rewritten:

$$\frac{\partial (C_s/C_{s\infty})}{\partial t} = kC_0\left[\frac{C}{C_0}\left(1 - \frac{C_s}{C_{s\infty}}\right) - \frac{1}{K_i}\frac{C_s}{C_{s\infty}}\left(1 - \frac{C}{C_0}\right)\right] \qquad \dots (7.63)$$

By defining time and distance parameters:

$$\tau = kC_0\left(t - \frac{z}{u}\right)$$

$$X = \frac{kC_{s\infty}z}{mu}$$

this equation and the conservation equation 7.51 can be simplified. If C is regarded as a function of X and τ:

$$dC = \left(\frac{\partial C}{\partial X}\right)_\tau dX + \left(\frac{\partial C}{\partial \tau}\right)_X d\tau$$

$$\left(\frac{\partial C}{\partial z}\right)_t = \left(\frac{\partial C}{\partial X}\right)_\tau \left(\frac{\partial X}{\partial z}\right)_t + \left(\frac{\partial C}{\partial \tau}\right)_X \left(\frac{\partial \tau}{\partial z}\right)_t$$

$$= \frac{kC_{s\infty}}{mu}\left(\frac{\partial C}{\partial X}\right)_\tau - \frac{kC_0}{u}\left(\frac{\partial C}{\partial \tau}\right)_X$$

Also

$$\left(\frac{\partial C}{\partial t}\right)_z = kC_0\left(\frac{\partial C}{\partial \tau}\right)_X$$

$$\left(\frac{\partial C_s}{\partial t}\right)_z = kC_0\left(\frac{\partial C_s}{\partial \tau}\right)_X$$

Substituting into equation 7.51:

$$u\frac{\partial C}{\partial z} + \frac{\partial C}{\partial t} = -\frac{1}{m}\frac{\partial C_s}{\partial t} \qquad \text{(equation 7.51)}$$

yields:

$$\left(\frac{\partial(C/C_0)}{\partial X}\right)_\tau = -\left(\frac{\partial(C_s/C_{s\infty})}{\partial \tau}\right)_X \qquad \dots\,(7.64)$$

which together with the rate equation (7.65) can be solved by the method of Thomas:

$$\frac{\partial(C_s/C_{s\infty})}{\partial \tau} = \frac{C}{C_0}\left(1 - \frac{C_s}{C_{s\infty}}\right) - \frac{1}{K_i}\frac{C_s}{C_{s\infty}}\left(1 - \frac{C}{C_0}\right) \qquad \dots\,(7.65)$$

The algebraic form of the solutions comprises complex functions of X, τ and K_i. For the purpose of design the solutions are more usefully presented in graphical form[69].

VERMEULEN and co-workers[57, 69] recognised the value of the solutions for systems other than those in which ion exchange kinetics control the rate.

(i) *Langmuir kinetics controlling.* If physical adsorption or chemisorption is controlling the rate, it can often be expressed in the form of Langmuir kinetics (page 543):

$$\frac{\partial C_s}{\partial t} = k\left[C(Q_s - C_s) - \frac{1}{K_a}C_s\right] \qquad \dots\,(7.66)$$

where Q_s is a measure of the ultimate capacity of the adsorbent. In this case a distinction must be made between Q_s and the attainable capacity $C_{s\infty}$, which depends upon the maximum fluid concentration with which equilibrium can be achieved. In adsorption, $C_{s\infty} = Q_s$ only when the relative concentration at the inlet to the bed approaches unity. For ion exchange, the total exchange capacity is realizable even when the relative concentration of the feed is low, and $C_{s\infty} = Q_s$ always.

At equilibrium, $C_s = C_{s\infty}$ and $C = C_0$. Also $\partial C_s/\partial t$ in equation 7.66 can be put equal to zero. Then:

$$Q_s = \left(\frac{1 + K_a C_0}{K_a C_0}\right)C_{s\infty}$$

Substituting for Q_s in equation 7.66 yields:

$$\frac{\partial(C_s/C_{s\infty})}{\partial t}$$
$$= k\left[\frac{1}{K_a} + C_0\right]\left[\frac{C}{C_0}\left(1 - \frac{C_s}{C_{s\infty}}\right) - \frac{1}{1 + K_a C_0}\frac{C_s}{C_{s\infty}}\left(1 - \frac{C}{C_0}\right)\right] \qquad \dots\,(7.67)$$

Equation 7.67 has the same form as equation 7.63. By defining an appropriate time parameter and a group containing the equilibrium constant, a rate expression for both ion exchange and Langmuir adsorption can be written in the same form:

$$\frac{\partial(C_s/C_{s\infty})}{\partial \tau} = \frac{C}{C_0}\left(1 - \frac{C_s}{C_{s\infty}}\right) - r^* \frac{C_s}{C_{s\infty}}\left(1 - \frac{C}{C_0}\right) \qquad \ldots (7.68)$$

For ion exchange:

Time parameter τ $= kC_0\left(t - \frac{z}{u}\right)$

Equilibrium parameter r^* $= \frac{1}{K_i}$

For adsorption:

Time parameter τ $= k\left[\frac{1}{K_a} + C_0\right]\left(t - \frac{z}{u}\right)$

Equilibrium parameter r^* $= \frac{1}{(1 + K_a C_0)}$

(ii) *External diffusion control.* Systems in which the kinetics of transfer are controlled wholly by the physical adsorption, the chemisorption or the ion exchange step are not common. If the solutions of Thomas are to be of general value, other rate controls, particularly those derived from physical diffusion steps, must also be interpreted. The simplest of these, though not the commonest, is controlled by the transfer through the boundary film. Thus:

$$\frac{\partial C_s}{\partial t} = k_g \frac{a_z}{1 - \varepsilon}(C - C_i^*) \qquad \ldots (7.69)$$

where a_z is the external surface of the solid in unit volume of column, and
       C_i^* is the fluid concentration at the exterior surface of the solid. In the case under discussion, the concentration of sorbate throughout the solid is uniform and is related to C_i^* through the sorption isotherm.

In equation 7.68, the right-hand side represents a driving force, since the velocity constant has been incorporated in the time parameter. In the case under consideration, external diffusion controls the overall rate of sorption, so the driving force of the sorption step itself approaches zero. Thus C_i^* can be expressed in other terms:

$$\frac{C_i^*}{C_0}\left(1 - \frac{C_s}{C_{s\infty}}\right) = r^* \frac{C_s}{C_{s\infty}}\left(1 - \frac{C_i^*}{C_0}\right)$$

$$C_i^* = \frac{C_s}{C_{s\infty}}\left[\frac{C_0}{1 + (C_s/C_{s\infty})(r^* - 1)}\right]$$

Substituting in equation 7.69 and rearranging gives:

$$\frac{\partial C_s/C_{s\infty}}{\partial t} = \frac{k_g a_z C_0}{(1-\varepsilon)[1+(r^*-1)(C_s/C_{s\infty})]C_{s\infty}} \times$$

$$\times \left[\frac{C}{C_0}\left(1-\frac{C_0}{C_{s\infty}}\right) - r^*\frac{C_s}{C_{s\infty}}\left(1-\frac{C}{C_0}\right)\right] \quad \ldots\ (7.70)$$

If a mean value is taken for the $C_s/C_{s\infty}$ term outside the square bracket, equation 7.70 has the same form as equation 7.68. Similar solutions can be used if the time parameter is redefined:

$$\tau = \frac{k_g a_z C_0}{(1-\varepsilon)\left[1+(r^*-1)(C_s/C_{s\infty})\right]C_{s\infty}}\left(t-\frac{z}{u}\right)$$

The choice of a suitable mean for $C_s/C_{s\infty}$ is discussed by HIESTER and VERMEULEN[57].

(iii) *Internal diffusion control.* Systems in which the rate of transfer is controlled by diffusion within the pellet are common. Thomas' equations can be used to give an approximate solution. The rate equation can be expressed in terms of a hypothetical solid film coefficient k_p and a contrived driving force $(C_s^* - C_s)$:

$$\frac{\partial C_s}{\partial t} = k_p \frac{a_z}{1-\varepsilon}(C_s^* - C_s) \quad \ldots\ (7.71)$$

where C_s^* is the concentration of the sorbed phase in equilibrium with C, and C_s is the mean concentration of sorbed phase averaged over the pellet.

Equilibration of the adsorption step itself gives an expression for C_s^*. From equation 7.68:

$$C_s^* = \frac{C_{s\infty}}{(1-r^*) + r^*(C_0/C)}$$

Substituting for C_s^* in equation 7.71 gives:

$$\frac{\partial(C_s/C_{s\infty})}{\partial t} = \frac{k_p a_z}{[C/C_0(1-r^*)+r^*](1-\varepsilon)} \times$$

$$\times \left[\frac{C}{C_0}\left(1-\frac{C_s}{C_{s\infty}}\right) - r^*\frac{C_s}{C_{s\infty}}\left(1-\frac{C}{C_0}\right)\right] \quad \ldots\ (7.72)$$

For a mean value of C/C_0 outside the square bracket or several mean values corresponding to different ranges of concentration, equation 7.72 has the form of equation 7.68 with a new time parameter:

$$\tau = \frac{k_p a_z}{[(C/C_0)(1-r^*)+r^*](1-\varepsilon)}\left(t-\frac{z}{u}\right)$$

(iv) *External and internal diffusion significant.* At different times during the period when the pellet is exposed to a sorption wave, both the external film and the internal diffusional resistance are important in determining the overall rate of transfer. When the sorption wave reaches a pellet and the fluid concentration around it begins to increase, the external film is important. The neglect of the external film in calculations results in the prediction of sharper break-through curves than in fact occur.

When the pellet becomes well immersed in the sorption wave, the transfer rate is usually controlled by internal diffusion, so a method of adding both internal and external effects is required. VERMEULEN[69] has suggested that the reciprocals of the rate coefficients be added.

(b) *Single pellet equation*

All the rate expressions used in the previous section, except those relating to ion-exchange and adsorption control, involve some degree of approximation. The approximations were chosen so that the rate equation conformed to the kind required for the solutions of Thomas to be valid. It is a step, only in degree rather than in principle, to extend the argument to the point of saying that any equation which satisfies the observed rate data can be used if it yields a convenient solution. In its limit, the argument applies to an empirical curve fit of a plot of C_s against t obtained for discrete pellets of solid exposed to a known fluid concentration. Let the relationship be expressed as:

$$t = j_0 + \sum_{n=1}^{n} j_n C_s^n \qquad \ldots (7.73)$$

where j_n, etc. are numerical coefficients.

Then:

$$\frac{dt}{dC_s} = \sum_{1}^{n} n j_n C_s^{n-1}$$

Implied in this equation is a dependence on the fluid concentration C_0 outside the pellet. Let the rate at any point in a fixed bed be written:

$$\left(\frac{\partial C_s}{\partial t}\right)_z = \frac{C}{C_0} \frac{1}{\sum_{1}^{n} n j_n C_s^{n-1}}$$

Or, in general:

$$\left(\frac{\partial C_s}{\partial t}\right)_z = \frac{bC}{F'(C_s)} \qquad \ldots (7.74)$$

where b is a constant, in this case $1/C_0$, and

$F'(C_s)$ is the derivative with respect to C_s of a function of C_s only.

The conservation equation for a fixed bed can be written:

$$\left(\frac{\partial C}{\partial x_1}\right)_{t_1} = -\left(\frac{\partial C_s}{\partial t_1}\right)_{x_1} \qquad \ldots (7.75)$$

where $t_1 = t - z/u$, and
$$x_1 = z/mu$$

For a bed initially free of sorbate and subjected to a constant inlet concentration of sorbate in the fluid, equations 7.73, 7.74 and 7.75 can be solved analytically[99] to give:

$$\frac{C}{C_0} = \frac{C_s}{C_s(0, t_1)} \qquad \qquad ....(7.76)$$

The concentration in the fluid leaving the fixed bed is given by:

$$j_1 \ln (C/C_0) - \sum_2^n \left\{ \frac{n}{(n-1)} j_n [C_s(0, t_1)]^{n-1} \left[1 - \left(\frac{C}{C_0} \right)^{n-1} \right] \right\} = -bx_1. \quad ....(7.77)$$

In equations 7.76 and 7.77 the measure of time is $C_s(0, t_1)$, the change in sorbate concentration of pellets at the inlet to the bed. Such pellets are exposed to a constant fluid concentration C_0 and this behaviour can be easily simulated in single pellet experiments leading to equation 7.73.

Equations 7.73, 7.76 and 7.77 can be evaluated using a digital computer which is programmed to receive the single pellet data and to print out C/C_0 as a function of time and bed length. There are dangers of misinterpretation if an empirical curve fit is used outside the range for which it was obtained, so it is important that the single pellet experiments should be carried out for a time that equals or exceeds the duration of the fixed-bed operation.

(c) Quadratic driving force equation

Another convenient method of correlating single pellet data was proposed by VERMEULEN[61]. The conservation equation for the diffusion of sorbate into a spherical pellet can be written:

$$D_e \left(\frac{\partial^2 C_{sr}}{\partial r^2} + \frac{2}{r} \frac{\partial C_s}{\partial r} \right) = \frac{\partial C_{sr}}{\partial t}$$

The equation has been solved[3, 44, 47, 52] for a constant fluid concentration outside the pellet and negligible resistance in the external film. The solution can be written:

$$Y = \frac{C_s - C_{s0}}{C_s^* - {}^*C_{s0}} = 1 - \frac{6}{\pi^2} \sum_{n=1}^{\infty} \frac{1}{n^2} \exp [-(D_e \pi^2 t)/(n^2 r_i^2)]$$

At large times the solution reduces to one term:

$$Y = 1 - \frac{6}{\pi^2} \exp [-(D_e \pi^2 t)/(r_i^2)]$$

The corresponding rate equations can be found by obtaining the derivative with respect to time:

$$\frac{dY}{dt} = \frac{\pi^2 D_e}{r_i^2} (1 - Y)$$

U

or

$$\frac{dC_s}{dt} = \frac{\pi^2 D_e}{r_i^2} (C_s^* - C_s) \qquad \qquad \dots (7.78)$$

where C_s^* is the concentration of the sorbed phase in equilibrium with the external fluid concentration. Equation 7.78 is an approximation due to GLUECKAUF and COATES[49], known as the linear driving force equation. Vermeulen found that the single pellet data were better described by an equation of the form:

$$\frac{dC_s}{dt} = \frac{\kappa D_e}{r_i^2} \frac{C_s^{*2} - C_s^2}{2(C_s - C_{s0})} \qquad \qquad \dots (7.79)$$

where κ depends on the equilibrium parameter (page 566) and in general, therefore, upon the concentrations outside the pellet.

Equation 7.79 is assumed to apply when the concentration outside a pellet is varying, that is at every point in a bed. Then:

$$\frac{\partial C_s}{\partial t} = \frac{\kappa D_e}{r_i^2} \frac{C_s^{*2} - C_s^2}{2(C_s - C_{s0})} \qquad \qquad \dots (7.80)$$

C_s^* is no longer constant since it varies with the concentration in the fluid outside the pellet.

For single pellets exposed to a constant concentration, equation 7.80 can be integrated for $C_{s0} = 0$:

$$C_s = C_s^* \left[1 - \exp(- \kappa D_e t / r_i^2)\right]^{\frac{1}{2}}$$

Hence a plot of

$$\ln \left[1 - \left(\frac{C_s}{C_s^*}\right)^2\right] \text{ against } t$$

should give a straight line, from the slope of which the diffusion factor $\kappa D_e / r_i^2$ can be obtained.

Equation 7.80 can be combined with the conservation equation to give numerical solutions of the fixed-bed condition.

(d) *Constant wave form*

It has been stated that favourable isotherms lead, in non-equilibrium adsorption, to a wave of constant shape moving down the bed. This fact can be used to give simplified solutions valid at times which are great enough for error incurred in the period during which the wave is being formed to be small by comparison[62]. A mass balance for a movement dz of the sorption wave gives:

$$\varepsilon u A C_0 \, dt = A[(1 - \varepsilon) C_{s\infty} + \varepsilon C_0] \, dz$$

$$\frac{dz}{dt} = \frac{\varepsilon u}{[(1 - \varepsilon)(C_{s\infty}/C_0) + \varepsilon]}$$

The left-hand term is the velocity of the zone; it is constant for the condition indicated.

A mass balance can be applied at any level of concentration into the zone and leads to:

$$\left(\frac{\partial z}{\partial t}\right)_c = \frac{\varepsilon u}{[(1 - \varepsilon)(C_s/C) + \varepsilon]} = \text{constant}$$

Therefore a constant wave form leads to the condition that:

$$\frac{C_s}{C} = \frac{C_{s\infty}}{C_0} \qquad \qquad \dots (7.81)$$

For constant pattern operation, equation 7.80 can be solved by the method given on page 568, as indeed can many of the rate equations on pages 564 to 568.

Constant wave solutions can be obtained for a more general rate equation than equation 7.74:

$$\frac{\partial C_s}{\partial t_1} = G_1(C, C_s) \qquad \qquad \dots (7.82)$$

where G_1 is any function of the sorbate concentration in the bulk fluid at a point and the mean sorbate concentration on the solid at that point.

For a constant wave condition:

$$C_s = \frac{C_{s\infty}}{C_0} C \qquad \qquad \text{(equation 7.81)}$$

Therefore, from equation 7.82:

$$\frac{\partial C_s}{\partial t_1} = G_1(C) = -\frac{\partial C}{\partial x_1}$$

Hence:

$$\int \frac{dC}{G_1(C)} = \left[\int \frac{dC}{G_1(C)}\right]_{C_0} - x_1$$

since $C = C_0$ at $x_1 = 0$ for all times.

Because constant wave or constant pattern solutions are generally not valid at short times, they are sometimes referred to as asymptotic solutions[62].

Deductive approach

Rate equations, such as those used in the previous sections in which little attempt is made to distinguish mechanisms of transfer within a pellet, are limited to the range of conditions for which rate data have been obtained. A more fundamental approach is to pick out the important mechanisms of transfer within a pellet and to combine them to form a rate equation, without regard to the mathematical convenience of the equation. Except however for some simple limiting cases, numerical solutions must be obtained for the resulting equations.

When the rate of sorption is controlled by diffusion through the external film the conservation equation can be written:

$$u\frac{\partial C}{\partial z} + \frac{\partial C}{\partial t} = -\frac{k_g a_z}{\varepsilon}(C - C_i^*) = -\frac{1}{m}\frac{\partial C_s}{\partial t} \qquad \dots(7.83)$$

This equation has been solved analytically for a linear isotherm[46].

If there is resistance to mass transfer within the pellet, sorbate concentrations are not constant throughout. The rate equation can still be written in terms of a concentration difference across the external film:

$$\frac{\partial C_s}{\partial t} = \frac{k_g a_z}{(1 - \varepsilon)}(C - C_i)$$

but C_i is related through the isotherm to the concentration at the external surface, rather than to the mean concentration C_s.

For the case of a linear isotherm and constant feed concentration, solutions have been found by ROSEN[59, 63].

The general solution has the form:

$$\frac{C}{C_0} = \frac{1}{2} + \psi(\lambda', \tau, \chi) \qquad \dots(7.84)$$

where:

length parameter $\qquad \lambda' = \dfrac{3D_e K_a z}{m u r_i^2}$

time parameter $\qquad \tau = \dfrac{2D_e}{r_i^2}\left(t - \dfrac{z}{u}\right)$

resistance parameter $\qquad \chi = \dfrac{D_e K_a}{r_i k_g}$

Except for small values of λ', the solution has the following convenient form:

$$\frac{C}{C_0} = \frac{1}{2}\left[1 + \text{erf}\left\{\frac{(3\tau/2\lambda' - 1)}{2\sqrt{[(1 + 5\chi)/5\lambda']}}\right\}\right] \qquad \dots(7.85)$$

Other solutions have been tabulated and graphed by ROSEN[59, 63].

The case of three resistances controlling (or any combination of one to three resistances) has been solved for a linear isotherm[85]. Inevitably the solutions are cumbersome and of little direct value in design. They can be of value in predicting the effect of a particlar variable.

However, one has now reached a stage in the development of deductive solutions when one should ask whether further refinements are justified until some of the grosser approximations have been removed. There is the common danger of evolving solutions which work because they have a sufficient number of constants that can be arbitrarily assigned, rather than because the constants have a fundamental significance.

Example†

A bed is packed with dry silica gel beads of mean diameter 1·72 mm to a density of 671 kg of gel per m³ of bed. The density of a bead is 1266 kg/m³ and the depth of packing is 0·305 m. Humid air containing 0·00267 kg of water per kg of dry air enters the bed at the rate of 0·1292 kg of dry air/m² s. The temperature of the air is 300 K and the pressure 1·024 × 10⁵ N/m². The bed is assumed to operate isothermally. Use the method of Rosen to find the effluent concentration as a function of time. Equilibrium data for the silica gel are given by the curve in Fig. 7.38.

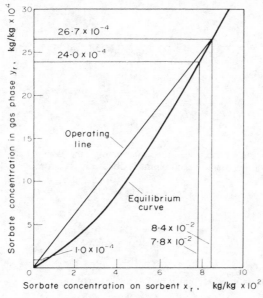

FIG. 7.38. Sorption isotherm for water vapour in air on silica gel

Solution

The length parameter λ' is calculated from the definition on page 572:

$$\lambda' = \frac{3D_e K_a z}{m u r_i^2}$$

$$K_a = \frac{0{\cdot}084}{0{\cdot}00267} \times \frac{1266}{1{\cdot}186} = 3{\cdot}36 \times 10^4$$

where K_a is obtained from the mean slope of the isotherm between its origin and the point corresponding to the inlet concentration. This slope has been multiplied by the ratio of bead to gas densities.

$$m = \frac{\varepsilon}{1-\varepsilon} = \frac{1266 - 671}{671} = 0{\cdot}89$$

where ε is the volume (and area) voidage between beads.

† This example is based on one given by CARTER[74].

$$\varepsilon = \frac{1266 - 671}{1266} = 0.47$$

$$u = \frac{0.1297}{0.47 \times 1.186} = 0.233 \text{ m/s}$$

where u is the interstitial velocity of the air.

$$r_i = 0.86 \text{ mm}$$
$$z = 0.305 \text{ m}$$

D_e is the diffusivity of sorbate referred to the sorbed phase (its value is between about 10^{-10} and 10^{-11} m^2/s).

Substituting values gives $\lambda' = 18.500$ to 1850. It is correct to use equation 7.85, valid for large values of λ'.

Now:

$$\frac{\tau}{\lambda'} = \frac{2\left(t - \dfrac{z}{u}\right)mu}{3 K_a z} = \left(\frac{2 \times 0.89}{3 \times 3.36 \times 10^4}\right)\left(\frac{0.233\,t}{0.305} - 1\right)$$

(for t expressed in hours)

$$t = 20.55\frac{\tau}{\lambda'} + \frac{1}{2750}$$

Again:

$$\frac{\chi}{\lambda'} = \frac{mur_i}{3zk_g}$$

The unknown on the right-hand side of this equation is k_g, the gas film mass transfer coefficient. From a correlation of the j_d factor (see Volume 1, Chapter 8) it can be shown that an appropriate value is:

$$k_g = 0.0833 \text{ m/s}$$

Therefore:

$$\frac{\chi}{\lambda'} = \frac{0.89 \times 0.233 \times 0.00086}{3 \times 0.305 \times 0.0833}$$

$$= 2.36 \times 10^{-3}$$

For the relative values of λ' and χ/λ', equation 7.85 can be rewritten:

$$\frac{C}{C_0} = \frac{1}{2}\left[1 + \text{erf}\left\{\frac{(3\tau/2\lambda' - 1)}{2\sqrt{(\chi/\lambda')}}\right\}\right]$$

$$= \frac{1}{2}[1 + \text{erf } E]$$

where:

$$E = \frac{(3\tau/2\lambda' - 1)}{2\sqrt{(\chi/\lambda')}}$$

For selected values of C/C_0, the value of E can be found from tables of error functions. From E a ratio τ/λ' can be calculated and hence a corresponding time. The calculations are summarised in the table.

$\dfrac{C}{C_0}$	erf E	E	$\dfrac{\tau}{\lambda'}$	t (h)
0.024	−0.952	−1.40	0.576	11.8
0.045	−0.910	−1.20	0.589	12.1
0.079	−0.842	−1.00	0.602	12.4
0.24	−0.520	−0.50	0.635	13.0
0.50	0	0	0.667	13.7
0.715	0.430	0.40	0.693	14.2
0.92	0.840	1.00	0.732	15.0

Staged operation

The idea of an ideal equilibrium stage[6] is familiar to the chemical engineer because of its occurrence in other separation processes such as absorption, distillation and solvent extraction. In the present context, as in the others, from an ideal stage streams of the contacting phases leave in equilibrium.

In its simplest form the equilibrium stage for a sorption process could be a vessel containing sorbent to which a volume of liquid containing a sorbate is added. The solid and liquid are left in contact until equilibrium is reached.

The conservation of sorbate then yields the following equality:

$$V_w(C_{s1}^* - C_{s0}) = V(C_0 - C_1)$$

where C_{s1}^* and C_1 are concentrations in equilibrium, and
V. V_w represent total volumes of fluid and of sorbent.

For differential changes of concentration:

$$V_w \, dC_s = -V \, dC$$

Writing:

$$y = C_s/C_{s\infty}$$

and

$$x = C/C_0$$

then:

$$C_{s\infty} V_w \, dy = -C_0 V \, dx$$

or

$$dx = -D^* \, dy$$

where:

$$D^* = \frac{C_{s\infty} V_w}{C_0 V} \qquad \text{is a } distribution \ coefficient$$

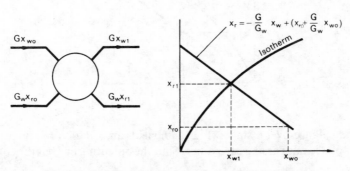

FIG. 7.39. Graphical solution for a single equilibrium stage

If sorbent and fluid are added and removed at the correct rates, the batch process described can become a steady state flow process. For small concentrations of sorbate in the fluid phase, a mass balance can be written:

$$G_w(x_{r1} - x_{r0}) = G(x_{w0} - x_{w1}) \qquad \qquad \dots (7.86)$$

where x_r is the mass of sorbate on unit mass or sorbate-free sorbent,
   x_w is the mass ratio of sorbate to carrier in the fluid,
   G is the mass flowrate of fluid, and
   G_w is the mass flowrate of sorbent.

If the equilibrium relationship is known, the effluent values x_{r1} and x_{w1} can be found graphically or by trial and error, in the general case (Fig. 7.39). From equation 7.86, it follows that the point (x_{r1}, x_{w1}) lies on a straight line through the point (x_{r0}, x_{w0}) with a slope $- G/G_w$. It is assumed that conditions are isothermal.

Multistage countercurrent operations. When several equilibrium stages are connected in series, the separation that results from the countercurrent contacting of fluid and sorbent can be calculated in a way similar to that used for other separation processes.

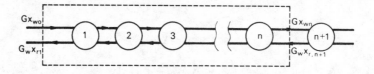

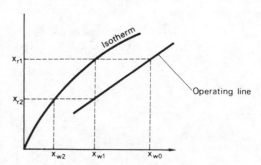

FIG. 7.40. Multistage countercurrent sorption

A mass balance over n stages (Fig. 7.40) yields an equation relating the compositions of streams passing each other. Thus the operating line has the equation:

$$x_{r,n+1} = \frac{G}{G_w} x_{wn} + \left(x_{r1} - \frac{G}{G_w} x_{w0} \right)$$

The number of stages necessary to achieve the required separation can be found by stepping off on the diagram. Pinch conditions can occur, and these set a minimum to the sorbent: fluid ratio for a particular isotherm. The situation is analogous to that found in absorption or distillation.

Moving bed—solids in plug flow. If the multiple stages of the previous section are merged into a continuously moving bed, the equivalent number of equilibrium stages can be found by stepping off between equilibrium and operating lines. But since little is known about the height of a sorption equilibrium stage, it is preferable to use the transfer unit concept (Volume 2, Chapter 11). A mass balance across an increment of the bed can be written:

$$u\varepsilon A \, dC = -k_g A(1 - \varepsilon) a_p (C - C_i) \, dz \qquad \ldots (7.87)$$

and rearranged to give:

$$\frac{k_g a_p z_e}{mu} = \int_{C_B}^{C_E} \frac{dC}{C - C_i} \qquad \ldots (7.88)$$

where C_E and C_B are the fluid sorbate concentrations into and out of the moving bed.

The integral in equation 7.88 represents the number of transfer units. In general the interface concentration C_i will not be known. It may be possible, however, to express the rate of transfer in terms of an overall driving force $(C - C^*)$ where C^* is the fluid concentration in equilibrium with the mean sorbed phase concentration. Then:

$$z_e = \frac{mu}{k_0 a_p} \int_{C_B}^{C_E} \frac{dC}{C - C^*} \qquad \ldots (7.89)$$

where k_0 is the overall mass transfer coefficient, and
z_e is the length of the moving bed which equals the length of the sorption zone.

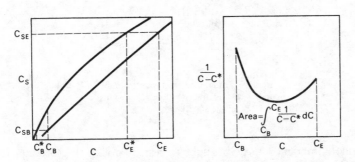

FIG. 7.41. Graphical calculations for the number of transfer units

In equation 7.89 the integral can be found graphically, as illustrated in Fig. 7.41. The factor outside the integral sign represents the height of a transfer unit, which, if known, enables the length of the bed for a given duty to be calculated. The ratio of fluid to solid flow can be obtained from the slope of the operating line in Fig. 7.41.

Points within the bed at a distance z can be related to the total distance z_e by the relationship:

$$\frac{z}{z_e} = \frac{\int_{C_B}^{C} \frac{dC}{(C - C^*)}}{\int_{C_E}^{C_E} \frac{dC}{(C - C^*)}} \qquad \dots (7.90)$$

Again, equation 7.90 can be solved graphically and z/z_e plotted as a function of C/C_0.

Fixed bed—transfer units. The transfer unit approach can be used for fixed beds operating with a constant sorption zone. Unlike moving beds, fixed beds must be long enough not only to contain the sorption zone but also to allow it a convenient residence time, which may be several hours. By definition, the sorption zone is that part of the bed in which mass transfer occurs, but, because the approach to equilibrium at the high and low concentration ends of the zone will be gradual, it is convenient to impose arbitrary limits on the extent of the zone. Let $C_B = 0.03C_0$, and $C_E = 0.97C_0$, or some such convenient fractions.

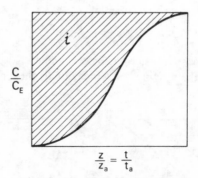

FIG. 7.42. Dimensionless breakthrough curve showing fractional unsaturation of the sorption zone

The zone is imagined to be brought to rest by giving the sorbent an equal and opposite velocity. Its length can be calculated as described for moving beds, and its fractional unsaturation i from a plot of C/C_0 vs. z/z_a, similar to Fig. 7.42.

The time to the breakpoint t_b of the bed can be found from a mass balance:

$$u A \varepsilon C_0 t_b = (z' - i z_a) C_{s\infty} A (1 - \varepsilon)$$

The shape of the breakthrough wave subsequent to breakpoint can be found from the proportionality between z' and t:

$$z' = \frac{(t - t_b)}{t_a} z_a$$

where t_a is the time for the sorption zone to move its own length, and
$\qquad z'$ is measured from the inlet of the sorption zone.

The method is described fully by TREYBAL[6] and illustrated in the example.

Example †

A bed is packed with silica gel as described in the previous example. Using the transfer unit concept, obtain a relationship between the effluent concentration and time. It can be assumed that the air leaves the bed in equilibrium with the sorbent and that the sorbent at the inlet to the bed is in equilibrium with the feed humidity. Experiments to find the relative resistances of the external gas film and of the sorbent gave the transfer coefficients per unit volume, referred to driving forces expressed as a difference in mass ratios, as $k'_p a_z = 0.964 \text{ kg/m}^3 \text{ s}$ and $k'_g a_z = 31.48\, G^{0.55} \text{ kg/m}^3 \text{ s}$ where G' is the mass flowrate of dry air per unit cross-section of the bed.

Solution

The sorption zone will be arbitrarily assumed to stretch from $y_r = 0.0001$ to 0.0024, where y_r is the mass ratio of moisture to dry air. The steady state velocity of the zone can be calculated by regarding the sorbent as having a velocity u_w sufficient to bring the zone to rest. Then:

$$671 \times u_w \times 0.078 = 0.1292 \times 0.0024$$

$$u_w = 5.93 \times 10^{-6} \text{ m/s}$$

To estimate the height of a sorption zone, it is necessary to calculate both the height and the number of transfer units. By analogy with the derivation on page 577, the number of transfer units is:

$$\int_{0.0001}^{0.0024} \frac{dy_r}{y_r - y_r^*}$$

where y_r^* is the humidity in equilibrium with the sorbed phase concentration corresponding to y_r on the operating line (Fig. 7.38). The integral is evaluated graphically from a plot of y_r against $\dfrac{1}{y_r - y_r^*}$. The values are tabulated below. The zone contains 10.95 transfer units. In terms of the overall mass transfer coefficient k'_0, the height of a transfer unit is $G'/k'_0 a_z$. The overall coefficient can be regarded as comprising contributions from a gas film coefficient k'_g and a hypothetical solid film coefficient k'_p. Now:

$$\frac{1}{k'_0 a_z} = \frac{1}{k'_g a_z} + \frac{m''}{k'_p a_z}$$

where m'' is the mean slope of the equilibrium curve over the concentration change of interest. In this case it is also the slope of the operating line $(0.00267/0.084)$. Substituting in the above equation when $G' = 0.1292 \text{ kg/m}^2 \text{ s}$

$$k'_0 a_z = 7.65 \text{ kg/m}^3 \text{ s}$$

Then:

$$\frac{G'}{k'_0 a_z} = \frac{0.1292}{7.65} = 0.0168 \text{ m}$$

The length z_a of the sorption zone $= 0.0168 \times 10.95 = 0.184$ m. The shape of the sorption

† This example is based on one given by TREYBAL[6].

curve can be found by plotting z'/z_a against y_e/y_{e0}, where y_e is the humidity at a distance z' into the sorption zone, and y_{e0}, the humidity at the inlet of the bed.

$$\frac{z'}{z_a} = \int_{y_{rB}}^{y_r} \frac{dy_r}{y_r - y_r^*} \bigg/ \int_{y_{rB}}^{y_{rE}} \frac{dy_r}{y_r - y_r^*}$$

where subscripts E and B denote conditions at the inlet to and outlet from the sorption zone. This ratio is tabulated below for values of y_r. When plotted the ratio shows that the unsaturated fraction of the sorption zone is 0·55. The time to the breakpoint of the bed can now be calculated from a mass balance over the whole bed.

$$0\cdot1292 \times 0\cdot00267 t_b = (1 - 0\cdot184 \times 0\cdot55)\,671 \times 0\cdot084$$

$$t_b = 33,120 \text{ s} = 9\cdot2 \text{ h}$$

From breakpoint to saturation the sorption zone moves its own length in a time t_a, where:

$$t_a = \frac{z_a}{u_w} = 8\cdot6 \text{ h}$$

Between breakpoint and saturation, times are given by:

$$t = t_b + \frac{z'}{z_a} t_a$$

Values are tabulated.

y_r	y_r^*	$\dfrac{1}{y_r - y_r^*}$	$\displaystyle\int_{y_{rB}}^{y_r} \dfrac{dy_r}{y_r - y_r^*}$	$\dfrac{z'}{z_a}$	$\dfrac{y_r}{y_{r0}}$	t (h)
$y_{rB} = 0\cdot0001$	0·00005	20,000			0·038	9·2
0·0002	0·00010	10,000	1·50	0·137	0·075	10·4
0·0006	0·00032	3,570	4·00	0·362	0·225	12·3
0·0010	0·00062	2,630	5·18	0·473	0·374	13·3
0·0014	0·00100	2,500	6·13	0·560	0·525	13·7
0·0018	0·00153	3,700	7·38	0·674	0·674	15·0
0·0022	0·00204	6,250	9·33	0·852	0·825	16·3
$y_{rE} = 0\cdot0024$	0·00230	10,000	10·95	1·000	0·899	16·9

Elution chromatography. A staged approach has been used to describe the operation of a chromatographic column. It assumes that equilibrium is attained between the phases in contact on a stage, and that the equilibrium relationship is linear. Longitudinal diffusion is considered to be negligible. The column is regarded as being composed of a number of stages, and the flow of mobile phase is considered as a series of discontinuous additions, each equal in volume to the free volume per stage.

When unit volume of sorbate is placed on the first plate, the sorbate becomes distributed so that a volume fraction x_v goes to the sorbed phase and $1 - x_v$ to the fluid. A volume of carrier fluid Δv, equal to the free volume of a stage, is added to the column. It sweeps the sorbate in the fluid phase of the first plate into the second. On plate 1 the sorbate that remains in the sorbed phase becomes redistributed in the same ratio as before. x_v^2 remains on the sorbent and $x_v(1 - x_v)$ enters the fluid. On plate 2 the sorbate carried forward, $(1 - x_v)$, is distributed, giving $x_v(1 - x_v)$ to the sorbed phase and leaving $(1 - x_v)^2$ in the fluid. When a second increment of fluid Δv is added, sorbate is carried from plate 1, to and from plate 2, to plate 3. Since the equilibrium distribution ratio is the same at all concentrations, and since operation is assumed to be isothermal, the new condition can be worked out. Table 7.5 summarises the position regarding the total amount of sorbate on each plate after four additions of carrier

TABLE 7.5. *Binomial distribution of sorbate over the theoretical plates of a packed column*

Volume increments of carrier added	Plate number				
	1	2	3	4	5
0	1				
1	x_v	$1 - x_v$			
2	x_v^2	$2x_v(1 - x_v)$	$(1 - x_v)^2$		
3	x_v^3	$3x_v^2(1 - x_v)$	$3x_v(1 - x_v)^2$	$(1 - x_v)^3$	
4	x_v^4	$4x_v^3(1 - x_v)$	$6x_v^2(1 - x_v)^2$	$4x_v(1 - x_v)^3$	$(1 - x_v)^4$

fluid. The distribution after q additions of fluid is given by the terms of the binomial expansion:

$$[x_v + (1 - x_v)]^q$$

The amount of sorbate on the $(n + 1)$th plate after q additions of fluid is:

$$Q'_{n+1} = \frac{q!(x_v)^{q-n}(1 - x_v)^n}{n!(q - n)!} \qquad \ldots (7.91)$$

If q is large, $q!/(q - n)! \approx q^n$ and $x_v^{q-n} \approx [1 - (1 - x_v)]^q = e^{-q(1-x_v)}$. Stirling's approximation for a factorial is given by:

$$n! = e^{-n} n^n (2\pi n)^{\frac{1}{2}}$$

if n is greater than about 10.

Substituting in equation 7.91 for $n!$ gives:

$$Q'_{n+1} = \frac{[q(1 - x_v)]^n \exp [n - q(1 - x_v)]}{(2\pi n)^{\frac{1}{2}} n^n} \qquad \ldots (7.92)$$

The peak of the elution curve at any time is obtained by differentiating equation 7.92 at constant q, and equating to zero. The condition for a maximum is that $n = q(1 - x_v)$. Substituting for this value gives $Q'_{max} = (2\pi n)^{-\frac{1}{2}}$. Since the argument so far refers to unit volume of sorbate initially, an initial volume V will result in:

$$Q'_{max} = \frac{V}{(2\pi n)^{\frac{1}{2}}} \qquad \ldots (7.93)$$

A transfer unit approach which allows for various transport resistances has been described by van Deemter *et al.*[65]

Non-isothermal operation

The bulk of the theoretical work reported for fixed-bed sorption applies to isothermal systems. Sorption from the gas phase may approach the isothermal condition when the sorbate concentration in the gas is low. Isothermal operation can also occur if the bed initially contains a sorbate which is displaced by a more strongly sorbed component in the gas. The net heat of the sorption is then the difference between the heat of sorption of the one component and the heat of desorption of the other. Sorption from the liquid phase is more likely to be isothermal than that from the gas phase, because of the lower heats of sorption and because of the greater heat capacity of the liquid phase.

Truly isothermal operation requires that the temperature throughout fluid and solid be constant; it can be approximated in laboratory equipment. Truly adiabatic operation requires that the heat generated in the bed by the sorption process be conserved in the fluid and solid phases, leaving neither to warm the walls of the containing vessel nor as heat loss through the walls to the surroundings. The latter condition can be approximated in large scale industrial plant if the thermal capacity of the containing vessel is small compared with that of the sorbent. The sorbent itself is a good insulator, so heat losses from large diameter beds will be a small proportion of that generated.

Quasi-isothermal operation

A quasi-isothermal condition of operation is commonly found in narrow laboratory columns which are cooled to remove the heat of sorption. Thermocouples inserted in the voids between granules indicate only small changes in temperature, so the operation is often thought to be isothermal. It is in the solid that the biggest temperature changes occur, however. Sorption occurs in the solid, and the temperature increases until heat is transferred to the surrounding fluid at the same rate as it is generated. The biggest resistance to heat transfer and, therefore, the biggest temperature gradient occurs in the laminar film surrounding a granule or adjacent to the walls of the containing vessel. By comparison, temperature gradients through the solid are small. Experiments on the sorption of water vapour in discrete spherical pellets of alumina have shown temperature rises within the pellets that varied from 4°C when the external relative humidity was constant at 9·3 per cent to 20°C when it was 96·7 per cent[104]. The same pellets used in a fixed bed would lose heat less readily because of the proximity of other pellets. But heat would be generated more gradually because of the slower exposure to the sorption wave experienced by all pellets except those at the bed inlet.

The quasi-isothermal operation of a column of activated alumina exposed to high humidity has been studied by Bowen and Donald[80]. A sectioned bed was weighed before and after operation, and the profile of the sorption wave plotted. It was found to consist of three zones corresponding to the three sections of the type IV isotherm of the system. The leading zone of the wave corresponded to the section of the isotherm associated with the formation of monomolecular layers; the second zone to the unfavourable section of the isotherm, associated with multilayer formation. A third zone corresponded to

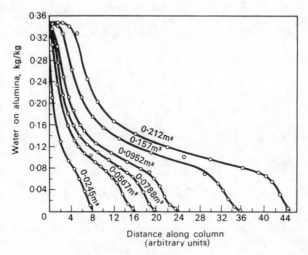

FIG. 7.43. Distribution of sorbate along a column in quasi-isothermal operation with a Type IV isotherm
Air flow $8·8 \times 10^{-6}$ m³/s. Temperature 303 K. Relative humidity of feed 95 per cent. The volumes to breakpoint for each bed are indicated on the curves

the capillary condensation section of the isotherm and merged into the saturated sorbent at the inlet end of the bed The three zones are shown clearly in Fig. 7.43. In this system the leading zone is self-sharpening, and soon reaches a constant profile. It was observed experimentally that the volume to breakpoint of the bed did not change significantly once the inlet concentration exceeded that corresponding to the upper limit of the first favourable section of the isotherm. It follows from the self-contained nature of the leading zone that it is possible to predict the breakpoint and the first part of the breakthrough curve for the system, if the analysis be confined to the leading zone. An effective feed humidity is used, equal to the upper limit referred to above.

Using the quadratic driving force equation to correlate single pellet data, RIMMER[104] has obtained close agreement between prediction and experiment for the quasi-isothermal case. Figure 7.44 shows breakthrough curves which are noteworthy for the agreement at breakpoint. Agreement in the region of $C/C_0 = 0·5$ is more readily obtained.

The work with quasi-isothermal columns indicates that isothermal theory can be applied to their solution if suitable mean properties of the system are found by experiment. But the design of large beds is not possible by this technique because of the appreciable temperature changes that occur both in the pellets and in the fluid. In these latter circumstances a complete analysis of mass and thermal fluxes is necessary.

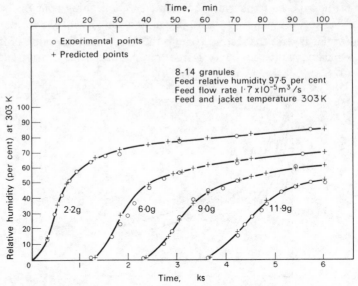

FIG. 7.44. Breakthrough curves for quasi-isothermal operation calculated using the quadratic driving force equation. The mass of sorbent in a bed is indicated against each curve

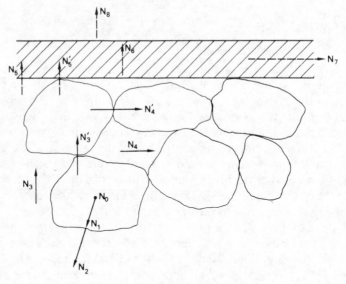

FIG. 7.45. Transport fluxes in a packed bed

The general condition of non-isothermal operation

The transport fluxes N that are possible in a fixed bed in which sorption occurs are shown in Fig. 7.45. The full arrows indicate heat and mass fluxes, the broken arrows heat flux only.

N_0, N_1 — are the intragranular processes. The sorption step itself and the associated liberation of the heat of sorption are subscripted 0. Both are assumed to be instantaneous, so equilibrium is assumed to exist between fluid and solid at the point of sorption. The second intragranular process, subscript 1, is the diffusion of sorbate molecules into the granule to the point of sorption, and the counter-movement of heat by conduction.

N_2 — is the transport of heat and mass between fluid and granule across the boundary film.

N_3, N'_3 — are the radial dispersions of heat and mass. N_3 occurs in the fluid and is assumed to be sufficiently fast for there to be no radial temperature or concentration gradients across the bed. N'_3 will be small because of the point contact between particles and the small area of stagnant fluid surrounding the point.

N_4, N'_4 — The convective flow N_4 of heat and mass in an axial direction is the main component of transport in that direction. By comparison, N'_4, the transfer by conduction and diffusion will be small.

N_5, N'_5 — Heat N_5 will be transferred to the wall of the vessel across the boundary film. By comparison, heat conducted through the point contact between wall and granule is likely to be small.

N_6, N_7 — Heat will be conducted through the walls in a radial and in an axial direction. It can be assumed that there is no radial variation in temperature in the wall. N_7 is likely to be small in thin-walled vessels.

N_8 — is the heat lost to the surroundings.

Radiant heat effects are normally neglected.

One of the most comprehensive analyses to the present time, by MEYER and WEBER[88], has taken into account fluxes N_1, N_2, N_4, N_5, N_8. When $N_8 = 0$, the condition is one of quasi-, or damped, adiabatic operation. It is a condition likely to be achieved in a laboratory column designed to simulate adiabatic operation. The independent variables are distance along the bed, time and radial position within a granule which is assumed to be spherical.

The dependent variables of interest are:

> the sorbate concentration in the bulk stream,
> the sorbate concentration in the pore fluid at a point in the granule,
> the interstitial velocity of the fluid,
> the fluid temperature,
> the temperature at a point in the granule, and
> the temperature of the wall.

The six unknowns can be found from six independent equations. Six such equations connecting these variables can be obtained from:

> (a) a component balance and heat balance in a single pellet,
> (b) a total balance, a component balance and a heat balance along a fixed bed, and
> (c) a balance of heat leaving the column.

W

Numerical solutions of such equations have been obtained by MEYER[103], MEYER and WEBER[88] and LEE and WEBER[102] using the method of characteristics[64].

Only when fluxes N_5 to N_8 are zero, will truly adiabatic conditions exist. The conditions are as unlikely to be obtained as are truly isothermal. The problem can be simplified somewhat by neglecting temperature differences through a granule. The temperature difference between a granule and the surrounding fluid at a point in the bed is then assumed to be confined to the external film. CARTER[86, 91] has given adiabatic solutions for this case.

When adiabatic operation occurs under equilibrium conditions, analytical solutions can be obtained. The work of DE VAULT[41] and more recently that of AMUNDSON et al.[84] is relevant. The conditions to which these solutions refer are rarely found in a practical situation.

The regeneration of a fixed bed by a flow of hot gas is mathematically similar to the sorption process[94]. For this reason and because it is more difficult to devise apparatus which will give a convenient initial condition for the desorption equations, little experimental work has been done.

Empirical design

Nowhere in chemical engineering is the split between theory and practice more apparent than in the process design of industrial sorption equipment. The complexity of the phenomena occurring in non-isothermal equipment has led to the widespread use of empirical methods of design. The methods often have little or no fundamental basis but their continued use in particular areas has led to an accumulation of reliable design data.

Most empirical methods involve use of the concept of a *capacity at breakpoint* which does not change appreciably with operating conditions other than temperature. The capacity at breakpoint is defined as the mass per cent or fraction of sorbate retained by the sorbate-free bed up to the breakpoint of the fluid stream leaving the bed. These data have been found to plot linearly against the *bed relative concentration* at inlet[68]. The latter is defined as the ratio of sorbate concentration in the feed to the concentration of sorbate that would saturate the feed at the mean temperature of the bed during operation.

Algebraically, the capacity at breakpoint x_{rb} can be written:

$$x_{rb} = \frac{(z - z_a) A(1 - \varepsilon) C_{s\infty} M_w + z_a(1 - i) A(1 - \varepsilon) C_{s\infty} M_w}{zA(1 - \varepsilon) \rho_a}$$

where z_a is the length of the sorption zone,
         i is the unsaturated fraction of the zone,
      M_w is the molecular weight of the sorbate, and
      ρ_a is the density of the sorbent granules.

Hence:

$$x_{rb} = \left[1 - \left(\frac{z_a}{z} \right) i \right] \left(C_{s\infty} \frac{M_w}{\rho_a} \right) \qquad \ldots (7.94)$$

It is clear from equation 7.94 that x_{rb} will be independent of bed length and of fluid velocity (which affects z_a and i) only if $(z_a/z)i \ll 1$.

A linear relationship between x_{rb} and the relative concentrations of the feed $\mathcal{H}_0$ can be predicted for the isothermal operation of some systems. The capacity at breakpoint can be written:

$$x_{rb} = \frac{uA\varepsilon C^* M_w \mathcal{H}_0 t_b}{W_0}$$

where C^* is the saturation concentration of sorbate in the feed,

$\mathcal{H}_0$ is the ratio of the actual concentration of sorbate in the feed to the saturation value,

t_b is the time to breakpoint, and

W_0 is the mass of the sorbate free sorbent.

Now:

$$\frac{\partial x_{rb}}{\partial \mathcal{H}_0} = \frac{uA\varepsilon C^* M_w}{W_0}\left(t_b + \mathcal{H}_0\frac{\partial t_b}{\partial \mathcal{H}_0}\right)$$

At constant temperature, fluid velocity and bed length, it has been shown experimentally[80] that t_b is constant for medium and high values of $\mathcal{H}_0$. It follows that in this range of $\mathcal{H}_0$, there is a linear relationship between x_{rb} and $\mathcal{H}_0$.

A mean temperature for the bed during adiabatic operation can be found from an overall heat balance:

$$uA\varepsilon C_0 \lambda t_b = zA\varepsilon C_M c_p(T - T_0) + zA(1 - \varepsilon)\rho_a S_c(T - T_0)$$
$$+ zA(1 - \varepsilon)\rho_a x_{rb} S_s(T - T_0) + uA\varepsilon C_M c_p(T - T_0)t_b \qquad \ldots(7.95)$$

where T_0 is the initial temperature of the bed,

c_p is the mean molar heat capacity of the fluid,

S_c, S_s are the specific heats of the sorbent and sorbate, and

C_M is the mean molar density of the mixture.

Now:

$$t_b = \frac{x_{rb} z A(1 - \varepsilon)\rho_a}{uA\varepsilon C_0 M_w}$$

Substituting for t_b in equation 7.95 and rearranging gives:

$$T = T_0 + \frac{\lambda x_{rb}}{\dfrac{mC_M c_p M_w}{\rho_a} + S_c M_w + x_{rb}S_s M_w + \dfrac{C_M c_p x_{rb}}{C_0}}$$

or

$$T \approx T_0 + \frac{\lambda C_0}{C_M c_p}$$

Fixed-bed arrangement. Capacity at breakpoint can be a useful concept when sorber beds and associated regenerating equipment are being examined for overall thermal requirements. A unit such as that shown in Fig. 7.8 consists of two beds, one of which is on-line and one regenerating, a heater and a cooler condenser. If the mode of operation is such that the full flow of process gas is available for regeneration, one criterion for continuous operation, namely:

$$\text{Time sorbing} \geqslant \text{Time heating} + \text{Time cooling}$$

can be given an algebraic significance.

(a) *Time sorbing*

$$t_s = \frac{W_0 x_{rw}}{u A \varepsilon C_0 M_w} \qquad \dots (7.96)$$

where x_{rw} is the working capacity of the sorber, the mass fraction of sorbate in the bed after time t_s.

(b) *Time heating*

$$\text{Latent heat requirement} = u A \varepsilon \, C_0 t_s \lambda$$

$$\text{Sensible heat requirement} = W' S_c \Delta T_h$$

where W' is the mass of the sorbent plus the equivalent mass of the sorbed phase and the containing vessel; i.e.:

$$W' = W_0 + W_0 x_{rw} \frac{S_s}{S_c} + W_c \frac{S_w}{S_c}$$

S_c is the specific heat of the sorbent,
S_s is the specific heat of the sorbed phase,
S_w is the specific heat of the containing vessel,
ΔT_h is the increase in temperature of the bed during regeneration, and
W_c is the mass of the containing vessel.

$$\text{Heat supplied by the generating gas} = u A \varepsilon C_M c_p \Delta T'_h t_h$$

where t_h is the heating time, and
$\Delta T'_h$ is the mean temperature difference between the inlet and outlet regenerating gas during heating.

Equating the heat supplied to the heat required, and substituting for $u A \varepsilon$ from equation 7.96 yields:

$$t_h = \left[\frac{W' S_c \Delta T_h M_w}{W_0 C_M c_p \Delta T'_h x_{rw}} + \frac{\lambda}{C_M c_p \Delta T'_h} \right] C_0 t_s \qquad \dots (7.97)$$

(c) *Time cooling*

$$\text{Heat to be removed from the bed} = W' S_c \Delta T_c$$

$$\text{Heat gained by the cooling gas} = u A \varepsilon C_M c_p \Delta T'_c t_c$$

where ΔT_c is the decrease in temperature of the bed during cooling, and
$\Delta T'_c$ is the mean temperature difference between the inlet and outlet
regenerating gas during cooling.

Equating the heat lost from the bed to that gained by the cooling gas, and
substituting for $uA\varepsilon$ from equation 7.96 gives:

$$t_c = \frac{W'S_c\Delta T_c C_0 M_w t_s}{W_0 C_M c_p \Delta T'_c x_{rw}} \qquad \dots (7.98)$$

The object of the cooling period is to return the bed to its temperature before
regeneration. It will be assumed that $\Delta T_h = \Delta T_c$ and also that $\Delta T'_h = \Delta T'_c$ for
continuous operation. Also:

$$t_s \geqslant t_h + t_c$$

Substituting for times from equations 7.96–7.98 and rearranging gives:

$$x_{rw} \geqslant \frac{2(W'/W_0)S_c\Delta T_h M_w}{C_M c_p \Delta T'_h/C_0 - \lambda} = x_{rl} \qquad \dots (7.99)$$

where x_{rl} is the minimum loading the arrangement can tolerate for continuous
operation.

But there is a second restriction on x_{rw}; that it shall not exceed x_{rb}, the capacity
at breakpoint. The condition for the satisfactory operation of the two bed full-
flow regeneration arrangement becomes:

$$x_{rb} \geqslant x_{rw} \geqslant x_{rl} \qquad \dots (7.100)$$

The argument can be generalised for a heating flow which is n' times the full
flow of process gas and a cooling flow which is m' times. A factor J', greater than
unity, is included to allow for heat losses during heating. Then:

$$x_{rl} = \frac{(W'/W_0)M_w S_c\Delta T_h [1 + n'/m'J']}{\dfrac{n'C_M c_p \Delta T'_h}{J'C_0} - \lambda} \qquad \dots (7.101)$$

It can also be shown that the ratio of the heating to cooling times can be given by:

$$\frac{t_h}{t_c} = \frac{J'm'}{n'}\left[1 + \frac{\lambda x_{rw}}{(W'/W_0)M_w S_c\Delta T_h}\right] \qquad \dots (7.102)$$

In the equations developed, the quantities ΔT_h and $\Delta T'_h$ must be given values.
There is no ambiguity about ΔT_h; it is the difference between the temperature
of the bed at the end of the heating period and its temperature at the beginning.
The value to be given to $\Delta T'_h$ is not equally apparent. An estimate can be obtained if it is assumed that a temperature wave moves through the bed during
regeneration in a manner analogous to the sorption wave. If the time taken for
the temperature zone to move its own length is t_a, then for a time $(t_h - t_a)$ the
heating stream will emerge at approximately the initial temperature of the bed.

During this time, the temperature difference is $T_h - T_0$, the suffixes referring to the temperatures of the hot regenerating gas and the temperature of the cold bed. For the remaining time t_a, the temperature difference will vary from $T_h - T_0$ to zero. An arithmetic mean would seem to be appropriate. Hence:

$$\Delta T'_h = \frac{(t_h - t_a)(T_h - T_0) + \frac{1}{2}t_a(T_h - T_0)}{t_h}$$

$$= (1 - \tfrac{1}{2}t_a/t_h)(T_h - T_0)$$

In general t_a is not known, but it will lie between 0 and t_h. Therefore:

$$(T_h - T_0) > \Delta T'_h > \tfrac{1}{2}(T_h - T_0)$$

In the absence of more accurate information, it can be assumed that:

$$\Delta T'_h = \tfrac{3}{4}(T_h - T_0) \qquad\qquad \dots\,(7.103)$$

It may be difficult in a two-bed system to satisfy the general requirement of equation 7.99. In that event, a three-bed system would be preferable. Such an arrangement consists of one bed sorbing, one being heated and one cooled at any instant. Of the last two operations, the time heating is the longer, so the criterion for continuous operation is $t_h \leqslant t_s$ which is more easily satisfied than the corresponding condition for the two-bed system[77].

Empirical methods that use the concept of capacity at breakpoint have served and still serve industry as a basis for design. This is not so much because of the reliability of such methods, but rather the formidable nature of the alternatives. But even the latter are only approximations of the real situation, making no allowances for radial variations of velocity, concentration or temperature; diffusivity is assumed constant. As long as these approximations remain, the most complex solutions can be no better than models of real systems.

Examples

A. In the first two columns of the table on page 592 are shown equilibrium data for the sorption of nitrogen on activated alumina at the normal boiling point of nitrogen. Test the applicability to these data of the following equilibrium theories: (a) Langmuir, (b) infinite BET, and (c) Harkins–Jura. For (a) and (b), obtain estimates of the surface area of the alumina and compare the values with that given by the point B method on Fig. 7.32.

B. Acetylene at 156·5 kN/m² is to be continuously dried from a dew-point of 303 K to a frost-point of less than 253 K using a fixed bed of activated alumina. If the acetylene may not be heated to a temperature greater than 433 K, would it be feasible to use a two-bed drying system with regeneration being carried out by the full flow of hot acetylene?

Assume the gas to be at 303 K at the inlet of the bed on-line and 433 K at the inlet of the bed being heated. The plant is well lagged so that only 20 per cent of the heat supplied for regeneration is lost extraneously. A suitable breakpoint capacity for the alumina in these conditions is 0·047[68]. The mass of the containing vessel can be taken as equal to that of the alumina contents.

For acetylene:	specific heat	= 1·843 kJ/kg K
	density at 293 K and 1 bar = 1·08 kg/m³	
For activated alumina:	specific heat	= 0·880 kJ/kg K
For mild steel:	specific heat	= 0·439 kJ/kg K
For water:	specific heat	= 4·187 kJ/kg K
	heat of sorption	= 2·51 MJ/kg

Solutions

A (a) Equation 7.12 can be rearranged and expressed in gas volumes:

$$\frac{(P/P^0)}{V} = \frac{(P/P^0)}{V^1} + \frac{1}{BV^i} \qquad \text{(equation 7.12)}$$

The equilibrium data are plotted as $\dfrac{(P/P^0)}{V}$ against (P/P^0) as shown below (Fig. 7.46). The equation is followed only at low relative pressures, suggesting that more than one sorbed layer is formed.

From the slope of the curve at low relative pressures, $V^1 = 7.96 \times 10^{-5}$ m³ nitrogen at 293 K, 10^2 kN/m². The corresponding surface area per kg of sorbent can be calculated from the area known to be associated with one adsorbed molecule in these conditions, 0.162 nm².

$$\text{Surface area} = \frac{79.6}{22400} \times 6.02 \times 10^{23} \times 16.2 \times 0.162 \times 10^{-18} = 3.46 \times 10^5 \text{ m}^2/\text{kg}$$

(b) The data conform to equation 7.14 to a higher value of relative pressures than in (a) (Fig. 7.46). From the slope of the line,

$$\frac{B-1}{V^1 B} = 0.014$$

and the intercept

$$\frac{1}{V^1 B} = 0.2 \times 10^{-3}$$

Hence $V^1 = 7.14 \times 10^{-5}$ m³ at 293 K, 10^2 kN/m². The corresponding area is 3.1×10^5 m²/kg sorbent.

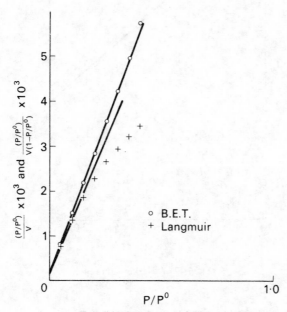

FIG. 7.46. Langmuir and BET plots

(c) From equation 7.21 a plot of $\ln P/P^0$ against $1/V^2$ should be linear. The plot is shown in Fig. 7.47.
Agreement in the middle range of relative pressures is good.
From Fig. 7.32 point B corresponds to about 7.5×10^{-5} m³ nitrogen or 3.26×10^5 m²/kg.

P/P^0	V gas at NTP 293 K, 10^2 kN/m²		$(P/P^0)\dfrac{1}{V} \times 10^3$	$(P/P^0)\dfrac{1}{V[1-(P/P^0)]} \times 10^5$	$\dfrac{1}{V^2} \times 10^4$
	cm³	10^5 m³			
0·05	66	6·6	0·76	0·80	2·30
0·10	74	7·4	1·35	1·50	1·83
0·15	81	8·1	1·85	2·18	1·52
0·20	88	8·8	2·27	2·83	1·29
0·25	94	9·4	2·66	3·55	1·13
0·30	102	10·2	2·94	4·20	0·96
0·35	109	10·9	3·21	4·94	0·84
0·40	117	11·7	3·42	5·73	0·73
0·50	138	13·8			0·53
0·60	165	16·5			0·37
0·70	196	19·6			0·26
0·80	221	22·1			0·20

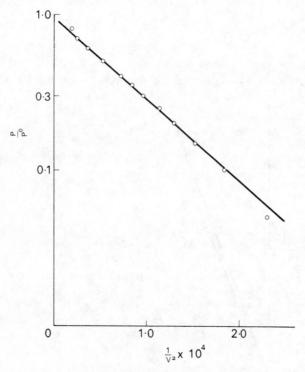

FIG. 7.47. A Harkins–Jura plot

B. For continuous operation the system must satisfy the criterion $x_{rb} \geqslant x_{rw} \geqslant x_{rl}$.
From equation 7.101 for $n' = m' = 1$ and $J' = 1·2$:

$$x_{rl} = \frac{(W'/W_0)S_c\Delta T_h M_w\left[1 + \dfrac{1}{1·2}\right]}{\dfrac{C_M c_p \Delta T_h'}{1·2\,C_0} - \lambda} \qquad \text{(equation 7.101)}$$

Now:

$$\frac{W'}{W_0} = 1 + x_{rw}\frac{S_s}{S_c} + \frac{W_c}{W_0}\frac{S_w}{S_c}$$

$$= 1 + x_{rw}\frac{4187}{880} + 1 \times \frac{439}{880}$$

To evaluate W'/W_0 assume the working capacity $x_{rw} = 0.04$.
Then:

$$W'/W_0 = 1.69.$$

Now:

$$\Delta T_n = 433 - 303 = 130 \text{ K}; \quad \Delta T'_n = \tfrac{3}{4} \times 130 = 97.5 \text{ K}$$

$$C_M = \frac{1.08 \times 10^{-6}}{26} \text{kmol/cm}^3; \quad c_p = 1.843 \times 26 \text{ kJ/kmol K}$$

At 303 K the vapour pressure of water is 4242 N/m². Therefore:

$$C_0 = \frac{4242}{156,500 - 4242} \times \frac{1}{22.4} \times \frac{273}{293} = 1.156 \times 10^{-3} \text{ kmol/m}^3 \text{ dry air at 293 K and 1 bar.}$$

The air leaving the bed on line is essentially free of water.
Substituting the above values into equation 7.101 gives:

$$x_{rl} = 0.068$$

Since $x_{rb} = 0.047$, the criterion for continuous operation cannot be satisfied by the full-flow regeneration arrangement outlined in the problem.

If a fan were used to circulate regenerating gas at twice the rate of the process flow, $n' = m' = 2$ and $x_{rl} = 0.027$. Such an arrangement would be workable.

If the number of beds were increased to three, the criterion for continuous operation would become $t_s \geqslant t_h$. The criterion would lead to a low value of x_{rl} that could be readily satisfied.

FURTHER READING

Necessarily, a chapter of this kind concentrates on classical sorption processes. New techniques and applications are continually being assessed for their commercial viability but these have not been discussed. In a recent study of research priorities, carried out for the Science Research Council, PRATT[1] comments on adsorption and ion exchange. He draws attention to the need for more work to be done on the development of continuous sorption columns, including ion exchange and chromatography. The process known as cycling zone adsorption,[2-5] in which bands of concentrated sorbate are generated within a bed by cyclic variations of temperature or pressure, has found few industrial applications so far. Advances in the field of "Adsorption and Ion Exchange" are regularly highlighted in the Symposium Series of that name published by the American Institute of Chemical Engineers.[6]

[1] PRATT, H. R. C.: Separation processes—research and development priorities (*SRC Report* 1977).

[2] BAKER, B., PIGFORD, R. L., and BLUM, D. E.: *Ind. Eng. Chem. (Fund.)* **8** (1969) 604. Equilibrium theory of parametric pumping.

[3] BAKER, B.: *Ind. Eng. Chem. (Fund.)* **9** (1970) 304. An equilibrium theory of the parametric pump.

[4] BAKER, B.: *Ind. Eng. Chem. (Fund.)* **9** (1970) 686. Parametric pumping and cycling zone adsorption.

[5] BAKER, B. and PIGFORD, R. L.: *Ind. Eng. Chem. (Fund.)* **10** (1971) 283. Cycling zone adsorption. Quantitative theory and experimental results.

[6] *Adsorption and Ion Exchange. A.I.Ch.E. Symposium Series* No. 152, **71** (1975) (also see numbers 14, 24, 74, 96, 117, 134).

REFERENCES TO CHAPTER 7

[1] YOUNG, T.: *Miscellaneous Works* (J. Murray 1855).

[2] FREUNDLICH, H.: *Colloid and Capillary Chemistry* (Methuen 1926).

[3] BARRER, R. M.: *Diffusion in and through Solids* (Cambridge U.P. 1941).

[4] BRUNAUER, S.: *The Adsorption of Gases and Vapours* (Oxford U.P. 1945).

[5] MANTELL, C. L.: *Adsorption* (McGraw-Hill 1951).

[6] TREYBAL, R. E.: *Mass Transfer Operations* (McGraw-Hill 1955).

[7] CARMAN, P. C.: *Flow of Gases through Porous Media* (Butterworth 1956).

[8] DE BOER, J. H. (ed.): *Second International Congress of Surface Activity*, Volume II (Butterworth 1957).

[9] EVERETT, D. H. and STONE, F. S. (ed.): *The Structure and Properties of Porous Materials* (Butterworth 1958).

[10] ADAMSON, A. W.: *Physical Chemistry of Surfaces* (Interscience 1960).

[11] AMBROSE, D. and AMBROSE, B. A.: *Gas Chromatography* (Newnes 1961).

[12] KNOX, J. H.: *Gas Chromatography* (Methuen 1962).

[13] NOGARE, S. D. and JUVET, R. S.: *Gas–Liquid Chromatography* (Interscience 1962).

[14] SCHOEN, H. M. (ed.): *New Chemical Engineering Separation Techniques* (Interscience 1962).

[15] PURNELL, H.: *Gas Chromatography* (Wiley 1962).

[16] HELFFERICH, F.: *Ion Exchange* (McGraw-Hill 1962).

[17] YOUNG, D. M., and CROWELL, A. D.: *Physical Adsorption of Gases* (Butterworth 1962).

[18] PERRY, J. H.: *Chemical Engineers' Handbook*, 4th edition (McGraw-Hill 1963).

[19] HAYWARD, D. O. and TRAPNELL, B. M. W.: *Chemisorption* (Butterworth 1964).

[20] ROSS, S. and OLIVIER, J. P.: *On Physical Adsorption* (Interscience 1964).

[21] KIPLING, J. J.: *Adsorption from Solutions of Non-electrolytes* (Academic Press 1965).

[22] DENBIGH, K. G.: *Principles of Chemical Equilibrium* (Cambridge U.P. 1966).

[23] GREGG, S. J. and SING, K. S. W.: *Adsorption Surface Area and Porosity* (Academic Press 1967).

[24] ARDEN, T. V.: *Water Purification by Ion Exchange* (Butterworth 1968).

[25] TSWETT, M.: *Ber. deut. botan. Ges.* **24** (1906) 316: Physikalische-chemische Studien über das Chlorophyll. Die Adsorptionen; 384. Adsorptionsanalyse und chromatographische Methode. Anwendung auf die Chemie des Chlorophylls.

[26] McBAIN, J. W.: *Phil. Mag.* **18** (1909) 916. The mechanism of adsorption ("sorption") of hydrogen by carbon.

[27] ZSIGMONDY, A.: *Z. anorg. Chem.* **71** (1911) 356. Über die Struktur des Gels der Kieselsäure.

[28] GURVITSCH, L.: *J. Phys. Chem. Russ.* **47** (1915) 805. O fiziko-khimicheskoi sil prityazheniya.

[29] LANGMUIR, I.: *J. Am. Chem. Soc.* **40** (1918) 1361. The adsorption of gases on plane surfaces of glass, mica and platinum.

[30] ANZELIUS, A.: *Z. angew. Math. und Mech.* **6** (1926) 291. Über Erwärmung vermittels durchströmender Medien.

[31] McBAIN, J. W. and BAKR, A. M.: *J. Am. Chem. Soc.* **48** (1926) 690. A new sorption balance.

[32] LOWRY, H. H. and OLMSTEAD, P. S.: *J. Phys. Chem.* **31** (1927) 1601. The adsorption of gases by solids with special reference to the adsorption of carbon dioxide by charcoal.

[33] EMMETT, P. H. and BRUNAUER, S.: *J. Am. Chem. Soc.* **59** (1937) 1553. The use of low temperature van der Waals adsorption isotherms in determining the surface area of iron synthetic ammonia catalysts.

[34] COHAN, L. H.: *J. Am. Chem. Soc.* **60** (1938) 433. Sorption hysteresis and vapour pressure of concave surfaces.

[35] BRUNAUER, S., EMMETT, P. H. and TELLER, E.: *J. Am. Chem. Soc.* **60** (1938) 309. Adsorption of gases in multimolecular layers. (Errata, see ref. 38.)

[36] MARGENAU, H.: *Rev. mod. Phys.* **11** (1939) 1. Van der Waals forces.

[37] BRUNAUER, S., DEMING, L. S., DEMING, W. E. and TELLER, E.: *J. Am. Chem. Soc.* **62** (1940) 1723. On a theory of the van der Waals adsorption of gases.

[38] EMMETT, P. H. and DE WITT, T.: *Ind. Eng. Chem.* (Anal.) **13** (1941) 28. Determination of surface areas.

[39] MARTIN, A. J. P. and SYNGE, R. L. M.: *Biochem. J.* **35** (1941) 1358. A new form of chromatogram employing two liquid phases.

[40] EMMETT, P. H.: *Advances in Colloid Science* **1** (1942) 1. The measurement of the surface areas of finely divided or porous solids by low temperature adsorption isotherms.

[41] DE VAULT, D.: *J. Am. Chem. Soc.* **65** (1943) 532. The theory of chromatography.

[42] HARKINS, W. D. and JURA, G.: *J. Chem. Phys.* **11** (1943) 431. An adsorption method for the determination of the area of a solid without the assumption of a molecular area and the area occupied by nitrogen molecules on the surface of solids. *J. Amer. Chem. Soc.* **66** (1944) 1366. Surface of solids. Part XIII.

[43] THOMAS, H. C.: *J. Am. Chem. Soc.* **66** (1944) 1664. Heterogeneous ion exchange in a flowing system.

[44] GEDDES, R. L.: *Trans. Am. Inst. Chem. Eng.* **42** (1946) 79. Local efficiencies of bubble plate fractionation.

[45] HILL, T. L.: *J. Chem. Phys.* **14** (1946) 441. Statistical mechanics of multimolecular adsorption: II. Localised and mobile adsorption and absorption.

[46] HOUGEN, O. A. and MARSHALL, W. R.: *Chem. Eng. Prog.* **43** (1947) 197. Adsorption from a fluid stream flowing through a stationary granular bed.

[47] BOYD, G. E., SCHUBERT, J., ADAMSON. A. W. and MYERS, L. S.: *J. Am. Chem. Soc.* **69** (1947) 2818, 2836, 2849. The exchange adsorption of ions from aqueous solutions by organic zeolites: Pts. I, II, III.

[48] HILL, T. L.: *J. Chem. Phys.* **15** (1947) 767. Statistical mechanics of multimolecular adsorption: III. Introductory treatment of horizontal interactions. Capillary condensation and hysteresis.

[49] GLUECKAUF, E., and COATES, J. I.: *J. Chem. Soc.* (1947) 1315. Theory of chromatography.

[50] DUBININ, M. M. and TIMOFEEV, D. P.: *J. Phys. Chem.* (USSR) **22** (1948) 133. Adsorbability and physicochemical properties of vapours.

[51] HILL, T. L.: *J. Chem. Phys.* **17** (1949) 762. Statistical mechanics of adsorption: VI. Localised unimolecular adsorption on a heterogeneous surface.

[52] EAGLE, S. and SCOTT, J. W.: *Ind. Eng. Chem.* **42** (1950) 1287. Liquid phase adsorption equilibrium and kinetics.

[53] LAPIDUS, L. and AMUNDSON, N. R.: *J. Phys. Coll. Chem.* **54** (1950) 821. Mathematics of adsorption in beds—radial flow.

[54] LEWIS, W. K., GILLILAND, E. R., CHERTOW, B. and CADOGAN, W. P.: *Ind. Eng. Chem.* **42** (1950) 1326. Adsorption equilibria. Pure gas isotherms.

[55] WHEELER, A.: *Advances in Catalysis* **3** (1951) 249. Reaction rates and selectivity in catalyst pores.

[56] McMILLAN, W. G. and TELLER, E.: *J. Chem. Phys.* **19** (1951) 25. The role of surface tension in multilayer gas adsorption.

[57] HIESTER, N. K. and VERMEULEN, T.: *Chem. Eng. Prog.* **48** (1952) 505. Saturation performance of ion exchange and adsorption columns.

[58] LAPIDUS, L. and AMUNDSON, N. R.: *J. Phys. Chem.* **56** (1952) 373. Mathematics of adsorption in fixed beds—The rate determining steps in radial adsorption analysis; **56** (1952) 984. The effect of longitudinal diffusion in ion exchange and chromatographic columns.

[59] ROSEN, J. B.: *J. Chem. Phys.* **20** (1952) 387. Kinetics of a fixed bed system for solid diffusion into spherical particles.

[60] HILL, T. L.: *Advances in Catalysis* **4** (1952) 211. Theory of physical adsorption.

[61] VERMEULEN, T.: *Ind. Eng. Chem.* **45** (1953) 1664. Theory for irreversible and constant pattern solid diffusion.

[62] LAPIDUS, L. and ROSEN, J. B.: *Chem. Eng. Prog.* Symp. Ser. No. 14, **50** (1954) 47. Experimental investigations of ion exchange mechanisms in fixed beds by means of an asymptotic solution.

[63] ROSEN, J. B.: *Ind. Eng. Chem.* **46** (1954) 1590. General numerical solutions for solid diffusion in fixed beds.

[64] ACRIVOS, A.: *Ind. Eng. Chem.* **45** (1956) 703. Method of characteristics technique. Application to heat and mass transfer problems.

[65] VAN DEEMTER, J. J., ZUIDERWEG, F. J. and KLINKENBERG, A.; *Chem. Eng. Sci.* **5** (1956) 271. Longitudinal diffusion and resistance to mass transfer as causes of nonideality in chromatography.

[66] CLUNIE, A. and GILES, C. H.: *Chem. and Ind.* No. 16 (1957) 481. Tumbling apparatus for liquid phase adsorption experiments.

[67] CRANSTON, R. W. and INKLEY, F. A.: *Advances in Catalysis* **9** (1957) 143. The determination of pore structures from nitrogen adsorption isotherms.

[68] MILLER A. W. and ROBERTS, C. W.: *Ind. Chemist* **34** (1958) 141. The drying of gases with activated alumina.

[69] VERMEULEN, T.: *Advances in Chemical Engineering* **2** (1958) 147. Separation by adsorption methods.

[70] BARRER, R. M.: *Brit. Chem. Eng.* **4** (1959) 267. New selective sorbents: porous crystals as molecular filters.

[71] KAGANER, M. G.: *J. Phys. Chem.* (USSR) **33** (1959) 2202. A new method for the determination of the specific adsorption surface of adsorbents and other finely dispersed substances.

[72] SETTER, N. J., GOOGIN, J. M. and MARROW, G. B.: *U.S.A.E.C. Report* Y–1257 (July 9th 1959). The recovery of uranium from reduction residues by semi-continuous ion exchange.

[73] SKARSTROM, C. W.: *Annals NY Acad. Sci.* **72** (1959) 751. Use of adsorption phenomena in automatic plant-type gas analysis.

[74] CARTER, J. W.: *Brit. Chem. Eng.* **5** (1960) 627. Adsorption drying of gases.

[75] JOHSWICH, F.: *Brennstoff-Wärme-Kraft* **14** (1962) 105. Abgasentschwefelung—Bedeutung und praktische Moeglichkeiten.

[76] BALLARD, D.: *Oil and Gas Journal* (June 3rd 1963) 97. How to get the most out of a short-cycle unit.

[77] BOWEN, J. H.: *Aust. Chem. Eng.* (Aug. 1963) 9. The arrangement of a fixed bed adsorption unit.

[78] ROWSON, H. M.: *Brit. Chem. Eng.* **8** (1963) 180. Fluid bed adsorption of carbon disulphide.

[79] STEELE, W. A.: *J. Phys. Chem.* **67** (1963) 2016. Monolayer adsorption with lateral interaction on heterogeneous surfaces.

[80] BOWEN, J. H. and DONALD, M. B.: *Chem. Eng. Sci.* **18** (1963) 599. Fixed bed adsorption from a high concentration feed; *Trans. Inst. Chem. Eng.* **42** (1964) T 259. Sizing adsorbers to take a high concentration feed.

[81] LIPPENS, B. C., LINSEN, B. G. and DE BOER, J. H.: *J. Catalysis* **3** (1964) 32. Studies on pore systems in catalysts I.

[82] FLEMING, J. B., GETTY, R. J. and TOWNSEND, F. M.: *Chem. Eng., Albany* **71** (Aug. 31st 1964) 69. Drying with fixed bed desiccants.

[83] BALLARD, D.: *Hydrocarbon Proc.* **44** (April 1965) 131. How to operate quick cycle plants.

[84] AMUNDSON, N. R., ARIS, R. and SWANSON, R.: *Proc. Roy. Soc.* **286** A (1965) 129. On simple exchange waves in fixed beds.

[85] MASAMUNE, S. and SMITH, J. M.: *A.I.Ch. E. Jl.* **11** (1965) 34. Adsorption rate studies—interaction of diffusion and surface processes; **11** (1965) 41. Adsorption of ethyl alcohol on silica gel.

[86] CARTER, J. W.: *Trans. Inst. Chem. Eng.* **44** (1966) T 253. A numerical method for the prediction of adiabatic adsorption in fixed beds.

[87] KATELL, S.: *Chem. Eng. Prog.* **62** (Oct. 1966) 67. Removing sulphur dioxide from flue gases.

[88] MEYER, O. A. and WEBER, T. W.: *A.I.Ch. E. Jl.* **13** (1967) 457. Non-isothermal adsorption in fixed beds.

[89] BOWEN, J. H., BOWREY, R. and MALIN, A. S.: *J. Catalysis* **7** (March 1967) 457. A study of the surface area and structure of activated alumina by direct observation.

[90] BROUGHTON, D. B.: *Chem. Eng. Prog.* **64** (Aug. 1968) 60. Molex. history of a process.

[91] CARTER, J. W.: *Trans. Inst. Chem. Eng.* **46** (1968) T 213. Isothermal case and adiabatic adsorption in fixed beds.

[92] COLLINS, J. J.: *Chem. Eng. Prog.* **64** (Aug. 1968) 66. Where to use molecular sieves

[93] CONDER, J. R.: *Advances in Analytical Chemistry and Instrumentation* **6** (1968) 209. Physical measurement by gas chromatography.

[94] OLSON, K. E., LUSS, D. and AMUNDSON, N. R.: *Ind. Eng. Chem. Process Design and Development* **7** (1968) 96. Regeneration of adiabatic fixed beds.

[95] RYAN, J. M., TIMMINS, R. S. and O'DONNELL, J. F.: *Chem. Eng. Prog.* **64** (Aug. 1968) 53 Production scale chromatography.

[96] SAWYER, D. T. and HARGROVE, G. L.: *Advances in Analytical Chemistry and Instrumentation* **6** (1968) 325. Preparative gas chromatography.

[97] TIMMINS, R. S., MIR, L. and RYAN, J. M.: *Chem. Eng., Albany* **75** (May 19th 1968) 170. Large scale chromatography: New separation tool.

[98] WEN, C. Y.: *Ind. Eng. Chem.* **60** (Sept. 1968) 34. Noncatalytic heterogeneous solid fluid reaction models.

[99] BOWEN, J. H. and LACEY, D. T.: *Chem. Eng. Sci.* **24** (1969) 965. A single pellet prediction of fixed bed behaviour.

[100] CORTELYOU, C. G.: *Chem. Eng. Prog.* **65** (Sept. 1969) 69. Commercial processes for SO_2 removal.

[101] SCHWARTZ, A. M.: *Ind. Eng. Chem.* **61** (Jan. 1969) 10. Capillarity—theory and practice.

[102] LEE, R. G. and WEBER, T. W.: *Can. J. Chem. Eng.* **47** (1969) 54. Isothermal adsorption in fixed beds; **47** (1969) 60. Interpretation of methane adsorption on activated carbon by non-isothermal and isothermal calculations.

[103] MEYER, O. A.: University of New York, Buffalo, Ph.D. Dissertation (1966). Dynamics of non-isothermal fixed bed adsorption.

[104] RIMMER, P. G.: University of Wales, Ph.D. Thesis (1970). A study of the effects of high concentration feeds to a fixed bed adsorber.

LIST OF SYMBOLS USED IN CHAPTER 7

A	Gross cross-sectional area of a bed	L^2
A_p	Surface area of a pore occupied by condensate	L^2
A_s	Area of a sorbed film	L^2
a	Fraction of available surface covered with sorbed molecules	—
a_A, a_B	Fractions of sorption surface occupied by sorbates $\mathbf{A}, \mathbf{B}$	—
a_m	Area occupied by one molecule in a sorbate film	L^2
a_0, a_1, a_2	Fraction of sorption surface which contains no sorbate, one layer, two layers of sorbed molecules	—
a_p	External area of granule per unit volume of granule	L^{-1}
a_z	External area of granule per unit volume of fixed bed	L^{-1}
a_{MA}, a_{MB}	Area occupied by one mol of $\mathbf{A}, \mathbf{B}$ in the sorbed state	$M^{-1}L^2$
a^1	Area of a monolayer per unit mass of sorbent	$M^{-1}L^2$
B	Constant for the BET equation defined by equation 7.9	—
b	Constant defined by equation 7.74	—
b'	Constant in Poiseuille's equation	L^2
b_i	Ratio of velocity constants $k_i \, k_i'$ for component	—
C	Concentration in mols per unit volume of sorbate in the fluid	ML^{-3}
C^*	Concentration in the fluid in equilibrium with the sorbed phase	ML^{-3}
C_A	Concentration of $\mathbf{A}$	ML^{-3}
C_{As}	Equivalent concentration of $\mathbf{A}$ in sorbed phase	ML^{-3}
C_B	Finite breakthrough concentration, e.g. $0.03\,C_0$	ML^{-3}
C_E	Finite saturation concentration, e.g. $0.97\,C_0$	ML^{-3}
C_i	Concentration of sorbate in the fluid at the external interface between granule and fluid	ML^{-3}
C_i^*	Interface concentration in equilibrium with the sorbed phase	ML^{-3}
C_M	Molar density	ML^{-3}
C_s	Concentration of the sorbed phase	ML^{-3}
C_{s0}	Initial concentration of sorbate on the sorbent	ML^{-3}
$C_{s\infty}$	Concentration of sorbed phase in equilibrium with C_0 in gas	ML^{-3}
C_s^*	Concentration of sorbed phase in equilibrium with the external fluid	ML^{-3}
C_0	Initial concentration of sorbate in the fluid entering the bed	ML^{-3}
c_p	Molar heat capacity	$L^2T^{-2}\theta^{-1}$
c_s'	Mass of solute per unit mass of sorbent	
D	Diffusivity for the bulk fluid	L^2T^{-1}
D'	Approximate diffusivity combining Knudsen and bulk diffusion effects	L^2T^{-1}
D_e	Effective diffusivity for a granule referred to fluid concentrations	L^2T^{-1}
$D_{e'}$	Effective diffusivity for a granule referred to the sorbed phase concentrations	L^2T^{-1}
D_K	Knudsen diffusivity	L^2T^{-1}
D_L	Longitudinal diffusivity	L^2T^{-1}
D^*	Distribution coefficient	—
E_0	Activation energy of desorption from empty surface	L^2T^{-2}
E_1, E_2	Activation energies of desorption from monolayers, bilayers	L^2T^{-2}
F'	Derivative of a function F defined by equation 7.74	—
f	Function defined by equation 7.25	—
f_1, f_2	Isotherm functions	—
$f'(C)$	Slope of the isotherm at a concentration C	—
f, f^0	Fugacities	$ML^{-1}T^{-2}$
G	Mass flowrate of fluid	MT^{-1}
G	Gibbs free energy of a sorbed film	ML^2T^{-2}
G_w	Mass flowrate of sorbent	MT^{-1}
$\mathscr{H}_0$	Relative concentration in the feed to a bed	—
i	Unsaturated fraction of a sorption zone	—
J	Geometric factor in equation 7.46	—

J'	Factor in equation 7.101	—
j	Factor relating mean and exit pressures	—
j_n	Numerical coefficient in equation 7.73	*
K	Equilibrium constant based on activities	*
K'	Constant in equation 7.22	$ML^{-1}T^{-2}$
K_a	Equilibrium constant for adsorption	*
K_{ax}	Equilibrium constant for adsorption based on mol fractions	—
K_c	Equilibrium constant based on concentrations	*
K_i	Equilibrium constant for ion exchange	*
K_x	Equilibrium constant based on mol fractions	—
k'	Desorption velocity constant	MT^{-1}
k''	Constant defined by equation 7.32	$M^{-1}LT^2$
k, k_0, k_1	Sorption velocity constant, general value, on empty surface, on monolayer surface, etc.	*
k_0	Film mass transfer coefficient allowing for internal and external diffusion	$M^{-1}L^3T^{-1}$
k_g	External film mass transfer coefficient	LT^{-1}
k_p	Hypothetical solid film mass transfer coefficient	LT^{-1}
k_N	Boltzmann's constant	$ML^2T^{-2}\theta^{-1}$
L	Constant in equation 7.21	—
L_p	Length of pore filled with condensate	L
M	Constant in equation 7.21	L^6
M_w	Molecular weight	—
m	Intergranular void ratio $\varepsilon/(1 - \varepsilon)$	—
m'	Factor in equation 7.101	—
m_m	Mass of a molecule	M
N	Avogadro number	M^{-1}
$N_0, N_1,$ etc.	Heat or mass fluxes referred to granule external surface	MT^{-3} or $ML^{-2}T^{-1}$
n	Number of layers of sorbate	—
n	Index in the Freundlich equation 7.33	—
n'	Factor in equation 7.101	—
n_0	Mols in the fluid phase initially	M
n_p	Number of pores	—
n_s	Mols of sorbate in a film	M
n_s^1	Mols of sorbate per unit mass of sorbent to complete monolayer	—
n_{sA}^1, n_{sB}^1	Mols of A or B to fill a monolayer on unit mass of sorbent	—
P	Total pressure of sorbate in a single sorbate system	$ML^{-1}T^{-2}$
$\bar{P}$	Mean total pressure	$ML^{-1}T^{-2}$
P_A	Partial pressure of A in a multi-sorbate system	$ML^{-1}T^{-2}$
P_e	Pressure at exit of bed	$ML^{-1}T^{-2}$
P_m	Vapour pressure over meniscus	$ML^{-1}T^{-2}$
P^0	Vapour pressure of sorbed phase	$ML^{-1}T^{-2}$
P_0	Pressure at the inlet of a column	$ML^{-1}T^{-2}$
Q_n'	Amount of sorbate on the nth plate	L^3
Q_s	Ultimate capacity of a sorbent	ML^{-3}
q	Number of additions in staged development	—
R	Gas constant	$L^2T^{-2}\theta^{-1}$
r	Radius within a spherical granule or a pore	L
r^*	Equilibrium parameter	—
r_i	Outer radius of a spherical granule	L
S_c	Specific heat of the sorbent	$L^2T^{-2}\theta^{-1}$
S_s	Specific heat of the sorbate	$L^2T^{-2}\theta^{-1}$
S_w	Specific heat of the containing vessel	$L^2T^{-2}\theta^{-1}$
T	Absolute temperature	θ
T_h	Temperature of hot regenerating gas	θ
T_0	Ambient or reference temperature	θ
$\Delta T_h, \Delta T_c$	Temperature differences of the sorbent during the heating and cooling stages of regeneration	θ

* Dimensions depend on order of process

$\Delta T'_h, \Delta T'_c$	Mean temperature differences of the gas during the heating and cooling stages of regeneration	θ
t	Time	T
t_a	Time for a sorption or temperature zone to move its own length	T
t_b	Time to breakpoint	T
t_c	Time of the cooling stage	T
t_h	Time of the heating stage	T
t_s	Time of the sorption stage	T
t_1	$= t - z/u$	T
u	Fluid intergranular velocity	LT^{-1}
$\bar{u}$	Mean velocity	LT^{-1}
u_m	Mean velocity of a molecule (RMS)	LT^{-1}
u_w	Velocity of sorption wave	LT^{-1}
u_0	Fluid velocity at the inlet to a column	LT^{-1}
V	Volume of fluid	L^3
V'	Volume flowrate of fluid	L^3T^{-1}
V''	Total volume of fluid that has entered unit cross-section of bed	L
V_c	Total volume of bed	L^3
V_g	Specific retention volume	$M^{-1}L^3$
$V''_{\min}$	Minimum volume of fluid to saturate a bed of unit cross-section with sorbate	L
V_0	Initial volume of solution	L^3
V_p	Volume of a pore filled with condensate	L^3
V_s	Volume of a sorbate	L^3
V_w	Total volume of sorbent	L^3
V_M	Molar volume of sorbate in the gas phase	$M^{-1}L^3$
V_T	Total volume of sorption space	L^3
V^1	Equivalent gas volume of sorbate in a complete monolayer	L^3
V_s^1	Monolayer volume for sorbate	L^3
$\bar{V}_s^1$	Mean monolayer volume for several components	L^3
V_{sA}, V_{sB}	Volumes sorbed of A and B	L^3
V_M	Liquid molar volume	$M^{-1}L^3$
v_s	Volume of sorbate on unit area of surface	L
v_s^1	Volume of sorbate in unit area of the monolayer	L
W	Total mass of bed	M
W'	Total mass of bed + equivalent mass of the containing vessel	M
W_c	Mass of containing vessel	M
W_0	Mass of empty sorbent in a bed	M
w_A, w_B	Mass of A, B in solution	M
w_{sA}, w_{sB}	Mass of A, B in the sorbed phase	M
w_0	Total mass of sorbates A + B	M
X	Distance parameter $kC_{s\infty}z/mu$	—
x	$= C/C_0$	—
x_1	$= z/mu$	T
x_A, x_B	Mol ratio of A, B to total sorbate in solution	—
x_m	Mol fraction	—
x_r	Mass ratio of sorbed phase to empty sorbent	—
x_{rb}	The value of x_r for a bed at its breakpoint—upper limit of bed working capacity	—
x_{rl}	Lower limit of bed working capacity	—
x_{rw}	Working capacity of sorber	—
x_v	Volume fraction	—
x_w	Mass fraction in fluid phase	—
x_{w0}	Mass fraction initially	—
Y	$(C_s - C_{so})/(C_s^* - C_{so})$	—
y	$C_s/C_{s\infty}$	—
y_A, y_B	Mol ratio of A, B to total exchangeable ion in a resin	—
z	Distance along a bed	L
z_a	Length of sorption zone	L
z_c	Distance to bed exit	L
z_0	Position at which concentration is initially C_s	L

Symbol	Description	Dimensions
z'	Distance measured from inlet of sorption zone	**L**
α, α_{1-9}	Constants in various equations	—
α_{10}	Separation factor	—
$\beta, \beta_{1-2}, \beta_{4-7}$	Constants in various equations	—
β_3	Coefficient of affinity	—
Γ	Two-dimensional or spreading pressure	$\mathbf{MT^{-2}}$
γ, γ_s	Activity coefficient for a component in the fluid, sorbed phase	—
Δ	Change in property	—
ε	Intergranular void fraction	—
ε_p	Sorption potential	$\mathbf{L^2T^{-2}}$
θ_1	Constant in equation 7.13	—
κ	Constant in equation 7.79	—
λ	Latent heat per mol	$\mathbf{L^2T^{-2}}$
λ'	Length parameter in equation 7.84	—
μ	Viscosity	$\mathbf{ML^{-1}T^{-1}}$
μ_g	Chemical potential of gas	$\mathbf{L^2T^{-2}}$
μ_s	Chemical potential of a sorbed film	$\mathbf{L^2T^{-2}}$
ν	Stoichiometric number	—
ν	Constant characteristic of pore structure (equation 7.49)	—
ρ	Mass density of the sorbed phase	$\mathbf{ML^{-3}}$
ρ_a	Mass density of the sorbent granules	$\mathbf{ML^{-3}}$
ρ_g	Mass density of the gas phase	$\mathbf{ML^{-3}}$
$\sigma_{sl}, \sigma_{sv}, \sigma_{lv}$	Surface tension of the three interfaces in a liquid–solid–vapour system	$\mathbf{MT^{-1}}$
τ	Time parameter	—
ϕ	Interfacial angle between liquid and solid	—
ϕ_1	Constant in equation 7.13	—
χ	Resistance parameter in equation 7.84	—
ψ	Function defined in equation 7.84	—

Appendix

TABLE A.1. *Laplace transforms*[a]

No.	Transform $\bar{f}(p) = \int_0^\infty e^{-pt} f(t)\, dt$	Function $f(t)$
1	$\dfrac{1}{p}$	1
2	$\dfrac{1}{p^2}$	t
3	$\dfrac{1}{p^n} \quad n = 1, 2, 3, \ldots$	$\dfrac{t^{n-1}}{(n-1)!}$
4	$\dfrac{1}{\sqrt{p}}$	$\dfrac{1}{\sqrt{(\pi t)}}$
5	$\dfrac{1}{p^{\frac{3}{2}}}$	$2\sqrt{\dfrac{t}{\pi}}$
6	$\dfrac{1}{p^{n+\frac{1}{2}}} \quad n = 1, 2, 3, \ldots$	$\dfrac{2^n t^{n-\frac{1}{2}}}{[1 \cdot 3 \cdot 5 \ldots (2n-1)]\sqrt{\pi}}$
7	$\dfrac{\Gamma(k)}{p^k} \quad k > 0$	t^{k-1}
8	$\dfrac{1}{p-a}$	e^{at}
9	$\dfrac{1}{(p-a)^2}$	$t e^{at}$
10	$\dfrac{1}{(p-a)^n} \quad n = 1, 2, 3, \ldots$	$\dfrac{1}{(n-1)!} t^{n-1} e^{at}$
11	$\dfrac{\Gamma(k)}{(p-a)^k} \quad k > 0$	$t^{k-1} e^{at}$

[a] By permission from *Operational Mathematics* by R. V. CHURCHILL, McGraw-Hill 1958.

TABLE A.1. *Laplace transforms (cont.)*

No.	Transform $\overline{f}(p) = \int\limits_0^\infty e^{-pt} f(t)\, dt$	Function $f(t)$
12	$\dfrac{1}{(p-a)(p-b)} \qquad a \neq b$	$\dfrac{1}{a-b}(e^{at} - e^{bt})$
13	$\dfrac{p}{(p-a)(p-b)} \qquad a \neq b$	$\dfrac{1}{a-b}(ae^{at} - be^{bt})$
14	$\dfrac{1}{(p-a)(p-b)(p-c)} \qquad a \neq b \neq c$	$-\dfrac{(b-c)e^{at} + (c-a)e^{bt} + (a-b)e^{ct}}{(a-b)(b-c)(c-a)}$
15	$\dfrac{1}{p^2 + a^2}$	$\dfrac{1}{a}\sin at$
16	$\dfrac{p}{p^2 + a^2}$	$\cos at$
17	$\dfrac{1}{p^2 - a^2}$	$\dfrac{1}{a}\sinh at$
18	$\dfrac{p}{p^2 - a^2}$	$\cosh at$
19	$\dfrac{1}{p(p^2 + a^2)}$	$\dfrac{1}{a^2}(1 - \cos at)$
20	$\dfrac{1}{p^2(p^2 + a^2)}$	$\dfrac{1}{a^3}(at - \sin at)$
21	$\dfrac{1}{(p^2 + a^2)^2}$	$\dfrac{1}{2a^3}(\sin at - at\cos at)$
22	$\dfrac{p}{(p^2 + a^2)^2}$	$\dfrac{t}{2a}\sin at$
23	$\dfrac{p^2}{(p^2 + a^2)^2}$	$\dfrac{1}{2a}(\sin at + at\cos at)$
24	$\dfrac{p^2 - a^2}{(p^2 + a^2)^2}$	$t\cos at$

TABLE A.1. *Laplace transforms (cont.)*

No.	Transform $\bar{f}(p) = \int\limits_{0}^{\infty} e^{-pt}f(t)\,dt$	Function $f(t)$
25	$\dfrac{p}{(p^2 + a^2)(p^2 + b^2)} \qquad a^2 \neq b^2$	$\dfrac{\cos at - \cos bt}{b^2 - a^2}$
26	$\dfrac{1}{(p - a)^2 + b^2}$	$\dfrac{1}{b} e^{at} \sin bt$
27	$\dfrac{p - a}{(p - a)^2 + b^2}$	$e^{at} \cos bt$
28	$\dfrac{3a^2}{p^3 + a^3}$	$e^{-at} - e^{at/2}\left(\cos\dfrac{at\sqrt{3}}{2} - \sqrt{3}\sin\dfrac{at\sqrt{3}}{2}\right)$
29	$\dfrac{4a^3}{p^4 + 4a^4}$	$\sin at \cosh at - \cos at \sinh at$
30	$\dfrac{p}{p^4 + 4a^4}$	$\dfrac{1}{2a^2} \sin at \sinh at$
31	$\dfrac{1}{p^4 - a^4}$	$\dfrac{1}{2a^3}(\sinh at - \sin at)$
32	$\dfrac{p}{p^4 - a^4}$	$\dfrac{1}{2a^2}(\cosh at - \cos at)$
33	$\dfrac{8a^3 p^2}{(p^2 + a^2)^3}$	$(1 + a^2 t^2)\sin at - at \cos at$
34	$\dfrac{1}{p}\left(\dfrac{p-1}{p}\right)^n$	$\dfrac{e^t}{n!}\dfrac{d^n}{dt^n}(t^n e^{-t})$ = Laguerre polynomial of degree n
35	$\dfrac{p}{(p - a)^{\frac{3}{2}}}$	$\dfrac{1}{\sqrt{(\pi t)}} e^{at}(1 + 2at)$
36	$\sqrt{(p - a)} - \sqrt{(p - b)}$	$\dfrac{1}{2\sqrt{(\pi t^3)}}(e^{bt} - e^{at})$

TABLE A.1. *Laplace transforms (cont.)*

No.	Transform $\bar{f}(p) = \int_0^\infty e^{-pt}f(t)\,dt$	Function $f(t)$
37	$\dfrac{1}{\sqrt{p}+a}$	$\dfrac{1}{\sqrt{(\pi t)}} - a\,e^{a^2 t}\mathrm{erfc}(a\sqrt{t})$
38	$\dfrac{\sqrt{p}}{p-a^2}$	$\dfrac{1}{\sqrt{(\pi t)}} + a\,e^{a^2 t}\mathrm{erf}(a\sqrt{t})$
39	$\dfrac{\sqrt{p}}{p+a^2}$	$\dfrac{1}{\sqrt{(\pi t)}} - \dfrac{2a\,e^{-a^2 t}}{\sqrt{\pi}}\displaystyle\int_0^{a\sqrt{t}} e^{\lambda^2}\,d\lambda$
40	$\dfrac{1}{\sqrt{p}(p-a^2)}$	$\dfrac{1}{a}e^{a^2 t}\mathrm{erf}(a\sqrt{t})$
41	$\dfrac{1}{\sqrt{p}(p+a^2)}$	$\dfrac{2e^{-a^2 t}}{a\sqrt{\pi}}\displaystyle\int_0^{a\sqrt{t}} e^{\lambda^2}\,d\lambda$
42	$\dfrac{b^2-a^2}{(p-a^2)(b+\sqrt{p})}$	$e^{a^2 t}[b - a\,\mathrm{erf}(a\sqrt{t})] - be^{b^2 t}\mathrm{erfc}(b\sqrt{t})$
43	$\dfrac{1}{\sqrt{p}(\sqrt{p}+a)}$	$e^{a^2 t}\mathrm{erfc}(a\sqrt{t})$
44	$\dfrac{1}{(p+a)\sqrt{(p+b)}}$	$\dfrac{1}{\sqrt{(b-a)}}e^{-at}\mathrm{erf}[\sqrt{(b-a)}\sqrt{t}]$
45	$\dfrac{b^2-a^2}{\sqrt{p}(p-a^2)(\sqrt{p}+b)}$	$e^{a^2 t}\left[\dfrac{b}{a}\mathrm{erf}(a\sqrt{t}) - 1\right] + e^{b^2 t}\mathrm{erfc}(b\sqrt{t})$
46	$\dfrac{(1-p)^n}{p^{n+\frac{1}{2}}}$	$\dfrac{n!}{(2n)!\sqrt{(\pi t)}}H_{2n}(\sqrt{t})$ where $H_n(t) = e^{t^2}\dfrac{d^n}{dt^n}e^{-t^2}$ is the Hermite polynomial
47	$\dfrac{(1-p)^n}{p^{n+\frac{3}{2}}}$	$-\dfrac{n!}{\sqrt{\pi}\,(2n+1)!}H_{2n+1}(\sqrt{t})$

TABLE A.1. *Laplace transforms (cont.)*

No.	Transform $\bar{f}(p) = \int_0^\infty e^{-pt}f(t)\,dt$	Function $f(t)$
48	$\dfrac{\sqrt{(p + 2a)}}{\sqrt{p}} - 1$	$ae^{-at}[I_1(at) + I_0(at)]$
49	$\dfrac{1}{\sqrt{(p + a)}\sqrt{(p + b)}}$	$e^{-\frac{1}{2}(a+b)t}I_0\left(\dfrac{a - b}{2}t\right)$
50	$\dfrac{\Gamma(k)}{(p + a)^k (p + b)^k} \qquad k > 0$	$\sqrt{\pi}\left(\dfrac{t}{a - b}\right)^{k-\frac{1}{2}} e^{-\frac{1}{2}(a+b)t}I_{k-\frac{1}{2}}\left(\dfrac{a - b}{2}t\right)$
51	$\dfrac{1}{\sqrt{(p + a)}(p + b)^{\frac{1}{2}}}$	$te^{-\frac{1}{2}(a+b)t}\left[I_0\left(\dfrac{a - b}{2}t\right)\right.$ $\left. + I_1\left(\dfrac{a - b}{2}t\right)\right]$
52	$\dfrac{\sqrt{(p + 2a)} - \sqrt{p}}{\sqrt{(p + 2a)} + \sqrt{p}}$	$\dfrac{1}{t}e^{-at}I_1(at)$
53	$\dfrac{(a - b)^k}{[\sqrt{(p + a)} + \sqrt{(p + b)}]^{2k}} \quad k > 0$	$\dfrac{k}{t}e^{-\frac{1}{2}(a+b)t}I_k\left(\dfrac{a - b}{2}t\right)$
54	$\dfrac{[\sqrt{(p + a)} + \sqrt{p}]^{-2j}}{\sqrt{p}\sqrt{(p + a)}} \quad j > -1$	$\dfrac{1}{a^j}e^{-\frac{1}{2}at}\,I_j\left(\dfrac{1}{2}at\right)$
55	$\dfrac{1}{\sqrt{(p^2 + a^2)}}$	$J_0(at)$
56	$\dfrac{[\sqrt{(p^2 + a^2)} - p]^j}{\sqrt{(p^2 + a^2)}} \quad j > 1$	$a^j J_j(at)$
57	$\dfrac{1}{(p^2 + a^2)^k} \qquad k > 0$	$\dfrac{\sqrt{\pi}}{\Gamma(k)}\left(\dfrac{t}{2a}\right)^{k-\frac{1}{2}} J_{k-\frac{1}{2}}(at)$
58	$[\sqrt{(p^2 + a^2)} - p]^k \quad k > 0$	$\dfrac{ka^k}{t}J_k(at)$
59	$\dfrac{[p - \sqrt{(p^2 - a^2)}]^j}{\sqrt{(p^2 - a^2)}} \quad j > -1$	$a^j I_j(at)$

TABLE A.1. *Laplace transforms (cont.)*

No.	Transform $\bar{f}(p) = \int_0^\infty e^{-pt} f(t)\,dt$	Function $f(t)$
60	$\dfrac{1}{(p^2 - a^2)^k} \qquad k > 0$	$\dfrac{\sqrt{\pi}}{\Gamma(k)} \left(\dfrac{t}{2a}\right)^{k-\frac{1}{2}} I_{k-\frac{1}{2}}(at)$
61	$\dfrac{e^{-kp}}{p}$	$S_k(t) = \begin{cases} 0 & \text{when } 0 < t < k \\ 1 & \text{when } t > k \end{cases}$
62	$\dfrac{e^{-kp}}{p^2}$	$\begin{cases} 0 & \text{when } 0 < t < k \\ t - k & \text{when } t > k \end{cases}$
63	$\dfrac{e^{-kp}}{p^j} \qquad j > 0$	$\begin{cases} 0 & \text{when } 0 < t < k \\ \dfrac{(t - k)^{j-1}}{\Gamma(j)} & \text{when } t > k \end{cases}$
64	$\dfrac{1 - e^{-kp}}{p}$	$\begin{cases} 1 & \text{when } 0 < t < k \\ 0 & \text{when } t > k \end{cases}$
65	$\dfrac{1}{p(1 - e^{-kp})} = \dfrac{1 + \coth \frac{1}{2}kp}{2p}$	$S(k, t) = n \quad \text{when } (n - 1)k < t < nk \\ \hspace{3cm} n = 1, 2, 3, \ldots$
66	$\dfrac{1}{p(e^{kp} - a)}$	$\begin{cases} 0 & \text{when } 0 < t < k \\ 1 + a + a^2 + \ldots + a^{n-1} \\ \hspace{1cm} \text{when } nk < t < (n + 1)k \\ \hspace{2cm} n = 1, 2, 3, \ldots \end{cases}$
67	$\dfrac{1}{p} \tanh kp$	$M(2k, t) = (-1)^{n-1} \\ \hspace{1cm} \text{when } 2k(n - 1) < t < 2kn \\ \hspace{2cm} n = 1, 2, 3, \ldots$
68	$\dfrac{1}{p(1 + e^{-kp})}$	$\dfrac{1}{2} M(k, t) + \dfrac{1}{2} = \dfrac{1 - (-1)^n}{2} \\ \hspace{1cm} \text{when } (n - 1)k < t < nk$
69	$\dfrac{1}{p^2} \tanh kp$	$H(2k, t) = \\ \begin{cases} t & \text{when } 0 < t < 2k \\ 4k - t & \text{when } 2k < t < 4k \end{cases}$
70	$\dfrac{1}{p \sinh kp}$	$2S(2k, t + k) - 2 = 2(n - 1) \\ \hspace{0.5cm} \text{when } (2n - 3)k < t < (2n - 1)k \\ \hspace{3cm} t > 0$

TABLE A.1. *Laplace transforms (cont.)*

No.	Transform $\bar{f}(p) = \int\limits_{0}^{\infty} e^{-pt} f(t)\, dt$	Function $f(t)$
71	$\dfrac{1}{p \cosh kp}$	$M(2k, t+3k) + 1 = 1 + (-1)^n$ when $(2n-3)k < t < (2n-1)k$ $t > 0$
72	$\dfrac{1}{p} \coth kp$	$2S(2k, t) - 1 = 2n - 1$ when $2k(n-1) < t < 2kn$
73	$\dfrac{k}{p^2 + k^2} \coth \dfrac{\pi p}{2k}$	$\lvert \sin kt \rvert$
74	$\dfrac{1}{(p^2 + 1)(1 - e^{-\pi p})}$	$\begin{cases} \sin t \text{ when } (2n-2)\pi < t < (2n-1)\pi \\ 0 \quad \text{when } (2n-1)\pi < t < 2n\pi \end{cases}$
75	$\dfrac{1}{p} e^{-k/p}$	$J_0[2\sqrt{(kt)}]$
76	$\dfrac{1}{\sqrt{p}} e^{-k/p}$	$\dfrac{1}{\sqrt{(\pi t)}} \cos 2\sqrt{(kt)}$
77	$\dfrac{1}{\sqrt{p}} e^{k/p}$	$\dfrac{1}{\sqrt{(\pi t)}} \cosh 2\sqrt{(kt)}$
78	$\dfrac{1}{p^{\frac{3}{2}}} e^{-k/p}$	$\dfrac{1}{\sqrt{(\pi k)}} \sin 2\sqrt{(k\mathbf{y})}$
79	$\dfrac{1}{p^{\frac{3}{2}}} e^{k/p}$	$\dfrac{1}{\sqrt{(\pi k)}} \sinh 2\sqrt{(kt)}$
80	$\dfrac{1}{p^j} e^{-k/p} \quad j > 0$	$\left(\dfrac{t}{k}\right)^{(j-1)/2} J_{j-1}[2\sqrt{(kt)}]$
81	$\dfrac{1}{p^j} e^{k/p} \quad j > 0$	$\left(\dfrac{t}{k}\right)^{(j-1)/2} I_{j-1}[2\sqrt{(k\,t)}]$
82	$e^{-k\sqrt{p}} \quad k > 0$	$\dfrac{k}{2\sqrt{(\pi t^3)}} \exp\left(-\dfrac{k^2}{4t}\right)$
83	$\dfrac{1}{p} e^{-k\sqrt{p}} \quad k \geqslant 0$	$\mathrm{erfc}\left(\dfrac{k}{2\sqrt{t}}\right)$

TABLE A.1. *Laplace transforms (cont.)*

No.	Transform $\bar{f}(p) = \int\limits_0^\infty e^{-pt} f(t)\, dt$	Function $f(t)$
84	$\dfrac{1}{\sqrt{p}} e^{-k\sqrt{p}} \qquad k \geqslant 0$	$\dfrac{1}{\sqrt{(\pi t)}} \exp\left(-\dfrac{k^2}{4t}\right)$
85	$p^{-\frac{3}{2}} e^{-k\sqrt{p}} \qquad k \geqslant 0$	$2\sqrt{\dfrac{t}{\pi}}\left[\exp\left(-\dfrac{k^2}{4t}\right)\right] - k\,\mathrm{erfc}\left(\dfrac{k}{2\sqrt{t}}\right)$
86	$\dfrac{a e^{-k\sqrt{p}}}{p(a+\sqrt{p})} \qquad k \geqslant 0$	$-\exp(ak)\exp(a^2 t)\mathrm{erfc}\left(a\sqrt{t} + \dfrac{k}{2\sqrt{t}}\right)$ $+\mathrm{erfc}\left(\dfrac{k}{2\sqrt{t}}\right)$
87	$\dfrac{e^{-k\sqrt{p}}}{\sqrt{p}(a+\sqrt{p})} \qquad k \geqslant 0$	$\exp(ak)\exp(a^2 t)\,\mathrm{erfc}\left(a\sqrt{t} + \dfrac{k}{2\sqrt{t}}\right)$
88	$\dfrac{e^{-k\sqrt{[p(p+a)]}}}{\sqrt{[p(p+a)]}}$	$\begin{cases} 0 & \text{when } 0 < t < k \\ \exp\left(-\tfrac{1}{2}at\right) I_0\left[\tfrac{1}{2}a\sqrt{(t^2-k^2)}\right] \\ & \text{when } t > k \end{cases}$
89	$\dfrac{e^{-k\sqrt{(p^2+a^2)}}}{\sqrt{(p^2+a^2)}}$	$\begin{cases} 0 & \text{when } 0 < t < k \\ J_0[a\sqrt{(t^2-k^2)}] & \text{when } t > k \end{cases}$
90	$\dfrac{e^{-k\sqrt{(p^2-a^2)}}}{\sqrt{(p^2-a^2)}}$	$\begin{cases} 0 & \text{when } 0 < t < k \\ I_0[a\sqrt{(t^2-k^2)}] & \text{when } t > k \end{cases}$
91	$\dfrac{e^{-k[\sqrt{(p^2+a^2)}-p]}}{\sqrt{(p^2+a^2)}} \qquad k \geqslant 0$	$J_0[a\sqrt{(t^2+2kt)}]$
92	$e^{-kp} - e^{-k\sqrt{(p^2+a^2)}}$	$\begin{cases} 0 & \text{when } 0 < t < k \\ \dfrac{ak}{\sqrt{(t^2-k^2)}} J_1[a\sqrt{(t^2-k^2)}] \\ & \text{when } t > k \end{cases}$
93	$e^{-k\sqrt{(p^2-a^2)}} - e^{-kp}$	$\begin{cases} 0 & \text{when } 0 < t < k \\ \dfrac{ak}{\sqrt{(t^2-k^2)}} I_1[a\sqrt{(t^2-k^2)}] \\ & \text{when } t > k \end{cases}$
94	$\dfrac{a^j e^{-k\sqrt{(p^2+a^2)}}}{\sqrt{(p^2+a^2)}[\sqrt{(p^2+a^2)}+p]^j}$ $j > -1$	$\begin{cases} 0 & \text{when } 0 < t < k \\ \left(\dfrac{t-k}{t+k}\right)^{\frac{1}{2}j} J_j[a\sqrt{(t^2-k^2)}] \\ & \text{when } t > k \end{cases}$

TABLE A.1. *Laplace transforms (cont.)*

No.	Transform $\bar{f}(p) = \int_0^\infty e^{-pt}f(t)\,dt$	Function $f(t)$
95	$\dfrac{1}{p}\ln p$	$\lambda - \ln t \qquad \lambda = -0.5772\ldots$
96	$\dfrac{1}{p^k}\ln p \qquad k > 0$	$t^{k-1}\left\{\dfrac{\lambda}{[\Gamma(k)]^2} - \dfrac{\ln t}{\Gamma(k)}\right\}$
97[b]	$\dfrac{\ln p}{p - a} \qquad a > 0$	$(\exp at)\,[\ln a - \mathrm{Ei}(-at)]$
98[c]	$\dfrac{\ln p}{p^2 + 1}$	$\cos t\,\mathrm{Si}(t) - \sin t\,\mathrm{Ci}(t)$
99[c]	$\dfrac{p\ln p}{p^2 + 1}$	$-\sin t\,\mathrm{Si}(t) - \cos t\,\mathrm{Ci}(t)$
100[b]	$\dfrac{1}{p}\ln(1 + kp) \qquad k > 0$	$-\mathrm{Ei}\left(-\dfrac{t}{k}\right)$
101	$\ln\dfrac{p - a}{p - b}$	$\dfrac{1}{t}(e^{bt} - e^{at})$
102[c]	$\dfrac{1}{p}\ln(1 + k^2 p^2)$	$-2\mathrm{Ci}\left(\dfrac{t}{k}\right)$
103[c]	$\dfrac{1}{p}\ln(p^2 + a^2) \qquad a > 0$	$2\ln a - 2\mathrm{Ci}(at)$
104[c]	$\dfrac{1}{p^2}\ln(p^2 + a^2) \qquad a > 0$	$\dfrac{2}{a}[at\ln a + \sin at - at\mathrm{Ci}(at)]$
105	$\ln\dfrac{p^2 + a^2}{p^2}$	$\dfrac{2}{t}(1 - \cos at)$

[b] $\mathrm{Ei}(-t) = -\displaystyle\int_t^\infty \dfrac{e^{-x}}{x}\,dx$ (for $t > 0$) = exponential integral function.

[c] $\mathrm{Si}(t) = \displaystyle\int_0^t \dfrac{\sin x}{x}\,dx$ = sine integral function.

$\mathrm{Ci}(t) = -\displaystyle\int_t^\infty \dfrac{\cos x}{x}\,dx$ = cosine integral function.

These functions are tabulated in M. ABRAMOWITZ and I. A. STEGUN, *Handbook of Mathematical Functions*, Dover Publications, New York 1965.

TABLE A.1. *Laplace transforms (cont.)*

No.	Transform $\bar{f}(p) = \int_0^\infty e^{-pt}f(t)\,dt$	Function f(t)
106	$\ln\dfrac{p^2 - a^2}{p^2}$	$\dfrac{2}{t}(1 - \cosh at)$
107	$\tan^{-1}\dfrac{k}{p}$	$\dfrac{1}{t}\sin kt$
108^c	$\dfrac{1}{p}\tan^{-1}\dfrac{k}{p}$	$\mathrm{Si}(kt)$
109	$\exp(k^2 p^2)\,\mathrm{erfc}(kp)$ $\qquad k > 0$	$\dfrac{1}{k\sqrt{\pi}}\exp\left(-\dfrac{t^2}{4k^2}\right)$
110	$\dfrac{1}{p}\exp(k^2 p^2)\,\mathrm{erfc}(kp)$ $\qquad k > 0$	$\mathrm{erf}\left(\dfrac{t}{2k}\right)$
111	$\exp(kp)\,\mathrm{erfc}\left[\sqrt{(kp)}\right]$ $\quad k > 0$	$\dfrac{\sqrt{k}}{\pi\sqrt{t(t+k)}}$
112	$\dfrac{1}{\sqrt{p}}\,\mathrm{erfc}\left[\sqrt{(kp)}\right]$	$\begin{cases} 0 & \text{when } 0 < t < k \\ (\pi t)^{-\frac{1}{2}} & \text{when } t > k \end{cases}$
113	$\dfrac{1}{\sqrt{p}}\exp(kp)\,\mathrm{erfc}\left[\sqrt{(kp)}\right]$ $\quad k > 0$	$\dfrac{1}{\sqrt{[\pi(t+k)]}}$
114	$\mathrm{erf}\left(\dfrac{k}{\sqrt{p}}\right)$	$\dfrac{1}{\pi t}\sin(2k\sqrt{t})$
115	$\dfrac{1}{\sqrt{p}}\exp\left(\dfrac{k^2}{p}\right)\mathrm{erfc}\left(\dfrac{k}{\sqrt{p}}\right)$	$\dfrac{1}{\sqrt{(\pi t)}}\exp(-2k\sqrt{t})$
116^d	$\mathrm{K}_0(kp)$	$\begin{cases} 0 & \text{when } 0 < t < k \\ (t^2 - k^2)^{-\frac{1}{2}} & \text{when } t > k \end{cases}$
117^d	$\mathrm{K}_0(k\sqrt{p})$	$\dfrac{1}{2t}\exp\left(-\dfrac{k^2}{4t}\right)$
118^d	$\dfrac{1}{p}\exp(kp)\,\mathrm{K}_1(kp)$	$\dfrac{1}{k}\sqrt{[t(t+2k)]}$

^{d} $\mathrm{K}_n(x)$ denotes the Bessel function of the second kind for the imaginary argument.

TABLE A.1. *Laplace transforms (cont.)*

No.	Transform $\overline{f}(p) = \int\limits_0^\infty e^{-pt} f(t)\, dt$	Function f(t)
119[d]	$\dfrac{1}{\sqrt{p}} K_1(k\sqrt{p})$	$\dfrac{1}{k} \exp\left(-\dfrac{k^2}{4t}\right)$
120[d]	$\dfrac{1}{\sqrt{p}} \exp\left(\dfrac{k}{p}\right) K_0\left(\dfrac{k}{p}\right)$	$\dfrac{2}{\sqrt{(\pi t)}} K_0[2\sqrt{(2kt)}]$
121[e]	$\pi \exp(-kp)\, I_0(kp)$	$\begin{cases} [t(2k-t)]^{-\frac{1}{2}} & \text{when } 0 < t < 2k \\ 0 & \text{when } t > 2k \end{cases}$
122[e]	$\exp(-kp)\, I_1(kp)$	$\begin{cases} \dfrac{k-t}{\pi k \sqrt{[t(2k-t)]}} & \text{when } 0 < t < 2k \\ 0 & \text{when } t > 2k \end{cases}$
123	unity	unit impulse

[e] $I_n(x)$ denotes the Bessel function of the first kind for the imaginary argument.

TABLE A.2. *Error function and its derivative*

x	$\frac{2}{\sqrt{\pi}}e^{-x^2}$	erf x	x	$\frac{2}{\sqrt{\pi}}e^{-x^2}$	erf x
0·00	1·12837	0·00000	0·45	0·92153	0·47548
0·01	1·12826	0·01128	0·46	0·91318	0·48465
0·02	1·12792	0·02256	0·47	0·90473	0·49374
0·03	1·12736	0·03384	0·48	0·89617	0·50274
0·04	1·12657	0·04511	0·49	0·88752	0·51166
0·05	1·12556	0·05637	0·50	0·87878	0·52049
0·06	1·12432	0·06762	0·51	0·86995	0·52924
0·07	1·12286	0·07885	0·52	0·86103	0·53789
0·08	1·12118	0·09007	0·53	0·85204	0·54646
0·09	1·11927	0·10128	0·54	0·84297	0·55493
0·10	1·11715	0·11246	0·55	0·83383	0·56332
0·11	1·11480	0·12362	0·56	0·82463	0·57161
0·12	1·11224	0·13475	0·57	0·81536	0·57981
0·13	1·10946	0·14586	0·58	0·80604	0·58792
0·14	1·10647	0·15694	0·59	0·79666	0·59593
0·15	1·10327	0·16799	0·60	0·78724	0·60385
0·16	1·09985	0·17901	0·61	0·77777	0·61168
0·17	1·09623	0·18999	0·62	0·76826	0·61941
0·18	1·09240	0·20093	0·63	0·75872	0·62704
0·19	1·08837	0·21183	0·64	0·74914	0·63458
0·20	1·08413	0·22270	0·65	0·73954	0·64202
0·21	1·07969	0·23352	0·66	0·72992	0·64937
0·22	1·07506	0·24429	0·67	0·72027	0·65662
0·23	1·07023	0·25502	0·68	0·71061	0·66378
0·24	1·06522	0·26570	0·69	0·70095	0·67084
0·25	1·06001	0·27632	0·70	0·69127	0·67780
0·26	1·05462	0·28689	0·71	0·68159	0·68466
0·27	1·04904	0·29741	0·72	0·67191	0·69143
0·28	1·04329	0·30788	0·73	0·66224	0·69810
0·29	1·03736	0·31828	0·74	0·65258	0·70467
0·30	1·03126	0·32862	0·75	0·64293	0·71115
0·31	1·02498	0·33890	0·76	0·63329	0·71753
0·32	1·01855	0·34912	0·77	0·62368	0·72382
0·33	1·01195	0·35927	0·78	0·61408	0·73001
0·34	1·00519	0·36936	0·79	0·60452	0·73610
0·35	0·99828	0·37938	0·80	0·59498	0·74210
0·36	0·99122	0·38932	0·81	0·58548	0·74800
0·37	0·98401	0·39920	0·82	0·57601	0·75381
0·38	0·97665	0·40900	0·83	0·56659	0·75952
0·39	0·96916	0·41873	0·84	0·55720	0·76514
0·40	0·96154	0·42839	0·85	0·54786	0·77066
0·41	0·95378	0·43796	0·86	0·53858	0·77610
0·42	0·94590	0·44746	0·87	0·52934	0·78143
0·43	0·93789	0·45688	0·88	0·52016	0·78668
0·44	0·92977	0·46622	0·89	0·51103	0·79184

TABLE A.2 (*cont.*)

x	$\frac{2}{\sqrt{\pi}} e^{-x^2}$	erf x	x	$\frac{2}{\sqrt{\pi}} e^{-x^2}$	erf x
0·90	0·50196	0·79690	1·35	0·18236	0·94376
0·91	0·49296	0·80188	1·36	0·17749	0·94556
0·92	0·48402	0·80676	1·37	0·17271	0·94731
0·93	0·47515	0·81156	1·38	0·16802	0·94901
0·94	0·46635	0·81627	1·39	0·16343	0·95067
0·95	0·45761	0·82089	1·40	0·15894	0·95228
0·96	0·44896	0·82542	1·41	0·15453	0·95385
0·97	0·44037	0·82987	1·42	0·15022	0·95537
0·98	0·43187	0·83423	1·43	0·14600	0·95685
0·99	0·42345	0·83850	1·44	0·14187	0·95829
1·00	0·41510	0·84270	1·45	0·13783	0·95969
1·01	0·40684	0·84681	1·46	0·13387	0·96105
1·02	0·39867	0·85083	1·47	0·13001	0·96237
1·03	0·39058	0·85478	1·48	0·12623	0·96365
1·04	0·38257	0·85864	1·49	0·12254	0·96489
1·05	0·37466	0·86243	1·50	0·11893	0·96610
1·06	0·36684	0·86614	1·51	0·11540	0·96727
1·07	0·35911	0·86977	1·52	0·11195	0·96841
1·08	0·35147	0·87332	1·53	0·10859	0·96951
1·09	0·34392	0·87680	1·54	0·10531	0·97058
1·10	0·33647	0·88020	1·55	0·10210	0·97162
1·11	0·32912	0·88353	1·56	0·09898	0·97262
1·12	0·32186	0·88678	1·57	0·09593	0·97360
1·13	0·31470	0·88997	1·58	0·09295	0·97454
1·14	0·30764	0·89308	1·59	0·09005	0·97546
1·15	0·30067	0·89612	1·60	0·08722	0·97634
1·16	0·29381	0·89909	1·61	0·08447	0·97720
1·17	0·28704	0·90200	1·62	0·08178	0·97803
1·18	0·28037	0·90483	1·63	0·07917	0·97884
1·19	0·27381	0·90760	1·64	0·07662	0·97962
1·20	0·26734	0·91031	1·65	0·07414	0·98037
1·21	0·26097	0·91295	1·66	0·07173	0·98110
1·22	0·25471	0·91553	1·67	0·06938	0·98181
1·23	0·24854	0·91805	1·68	0·06709	0·98249
1·24	0·24248	0·92050	1·69	0·06487	0·98315
1·25	0·23652	0·92290	1·70	0·06271	0·98379
1·26	0·23065	0·92523	1·71	0·06060	0·98440
1·27	0·22489	0·92751	1·72	0·05856	0·98500
1·28	0·21923	0·92973	1·73	0·05657	0·98557
1·29	0·21367	0·93189	1·74	0·05464	0·98613
1·30	0·20820	0·93400	1·75	0·05277	0·98667
1·31	0·20284	0·93606	1·76	0·05095	0·98719
1·32	0·19757	0·93806	1·77	0·04918	0·98769
1·33	0·19241	0·94001	1·78	0·04747	0·98817
1·34	0·18734	0·94191	1·79	0·04580	0·98864

APPENDIX

TABLE A.2 (*cont.*)

x	$\frac{2}{\sqrt{\pi}}e^{-x^2}$	erf x	x	$\frac{2}{\sqrt{\pi}}e^{-x^2}$	erf x
1·80	0·04419	0·98909	1·90	0·03052	0·99279
1·81	0·04262	0·98952	1·91	0·02938	0·99308
1·82	0·04110	0·98994	1·92	0·02827	0·99337
1·83	0·03963	0·99034	1·93	0·02721	0·99365
1·83	0·03820	0·99073	1·94	0·02617	0·99392
1·85	0·03681	0·99111	1·95	0·02517	0·99417
1·86	0·03547	0·99147	1·96	0·02421	0·99442
1·87	0·03417	0·99182	1·97	0·02328	0·99466
1·88	0·03292	0·99215	1·98	0·02237	0·99489
1·89	0·03170	0·99247	1·99	0·02150	0·99511
			2·00	0·02066	0·99532

Problems

1.1. A preliminary assessment of a process for the hydrodealkylation of toluene is to be made:

$$C_6H_5 \cdot CH_3 + H_2 \rightleftharpoons C_6H_6 + CH_4$$

The feed to the reactor will consist of hydrogen and toluene in the ratio $2H_2 : 1\ C_6H_5 \cdot CH_3$.

(a) Show that with this feed and an outlet temperature of 900 K, the maximum conversion attainable, i.e. the equilibrium conversion, is 0·996 based on toluene. The equilibrium constant of the reaction at 900 K, $K_p = 227$.

(b) Calculate the temperature rise which would occur with this feed if the reactor were operated adiabatically and the products left at equilibrium. For the above reaction at 900 K, $-\Delta H = 50,000$ kJ/kmol.

(c) If the maximum permissible temperature rise for this process is 100 degrees K (i.e. just over one-half the answer to (b)), suggest a suitable design for the reactor.

Heat capacities at 900 K (kJ/kmol K): C_6H_6 198; $C_6H_5CH_3$ 240; CH_4 67; H_2 30

1.2. In a process for producing hydrogen which is required for the manufacture of ammonia, natural gas (methane) is to be reformed with steam according to the equations:

$$CH_4 + H_2O \rightleftharpoons CO + 3H_2; \quad K_p(\text{at } 1173\ \text{K}) = 1 \cdot 43 \times 10^{13}\ N^2/m^4$$
$$CO + H_2O \rightleftharpoons CO_2 + H_2; \quad K_p(\text{at } 1173\ \text{K}) = 0 \cdot 784$$

The natural gas is mixed with steam in the mol ratio $1CH_4 : 5H_2O$ and passed into a catalytic reactor which operates at a pressure of 30 bar. The gases leave the reactor virtually at equilibrium at 1173 K.

(a) Show from the values of the equilibrium constants given above that for every 1 mol of CH_4 entering the reactor, 0·950 mol reacts, and 0·44 mol of CO_2 is formed.

(b) Explain why other reactions such as:

$$CH_4 + 2H_2O \rightleftharpoons CO_2 + 4H_2$$

need not be considered.

(c) By considering the reaction:

$$2CO \rightleftharpoons CO_2 + C$$

for which $K_p' = P_{CO_2}/P_{CO}^2 = 2 \cdot 76 \times 10^{-7}\ m^2/N$ at 1173 K show that carbon deposition on the catalyst is unlikely to occur under the operating conditions.

(d) Qualitatively, what will be the effect on the composition of the exit gas of increasing the total pressure in the reformer? Can you suggest why for ammonia manufacture the reformer is operated at 30 bar instead of at a considerably lower pressure? The reforming step is followed by a shift converter:

$$CO + H_2O \rightleftharpoons CO_2 + H_2$$

absorption of the CO_2, and ammonia synthesis:

$$N_2 + 3H_2 \rightleftharpoons 2NH_3$$

1.3. An aromatic hydrocarbon feedstock consisting mainly of m-xylene is to be isomerised catalytically in a process for making p-xylene. The product from the reactor consists of a mixture of p-xylene, m-xylene, o-xylene and ethylbenzene. As part of a preliminary assessment of the process, calculate the composition of this mixture if equilibrium were established over the catalyst at 730 K.

Equilibrium constants at 730 K are as follows:

$$m\text{-xylene} \rightleftharpoons p\text{-xylene}; \qquad K_p = 0.45$$
$$m\text{-xylene} \rightleftharpoons o\text{-xylene}; \qquad K_p = 0.48$$
$$m\text{-xylene} \rightleftharpoons \text{ethylbenzene}; \quad K_p = 0.19$$

Why is it unnecessary to consider also reactions such as:

$$o\text{-xylene} \underset{k_r}{\overset{k_f}{\rightleftharpoons}} p\text{-xylene}?$$

1.4.. The alkylation of toluene with acetylene in the presence of sulphuric acid is carried out in a batch reactor. 6000 kg of toluene is charged in each batch, together with the required amount of sulphuric acid, and the acetylene is fed continuously to the reactor under pressure. Under circumstances of intense agitation, it may be assumed that the liquid is always saturated with acetylene, and that the toluene is consumed in a simple pseudo-first-order reaction with a rate constant of $0.0011 \, s^{-1}$.

If the reactor is shut down for a period of 15 min between batches, determine the optimum reaction time for the maximum rate of production of alkylate, and calculate this maximum rate in terms of kg toluene consumed per hour.

1.5. Methyl acetate is hydrolysed by water in accordance with the following equation

$$CH_3 \cdot COOCH_3 + H_2O \underset{k_r}{\overset{k_f}{\rightleftharpoons}} CH_3 \cdot COOH + CH_3OH$$

A rate equation is required for this reaction taking place in dilute solution. It is expected that reaction will be pseudo-first order in the forward direction and second order in reverse. The reaction is studied in a laboratory batch reactor starting with a solution of methyl acetate and with no products present. In one experiment, the initial concentration of methyl acetate was $0.05 \, kmol/m^3$ and the fraction hydrolysed at various times subsequently was as follows:

Time (s)	0	1350	3060	5340	7740	∞
Fractional conversion	0	0.21	0.43	0.60	0.73	0.90

(a) Write the rate equation for the reaction and develop its integrated form applicable to a batch reactor.

(b) Plot the data in the manner suggested by the integrated rate equation, confirm the order of reaction and evaluate the forward and reverse rate constants, k_f and k_r.

1.6. Styrene is to be made by the catalytic dehydrogenation of ethyl benzene:

$$C_6H_5 \cdot CH_2 \cdot CH_3 \rightleftharpoons C_6H_5 \cdot CH:CH_2 + H_2$$

The rate equation for this reaction has been reported as follows:

$$\mathcal{R} = k \left(P_{Et} - \frac{1}{K_p} P_{St} P_H \right)$$

where P_{Et}, P_{St} and P_H are partial pressures of ethyl benzene, styrene and hydrogen respectively.

The reactor will consist of a number of tubes each of 80 mm diameter packed with catalyst having a bulk density of $1440 \, kg/m^3$. The ethyl benzene will be diluted with steam, the feed rates per unit cross-sectional area being ethyl benzene $1.6 \times 10^{-3} \, kmol/m^2 \, s$, steam $29 \times 10^{-3} \, kmol/m^2 \, s$. The reactor will be operated at an average pressure of 1.2 bar and the temperature will be maintained at $560°C$ throughout. If the fractional conversion of ethyl benzene is to be 0.45, estimate the length and number of tubes required to produce 20 tonne styrene per day.

At $560°C$ (833 K) $k = 6.6 \times 10^{-9} \, kmol \, m^2/Ns \, kg \, catalyst$, $K_p = 1.0 \times 10^4 \, N/m^2$.

1.7.. Ethyl formate is to be produced from ethanol and formic acid in a continuous flow tubular reactor operated at a constant temperature of $30°C$. The reactants will be fed to the reactor in the proportions 1 mol $HCOOH : 5$ mols C_2H_5OH at a combined flow rate of $0.72 \, m^3/h$. The reaction will be catalysed by a small amount of sulphuric acid. At the temperature, mol ratio, and catalyst

concentration to be employed, the rate equation determined from small-scale batch experiments
has been found to be: $\mathscr{R} = kC_F^2$

where $\mathscr{R}$ is kmol formic acid reacting$/(m^3/s)$

C_F is concentration of formic acid $kmol/m^3$, and

$k = 2.8 \times 10^{-4} \, m^3/kmol \, s$.

The density of the mixture is $820 \, kg/m^3$ and may be assumed constant throughout. Estimate the
volume of the reactor required to convert 70 per cent of the formic acid to the ester.

If the reactor consists of a pipe of 50 mm i.d. what will be the total length required? Determine
also whether the flow will be laminar or turbulent and comment on the significance of
your conclusion in relation to your estimate of reactor volume. The viscosity of the solution is
$1.4 \times 10^{-3} \, Ns/m^2$.

1.8. A batch reactor and a single continuous stirred tank reactor are being compared in relation to
their performance in carrying out the simple liquid phase reaction $A + B \rightarrow$ products. The reaction
is first order with respect to each of the reactants, i.e. second order overall. If the initial concentrations of the reactants are equal, show that the volume of the continuous reactor must be $1/(1 - \alpha)$
times the volume of the batch reactor for the same rate of production from each, where α is the
fractional conversion. Assume that there is no change in density associated with the reaction and
neglect the shutdown period between batches for the batch reactor.

In qualitative terms, what is the advantage of using a series of continuous stirred tanks for such
a reaction?

1.9. Two stirred tanks are available at a chemical works one of volume $100 \, m^3$, the other $30 \, m^3$.
It is suggested that these tanks be used as a two-stage CSTR for carrying out an irreversible liquid
phase reaction $A + B \rightarrow$ product. The two reactants will be present in the feed stream in equimolar
proportion, the concentration of each being $1.5 \, kmol/m^3$. The volumetric flowrate of the feed
stream will be $0.3 \times 10^{-3} \, m^3/s$. The reaction is irreversible and is of first order with respect to
each of the reactants A and B, i.e. second order overall, with a rate constant 1.8×10^{-4}
$m^3/kmol \, s$.

(a) Which tank should be used as the first stage of the reactor system, the aim being to effect as
high a conversion as possible?

(b) With this configuration, calculate the conversion obtained in the product stream leaving the
second tank after steady conditions have been reached.

(If in doubt regarding which tank should be used as the first stage, calculate the conversions
for both configurations and compare. Note that accurate calculations are required in order to
distinguish between the two.)

1.10. The kinetics of a liquid-phase chemical reaction are being investigated in a laboratory-scale
continuous stirred tank reactor. The stoichiometric equation for the reaction is $A \rightarrow 2P$ and it is
irreversible. The reactor is a single vessel which contains $3.25 \times 10^{-3} \, m^3$ of liquid when it is
filled just to the level of the outflow. In operation, the contents of the reactor are well stirred
and uniform in composition. The concentration of the reactant A in the feed stream is 0.5
$kmol/m^3$. Results of three steady-state runs are as follows:

Feed rate $m^3/s \times 10^5$	Temperature $°C$	Concentration of P in outflow $kmol \, P/m^3$
0.100	25	0.880
0.800	25	0.698
0.800	60	0.905

Hence determine the constants in the rate equation

$$\mathscr{R}_A = \mathscr{A} \exp(-E/RT) \, C_A^p.$$

1.11. A reaction $A + B \rightarrow P$ which is first order with respect to each of the reactants, with a rate
constant of $1.5 \times 10^{-5} \, m^3/kmol \, s$ is carried out in a single continuous flow stirred tank reactor.
This reaction is accompanied by a side reaction $2B \rightarrow Q$, where B is a waste product, the side
reaction being second order with respect to B, with a rate constant of $11 \times 10^{-5} \, m^3/kmol \, s$.

An excess of A is used for the reaction, the feed rates to the tank being $0.014 \, kmol/s$ of A and

0·0014 kmol/s of **B**; ultimately reactant **A** is recycled whereas **B** is not. Under these circum-stances the overflow from the tank is at the rate of $1·1 \times 10^{-3}$ m³/s, while the capacity of the tank is 10 m³.

Calculate (a) the fraction of **B** converted into the desired product **P**, and (b) the fraction of **B** converted into **Q**.

If a second tank of equal capacity becomes available, suggest with reasons in what manner it might be incorporated (i) if **A** but not **B** is recycled as above, and (ii) if both **A** and **B** are recycled.

1.12. Consider a model reaction scheme in which a substance **A** reacts with a second substance **B** to give a desired product **P**, but **B** also undergoes a simultaneous side reaction to give an unwanted product **Q**:

$$\mathbf{A} + \mathbf{B} \rightarrow \mathbf{P}; \quad \text{rate} = k_P C_A C_B$$
$$2\mathbf{B} \rightarrow \mathbf{Q}; \quad \text{rate} = k_Q C_B^2$$

The rate equations are given above where C_A and C_B are the concentrations of **A** and **B** respectively.

Let a single continuous stirred tank reactor be used to make these products. **A** and **B** are mixed in equimolar proportions such that each has the concentration C_0 in the combined stream fed at a volumetric flowrate v to the reactor. If the rate constants above are equal $k_P = k_Q = k$ and the total conversion of **B** is 0·95, i.e. the concentration of **B** in the outflow is $0·05 C_0$, shows that the volume of the reactor will be $69v/kC_0$ and that the relative yield of **P** will be 0·82 (i.e. case d, Fig. 1.27).

Do you consider that a simple tubular reactor would give a larger or smaller yield of **P** than the C.S.T.R. above? What is the essential requirement for a high yield of **P**? Can you suggest any alternative modes of contacting the reactants **A** and **B** which would give better yields than either a single C.S.T.R. or a simple tubular reactor?

2.1. An approximate design procedure for packed tubular reactors entails the assumption of plug flow conditions through the reactor. Discuss critically those effects which would

(a) invalidate plug flow assumptions, and

(b) enhance plug flow.

2.2. A first order chemical reaction occurs isothermally in a reactor packed with spherical catalyst pellets of radius r. If there is a resistance to mass transfer from the main fluid stream to the surface of the particle in addition to a resistance within the particle, show that the effectiveness factor for the pellet is given by:

$$\eta = \frac{3}{\lambda r} \left\{ \frac{\coth \lambda r - 1/\lambda r}{1 + (2\lambda r/Sh.')(\coth \lambda r - 1/\lambda r)} \right\}$$

where $\lambda = (k/D_e)^{\frac{1}{2}}$ and $Sh.' = \dfrac{h_D d_p}{D_e}$,

k is the first order rate constant per unit volume of particle,

D_e is the effective diffusivity, and

h_D is the external mass transfer coefficient.

Discuss the limiting cases pertaining to this effectiveness factor.

2.3. Two consecutive first order reactions;

$$\mathbf{A} \overset{k_1}{\rightarrow} \mathbf{B} \overset{k_2}{\rightarrow} \mathbf{C}$$

occur under isothermal conditions in porous catalyst pellets. Show that the rate of formation of **B** with respect to **A** at the exterior surface of the pellet is:

$$\frac{(k_1/k_2)^{\frac{1}{2}}}{1 + (k_1/k_2)^{\frac{1}{2}}} - \left(\frac{k_2}{k_1} \right)^{\frac{1}{2}} \frac{C_B}{C_A}$$

when the pellet size is large, and

$$1 - \frac{k_2}{k_1} \frac{C_B}{C_A}$$

when the pellet size is small. C_A and C_B represent the concentrations of **A** and **B** respectively at the exterior surface of the pellet, and k_1 and k_2 are the specific rate constants of the two reactions.

Comparing these results, what general conclusions can you deduce concerning the selective formation of **B** on large and small catalyst pellets?

2.4. A packed tubular reactor is used to produce a substance **D** at a total pressure of 1 atm utilising the exothermic equilibrium reaction:

$$A + B \rightleftharpoons C + D$$

600 mol min^{-1} of an equimolar mixture of **A** and **B** is fed to the reactor and plug flow conditions within the reactor may be assumed.

Find the optimal isothermal temperature for operation and the corresponding reactor volume. Is this the best way of operating the reactor?

The forward and reverse kinetics are second order with the velocity constants $k_1 = 4.4 \times 10^{13}$ exp $(-25,000/RT)$ and $k_2 = 7.4 \times 10^{14}$ exp $(-30,000/RT)$ respectively, both in units of (l. mol^{-1} sec^{-1}).

3.1. After being in use for some time, a three-term controller (Fig. 3.19) develops a significant leak in the partition between the integral bellows and the proportional bellows. It is known that the rate of change of pressure in the integral bellows due to the leak is half that due to air flow through the integral restrictor. Show that the leak does not affect the form of the output response of the controller and that the ratio of the gain of the controller with the leak to that of the same controller before the leak developed is given by:

$$\frac{3\tau_2 + \tau_1}{2\tau_2 + \tau_1}$$

where τ_1 and τ_2 are the time constants of the integral and derivative restrictors respectively.

3.2. An industrial controller is constructed in the following way. A change in input pressure P_1 is applied to a restrictor R_1 followed by a chamber of fixed volume C_1. The latter is in turn connected to a further restrictor R_2 followed by a bellows of initial volume C_2. A second bellows of initial volume C_3 is connected directly to C_1. The controller is arranged so that the output pressure P_0 at any instant is equal to the pressure in C_3 minus the pressure in C_2. If $R_1 = R_2$ and $C_1 = C_2 = C_3$, show that:

$$P_1 - P = 5P_0 + 1/\tau \int_0^t P_0 \, dt + 2\tau \frac{dP_0}{dt}$$

where P is the input steady-state pressure at $t = 0$ and $\tau = C_2 R_2$. (Assume that variations in the volumes of C_2 and C_3 are negligible and neglect volumes of connections.)

3.3. A mercury thermometer having first order dynamics with a time constant of 60 s is placed in a temperature bath at 35°C. After the thermometer reaches steady-state it is suddenly placed in a bath at 40°C at $t = 0$ and left there for 60 s, after which it is immediately returned to the bath at 35°C.

(a) Draw a sketch showing the variation of the thermometer reading with time.
(b) Calculate the thermometer reading at $t = 30$ s and at $t = 120$ s.
(c) What would be the reading at $t = 6$ s if the thermometer had only been immersed in the 40°C bath for less than 1 s before being returned to the 35°C bath?

3.4. A tank having a cross-sectional area of 0·2 m^2 is operating at steady-state with an inlet flowrate of 10^{-3} m^3/s. Between the liquid heads of 0·3 m and 0·09 m the flow-head characteristics are given by the equation:

$$Q_2 = 2Z + 0.0006$$

Determine the transfer functions relating (a) inflow and liquid level, (b) inflow and outflow.

If the inflow increases from 10^{-3} to 1·1 × 10^{-3} m^3/s according to a step change, calculate the liquid level 200 s after the change has occurred.

3.5. A continuous stirred tank reactor is fed at a constant rate F m^3/s. The reaction occurring is :

$$A \rightarrow B$$

and proceeds at a rate:

$$\mathscr{R} = kC_0$$

where $\mathcal{R}$ = mols **A** reacting/m³ (mixture in tank) s,
 $\quad k$ = reaction velocity constant, and
 $\quad C_0$ = concentration of **A** in reactor kmol/m³.

If the density and volume V of the reaction mixture in the tank are assumed to remain constant, derive the transfer function relating the concentration of **A** in the reactor at any instant to that in the feed stream C_i. Sketch the response of C_0 to an impulse in C_i.

3.6. Liquid flows into a tank at the rate of Q m³/s. The tank has three vertical walls and one sloping inwards at an angle β to the vertical. The base of the tank is a square with sides of length x m and the average operating level of liquid in the tank is z_0 m. If the relationship between liquid level and flow out of the tank at any instant is linear develop a formula for determining the time constant of the system.

3.7. Write the transfer function for a mercury manometer consisting of a glass U-tube 0·012 m i.d., with a total mercury-column length of 0·54 m, assuming that the actual frictional damping forces are four times greater than would be estimated from Poiseuille's equation. Sketch the response of this instrument when it is subjected to a step change in an air-pressure differential of 14,000 N/m² if the original steady differential was 5000 N/m². Draw the frequency-response characteristics of this system on a Bode diagram.

3.8. The response of an under-damped second order system to a unit step change may be shown to be:

$$Y(t) = 1 - \frac{1}{\sqrt{(1 - \zeta^2)}} \exp(-\zeta t/\tau) \left\{ \zeta \sin\left[\sqrt{(1 - \zeta^2)}\frac{t}{\tau}\right] + \sqrt{(1 - \zeta^2)} \cos\left[\sqrt{(1 - \zeta^2)}\frac{t}{\tau}\right] \right\}$$

Prove that the overshoot for such a response is given by:

$$\exp\{-\pi\zeta/\sqrt{(1 - \zeta^2)}\}$$

and that the decay ratio is equal to the (overshoot)².

A forcing function, whose transform is a constant, K, is applied to an under-damped second order system having a time constant of 0·5 min and a damping coefficient of 0·5. Show that the decay ratio for the resulting response is the same as that due to the application of a unit step function to the same system.

3.9. Air containing ammonia is contacted with fresh water in a two-stage counter-current bubble-plate absorber.

L_n and V_n are the molar flowrates of liquid and gas respectively leaving the nth plate. x_n and y_n are the mol fractions of NH_3 in liquid and gas respectively leaving the nth plate. H_n is the molar hold-up of liquid on the nth plate. Plates are numbered up the column.

(a) Assuming (i) temperature and total pressure throughout the column to be constant, (ii) no change in molar flowrates due to gas absorption, (iii) plate efficiencies to be 100 per cent, (iv) the equilibrium relation to be given by $y_n = mx_n^* + b$, (v) the holdup of liquid on each plate to be constant and equal to H, and (vi) the hold-up of gas between plates to be negligible, show that the variations of the liquid compositions on each plate are given by:

$$\frac{dx_1}{dt} = \frac{1}{H}(L_2 x_2 - L_1 x_1) + \frac{mV}{H}(x_0 - x_1)$$

$$\frac{dx_2}{dt} = \frac{mV}{H}(x_1 - x_2) - \frac{1}{H}L_2 x_2$$

where $V = V_1 = V_2$.

(b) If the inlet liquid flowrate remains constant, prove that the open-loop transfer function for the response of y_2 to a change in inlet gas composition is given by:

$$\frac{\bar{y}_2}{\bar{y}_0} = \frac{c^2/(a^2 - bc)}{\{1/(a^2 - bd)\}p^2 + \{2a/(a^2 - bc)\}p + 1}$$

where $\bar{\mathcal{Y}}_2$, $\bar{\mathcal{Y}}_0$ are the transforms of the appropriate deviation variables and

$$L = L_1 = L_2, \quad a = \frac{L}{H} + \frac{mV}{H}, \quad b = \frac{L}{H}, \quad c = \frac{mV}{H}.$$

Discuss the problems involved in determining the relationship between $\bar{\mathcal{Y}}_2$ and changes in inlet liquid flowrate.

3.10. In the theoretical prediction of the frequency response of a plate distillation column, a perturbation procedure is used, i.e. the instantaneous value of any particular variable is assumed to be the sum of the steady-state value of that variable and a small time-dependent quantity or perturbation, arising from an imposed disturbance. e.g.

$$L_n = L_n^* + l_n$$
$$V_{n-1} = V_{n-1}^* + v_{n-1}$$
$$H_n = H_n^* + h_n$$

where L_n, L_n^* and l_n are the instantaneous, steady-state and perturbation values of the liquid flowrate from the plate n; H_n, H_n^* and h_n are the corresponding values of the liquid hold-up on plate n; V_{n-1}, V_{n-1}^* and v_{n-1} refer to the vapour flowrate from plate $n - 1$. (Plates are numbered up the column.)

If the vapour hold-up in the column can be considered negligible, show that the liquid flow from any tray n can be described by:

$$\frac{dh_n}{dt} = l_{n+1} - l_n$$

(Assume adiabatic operation and constant molar overflow.)

3.11. A proportional controller is used to control a process which may be represented as two non-interacting first order lags each having a time constant of 10 min. The only other lag in the closed loop is the measuring unit which can be approximated by a distance/velocity lag equal to 1 min. Show that when the proportional controller is set such that the loop is on the limit of stability the frequency of the oscillation is given by:

$$\tan \omega = \frac{-20\omega}{1 - 100\omega^2}$$

3.12. A control loop consists of a proportional controller, a first order valve (time constant τ_v, gain K_v) and a first order process (time constant τ_1, gain K_1). Show that when the system is critically damped the controller gain is given by:

$$K_c = \frac{(E - 1)^2}{4EK_vK_1} \quad \text{where } E = \frac{\tau_v}{\tau_1}$$

If the desired value is suddenly changed by an amount ΔR when the controller is set to give critical damping, show that the error ε will be given by:

$$\frac{\varepsilon}{\Delta R} = \frac{4E}{(1 + E)^2} + \left\{ \left[\frac{(1 - E)^2}{2E(1 + E)} \right] \frac{t}{\tau_1} + \frac{(1 - E)^2}{(1 + E)^2} \right\} \exp \left[-\left(\frac{1 + E}{2E} \right) \frac{t}{\tau_1} \right]$$

3.13. A temperature controlled polymerisation process is estimated to have a transfer function:

$$G(p) = \frac{K}{(p - 40)(p + 80)(p + 100)}$$

Show by means of the Routh–Hurwitz criterion that two conditions of controller parameters define upper and lower bounds on the stability of the feedback system incorporating this process.

3.14. Make a critical assessment of the various methods of determining control system stability. In discussing each procedure make use of the following simple control loops (as Fig. 3.2) as examples:

(a)   Controller transfer function $= 4\left(1 + \dfrac{10}{p}\right)$

Process transfer function $= \dfrac{1}{p^2 + 0.2p + 2}$

(b)   Controller transfer function $= 4$

Process transfer function $= \dfrac{1}{p^2 + 0.2p + 2}$

All other transfer functions in the loops may be considered as being unity.

What conclusions can you draw concerning the essential difference between the two control systems?

3.15. A process is controlled by an industrial **P + I** controller having the transfer function:

$$G_c = \frac{\tau_I p^2 + (K_c + 2\tau_I)p + 2K_c}{2\tau_I^2 p}$$

The measuring and final control elements in the control loop are described by transfer functions which can be approximated by constants of unit gain, and the process has the transfer function:

$$G_2 = \frac{1/\tau_2}{\tau_1 \tau_2 p^2 - (2\tau_1 - \tau_2)p - 2}$$

If $\tau_1 = \tfrac{1}{2}$, show that the characteristic equation of the system is given by:

$$(p + 2)\{\tau_2^2 p^2 + (1/\tau_I - 2\tau_2)p + K_c/\tau_I^2\} = 0$$

Show also that the condition under which this control loop will be stable:

(a) when

$$K_c \geqslant \left(\frac{1}{2\tau_2} - \tau_I\right)^2$$

(b) when

$$K_c < \left(\frac{1}{2\tau_2} - \tau_I\right)^2$$

is that $1/\tau_I > 2\tau_2$, provided that K_c and $\tau_2 > 0$.

3.16. The following information is known concerning a control loop of the type shown in Fig. 3.2:

(a) the transfer function of the process is given by:

$$G_2 = \frac{1}{0.1p^2 + 0.3p + 0.2}$$

(b) the steady-state gains of valve and measuring elements are 0.4 and 0.6 units respectively;

(c) the time constants of both valves and measuring element may be considered negligible.

It is proposed to use one of two types of controller in this control loop:

either (i) a **P + D** controller whose action approximates to the relationship:

$$P = P_0 + K_c \varepsilon + K_D \frac{d\varepsilon}{dt}$$

where P is the output pressure at time t, P_0 is the output pressure at $t = 0$, ε is the error, and K_c (proportional gain) = 2 units and K_D = 4 units;

or (ii) an *inverse* rate controller which has the action:

$$P = P_0 + K_c\varepsilon - K_D\frac{d\varepsilon}{dt}$$

(P, P_0, ε, K_c and K_D having the same meaning and values as above.)

Basing your judgement only on the amount of offset obtained when a step change in load is made, which of these two controllers would you recommend? Would you change your mind if you took the system stability into account?

Having regard to the form of the equation describing the control action in each case, under what general circumstances do you think inverse rate control would be better than normal **P** + **D** control?

3.17. A proportional plus integral controller is used to control the level in the reflux accumulator of a distillation column by regulating the top product flowrate. At time $t = 0$ the desired value of the flow controller which is controlling the reflux is increased by 3×10^{-4} m³/s. If the integral action time of the level controller is half the value which would give a critically damped response and the proportional band is 50 per cent, obtain an expression for the resulting change in level.

The range of head covered by the level controller is 0.3 m, the range of top product flowrate is 10^{-3} m³/s and the cross-sectional area of the accumulator is 0.4 m². It may be assumed that the response of the flow controller is instantaneous and that all other conditions remain the same.

If there had been no integral action, what would have been the offset in the level in the accumulator?

3.18. Draw the Bode diagrams of the following transfer functions:

$$\text{(i)} \quad G(p) = \frac{1}{p(1 + 6p)}$$

$$\text{(ii)} \quad G(p) = \frac{(1 + 3p)\exp(-2p)}{p(1 + 2p)(1 + 6p)}$$

$$\text{(iii)} \quad G(p) = \frac{5(1 + 3p)}{p(p^2 + 0.4p + 1)}$$

Comment on the stability of the closed-loop systems having these transfer functions.

3.19. The transfer function of a process and measuring element connected in series is given by:
$$(2p + 1)^{-2}\exp(-0.4p)$$

(a) Sketch the open-loop Bode diagram of a control loop involving this process and measurement lag (but without the controller).

(b) Specify the maximum gain of a proportional controller to be used in this control system without instability occurring.

3.20. A closed-loop system contains a proportional controller, a process which can be represented by a first order transfer function of unit steady-state gain with a time constant of 1 min, and a measuring element having the same transfer function as the process.

Show graphically, or otherwise, that for phase margins $\leqslant 30°$ the phase margin for this control system is approximately related to its damping coefficient by the expression:

$$\frac{\text{phase margin}}{\text{damping coefficient}} = 115 \text{ degrees.}$$

3.21. The block diagram of a control system using a **P** + **I** + **D** pneumatic controller is shown in Fig. 3.2. The gain of the controller is 10 units change in output pressure/unit error, and the integral time is 1 min. The derivative time is adjusted such that the derivative action is negligible. The transfer function of the control valve may be represented by a constant $K_v = 0.5$ units change in M/unit change in input pressure. The process can be approximated by a first and second order system in series. The first order system has the transfer function $G(p) = 1/(5p + 2)$. The second order system has a time constant of 1 min, a damping coefficient of 0.5 and gives a steady-state gain of unity. The transfer function of the measuring element is given by the constant $K_1 = 0.5$ units change in B/unit change in C. The measurement of the controlled variable by the measuring element also introduces a distance-velocity lag of 0.8 min.

By the determination of gain and phase margins, compare the stability of this control system with that obtained when the derivative time is reset to a value of one minute.

3.22. A control system is made up of a process having a transfer function G_2, a measuring element H and a controller G_c.

If $G_2 = (3p + 1)^{-1} \exp(-0.5p)$ and $H = 4.8(1.5p + 1)^{-1}$, determine, using the method of Ziegler and Nichols, the controller settings for P, $P + I$, $P + I + D$ controllers.

3.23. Determine the open-loop response of the output of the measuring element in problem 3.22 to a unit step change in input to the process. Hence determine controller settings for the control loop by the Cohen–Coon and ITAE methods for P, $P + I$ and $P + I + D$ control actions. Compare the settings obtained with those in problem 3.22.

3.24. A continuous process consists of two sections A and B. Feed of composition X_1 enters section A where it is extracted with a solvent which is pumped at a rate L_1 to A. The raffinate is removed from A at a rate L_2 and the extract is pumped to a cracking section B. Hydrogen is added at the cracking stage at a rate L_3 whilst heat is supplied at a rate Q. Two products are formed having compositions X_3 and X_4. The feed-rate to A and L_1 and L_2 can easily be kept constant, but it is known that fluctuations in X_1 can occur. Consequently a feed-forward control system is proposed to keep X_3 and X_4 constant for variations in X_1 using L_3 and Q as controlling variables. Experimental frequency response analysis gave the following transfer functions:

$$\frac{\bar{x}_2}{\bar{x}_1} = \frac{1}{p + 1}, \qquad \frac{\bar{x}_3}{\bar{l}_3} = \frac{2}{p + 1}$$

$$\frac{\bar{x}_3}{\bar{q}} = \frac{1}{2p + 1}, \qquad \frac{\bar{x}_4}{\bar{l}_3} = \frac{p + 2}{p^2 + 2p + 1}$$

$$\frac{\bar{x}_4}{\bar{q}} = \frac{1}{p + 2}, \qquad \frac{\bar{x}_3}{\bar{x}_2} = \frac{2p + 1}{p^2 + 2p + 1}$$

$$\frac{\bar{x}_4}{\bar{x}_2} = \frac{1}{p^2 + 2p + 1}$$

where $\bar{x}_1$, $\bar{x}_2$, $\bar{l}_3$, $\bar{q}$, etc., represent the transforms of small time dependent perturbations in X_1, X_2, L_3, Q, etc.

Determine the transfer functions of the feed-forward control scheme assuming linear operation and negligible distance-velocity lag throughout the process. Comment on the stability of the feed-forward controllers you design.

4.1. Write a computer program to carry out the following task:

Read 30 numbers of format F7.1 and sort the numbers into two groups—one in which the integer part of the number is even and a second in which the integer part is odd. Store the odd integer numbers from the smallest upwards in an array, and the even from the largest down.

Find the sum of the fractional parts for each group separately. Print the results.

4.2. Write a computer program to carry out the following task:

Read in 30 numbers of format F7.1 within the range 10–100. Store those that lie within the range 60–70 in an array in ascending order, square each number and sum the squares. Store those that lie within the range 20–30 in an array in ascending order, multiply each number by -1 and add 25. Print out the two groups.

4.3. Write a computer program to carry out the following task:

Integrate the differential equation:

$$\frac{dy}{dx} = y^2(5x^2 + 6x)$$

using Euler's modified method for step lengths of 0·1 and 0·02. The initial condition is $y = -1$ at $x = 0$ and the integration is to be carried to $x = 10$.

Compare the results with the analytical solution.

4.4. Write a computer program to carry out the following task:

Integrate the differential equation:

$$\frac{d^2y}{dx^2} + \ln x \frac{dy}{dx} + xy = 0$$

when the initial conditions are $y = 1$ at $x = 1$,

$$\frac{dy}{dx} = -1 \text{ at } x = 1.$$

Use the method of Runge and Kutta and integrate to $x = 10$ using step lengths of 0·2, 0·05 and 0·01.

4.5. Write a computer program to carry out the following task:

Find the value of the definite integral:

$$y = \int_1^{10} (6x + 15x \ln x)\,dx$$

by means of Simpson's Rule. Find a value of the step length of integration so that when the step length is halved the integral does not change in value by more than 0·1 per cent.

4.6. Write a computer program to carry out the following task:

Find the value of the definite integral:

$$y = \int_1^5 \left(6x^3 + 25x^2 \ln x + \frac{72}{x}\right) dx$$

using Simpson's Rule and the trapezoidal rule. For both methods choose a value of step length so that when this length is halved the numerical value of the integral does not change by more than 0·05 per cent. What conclusions do you draw from the comparison?

4.7. Write a computer program to carry out the following task:

Integrate the differential equation:

$$\frac{dy}{dx} = 6x^3 + 25x^2 \ln x + \frac{72}{x}$$

by means of the modified Euler method using the initial condition $y = 0$ at $x = 1$ for a range of step lengths. Integrate to $x = 5$. Comment upon the comparison between this and the previous problem.

4.8. Write a computer program to carry out the following task:

Integrate the differential equation:

$$\frac{d^2y}{dx^2} + \ln x \frac{dy}{dx} + xy = 0$$

when the boundary conditions are:

$$y = 1 \text{ at } x = 1$$
$$y = -1 \text{ at } x = 10$$

Use step lengths of 0·2, 0·05 and 0·01 with the method of Runge and Kutta.

4.9. Write a computer program to carry out the following task:

Integrate the partial differential equation:

$$\frac{\partial^2 u}{\partial x^2} + \frac{\partial^2 u}{\partial y^2} = 0 \qquad \text{(Laplace's equation)}$$

when the boundary conditions over a square region are:

$$u = 0, y = 0, 0 < x < L$$
$$u = 1, y = L, 0 < x < L$$
$$u = 0, x = 0, 0 < y < L$$
$$u = 0, x = L, 0 < y < L.$$

Replace the partial differential equation by a finite difference expression. Divide each side into intervals of $L/5$ and find the values of u at the nodes of the grid so formed.

4.10. Write a computer program to carry out the following task:
Integrate the partial differential equation:

$$\frac{\partial^2 u}{\partial x^2} + \frac{\partial^2 u}{\partial y^2} = 1 \quad \text{(Poisson's equation)}$$

when the boundary conditions over a square region are:

$$u = 0, y = 0, 0 < x < L$$
$$u = 0, y = L, 0 < x < L$$
$$u = 0, x = 0, 0 < y < L$$
$$u = 0, x = L, 0 < y < L$$

Replace the partial differential equation by a finite difference expression. Divide each side into intervals of $L/5$ and find the values of u at the nodes of the grid so formed.

4.11. Write a computer program to carry out the following task:
Integrate the partial differential equation:

$$\frac{\partial^2 u}{\partial x^2} + \frac{\partial^2 u}{\partial y^2} = \frac{\partial u}{\partial t}$$

when the boundary conditions over a square region are:

$$u = 0, y = 0, 0 < x < L$$
$$u = 1, y = L, 0 < x < L$$
$$u = 0, x = 0, 0 < y < L$$
$$u = 0, x = L, 0 < y < L$$

and the initial condition is that $u = 0$ everywhere when $t = 0$. Replace the partial differential equation by a finite difference expression. Divide each side into intervals of $L/5$ and find the values of u at the nodes of the grid so formed. Carry the integration forward at intervals of $\Delta = 0.01$.

4.12. Write a computer program to carry out the following task:
Integrate the differential equation given in problem 4.4 by Hamming's method.

4.13. Write a computer program to carry out the following task:
Use the Gauss–Seidel method to solve the equations:

$$1.5w + 1.9x + 3.8y + z = 5.1$$
$$w + 2.9x + 0.7y + 0.5z = 3.6$$
$$4.0w + x + 1.2y + z = 2.9$$
$$0.5w + 0.8x + y + 3.5z = 1.8$$

4.14. Write a computer program to carry out the following task:
Solve the following system of equations by the method of *pivotal condensation*:

$$1\cdot2v + \quad 2w + \quad x + \quad y + z = \quad 6$$
$$2\cdot5v + 1\cdot2w + \quad 2x + \quad 3y + z = \quad 10$$
$$v + \quad w + \quad x + \quad 2y + z = \quad 3\cdot5$$
$$2v + \quad w + 1\cdot5x + 1\cdot5y + 2z = \quad 4$$
$$3v + 2\cdot5w + \quad x + \quad 6y - z = \quad 5$$

6.1. The effective viscosity of a non-Newtonian fluid can be expressed by the following relationship:

$$\mu_a = k'' \frac{du_x}{dr}$$

where k'' is constant.

Show that the volumetric flowrate of this fluid in a horizontal pipe of radius a under isothermal laminar flow conditions with a pressure gradient $-\Delta P/l$ per unit length is:

$$Q = \frac{2\pi}{7} a^{7/2} \left(\frac{-\Delta P}{2k''l} \right)^{1/2}$$

6.2. Determine the yield stress of a Bingham fluid of density 2×10^3 kg/m³ which will just flow out of an open-ended vertical tube of diameter 300 mm under the influence of its own weight.

6.3. A fluid of density $1\cdot2 \times 10^3$ kg/m³ flows down an inclined plane at 15° to the horizontal. If the viscous behaviour is described by the relationship:

$$R_{yx} = -k \left(\frac{du_x}{dy} \right)^n$$

where $k = 4\cdot0$ N s m$^{-1\cdot6}$, and
$$n = 0\cdot4$$

calculate the volumetric flowrate per unit width if the fluid film is 10 mm thick.

6.4. A fluid with a finite yield stress is sheared between two concentric cylinders, 50 mm long. The inner cylinder is 30 mm diameter and the gap is 20 mm. The outer cylinder is held stationary while a torque is applied to the inner. The moment required just to produce motion was $0\cdot01$ N m. Calculate the force needed to ensure all the fluid is flowing under shear if the plastic viscosity is $0\cdot1$ N s m^{-2}.

6.5. Experiments with a capillary viscometer gave the following results:

Capillary length 100 mm, diameter 2 mm.

Applied pressure N/m²	Volumetric flowrate m³/s
1×10^3	1×10^{-7}
2×10^3	$2\cdot8 \times 10^{-7}$
5×10^3	$1\cdot1 \times 10^{-6}$
1×10^4	3×10^{-6}
2×10^4	9×10^{-6}
5×10^4	$3\cdot5 \times 10^{-5}$
1×10^5	1×10^{-4}

Suggest a suitable model to describe the fluid properties.

6.6. Data obtained with a cone and plate viscometer were as follows:

cone half-angle 89°
cone radius 50 mm

cone speed rev/s	measured torque N m
0·1	4·6 × 10^{-2}
0·5	7 × 10^{-2}
1	1·0
5	3·4
10	6·4
50	3·0 × 10

Suggest a suitable model to describe the fluid properties.

6.7. Tomato purée of density 1300 kg/m³ is pumped through a 50 mm diameter factory pipeline at a flowrate of 1 m³/h. It is suggested that in order to double production either:

(a) a similar line with pump should be put in parallel to the existing one, or
(b) a large pump should force the material through the present line, or
(c) a large pump should supply the liquid through a line of twice the cross-sectional area.

Given that the flow properties of the purée can be described by the Casson equation:

$$(-R_{yx})^{1/2} = (-R_Y)^{1/2} + \left(-\mu_c \frac{du_x}{dy}\right)^{1/2}$$

where R_Y is a yield stress, here 20 N/m²,
    μ_c is a characteristic Casson plastic viscosity, 5 Ns/m², and
    $\dfrac{du_x}{dy}$ is the velocity gradient.

evaluate the relative pressure drops of the three suggestions, assuming laminar flow throughout.

7.1. Spherical particles of 15 nm diameter and a density of 2290 kg/m³ are pressed together to form a pellet. The following equilibrium data were obtained for the sorption of nitrogen at 77 K. Obtain estimates of the surface area of the pellet from the sorption isotherm and compare the estimates with the geometric surface. The density of liquid nitrogen at 77 K is 808 kg/m³.

P/P_0	0·1	0·2	0·3	0·4	0·5	0·6	0·7	0·8	0·9
m³ liq N_2 × 10^6/ kg solid	66·7	75·2	83·9	93·4	108·4	130·0	150·2	202·0	348·0

7.2. A volume of 1 m³ contains a mixture of air and acetone vapour. The temperature is 303 K and the total pressure 1 bar. If the relative saturation of the air by acetone is 40 per cent, how much activated carbon must be added to the space in order to reduce the value to 5 per cent at 303 K?

If, in fact, 1·6 kg carbon is added, what will the per cent relative saturation of the equilibrium mixture be, assuming the temperature to be unchanged?

The vapour pressure of acetone at 303 K is 37·9 kN/m².

Sorption equilibrium data for acetone on carbon at 303 K are:

Partial pressure acetone × 10^{-2} N/m²	0	5	10	30	50	90
kg acetone/kg carbon	0	0·14	0·19	0·27	0·31	0·35

7.3. A solvent contaminated with 0·03 kmol/m³ of a fatty acid is to be purified by passing it through a fixed bed of activated carbon which will adsorb the acid but not the solvent. If the operation is essentially isothermal and equilibrium is maintained between the liquid and the solid, calculate the length of a bed of 0·15 m diameter to give one hour's operation when the fluid is fed at 1×10^{-4} m³/s. The bed is free of sorbate initially; the intergranular voidage is 0·4. Use an equilibrium, fixed-bed theory to obtain an answer for three types of isotherm:

(a) $C_s = 10\ C$.
(b) $C_s = 3\cdot0\ C^{0\cdot3}$ (use the mean slope).
(c) $C_s = 10^4\ C^2$ (take the breakthrough concentration as 0·003 kmol/m³).

In each case the units of concentration are kmol/m³.

7.4. A single pellet of alumina is exposed to a flow of humid air at a constant temperature. The increase in mass of the pellet is followed automatically, yielding the following results:

t (min)	2	4	10	20	40	60	120
C_s, g. g	0·091	0·097	0·105	0·113	0·125	0·128	0·132

Assuming the effect of the external film is negligible, predict t vs. C_s values for a pellet of twice the radius.

7.5. For the problem of drying acetylene described in the example on page 590, show that a three-bed system using full-flow regeneration can satisfy the criterion for continuous operation $t_s \geqslant t_h$.

Index